Bacteriophage Ecology

Population Growth, Evolution, and Impact of Bacterial Viruses

Bacteriophages, or phages, are viruses that infect bacteria, and are believed to be the most abundant and most genetically diverse organisms on Earth. As such, their ecology is vast both in quantitative and qualitative terms. Their abundance, and their impact on bacteria, makes an understanding of phage ecology increasingly relevant to bacterial ecosystem ecology, bacterial genomics, and bacterial pathology. This volume is the first on phage ecology in over 20 years. Written by leading experts, it seeks to synthesize three key approaches toward studying phage ecology, namely determination of natural (*in situ*) phage abundance and diversity; experimentation in the laboratory as well as *in situ* experimentation on the dynamics of phage–phage, phage–bacterium, and phage–ecosystem interactions; and the development, using mathematical and computer models, of ecological and evolutionary theory based on phage populations and phage-containing bacterial communities. With strong emphasis on microbial population biology, and by distilling cutting-edge research into basic principles, this book will serve as an essential resource for graduate students and researchers, particularly those with an interest in phage ecology or phage evolutionary biology.

STEPHEN T. ABEDON is Associate Professor of Microbiology at the Ohio State University. He contributed to the editing of *The Bacteriophages* (2006) and founded the Bacteriophage Ecology Group at www.phage.org to encourage collaboration and to provide a central resource for the bacteriophage community.

Published titles

1. *Bacterial Adhesion to Host Tissues.* Edited by Michael Wilson 0521801079
2. *Bacterial Evasion of Host Immune Responses.* Edited by Brian Henderson and Petra Oyston 0521801737
3. *Dormancy in Microbial Diseases.* Edited by Anthony Coates 0521809401
4. *Susceptibility to Infectious Diseases.* Edited by Richard Bellamy 0521815258
5. *Bacterial Invasion of Host Cells.* Edited by Richard Lamont 0521809541
6. *Mammalian Host Defense Peptides.* Edited by Deirdre Devine and Robert Hancock 0521822203
7. *Bacterial Protein Toxins.* Edited by Alistair Lax 052182091X
8. *The Dynamic Bacterial Genome.* Edited by Peter Mullany 0521821576
9. *Salmonella Infections.* Edited by Pietro Mastroeni and Duncan Maskell 0521835046
10. *The Influence of Cooperative Bacteria on Animal Host Biology.* Edited by Margaret J. McFall Ngai, Brian Henderson and Edward Ruby 0521834651
11. *Bacterial Cell-to-Cell Communication.* Edited by Donald R. Demuth and Richard Lamont 0521846382
12. *Phagocytosis of Bacteria and Bacterial Pathogenicity.* Edited by Joel Ernst and Olle Stendahl 0521845696
13. *Bacterial-Epithelial Cell Cross-Talk: Molecular Mechanisms in Pathogenesis.* Edited by Beth A. McCormick 0521852447
14. *Dendritic Cell Interactions with Bacteria.* Edited by Maria Rescigno 9780521855860

Over the past decade, the rapid development of an array of techniques in the fields of cellular and molecular biology has transformed whole areas of research across the biological sciences. Microbiology has perhaps been influenced most of all. Our understanding of microbial diversity and evolutionary biology and of how pathogenic bacteria and viruses interact with their animal and plant hosts at the molecular level, for example, has been revolutionized. Perhaps the most exciting recent advance in microbiology has been the development of the interface discipline of Cellular Microbiology, a fusion of classical microbiology, microbial molecular biology and eukaryotic cellular and molecular biology. Cellular Microbiology is revealing how pathogenic bacteria interact with host cells in what is turning out to be a complex evolutionary battle of competing gene products. Molecular and cellular biology are no longer discrete subject areas but vital tools and an integrated part of current microbiological research. As part of this revolution in molecular biology, the genomes of a growing number of pathogenic and model bacteria have been fully sequenced, with immense implications for our future understanding of microorganisms at the molecular level.

Advances in Molecular and Cellular Microbiology is a series edited by researchers active in these exciting and rapidly expanding fields. Each volume will focus on a particular aspect of cellular or molecular microbiology and will provide an overview of the area; it will also examine current research. This series will enable graduate students and researchers to keep up with the rapidly diversifying literature in current microbiological research.

AMCM

ADVANCES IN MOLECULAR AND CELLULAR MICROBIOLOGY

Advances in Molecular and Cellular Microbiology 15

Bacteriophage Ecology

Population Growth, Evolution, and Impact of Bacterial Viruses

EDITED BY

STEPHEN T. ABEDON

The Ohio State University

CAMBRIDGE
UNIVERSITY PRESS

CAMBRIDGE
UNIVERSITY PRESS

Shaftesbury Road, Cambridge CB2 8EA, United Kingdom

One Liberty Plaza, 20th Floor, New York, NY 10006, USA

477 Williamstown Road, Port Melbourne, VIC 3207, Australia

314–321, 3rd Floor, Plot 3, Splendor Forum, Jasola District Centre, New Delhi – 110025, India

103 Penang Road, #05–06/07, Visioncrest Commercial, Singapore 238467

Cambridge University Press is part of Cambridge University Press & Assessment, a department of the University of Cambridge.

We share the University's mission to contribute to society through the pursuit of education, learning and research at the highest international levels of excellence.

www.cambridge.org
Information on this title: www.cambridge.org/9780521858458

First published 2008

A catalogue record for this publication is available from the British Library

ISBN 978-0-521-85845-8 Hardback

Contents

About the cover *page* ix
List of contributors xi
Foreword by Bruce R. Levin xiv
Preface xvii

1 Phages, ecology, evolution 1
Stephen T. Abedon

Part I Phage ecology

2 Bacteriophages: models for exploring basic principles of ecology 31
Benjamin Kerr, Jevin West, and Brendan J. M. Bohannan

3 Phage population growth: constraints, games, adaptation 64
Stephen T. Abedon

4 Impact of spatial structure on phage population growth 94
Stephen T. Abedon and John Yin

5 Contribution of lysogeny, pseudolysogeny, and starvation to phage ecology 114
Robert V. Miller and Martin J. Day

Part II Phage evolutionary biology

6 Phage evolutionary biology 147
Siobain Duffy and Paul E. Turner

7 Phage evolution 177
Roger W. Hendrix

8 Evolutionary ecology of multiple phage adsorption and infection 195
Paul E. Turner and Siobain Duffy

9 Patterns in phage experimental adaptation 217
J. J. Bull

Part III Phage ecology in environments

10 Aquatic phage ecology 251
T. Frede Thingstad, Gunnar Bratbak, and Mikal Heldal

11 Phage ecology of terrestrial environments 281
Martin J. Day and Robert V. Miller

12 Phages, bacteria, and food 302
Lawrence D. Goodridge

13 Interaction of bacteriophages with animals 332
Carl R. Merril

14 Phage ecology of bacterial pathogenesis 353
Paul Hyman and Stephen T. Abedon

Part IV Modeling phage ecology

15 Modeling bacteriophage population growth 389
David Stopar and Stephen T. Abedon

16 Modeling phage plaque growth 415
Stephen M. Krone and Stephen T. Abedon

17 Modeling of bacteriophage therapy 439
Jason J. Gill

Index 465

Color plate section appears between pages 46 and 47.

About the cover

T4 bacteriophage cover image created by science artist Steven McQuinn from diverse data sources.

ARTIST'S COMMENT

While it is tempting to say, "This is what bacteriophage T4 really looks like," such a statement makes no sense in the nanoscale microcosm where visible light washes over phages the way ocean swells move through plankton. Rather, phages must be probed using the severely short end of the electromagnetic spectrum. Electron microscopists and X-ray crystallographers measure energy–matter deflections scattered by hundreds of samples. Such datasets are visualized in various ways to illustrate protein morphology, even protein molecular structure. Data furnished to the eye seem intuitively comprehensible, although the caveats of method should never be slighted. We are not looking at a thing, we are looking at the result of a process. T4 thrives in a domain of scale where matter and energy function very differently from the way they do in the everyday world of our visual intuition. The image appears to be a portrait, but it is in fact a synthesis of data with constrained resolution, assembled from diverse sources.

Tools

UCSF Chimera, VMD, Carrara Studio Pro, Amapi 3D

DATA SOURCES

Cerritelli, M. E., J. S. Wall, M. N. Simon, J. F. Conway, and A. C. Steven. 1996. Stoichiometry and domainal organization of the long tail-fiber of bacteriophage T4: a hinged viral adhesin. *J. Mol. Biol.* **260**: 767–80.

Fokine, A., P. R. Chipman, P. G. Leiman, V. V. Mesyanzhinov, V. B. Rao, M. G. Rossmann. 2004. Molecular architecture of the prolate head of bacteriophage T4. *Proc. Natl. Acad. Sci. U.S.A.* **101**: 6003–8.

Fokine, A., V. A. Kostyuchenko, A. V. Efimov, et al. 2005. A three-dimensional cryo-electron microscopy structure of the bacteriophage ϕKZ head. *J. Mol. Biol.* **352**: 117–24.

Kanamaru, S., P. G. Leiman, V. A. Kostyuchenko, et al. 2002. Structure of the cell-puncturing device of bacteriophage T4. *Nature* **415**: 553–7.

Kostyuchenko, V. A., P. G. Leiman, P. R. Chipman, et al. 2003. Three-dimensional structure of bacteriophage T4 baseplate. *Nat. Struct. Biol.* **10**: 688–93.

Leiman, P. G., P. R. Chipman, V. A. Kostyuchenko, V. V. Mesyanzhinov, M. G. Rossmann. 2004. Three-dimensional rearrangement of proteins in the tail of bacteriophage T4 on infection of its host. *Cell* **118**: 419–29.

Leiman, P. G., S. Kanamaru, V. V. Mesyanzhinov, F. Arisaka, M. G. Rossmann. 2004. Structure and morphogenesis of bacteriophage T4. *Cell. Mol. Life. Sci.* **60**: 2356–70.

Lepault, J., and K. Leonard. 1985. Three-dimensional structure of unstained, frozen-hydrated extended tails of bacteriophage T4. *J. Mol. Biol.* **182**: 431–41.

Rossmann, M. G., V. V. Mesyanzhinov, F. Arisaka, P. G. Leiman. 2004. The bacteriophage T4 DNA injection machine. *Curr. Opin. Struct. Biol.* **14**: 171–80.

Contributors

Stephen T. Abedon
Department of Microbiology, The Ohio State University, Mansfield, OH 44906, USA

Brendan J. M. Bohannan
Center for Ecology and Evolutionary Biology, 309 Pacific Hall, 5289 University of Oregon, Eugene, OR 97403, USA

Gunnar Bratbak
Department of Biology, University of Bergen, Jahnebakken 5, PO Box 7800, 5020 Bergen, Norway

J. J. Bull
Section of Integrative Biology, University of Texas at Austin, 1 University Station #C0930, Austin, TX 78712, USA

Martin J. Day
Cardiff School of Biosciences, Cardiff University, Main Building, Park Place, PO Box 915, Cardiff CF10 3TL, United Kingdom

Siobain Duffy
Department of Ecology and Evolutionary Biology, Yale University, PO Box 208106, New Haven, CT 06520, USA

Jason J. Gill
Biochemistry & Biophysics, Texas A&M University, 2128 TAMU, Room 336, College Station, TX 77843, USA

Lawrence D. Goodridge
Department of Animal Sciences, Colorado State University, 350 West Pitkin Street, Fort Collins, CO 80523, USA

Mikal Heldal
Department of Biology, University of Bergen, Jahnebakken 5, PO Box 7800, 5020 Bergen, Norway

Roger W. Hendrix
Pittsburgh Bacteriophage Institute & Department of Biological Sciences, University of Pittsburgh, 4249 Fifth Avenue, Pittsburgh, PA 15260, USA

Paul Hyman
MedCentral College of Nursing, 335 Glessner Avenue, Mansfield, OH 44903, USA

Benjamin Kerr
Department of Biology, University of Washington, 106 Kincaid Hall, Box 351800, Seattle, WA 98195, USA

Stephen M. Krone
Department of Mathematics, University of Idaho, Moscow, ID 83844, USA

Robert V. Miller
Department of Microbiology and Molecular Genetics, Oklahoma State University, 307 Life Science East, Stillwater, OK 74078, USA

Carl R. Merril
National Institute of Mental Health, National Institutes of Health, 6840 Capri Place, Bethesda, MD 20817, USA

David Stopar
Biotechnical Faculty, Department of Food Technology, University of Ljubljana, Vecna pot 111, 1000 Ljubljana, Slovenia

T. Frede Thingstad
Department of Biology, University of Bergen, Jahnebakken 5, PO Box 7800, 5020 Bergen, Norway

Paul E. Turner
Department of Ecology and Evolutionary Biology, Yale University, PO Box 208106, New Haven, CT 06520, USA

Jevin West
Department of Biology, University of Washington, 106 Kincaid Hall, Box 351800, Seattle, WA 98195, USA

John Yin
Department of Chemical and Biological Engineering, University of Wisconsin–Madison, 3633 Engineering Hall, 1415 Engineering Drive, Madison, WI 53706, USA

Foreword

More than forty years ago the then major honchos of molecular biology published a festschrift for Max Delbrück, lauding, in a self-congratulatory but often delightful and enthusiastic way, the central role bacteriophages played in the development of molecular biology (Cairns, J., G. Stent, and J. Watson. 1966. *Phage and the Origins of Molecular Biology*. Cold Spring Harbor, NY: Cold Spring Harbor Laboratory). The prominence of bacteriophages as model systems for studying molecular biology has since waned. However, even in these days where genetics is rarely done by allowing organisms to move genes by themselves (horror, mating between the unmarried?), bacteriophages maintain a major rep as tools for studies of the genetics and molecular biology of bacteria.

Bacteriophages, however, rarely come to mind when one thinks of ecology and evolutionary biology. Despite increasing numbers of meetings devoted entirely or partially to the ecology and evolution of bacteriophages, and a growing constituency of young investigators participating in these meetings, ecology and evolutionary biology has been and remains dominated not only by eukaryotic practitioners but also by eukaryotes as objects of study. Of some 195 000 articles listed on PubMed with the word "evolution" in some field, less than 0.7% list "bacteriophage" in another field, and of approximately 23 000 articles with "ecology," less than 0.6% list "bacteriophage."

Why have bacteriophages played so small a role either as model systems or as the focus of study in ecology and evolutionary biology? Could it be because bacteriophages and their hosts are neither cute nor live in appealing places, or is it because they are hard for humans to identify with? Phages are certainly not rare: 10^{31} (see Chapter 1) is a substantial constituency. Their host bacteria are unquestionably important evolutionarily, ecologically, agriculturally, gastronomically, and clinically. If not for bacteria, humans and other eukaryotes wouldn't be, much less be publishing books – for a delightful excursion into the dominance of bacteria in evolution and the ecology of our world, see Stephen Jay Gould's 1996 essay on "Planet of the bacteria" (*Washington Post Horizon* **119** [334]: H1. www.stephenjaygould.org/library/gould˙bacteria.html). And then there are the bacterial diseases: plague, pneumonia, tuberculosis, anthrax, diphtheria, cholera and other diarrheas, etc., which collectively remain a major – and in some parts of the world are still *the* major – source of human morbidity and mortality. How important are bacteriophages in the ecology and evolution of the bacteria that are so fundamental to our lives and well-being, and to the well-being of our planet?

As I read through the chapters of this volume, I kept thinking of the question of why bacteriophages have not played a more prominent role in ecology and evolutionary biology. As model systems for studying these subjects, bacteriophages and bacteria are awesome. They can be used to actually test, rather than just champion, ecological and evolutionary hypotheses. What a concept! With bacteriophages, evolution is not just history that can only be studied retrospectively, nor is it "just a theory": you can watch evolution in action, whether you want to or not. With bacteriophages, mathematical modeling is not just a precious exercise but rather a useful (read: essential) tool for the design and interpretation of empirical studies and for generating testable hypotheses. Parameters can be directly estimated and the validity of predictions generated from mathematical and computer simulation models can be evaluated quantitatively.

Bacteriophages are not only useful as model systems for studying ecological and evolutionary problems, they raise oodles of delicious ecological and evolutionary questions in their own right, questions of importance and utility beyond the precious realm of academe. What is the contribution of bacteriophages to regulating the densities of bacterial populations and structuring their communities? How important are bacteriophages as vectors for the horizontal gene transfer that plays such a prominent role in bacterial evolution? Can we use these viruses that kill bacteria to control those that are inconvenient to our agriculture and impinge on our health and well-being? Why is the virulence of the bacteria responsible for diphtheria, hemolytic uremic syndrome, cholera, toxic shock, etc. determined by genes borne on bacteriophages rather than on the chromosomes of their host bacteria?

The chapters in this volume (I recommend starting by reading Chapter 1, which provides an overview) do not address all of these bacteria-phage specific questions, nor completely answer them (fortunately?), but they do address some, as well as raising others. These chapters, which can be read independently of each other, also provide wonderful illustrations of the power of bacteriophages as model systems for studying fundamental ecological and evolutionary questions and the utility of mathematical modeling in that enterprise. This collection is both a how-to guide and an up-to-date review of what's happening in the ecology and evolution of bacteriophages for students, for practicing ecologists, and for practicing evolutionary biologists with the good taste to be working with bacteriophages, or wanting to.

This is a very appealing time to study the ecology and evolution of bacteriophages. The tools of molecular biology, which bacteriophages played a prominent role in developing, are now available to the masses. These tools are particularly amenable to addressing and answering ecological and evolutionary questions with and about bacteriophages.

Enjoy!

Bruce R. Levin
Department of Biology, Emory University

Preface

This monograph introduces the biologically suave reader to the unseen but vibrant ecology of bacterial viruses, also known as phages, whose numbers potentially exceed those of all other organisms combined. Our focus is on phage population dynamics and community interactions (Part I) as well as various aspects of phage evolutionary biology (Part II). Also included are introductions to the phage ecology of a range of environments (Part III), as well as the mathematical modeling of phage (and bacteria) population growth and community interactions (Part IV). In short, this monograph provides a much-needed snapshot of phage ecology as reflected by the interests of the authors involved, and as seen especially from the perspective of the individual phage.

This book, along with my personal research interests, reflects my position at Ohio State University, which is on a regional campus in the town of Mansfield, Ohio. At OSU–Mansfield I lack access to fellow microbiologists, graduate students, and even upper-level undergraduates. Instead, I'm surrounded by a number of fabulous liberal-arts faculty who tend to publish without doing laboratory research (gasp!) in such disciplines as English and History and who otherwise write books for a living. My isolation, as well as my graduate and postdoctoral experience, led me early on to reach out to fellow phage ecologists by a then new-fangled medium called the World Wide Web. My web site on the subject, found at www.phage.org, was originally assembled, in 1996, as a catharsis following my mother's death. It took much time for the site to hit, but it was reviewed in *Science*, in 2000, mere weeks before my tenure packet was due. Its renown and/or my experience with it has resulted in a number of opportunities, most notably my contribution to the editing of *The Bacteriophages*, 2nd edition (Rich Calendar as main editor, Oxford University Press, 2006).

It was during the final stages of line-editing *The Bacteriophages* that I received an email from Mike Wilson, who is one of the editors of the series *Advances in Molecular and Cellular Microbiology*, published by Cambridge University Press. My guess is that Mike somehow came upon the idea of phage ecology, decided that it would make a nice contribution to the series, typed "phage ecology" into Google, and, *voila*, found me. He then invited me to write and submit a proposal, which of course I did, and, amazingly, it was accepted with minimal revision.

In addition to Mike Wilson, I want to thank my main mentors from graduate school: Harris Bernstein, Richard Michod, Stephen Zegura, and John Spizizen, all of whom were there, supporting me through my transition into a phage ecologist. Thank you to John Riedl, who, as my first Dean-Director at Ohio State University Mansfield, took a chance on me – apparently impressed especially by the writing in my teaching statement! John also put up with me while I created, against his best instincts, a web-based virtual community: the Bacteriophage Ecology Group.

Thank you to Hans-Wolfgang Ackermann, Richard Calendar, Elizabeth Kutter, and Ryland Young, who as members of the phage community were there and very supportive of both me and the field at various key moments in my career. Thank you to James Bull for his ongoing interest in my research and almost-from-the-beginning support of this project. An enormous thank you to Steven McQuinn for his ongoing phage artistic efforts and, especially, for producing the cover illustration for this volume. Thank you goes also to Bruce Levin, whose theoretical work with phages inspired much of mine.

Thank you to Cameron Thomas (now Thomas-Abedon), who raises my children, in her spare time does my laboratory research, and otherwise has been extremely supportive of my endeavors during the writing and editing of this book. Thank you, especially, for putting up with me while I suffered and complained through a month-plus devoted to indexing this volume; such dedication is above and beyond the call of duty!

To all of my various authors, who rose to the occasion both in writing their chapters and in putting up with my sometimes heavy editing, an enormous thanks. Thank you also to a number of individuals who have anonymously read and commented on various chapters of this book, and to Hugh Brazier for his sometimes heroic efforts copy-editing this volume. Finally, I would like to dedicate this volume to Gisela Mosig (1930–2003), whose interests were broad, and working knowledge even broader, but who ultimately was driven by the inspirational goal of understanding phages as organisms.

CHAPTER 1

Phages, ecology, evolution

Stephen T. Abedon

1.1 INTRODUCTION

Ecology is the study of the interactions between organisms and their environments. Within our planet's varied environments there exist 10^{30} or more individual bacteria (Whitman *et al.*, 1998), and perhaps 10 or more times as many bacterial viruses, or bacteriophages (phages) as they are more typically described. Whether these viruses are diffusing, decaying, or finding bacteria to adsorb and infect, what is certain is that phages must participate in enormous numbers of ecological interactions. From these numbers we can limit our considerations to approximately four perspectives: (1) how phages operate in the wild; (2) the impact of phages on bacteria, on bacterial communities, and on ecosystem nutrient flow; (3) the employment of phages as model systems for the study of basic principles of ecology, and (4) the diversity of phages in environments. In this monograph, *Bacteriophage Ecology: Population Growth, Evolution, and Impact of Bacterial Viruses*, we take on these approaches, emphasizing how an understanding of phages in the laboratory, both theoretically and empirically, can translate into an understanding of how they operate in the wild. In this chapter I introduce the concept of what phages are and provide overviews of phage biology, phage ecology, and phage evolutionary biology.

For general reviews especially of phage ecology see Anderson (1957), Levin and Lenski (1985), Goyal *et al.* (1987), Ackermann (1997), Paul and Kellogg (2000), Weinbauer (2004), Chibani-Chennoufi *et al.* (2004), Day (2004), Breitbart *et al.* (2005), Brüssow and Kutter (2005b), Abedon (2006a, 2007), and Weinbauer *et al.* (2007). Additional, more narrowly focused

Bacteriophage Ecology: Population Growth, Evolution, and Impact of Bacterial Viruses, ed. Stephen T. Abedon. Published by Cambridge University Press.

phage ecology reviews are listed at the "Phage ecology" entry of Wikipedia (wikipedia.org). A list of 100-plus phage monographs can also be found at Wikipedia under the heading "Phage monographs." See also Abedon (2006b).

1.2 WHAT IS A PHAGE?

For a phage ecologist a phage can be a concept, a fairly well-defined diversity of entities, or a less well-defined collection of ecosystem-modifying creatures. These we will consider in turn.

1.2.1 Phages as a concept

Just what a phage is was once a very contentious issue. Here I provide a brief outline of the history of thinking on phage biology, and then present my own broader sense of what a "phage" is as derived from a perspective of phage ecological theory.

1.2.1.1 Phages as viruses of bacteria

As we enter our tenth decade of the study of phage biology, we can look back upon a long tradition of celebrating the concept of phages as bacterial viruses. Following the approximate co-discovery of phages by Twort (1915) and d'Hérelle (1917) (an English translation of the latter can be found in Summers, 1999), there was much controversy – in those days prior to the invention of electron microscopy – over d'Hérelle's insistence upon the viral nature of phages ("I did not think, then, that a day would come when an electron microscope would be invented . . ."; d'Hérelle, 1949). This controversy over whether phages were indeed viruses was likely due, at least in part, to confusion over the nature of such phenomena as lysogeny (Chapter 5), pseudolysogeny (Chapter 5), or carrier states (Barksdale and Arden, 1974) (see also Section 1.2.2.4). History, of course, has long proven d'Hérelle correct, and today we understand that bacteriophages are viruses with bacterial or, more broadly, prokaryotic host ranges.

D'Hérelle called bacterial viruses bacteriophages, a description of their macro effects on bacterial cultures. At that time the characteristics of whole cultures were a common emphasis in microbiology – a contrast with a later preoccupation with the workings of individual bacteria (Summers, 1991): "a culture was conceptualized not in terms of the population dynamics of individual cells, but as an organism in itself" (Summers, 2005a). D'Hérelle (1949), in fact, provides a clearly whole-cultural description of the phage phenomenon:

> at various times I noticed an anomaly shown by some cultures of the coccobacillus which intrigued me greatly, although in fact the observation was ordinary enough, so banal indeed that many bacteriologists had certainly made it before on a variety of cultures . . . The anomaly consisted of *clear spots*, quite circular, two or three millimetres in diameter, speckling the cultures grown on agar. [emphasis his]

That is, turbid cultures – whether broth, bacterial lawns, or even individual bacterial colonies – can be dramatically reduced in turbidity (i.e., "eaten") as a consequence of phage exposure. The original concept of a phage thus was as an agent, seemingly produced by bacteria, that (1) was not filterable through devices normally capable of filtering bacteria from suspensions, (2) was capable of some sort of self-propagation given exposure to bacteria, and (3) "ate" bacterial cultures. Indeed, the word "phage" is derived from the Greek φαγειν, meaning to eat or devour (Stent, 1963).

D'Hérelle's 1949 retrospective on his 30-plus years studying bacteriophages is very phage-ecological: "The bacteriophage is everywhere . . . ceaseless struggle which goes on in nature between bacteriophage and bacteria . . . In nature every time that bacteria do something, a bacteriophage interferes and destroys the bacteria, or provokes a modification of their action." A phage-infested culture – serving simultaneously as population, community, and ecosystem – is a much more ecological notion than the more modern idea of the microorganism as predominantly a molecularly reducible individual. The study of phage ecology and phage biology thus are intimately entwined, possessing common origins and a common history, especially in terms of the whole-cultural analysis of the phage impact on bacteria. For more on the history of phage biology, start with the various recent reviews by Bill Summers (2005a; 2005b; 2006).

1.2.1.2 Phages as unicellular-organism parasites

"UOPs" (pronounced o͞ops) and "MOPs" (pronounced mŏps) are the parasites of unicellular organisms (unicellular-organism parasites) and multicellular organisms (multicellular-organism parasites), respectively. I first presented the concepts of UOPs and MOPs as part of the mission statement of the Bacteriophage Ecology Group (BEG; www.phage.org/beg_mission_statement.htm), as written, revised, and posted between 1996 and 1999. My intention in describing phages as UOPs was to broaden what one might think of as phage-like when considering the theory of phage ecological interactions (e.g., Abedon, 1999, 2006a).

Included among UOPs, of course, are bacteriophages – such as coliphages (Gerba, 2006), cyanophages (Abedon, 2004, 2006b; Mann, 2006), actinophages (Kurtboke, 2005), etc. – which, with the addition of archaeal viruses (Stedman *et al.*, 2006; Prangishvili *et al.*, 2006a), are viruses whose hosts are prokaryotes. However, there are also a number of non-bacterial unicellular organisms which can serve as viral hosts, plus bacteria that are (or, at least, appear to be) obligately intracellular parasites of unicellular organisms. Together these non-bacteriophage UOPs include the viruses of protozoa (Wang and Wang, 1991; Bruenn, 2000; Attoui *et al.*, 2006); the mycoviruses of yeasts (Wickner, 1989); the viruses of unicellular, eucaryotic algae (Takao *et al.*, 2006); and *Bdellovibrio* spp. along with additional bacterial species that parasitize other bacteria intracellularly (Shilo, 1984; Guerrero *et al.*, 1987; Martin, 2002; Sockett and Lambert, 2004). UOPs also include *Legionella* spp., which are bacteria that parasitize protozoa intracellularly (Fields *et al.*, 2002; Steinert *et al.*, 2002; Greub and Raoult, 2004), and viruses of multicellular creatures (such as animals) when infecting their quasi-unicellular counterparts grown in the laboratory as tissue cultures (e.g., Yin *et al.*, 2001; Cuevas *et al.*, 2003; see also Abedon, 1999).

UOPs may be distinguished from MOPs by a number of criteria. For example, UOPs typically are organisms which (1) are obligate intracellular parasites; (2) employ simpler infection strategies, ones in which within-host growth and intracellular growth are synonymous; and (3) infect morphologically simpler hosts. By contrast: (1) MOPs include obligate intracellular parasites but also many parasites which replicate extracellularly or even extra-organismally. (2) MOPs that are obligate intracellular parasites display four fundamental and differentiable phases: host acquisition (which itself may be broken into two distinct phases, host entry and cell adsorption); an intracellular phase; an extracellular dispersal-within-the-host phase; and finally an extracellular between-host phase. The simpler UOPs, by contrast, exist in just two fundamental and differentiable phases: an extracellular host acquisition phase (i.e., adsorption) and an intracellular phase. (3) MOP hosts tend to be more complex than unicellular organisms.

1.2.2 Phages in terms of their diversity

We can reflect on phages from the perspective of their diversity. This diversity may be considered in terms of those few phages which have been subject to extensive laboratory characterization or, alternatively, in terms of that diversity which exists in the wild. The latter approach I describe as phage environmental microbiology (Abedon, 2008). See Chapter 10 for a brief description of techniques employed for determining aquatic viral diversity as well as discussion of the extent of that diversity.

1.2.2.1 Diversity characterization involving phage culturing

The traditional means of characterizing phage environmental diversity is by means of plaque assays (Wilhelm and Poorvin, 2001). The strength of these viable counts lies in the potential for phenotypic characterization of phages, including host-range determination. On the other hand, requiring phages to produce plaques demands that a suitable host bacterium be used, or even exist as a pure culture, and that plaques form under the plaquing conditions employed. Thus, there are strong selective biases in determinations of phage environmental diversity if those determinations require plaque formation, or even phage culturing in broth (Snyder *et al.*, 2004).

Biases probably are even stronger given the employment of methods of pre-enrichment prior to plaquing. These biases exist especially in terms of phage virulence (Chapter 3) and perhaps also in terms of phage valence (which is the number of bacterial types that a phage is capable of infecting and which may be lower within enriched cultures than among wild phage populations; Jensen *et al.*, 1998). There also exists no guarantee that the indicator host or hosts one employs are a fair representation of the actual host types that a phage has encountered in its recent past. That is, only so long as phages are isolated from environmental samples as phage-infected bacteria, and then only so long as a second, effectively identical uninfected host bacterium is simultaneously isolated, can we be reasonably sure that we have isolated the (or, at least, a) true host of a given phage.

1.2.2.2 Diversity characterization without culturing

A second approach towards characterization of phage environmental diversity involves electron microscopy (Breitbart *et al.*, 2004; Williamson *et al.*, 2005) or genotype determination without first culturing (Suttle, 2002; Weinbauer, 2004; Breitbart *et al.*, 2005; Breitbart and Rohwer, 2005). These approaches hold out the promise of providing at least some characterization and identification of phages that otherwise cannot be easily cultivated. However, these advantages come at the expense of phenotypic characterization, though some inferences may be made based on virion morphology or from gene sequences obtained via metagenomic analysis (Breitbart and Rohwer, 2005; see also Chapter 7).

1.2.2.3 Phage types

There exist three basic means of characterizing phages into types: infection phenotype (including host range as well as additional culturing characteristics), morphology (including virion shape, dimensions, serology, physicochemical properties, and nucleic acid characteristics), and genome sequence. The order of presentation reflects the development of phage research. Early studies were limited to the phenotype of infection. The study of infection phenotype was subsequently complemented by serological and electron microscopic characterization. Serological characterization was then replaced by DNA–DNA hybridization, which, in turn, gave way to gene–gene sequence comparison as a means of analyzing phage diversity. Characterization of phage diversity today has seemingly reached a near pinnacle of sophistication through a combination of whole-genome sequencing, comparative phage genomics, and electron microscopic analysis. For recent review of phage virion classification by morphology and physicochemical properties, see Ackermann (2005, 2006). For consideration of phage genomic comparisons, see Rohwer and Edwards (2002) and Brüssow and Desiere (2006). For a compilation of detailed analyses of infection phenotypes of select phages (especially molecular details), see Calendar and Abedon (2006).

By both morphological and genome analysis, bacterial viruses generally may be divided into a number of basic types (after Ackermann, 2006, and the International Committee on Taxonomy of Viruses, www.ncbi.nlm.nih.gov/ICTVdb/Ictv). In terms of genome types these are ssDNA, dsDNA, ssRNA, and dsRNA. In terms of virion morphology there are tailed (also described as binary for their symmetry), isometric (generally icosahedral), helical (which are filamentous or rod-shaped), and pleomorphic phages. All tailed phages (order *Caudovirales*) have dsDNA genomes and may be differentiated into the families *Myoviridae* (long contractile tails), *Siphoviridae* (long noncontractile tails), and *Podoviridae* (short contractile tails). Among isometric phages are all four genome types, including segmented dsRNA (family *Cystoviridae*). Helical and pleomorphic phages are mostly represented by phages with dsDNA genomes, with one exception: family *Inoviridae*, which are filamentous helical viruses with ssDNA genomes. The most prevalent phages are tailed.

I link common phages with their families, morphologies, and genome types as follows:

phage T4 – family *Myoviridae*, tailed, dsDNA; also representing family *Myoviridae* are P1-, P2-, Mu-, and SPO1-like phages
phage λ – family *Siphoviridae*, tailed, dsDNA; also representing *Siphoviridae* are T1-, T5-, c2-, and L5-like phages
phages T3 and T7 – family *Podoviridae*, tailed, dsDNA; also representing family *Podoviridae* are φ29- and P22-like phages
phage φX174 – family *Microviridae*, icosahedral, ssDNA

phage ϕ6 – family *Cystoviridae*, icosahedral and enveloped, segmented dsRNA
phage MS2 – family *Leviviridae*, icosahedral, ssRNA; also phages F2 and Qβ
phages f1, fd, and M13 – family *Inoviridae*, filamentous, ssDNA

Phages f1, fd, and M13 are also distinct in that they adsorb to F (sex) pili and are released chronically by extrusion (i.e., continuously from infected bacteria rather than via lysis; see Section 1.2.2.4 and Russel and Model, 2006). Almost all *Leviviridae* adsorb to pili, too. Viruses of domain Archaea (often described simply as viruses rather than as phages) are reviewed by Prangishvili *et al.*, (2006b) and Stedman *et al.* (2006).

1.2.2.4 Types of phage infection

Several recent articles review some of the more common aspects of phage biology (Calendar and Inman, 2005; Guttman *et al.*, 2005; Kutter *et al.*, 2005). I direct readers to these references to gain a relatively quick, particularly molecular understanding of phage biology. There are also various recent phage books (e.g., Kutter and Sulakvelidze, 2005; Calendar and Abedon, 2006), which will provide greater detail on numerous areas of phage biology. In this section I instead discuss, and classify, general aspects of phage biology as viewed especially from an ecological perspective. I do this not to leave you with the impression that an understanding of molecular aspects of phage biology is unimportant. Far from that, it is always of vital importance for any biologist, working at any level of biological organization, to learn as much as possible about the organism with which he or she works, *especially those aspects that might directly impact what one might be studying* (Abedon, 2000a). Toward that end, I cannot stress too strongly the importance of determining (and determining well) at the very least the basic phage growth parameters – adsorption constant, burst size, latent periods – for each phage–bacterium–environment combination with which one works (Carlson, 2005; Hyman and Abedon, in press).

Rather than attempting to address the molecular and physiological details of any particular phage, my approach, instead, is to broadly differentiate phage infections into functional types. Traditionally (and simplistically) this has been done by distinguishing phage infections into two basic categories, lytic versus lysogenic, plus perhaps a third: chronic. Here, however, I will take a road untraveled, considering three higher-level and six lower-level categories. I differentiate each category in terms of bacterial survival (yes or no), phage survival (yes or no), phage genome replication (yes or no), phage virion production (yes or no), and phage progeny release (lysis vs. chronic vs. none) (Table 1.1).

Table 1.1 *Categories of phage infection types.*

Higher category[a]	Lower category	Infection type	Bacterial survival[b]	Phage survival	Phage genome replication	Virion production	Virion release
I	i	lytic	no	yes	yes	yes	lysis
I	ii	chronic[c]	yes	yes	yes	yes	chronic[d]
II	iii	lysogenic	yes/no[e]	yes	yes	no	no
III	iv	pseudolysogenic	yes/no[f]	yes	no[g]	no	no
III	v	restricted[h]	yes	no	no	no	no
III	vi	abortive[i]	no	no	no	no	no
n.a.	n.a.	resistant[j]	yes	yes	no	no	no
all	all	partial resistance[k]	yes/no	yes	yes/no	yes/no	yes/no

[a] I = productive, II = lysogenic, III = non-productive (or minimally-productive) infections.
[b] Survival at least over the short term, so long as the current infection type continues, with the caveat that one infection type may change with time into another, e.g., lysogenic infection followed by lytic infection.
[c] A "chronic" infection instead may be described as "continuous" (Ackermann and DuBow, 1987a).
[d] Virion release during chronic infection occurs via extrusion or budding across the cell envelope of the infected bacterium (Ackermann and DuBow, 1987b).
[e] Absent induction, bacteria survive.
[f] Some but not all bacteria progeny may survive (Chapter 5).
[g] Even if phage genome replication occurs, it is at a low level compared to during productive or even lyosgenic infection.
[h] Restrition renders bacteria "immune" (Levin and Lenski, 1983) to infecting or adsorbing phages. Retriction-like mechanisms involve blocking of genome injection by adsorbed phages (exclusion); targeted destruction of phage genomes within bacterial cytoplasms (restriction); blocking of the replication superinfecting phage genomes by prophage (immunity); or prophage eradication (curing). See Chapter 8 for further consideration of superinfection exclusion and immunity.
[i] An abortive infection (also known as phage exclusion; Chopin *et al.*, 2005) is halted after phage macromolecular synthesis has begun (Duckworth *et al.*, 1981). By definition, the host cell is killed (Klaenhammer and Fitzgerald, 1994).
[j] Resistance meaning that phage–bacterium encounters result in an absence of irreversible adsorption (Levin and Lenski, 1985). Note that Levin and Lenski (1983) do not distinguish between exclusion and resistance, instead considering bacterial resistance to irreversible phage adsorption to be a form of exclusion.
[k] The phrase "partially phage-resistant" bacteria is in the sense of Kerr *et al.* (Chapter 2). What this means is that bacteria display a reduced affinity for phage relative to some ideal (this ideal is $f = 1$ = fraction of all collisions that result in irreversible adsorption; Stent, 1963). That is, phage–bacterium encounters do not necessarily lead to infection and therefore some bacteria survive since the phage adsorption constant is reduced on these bacteria, but not to zero. Strictly, it is likely that all bacteria are at least partially resistant to all phages, in that in few cases would all phage–bacterium encounters result in phage irreversible adsorption. Even in the example presented by Stent (1963), by the statement "nearly every collision between a T4 phage and an *E. coli* bacterium seems to result in fixation of the virus particle" (p. 90), by "nearly" he means 0.5, e.g., rather than 0.001 (G. Stent, personal communication).

The six "lower-level" categories include:

(i) Lytic infections, which are infections terminated following phage virion production by a phage-induced lysis of the bacterial host (see Chapter 3 for discussion of the ecology of lytic-phage population growth and Fig. 2.1 for illustration, in broad overview, of a phage lytic cycle).

(ii) Chronic infections (Russel and Model, 2006), which involve the release of produced virions without immediate destruction of the host infection.

(iii) Lysogeny (Chapters 5 and 14), which involves phage genome replication but no virion production or release, all the while retaining bacterial viability.

(iv) Pseudolysogeny (Chapter 5), which involves no (or, at best, minimal) phage genome replication but retains some potential for future phage genome replication and subsequent virion production. That is, with pseudolysogens host bacteria do not die, at least not in the short term, and unlike in most cases of lysogeny, phage genomes do not integrate into the host chromosome.

(v) Phage restriction (e.g., Jeltsch, 2003), described as a kind of anti-phage immunity (Levin and Lenski, 1985; Lenski, 1988), involves loss of phage but not of bacterial viability. Prophage curing (conversion of lysogens to non-lysogens) also involves loss of phage without loss of bacterial viability.

(vi) Phage abortive infections (e.g., Bouchard *et al.*, 2002), which involve loss of both phage and bacterial viability.

Note that both abortive and restricted infection can be viewed as consequences of the deployment by bacteria of antiphage defenses (Weinbauer, 2004). Weinbauer also reviews mechanisms of phage adsorption inhibition which I describe here as a kind of phage resistance. Chapter 2 discusses "partial" phage resistance (see also Abedon, 2006a). These different categories of phage infection are summarized in Table 1.1.

At the proposed higher level of categorization we have (I) phage productive infections, which cover categories (i) and (ii) as discussed above (lytic as well as chronic infection) and which may be described as yes–yes infections in terms of phage genome replication and virion production; (II) phage lysogenic infections, which may or may not, depending on the phage, involve phage genome integration, but which in terms of phage genome replication and virion production may be described as yes–no infections (respectively); and (III) phage non-productive infections, which cover categories (iv) through (vi), and which represent no–no infections (even if only temporarily, as is the case with pseudolysogeny). An alternative means of classification, one that was formulated only as this chapter went to press, is to distinguish phage infections into "productive" (lytic or chronic), "reductive" (lysogenic or pseudolysogenic), or "destructive" (restrictive or abortive).

1.2.2.4.1 Another word on pseudolysogeny

Pseudolysogeny, as employed here (and in Chapter 5) is described by some authors as a phage "carrier state" (e.g., Ackermann and DuBow, 1987a). The alternative meaning of each of these terms is a culture containing mostly phage-resistant bacteria but also a small number of phage-sensitive bacteria that maintain a smaller number of phage *lytic* infections (this is a "carrier strain" as Lwoff [1953] employed the term, which he seems to have considered to be synonymous with pseudolysogenic). Both pseudolysogeny and the carrier state may be described as persistent phage infections, as distinguished from lysogeny or chronic infection (Ackermann and DuBow, 1987a), or "persistently contaminated" cultures (Barksdale and Arden, 1974). Persisting infections mimic lysogeny in that they can give rise to phages (i.e., they are *lyso-genic*) and, in the case of "carrier strains," they are dominated by phage-resistant bacteria. They are unlike lysogens, however, in that the persisting state is easily cured via application of anti-phage serum. Note that in their 1974 review on the subject of persisting infections (as well as lysogeny), Barksdale and Arden, like Lwoff before them, use "pseudolysogeny" and "carrier state" synonymously, apparently using the definition for both that is equivalent to Lwoff's carrier-strain definition.

Under conditions of partial host starvation the two phenomena – pseudolyosgeny and the carrier state – likely occur both simultaneously and distinctly, resulting in the formation, for example, of turbid plaques (Chapter 5). In that case bacteria would be resistant in a physiological sense, with phage adsorption resulting in pseudolysogeny. Other bacteria within the same lawn, by chance or due to their history or microenvironment, would be less physiologically resistant, resulting in these more phage-sensitive bacteria serving as carriers (*sensu* Chapter 2) of sufficient numbers of productive phage infections to give rise to plaque formation. In the case of the turbid plaques described in Chapter 5, and using the terms as employed in this monograph, plaque formation would be the signature of the carrier state (low-level productive phage infection and bacterial lysis) while at least some of the intact bacteria giving rise to plaque turbidity would be pseudolysogenically infected. That is, these latter, non-productively infected bacteria would contain a non-replicating phage genome, or preprophage as these are described in Chapter 5.

1.2.2.5 Virulence versus temperance

It is important to appreciate the following (from Abedon, 2000b): (1) There is no such thing as a "lysogenic phage." Bacteria can be lysogenic. Bacteria can possess prophages. But potentially lysogenizing phages are properly described as temperate rather than as lysogenic phages. (2) A lytic phage is a phage capable of going through a lytic cycle, but temperate phages (most of them, anyway; Chapter 14) are just as capable of going through a lytic cycle as a lysogenic cycle. Thus, the dichotomy of lytic versus temperate is a false one. (3) The prevalence of the above-noted dichotomy stems from a common equating of "lytic phage" with "virulent phage," which are, in turn, seen as opposites of "temperate phage." However, since temperate phages are also able to display lytic cycles (and, indeed, are thought to predominantly display lytic versus lysogenic cycles on a per-adsorption basis), it is again a false dichotomy to imply that temperate phages are not virulent (in the sense of being able to mostly wipe out bacterial cultures; d'Hérelle, 1949). The error in this latter case is one of overlooking the death and mayhem that can be caused by temperate phages, as exemplified by the origin of the term lysogenic, i.e., capable of generating the lysis of cultures of non-lysogens. Instead, undue emphasis – in terms of phage population growth – is placed on the small lack of death that occurs in these same cultures as a consequence of lysogenization. In other words, it is not true that temperate phages are *not* virulent, but instead that they are *less* virulent, and even "less virulent" deserves a qualifier of "potentially."

What temperate phages are not is obligately lytic, a phrase which I prefer to the more ambiguous "virulent" when describing otherwise lytic phages as not temperate. Studies (other than my own or those that I have edited) that employ the phrase "obligately lytic" include Matsuda and Barksdale (1967), Madsen *et al.* (2001), Durmaz *et al.* (2002), Mann (2003), Sullivan *et al.* (2003), Bailey *et al.* (2004), Sturino and Klaenhammer (2004), Paul and Sullivan (2005), Brüssow (2005), Trotter *et al.* (2006), and Bull *et al.* (2006). Note, though, that Barksdale and Arden (1974) are stricter in their definition of "obligately lytic" than simply indicating not temperate or not being capable of displaying a lysogenic cycle. Instead, they argue that "phages are *obligately lytic*, i.e. there is for them no finite probability of giving rise to mutants capable of lysogenizing host bacteria" (p. 276, emphasis theirs). Barksdale and Arden (1974) furthermore reserve the word "virulent" for those phages that in fact can mutate back to being capable of lysogenizing, i.e., otherwise temperate phages that have mutationally lost that ability. I would urge phage workers to follow their lead.

1.2.3 Phages as modifiers

There exist a variety of mechanisms whereby phages can modify ecosystems. Most of these mechanisms are mediated through the phage infection of bacteria. These effects are manifest at the level of bacterial communities, contributing to increases in bacterial diversity, and at the level of nutrient cycling within ecosystems. I consider these effects in reverse order.

1.2.3.1 Phages as ecosystem modifiers

During the 1990s, in reaction to electron microscopic observations of an unexpected plethora of aquatic virions (Bergh *et al.*, 1989), there was an explosion of interest in aquatic phage ecology (Chapter 10). A dominant theme throughout this literature is consideration of the phage impact on aquatic nutrient cycling. The phage impact in aquatic environments is a consequence of two phenomena. The first is that a large fraction of primary productivity (photosynthesis) in aquatic ecosystems is carried out by prokaryotes, i.e., cyanobacteria, and the resulting organic carbon becomes available to higher trophic levels as a consequence of grazing by eukaryotes, especially protozoa. The second is that most dissolved organic material within aquatic ecosystems becomes available to eukaryotic grazers only once it is assimilated by bacteria.

Because of phages, however, "6–26% of photosynthetically fixed organic carbon is recycled back to dissolved organic material by viral lysis" (Wilhelm and Suttle, 1999). That dissolved carbon may then be assimilated by heterotrophic bacteria. Heterotrophic bacteria, like photosynthetic bacteria, are susceptible to phage-induced lysis. The result is a diminishing of the pace of movement of organic carbon from photosynthesizers, or from heterotrophic bacteria, to higher trophic levels. In other words, phage-induced lysis of prokaryotes creates a "viral loop" (Thingstad, 2000) that "short-circuits" the movement of organic carbon through what is otherwise known as the aquatic "microbial loop" (Chapter 10).

Interestingly, lysis products may accumulate as "recalcitrant dissolved organic material in the deep oceans," material that is somewhat less available to heterotrophic bacteria and which consequently can serve as a crucial global carbon sink (Suttle, 2000). Phages may also interfere with terrestrial ecosystem functioning, such as by infecting nitrogen-fixing (Abebe *et al.*, 1992; Ahmad and Morgan, 1994; Hammad, 1998) or biocontrol-mediating

(Keel *et al.*, 2002) bacteria. Furthermore, phages, by disrupting normal-flora bacteria, are possible causes of bacterial vaginosis (Tao *et al.*, 2005) as well as inefficient food use by ruminants (Klieve, 1996). Phage-encoded genes, such as those encoding many bacterial exotoxins, can also modify ecosystems such as animal gastrointestinal tracts (Abedon and LeJeune, 2005, and Chapter 14).

1.2.3.2 Phages increase bacterial diversity

There exist a number of mechanisms whereby phages can modify bacterial communities, including by increasing bacterial diversity. These mechanisms can be differentiated into those associated with phage-mediated predation of bacteria, those associated with genetic transduction, and those associated with lysogenic conversion (Anderson, 1957). The impact of phages as predators of bacteria on bacterial diversity is considered under the general heading of "killing the winner," as presented in Chapter 10. The idea is that phages can impart frequency-dependent selection on bacteria whereby those bacteria that succeed in growing to higher population densities are also the bacteria which should be most susceptible to phage attack (both due to the higher densities of these more "successful" bacteria and because of trade-offs those bacteria may have made in terms of phage susceptibility in order to grow to these higher densities). See Chapter 2 for consideration of the consequence on bacterial phenotype of selection for bacterial resistance to phage attack.

Transduction is the transfer of genetic material (horizontal or lateral gene transfer) from bacterium to bacterium as mediated by phage virion particles and resulting in modification of bacterial genotypes. Therefore, transduction increases bacterial genetic diversity. See Chapter 11 for additional consideration of phage-mediated horizontal transfer. From the perspective especially of phage ecology, this transfer can occur in two basic guises: one in which otherwise viable phages are carrying DNA between bacteria and the other in which genetically inviable virions are doing the carrying (Breitbart *et al.*, 2005). The carrying of morons by phages (meaning "more" DNA, typically assumed to be of bacterial origin; Chapter 7; Hendrix *et al.*, 2000, 2003) is an example of DNA carried by otherwise viable phage virions. At the other extreme is generalized transduction in which phage virions carry bacterial DNA but not phage DNA, implying a lack of phage infection upon acquisition of a bacterial cell. In between these extremes exists specialized transduction, in which certain bacterial genes formerly found in close association with an integrated prophage genome become incorporated, upon imprecise prophage excision, into the resulting phage progeny. Virions displaying specialized transduction may or may not be able to support subsequent productive phage infection.

Lysogenic conversion represents the expression of prophage-encoded genes that results in changes in bacterial phenotype (Barksdale and Arden, 1974; Breitbart *et al.*, 2005; Abedon and LeJeune, 2005; see Chapters 5, 11, and 14). Thus, phage action can select among bacterial phenotypes, can modify bacteria genetically by mediating what is essentially bacterial sex, and can modify bacterial phenotypes by expressing genes over the course of otherwise latent phage infections.

1.2.3.3 Phage ubiquity and prevalence

Phages have an impact on bacterial communities and ecosystems only to the extent that phages are actually present in those communities. Consistent with their overall importance as modifiers of environments, phages have been observed or isolated in a wide variety of environments (Ackermann and DuBow, 1987c; Ackermann, 1997; Chapters 10–14). To have a significant impact, phages also must be present in these environments at relatively high densities (e.g., Chapters 5 and 10). Densities vary both spatially and temporally, ranging from pelagic lows of less than 10^4 virus-like particles per mL to highs that are in excess of 10^8 mL^{-1} (Wommack and Colwell, 2000), or even 10^{10} mL^{-1} in sediments (Maranger and Bird, 1996) (see Chapter 10 for discussion of the phage ecology of aquatic environments). Correlations of virus-particle density with bacterial density and activity suggest that a substantial fraction of these virus particles are phages (Weinbauer, 2004). Recently it has been argued that the phage density in soils also may be on the order of 10^8 g^{-1} (Ashelford *et al.*, 2003; Williamson *et al.*, 2003; see Chapter 11 for discussion of the phage ecology of terrestrial environments).

If we assume an average of 10^6 phages per mL within aquatic environments and multiply this number by the volume of the world's major oceans – 3, 3, and 7×10^{23} mL for the Atlantic, Indian, and Pacific oceans, respectively (Abedon, 2001) – then we can estimate worldwide total phage numbers of at least 10^{30} (or, as speculated, e.g., by Brüssow and Hendrix [2002], Rohwer and Edwards [2002], and Hendrix [2005], perhaps even more). 10^{30} phages is approximately one phage per prokaryotic cell (Whitman *et al.*, 1998). Lined up end to end, a string of 10^{30} phages would extend 5 or more million light-years into space (Weinbauer and Rassoulzadegan, 2004; Hendrix, 2005). More down to earth, assuming an average mass of 10^8 daltons per phage (Dubin *et al.*, 1970), then 10^{30} phages represents a combined mass of 10^{14} g ($\approx 10^8 \times 10^{30}$/Avogadro's number), which is equivalent to the weight of one million adult blue whales at 100 metric tons (10^8 g) per whale (Abedon, 2001).

Further suggestions of just what 1 000 000 000 000 000 000 000 000 000 000-plus might mean in "human" terms I discuss elsewhere (Abedon, 2001).

1.3 WHAT IS PHAGE ECOLOGY?

Ecology is the study of the interactions between organisms and their environments, and environments can be defined in terms of the biotic (other organisms, including conspecifics) or the abiotic (non-living things, including the remains of once-living things). Phage ecological thinking may be divided into the various subdisciplines (or areas of ecology) as generally employed by ecologists. I originally presented such categorizations of phage ecology research in the Bacteriophage Ecology Group's *BEG News* (Abedon, 2000c). This approach was elaborated upon in Breitbart *et al.* (2005) and Abedon (2006a, 2008). These various ecological subdisciplines include organismal ecology, physiological ecology, evolutionary ecology, behavioral ecology, population ecology, community ecology, ecosystem ecology, and landscape ecology, plus mathematical ecology and metapopulation ecology (Chapter 2).

Organismal ecology studies the adaptations that phages employ to survive as well as to acquire new bacteria to infect. We can consider phage organismal ecology to involve the evolution of various phage intracellular and extracellular adaptations to better fit current environmental conditions (see, for example, Chapter 5). At least some of these adaptations may be summarized under the heading of phage growth parameters (Chapter 15), with methods for the determination of these growth parameters reviewed elsewhere (Carlson, 2005; Hyman and Abedon, in press). This subdiscipline of phage ecology, along with the associated phage physiological ecology (the impact of environments on phage growth parameters), is most closely allied with the extensive phenotypic and molecular characterization of phages that has occurred over the past 90 or so years. Reviews with significant organismal ecological components include those by Lwoff, (1953), Kutter *et al.* (1994), Schrader *et al.* (1997), Robb and Hill (2000), Paul and Kellogg (2000), Weinbauer (2004), Chibani-Chennoufi *et al.* (2004), Breitbart *et al.* (2005), and Abedon (2006a, 2007).

Evolutionary ecology is the study of adaptations from an evolutionary perspective or, as DeFilippis and Villarreal (2000, p. 127) put it, "Evolutionary ecology is the study of how organisms respond and have responded evolutionarily to the selective environments in which they exist." This discipline, as it may be applied to phages, is considered especially in Chapters 2, 3, 4, 8, and 14, while phage behavioral ecology, to the degree that phages can be considered to display behaviors, is addressed in Chapter 2 (optimal foraging theory) and in Chapters 3 and 8 (game theory). Population ecology considers phage population growth and intraspecific interactions. These phenomena are considered especially in Chapters 3, 4, 8, 15, and 16. See also Bull *et al.* (2004), Abedon (2006a, 2007), and Abedon and Yin (in press).

Community ecology considers phage interactions with other organisms, including with other phages, with bacteria (including the bacterial host), and even with organisms harboring the bacterial host. See Chapters 2, 8, 10, 11, 13, 14, 15, and 17 and Section 1.2.3 (this chapter) for consideration of these various interspecific interactions. Note that these interactions may be classified as predatory, parasitic, parasitoid (more generally, victim-exploiter; Chapter 2), or even mutualistic (Levin and Lenski, 1983, 1985; Lenski, 1988; Weinbauer, 2004; Day, 2004; Forde *et al.*, 2004), depending upon the system and also, seemingly, on the mood of the researcher (see Martin [2002] for a similar discussion vis-à-vis *Bdellovibrio*). These interactions also may be better described as a process rather than as the trophic level that the terms predation or parasitism (etc.) would imply (Chapter 10). Chronic phage infections (Table 1.1) are better described as parasitic rather than predatory. Note that it is also possible to consider the interactions between different phages (i.e., phage–phage interactions) in terms of mutualism, predator–prey interactions, etc. (Weinbauer, 2004). See Chapters 5 and 14 for more on the concept of phages as mutualistic symbionts. For more on phage community ecology, see Anderson (1957), Barksdale and Arden (1974), Levin and Lenski (1983, 1985), Krüger and Bickle (1983), Lenski (1988), Miller and Sayler (1992), Suttle (1994), Bohannan and Lenski (2000), Levin and Bull (2004), Weinbauer and Rassoulzadegan (2004), Sutherland *et al.* (2004), Comeau and Krisch (2005), and Abedon (2006a, 2007).

Phage ecosystem ecology describes the phage impact on nutrient and energy flow within environments. This field has been extensively reviewed in recent years (see Section 1.2.3.1) and is at least touched upon by Chapter 10. Landscape ecology considers the interaction between ecosystems, and is an area of phage ecology that has been underappreciated except in association with the use of phages as markers for environmental fecal contamination (e.g., Gerba, 2006), but even then concepts from landscape ecology are not considered explicitly.

1.4 WHAT IS PHAGE EVOLUTIONARY BIOLOGY?

There exist at least five ways of considering phage evolution: phage evolutionary relatedness, phage origins (including evolution from the perspective of genomics), phage evolution in terms of mechanisms (e.g., as deviating from Hardy–Weinberg assumptions), phage evolutionary ecology, and the use of phages as model systems to study basic evolutionary principles (experimental evolution). Despite the title of this chapter, with its conspicuous use of the word "evolution," I will be somewhat limited in my discussion of phage evolution, concentrating mainly on cheerleading for the evolutionary biology chapters presented in this volume (especially Part II, Chapters 6–9).

1.4.1 Phage relatedness

Study of phage relatedness was once a province of the analysis of virion phenotypic properties (such as virion morphology) and host range (Ackermann, 2001, 2003, 2005, 2006; Ackermann and Abedon, 2001). Today, like the analysis of the evolutionary relatedness of essentially every other microorganism, determination of phage relatedness increasingly employs DNA sequence (Rohwer and Edwards, 2002; Brüssow and Desiere, 2006; Prangishvili *et al.*, 2006a; Liu *et al.*, 2006). Due to space limitations, as well as the general proclivity of this editor, emphasis in this volume will not be on issues of phage relatedness. Instead I direct the interested reader to the references listed above.

1.4.2 Phage origins and genomic evolution

The extensive sequencing of phage genomes has not only resulted in a potential to estimate phage relatedness, but also has allowed a gaining of insights into general principles of phage evolutionary pathways. See Chapter 7 for consideration of this important subject, as well as Brüssow *et al.* (2004), Casjens (2005), Hendrix (2005), Brüssow and Kutter (2005a), and Brüssow and Desiere (2006) for recent reviews.

1.4.3 Phage evolution in terms of mechanisms

The mechanisms of phage evolution, in addition to horizontal gene transfer, are processes that can give rise to changes in allele frequency within populations: mutation, selection, genetic drift, and migration. These are considered, using these deviations from Hardy–Weinberg assumptions as a guiding principle, in Chapters 6 and 9. Relevant issues include mutation rates, phage Darwinian fitness (including the concept of adaptive landscapes), and the impact of small population sizes on allele frequency (genetic drift).

1.4.4 Phage evolutionary ecology

I introduce phage evolutionary ecology in Section 1.3. Evolutionary ecology exists as the other "half" of the science of evolutionary biology, contrasting with issues of evolutionary mechanisms and relatedness. In essence, evolutionary ecology is evolutionary biology with an emphasis on the evolution of phenotype, whereas the rest of evolutionary biology, ultimately, has an emphasis on inheritance of genotype. Issues of phage evolutionary ecology are touched upon in this volume, especially in Chapters 2 and 3, where evolution of phage virulence and game theory are introduced, and then in Chapter 8, which considers the evolutionary implications of phage–phage interactions within infected bacteria.

1.4.5 Experimental evolution

The same relative phenotypic and genotypic simplicity that made phages key to the development of the science and techniques of molecular genetics and biology, make phages ideal for studying evolution as it can occur in the laboratory. Especially helpful are small viruses, which may be readily sequenced in full, as well as single-stranded and RNA viruses, which both display and are adapted to high mutation rates as compared with dsDNA-genomed organisms. Thus, phages can display high mutation rates, seemingly significant potential to vary genotypically, ease of propagation, and straightforward implementation of natural selection, such as by altering biotic (i.e., host) or abiotic (e.g., temperature) conditions. See Chapter 9 for an overview of phage experimental evolution. An abbreviated list of mostly recent phage experimental evolutionary biology references, sorted by category, and which is based on a table presented by Breitbart *et al.* (2005), can be found under the heading "Phage experimental evolution" via Abedon (2006b).

ACKNOWLEDGMENTS

Thank you to Hans Ackermann, Richard E. Herman, Benjamin Kerr, and Cameron Thomas for their help in editing this chapter.

REFERENCES

Abebe, H. M., M. J. Sadowsky, B. K. Kinkle, and E. L. Schmidt. 1992. Lysogeny in *Bradyrhizobium japonicum* and its effect on soybean nodulation. *Appl. Environ. Microbiol.* **58**: 3360–6.

Abedon, S. T. 1999. When grown *in vitro*, do parasites of multicellular organisms (MOPs) become unicellular organism parasites (UOPs)? *BEG News* **2**. www.phage.org/bgnws002.htm#editorial.

Abedon, S. T. 2000a. Bacteriophages as model systems. *BEG News* **3**. www.phage.org/bgnws003.htm#editorial.

Abedon, S. T. 2000b. Lytic, lysogenic, temperate, chronic, virulent, quoi? *BEG News* **5**. www.phage.org/bgnws005.htm#editorial.

Abedon, S. T. 2000c. Which ecology are you? *BEG News* **6**. www.phage.org/bgnws006.htm#editorial.

Abedon, S. T. 2001. How big is 10^{30}? *BEG News* **7**. www.phage.org/bgnws007.htm#submissions.

Abedon, S. T. 2004. The cyanophage literome. *BEG News* **22**. www.phage.org/bgnws022.htm#submissions.

Abedon, S. T. 2006a. Phage ecology. In R. L. Calendar and S. T. Abedon (eds.), *The Bacteriophages*, 2nd edn. Oxford: Oxford University Press, pp. 37–46.

Abedon, S. T. 2006b. Phage ecology on Wikipedia. *BEG News* **25**. www.phage.org/bgnws025.htm#abedon_editorial.

Abedon, S. T. 2008. Ecology of viruses infecting bacteria. In B. Mahy and M. van Regenmortel (eds.), *Encyclopedia of Virology*. Oxford: Elsevier.

Abedon, S. T., and J. T. LeJeune. 2005. Why bacteriophage encode exotoxins and other virulence factors. *Evol. Bioinf. Online* **1**: 97–110.

Abedon, S. T., and J. Yin, in press. Bacteriophage plaques: theory and analysis. In M. Clokie and A. Kropinski (eds.), *Bacteriophages: Methods and Protocols*. Totowa, NJ: Humana Press.

Ackermann, H.-W. 1997. Bacteriophage ecology. In M. T. Martins, M. I. Z. Sato, J. M. Tiedje, L. C. N. Hagler, J. Döbereiner, and P. S. Sanchez (eds.), *Progress in Microbial Ecology* (Proceedings of the Seventh International Symposium on Microbial Ecology). São Paulo: Brazilian Society for Microbiology/International Committee on Microbial Ecology, pp. 335–9.

Ackermann, H.-W. 2001. Frequency of morphological phage descriptions in the year 2000: brief review. *Arch. Virol.* **146**: 843–57.

Ackermann, H.-W. 2003. Bacteriophage observations and evolution. *Res. Microbiol.* **154**: 245–51.

Ackermann, H.-W. 2005. Bacteriophage classification. In E. Kutter and A. Sulakvelidze (eds.), *Bacteriophages: Biology and Application*. Boca Raton, FL: CRC Press, pp. 67–90.

Ackermann, H.-W. 2006. Classification of bacteriophages. In R. L. Calendar and S. T. Abedon (eds.), *The Bacteriophages*, 2nd edn. Oxford: Oxford University Press, pp. 8–16.

Ackermann, H.-W., and S. T. Abedon. 2001. Bacteriophage names 2000. www.phage.org/names.htm.

Ackermann, H.-W., and M. S. DuBow. 1987a. Lysogeny. In H.-W. Ackermann and M. S. DuBow (eds.), *Viruses of Prokaryotes. Volume I: General Properties of Bacteriophages.* Boca Raton, FL: CRC Press, pp. 87–101.

Ackermann, H.-W., and M. S. DuBow. 1987b. Phage multiplication. In H.-W. Ackermann and M. S. DuBow (eds.), *Viruses of Prokaryotes. Volume I: General Properties of Bacteriophages.* Boca Raton, FL: CRC Press, pp. 49–85.

Ackermann, H.-W., and M. S. DuBow. 1987c. *Viruses of Prokaryotes, Volume 1, General Properties of Bacteriophages.* Boca Raton, FL: CRC Press.

Ahmad, M. H., and V. Morgan. 1994. Characterization of a cowpea (*Vigna unguiculata*) rhizobiophage and its effects on cowpea nodulation and growth. *Biol. Fertil. Soils* **18**: 297–301.

Anderson, E. S. 1957. The relations of bacteriophages to bacterial ecology. In R. E. O. Williams and C. C. Spicer (eds.), *Microbial Ecology.* Cambridge: Cambridge University Press, pp. 189–217.

Ashelford, K. E., M. J. Day, and J. C. Fry. 2003. Elevated abundance of bacteriophage infecting bacteria in soil. *Appl. Environ. Microbiol.* **69**: 285–9.

Attoui, H., F. M. Jaafar, M. Belhouchet, P. de Micco, X. de Lamballerie, and C. P. Brussaard. 2006. Micromonas pusilla reovirus: a new member of the family Reoviridae assigned to a novel proposed genus (Mimoreovirus). *J. Gen. Virol.* **87**: 1375–83.

Bailey, S., M. R. J. Clokie, A. Millard, and N. H. Mann. 2004. Cyanophage infection and photoinhibition in marine cyanobacteria. *Res. Microbiol.* **155**: 720–5.

Barksdale, L., and S. B. Arden. 1974. Persisting bacteriophage infections, lysogeny, and phage conversions. *Annu. Rev. Microbiol.* **28**: 265–99.

Bergh, O., K. Y. Børsheim, G. Bratbak, and M. Heldal. 1989. High abundance of viruses found in aquatic environments. *Nature* **340**: 467–8.

Bohannan, B. J. M., and R. E. Lenski. 2000. Linking genetic change to community evolution: insights from studies of bacteria and bacteriophage. *Ecol. Lett.* **3**: 362–77.

Bouchard, J. D., E. Dion, F. Bissonnette, and S. Moineau. 2002. Characterization of the two-component abortive phage infection mechanism AbiT from *Lactococcus lactis. J. Bacteriol.* **184**: 6325–32.

Breitbart, M., and F. Rohwer. 2005. Here a virus, there a virus, everywhere the same virus? *Trends Microbiol.* **13**: 278–84.

Breitbart, M., L. Wegley, S. Leeds, T. Schoenfeld, and F. Rohwer. 2004. Phage community dynamics in hot springs. *Appl. Environ. Microbiol.* **70**: 1633–40.

Breitbart, M., F. Rohwer, and S. T. Abedon. 2005. Phage ecology and bacterial pathogenesis. In M. K. Waldor, D. I. Friedman, and S. L. Adhya (eds.), *Phages: Their Role in Bacterial Pathogenesis and Biotechnology.* Washington DC: ASM Press, pp. 66–91.

Bruenn, J. A. 2000. Viruses of fungi and protozoans: is everyone sick? In C. J. Hurst (ed.), *Viral Ecology*. San Diego, CA: Academic Press, pp. 297–317.

Brüssow, H. 2005. Phage therapy: the *Escherichia coli* experience. *Microbiology (Reading)* **151**: 2133–40.

Brüssow, H., and F. Desiere. 2006. Evolution of tailed phages: insights from comparative phage genomics. In R. L. Calendar and S. T. Abedon (eds.), *The Bacteriophages*, 2nd edn. Oxford: Oxford University Press, pp. 26–36.

Brüssow, H., and R. W. Hendrix. 2002. Phage genomics: small is beautiful. *Cell* **108**: 13–16.

Brüssow, H., and E. Kutter. 2005a. Genomics and the evolution of tailed phages. In E. Kutter and A. Sulakvelidze (eds.), *Bacteriophages: Biology and Application*. Boca Raton, FL: CRC Press, pp. 91–128.

Brüssow, H., and E. Kutter. 2005b. Phage ecology. In E. Kutter and A. Sulakvelidze (eds.), *Bacteriophages: Biology and Application*. Boca Raton, FL: CRC Press, pp. 129–64.

Brüssow, H., C. Canchaya, and W. D. Hardt. 2004. Phages and the evolution of bacterial pathogens: from genomic rearrangements to lysogenic conversion. *Microbiol. Mol. Biol. Rev.* **68**: 560–602.

Bull, J. J., D. W. Pfening, and I.-W. Wang. 2004. Genetic details, optimization, and phage life histories. *Trends Ecol. Evol.* **19**: 76–82.

Bull, J. J., J. Millstein, J. Orcutt, and H. A. Wichman. 2006. Evolutoinary feedback mediated through population density, illustrated with viruses in chemostats. *Am. Nat.* **167**: E39–51.

Calendar, R. L., and S. T. Abedon (eds.). 2006. *The Bacteriophages*, 2nd edn. Oxford: Oxford University Press.

Calendar, R. L., and R. Inman. 2005. Phage biology. In M. K. Waldor, D. I. Friedman, and S. L. Adhya (eds.), *Phages: Their Role in Bacterial Pathogenesis and Biotechnology*. Washington DC: ASM Press, pp. 18–36.

Carlson, K. 2005. Working with bacteriophages: common techniques and methodological approaches. In E. Kutter and A. Sulakvelidze (eds.), *Bacteriophages: Biology and Application*. Boca Raton, FL: CRC Press, pp. 437–94.

Casjens, S. R. 2005. Comparative genomics and evolution of the tailed-bacteriophages. *Curr. Opin. Microbiol.* **8**: 451–8.

Chibani-Chennoufi, S., A. Bruttin, M. L. Dillmann, and H. Brüssow. 2004. Phage–host interaction: an ecological perspective. *J. Bacteriol.* **186**: 3677–86.

Chopin, M. C., A. Chopin, and E. Bidnenko. 2005. Phage abortive infection in lactococci: variations on a theme. *Curr. Opin. Microbiol.* **8**: 473–9.

Comeau, A. M., and H. M. Krisch. 2005. War is peace: dispatches from the bacterial and phage killing fields. *Curr. Opin. Microbiol.* **8**: 488–94.

Cuevas, J. M., A. Moya, and S. F. Elena. 2003. Evolution of RNA virus in spatially structured heterogeneous environments. *J. Evol. Biol.* **16**: 456–66.

Day, M. 2004. Bacterial sensitivity to bacteriophage in the aquatic environment. *Sci. Prog.* **87**: 179–91.

DeFilippis, V. R., and L. P. Villarreal. 2000. An introduction to the evolutionary ecology of viruses. In C. J. Hurst (ed.), *Viral Ecology*. San Diego, CA: Academic Press, pp. 125–208.

d'Hérelle, F. 1917. Sur un microbe invisible antagoniste des bacilles dysentériques. *C. R. Acad. Sci. Ser. D* **165**: 373–5.

d'Hérelle, F. 1949. The bacteriophage. *Sci. News (Harmondsworth, London)* **14**: 44–59.

Dubin, S. B., G. B. Benedek, F. C. Bancroft, and D. Freifelder. 1970. Molecular weights of coliphages and coliphage DNA. II. Measurement of diffision coefficients using optical mixing spectroscopy, and measurement of sedimentation coefficients. *J. Mol. Biol.* **54**: 547–56.

Duckworth, D. H., J. Glenn, and D. J. McCorquodale. 1981. Inhibition of bacteriophage replication by extrachromosomal genetic elements. *Microbiol. Rev.* **45**: 52–71.

Durmaz, E., S. M. Madsen, H. Israelsen, and T. R. Klaenhammer. 2002. Lactococcus lactis lytic bacteriophages of the P335 group are inhibited by overexpression of a truncated CI repressor. *J. Bacteriol.* **184**: 6532–44.

Fields, B. S., R. F. Benson, and R. E. Besser. 2002. Legionella and Legionnaires' disease: 25 years of investigation. *Clin. Microbiol. Rev.* **15**: 506–26.

Forde, S. E., J. N. Thompson, and B. J. M. Bohannan. 2004. Adaptation varies through space and time in a coevolving host–parasitoid interaction. *Nature* **431**: 841–4.

Gerba, C. 2006. Bacteriophage as pollution indicators. In R. L. Calendar and S. T. Abedon (eds.), *The Bacteriophages*, 2nd edn. Oxford: Oxford University Press, pp. 695–701.

Goyal, S. M., C. P. Gerba, and G. Bitton. 1987. *Phage Ecology*. Boca Raton, FL: CRC Press.

Greub, G., and D. Raoult. 2004. Microorganisms resistant to free-living amoebae. *Clin. Microbiol. Rev.* **17**: 413–33.

Guerrero, R., I. Esteve, C. Pedrós-Alió, and N. Gaju. 1987. Predatory bacteria in prokaryotic communities. *Ann. N. Y. Acad. Sci.* **503**: 238–50.

Guttman, B., R. Raya, and E. Kutter. 2005. Basic phage biology. In E. Kutter and A. Sulakvelidze (eds.), *Bacteriophages: Biology and Application*. Boca Raton, FL: CRC Press, pp. 26–66.

Hammad, A. M. M. 1998. Evaluation of alginate-encapsulated *Azotobacter chroococcum* as a phage-resistant and an effective inoculum. *J. Basic Microbiol.* **38**: 9–16.

Hendrix, R. W. 2005. Bacteriophage evolution and the role of phages in host evolution. In M. K. Waldor, D. I. Friedman, and S. L. Adhya (eds.), *Phages: Their Role in Bacterial Pathogenesis and Biotechnology*. Washington DC: ASM Press, pp. 55–65.

Hendrix, R. W., J. G. Lawrence, G. F. Hatfull, and S. Casjens. 2000. The origins and ongoing evolution of viruses. *Trends Microbiol.* **8**: 504–8.

Hendrix, R. W., G. F. Hatfull, and M. C. M. Smith. 2003. Bacteriophages with tails: chasing their origins and evolution. *Res. Microbiol.* **154**: 253–7.

Hyman, P., and S. T. Abedon, in press. Practical methods for determining phage growth parameters. In M. Clokie and A. Kropinski (eds.), *Bacteriophages: Methods and Protocols*. Totowa, NJ: Humana Press.

Jeltsch, A. 2003. Maintenance of species identity and controlling speciation of bacteria: a new function for restriction/modification systems? *Gene* **317**: 13–16.

Jensen, E. C., H. S. Schrader, B. Rieland, *et al.* 1998. Prevalence of broad-host-range lytic bacteriophages of *Sphaerotilus natans, Escherichia coli*, and *Pseudomonas aeruginosa*. *Appl. Environ. Microbiol.* **64**: 575–80.

Keel, C., Z. Ucurum, P. Michaux, M. Adrian, and D. Haas. 2002. Deleterious impact of a virulent bacteriophage on survival and biocontrol activity of *Pseudomonas fluorescens* strain CHAO in natural soil. *Mol. Plant-Microbe Interact.* **15**: 567–76.

Klaenhammer, T. R., and G. F. Fitzgerald. 1994. Bacteriophages and bacteriophage resistance. In M. J. Gasson and W. M. de Vos (eds.), *Genetics and Biotechnology of Lactic Acid Bacteria*. London: Blackie, pp. 106–68.

Klieve, A. V. 1996. Bacteriophages in the rumen: types present, population size and implications for the efficiency of feed utilisation. *Proc. Aust. Soc. Anim. Prod.* **21**: 92–4.

Krüger, D. H., and T. A. Bickle. 1983. Bacteriophage survival: multiple mechanisms for avoiding deoxyribonucleic acid restriction systems of their hosts. *Microbiol. Rev.* **47**: 345–60.

Kurtboke, D. I. 2005. Actinophages as indicators of actinomycete taxa in marine environments. *Antonie van Leeuwenhoek* **87**: 19–28.

Kutter, E., and A. Sulakvelidze. 2005. *Bacteriophages: Biology and Application*. Boca Raton, FL: CRC Press.

Kutter, E., E. Kellenberger, K. Carlson, *et al.* 1994. Effects of bacterial growth conditions and physiology on T4 infection. In J. D. Karam (ed.), *The Molecular Biology of Bacteriophage T4*. Washington, DC: ASM Press, pp. 406–418.

Kutter, E., R. Raya, and K. Carlson. 2005. Molecular mechanisms of phage infection. In E. Kutter and A. Sulakvelidze (eds.), *Bacteriophages: Biology and Application.* Boca Raton, FL: CRC Press, pp. 165–222.

Lenski, R. E. 1988. Dynamics of interactions between bacteria and virulent bacteriophage. *Adv. Microbial. Ecol.* **10**: 1–44.

Levin, B. R., and J. J. Bull. 2004. Population and evolutionary dynamics of phage therapy. *Nature Rev. Microbiol.* **2**: 166–73.

Levin, B. R., and R. E. Lenski. 1983. Coevolution in bacteria and their viruses and plasmids. In D. J. Futuyama and M. Slatkin (eds.), *Coevolution.* Sunderland, MA: Sinauer, pp. 99–127.

Levin, B. R., and R. E. Lenski. 1985. Bacteria and phage: a model system for the study of the ecology and co-evolution of hosts and parasites. In D. Rollinson and R. M. Anderson (eds.), *Ecology and Genetics of Host–Parasite Interactions.* London: Academic Press, pp. 227–42.

Liu, J., G. Glazko, and A. Mushegian. 2006. Protein repertoire of double-stranded DNA bacteriophages. *Virus Res.* **117**: 68–80.

Lwoff, A. 1953. Lysogeny. *Bacteriol. Rev.* **17**: 269–337.

Madsen, S. M., D. Mills, G. Djordjevic, H. Israelsen, and T. R. Klaenhammer. 2001. Analysis of the genetic switch and replication region of a P335-type bacteriophage with an obligate lytic lifestyle on *Lactococcus lactis. Appl. Environ. Microbiol.* **67**: 1128–39.

Mann, N. H. 2003. Phages of the marine cyanobacterial picophytoplankton. *FEMS Microbiol. Rev.* **27**: 17–34.

Mann, N. H. 2006. Phages of cyanobacteria. In R. L. Calendar and S. T. Abedon (eds.), *The Bacteriophages*, 2nd edn. Oxford: Oxford University Press, pp. 517–33.

Maranger, R., and D. E. Bird. 1996. High concentrations of viruses in the sediments of Lac Gilbert, Quebec. *Microb. Ecol.* **31**: 141–51.

Martin, M. O. 2002. Predatory prokaryotes: an emerging research opportunity. *J. Mol. Microbiol. Biotechnol.* **4**: 467–77.

Matsuda, M., and L. Barksdale. 1967. System for the investigation of the bacteriophage-directed synthesis of diphtherial toxin. *J. Bacteriol.* **93**: 722–30.

Miller, R. V., and G. S. Sayler. 1992. Bacteriophage–host interactions in aquatic systems. In E. M. H. Wellington and J. D. van Elsas (eds.), *Genetic Interactions among Microorganisms in the Natural Environment.* Oxford: Pergamon Press, pp. 176–93.

Paul, J. H., and C. A. Kellogg. 2000. Ecology of bacteriophages in nature. In C. J. Hurst (ed.), *Viral Ecology.* San Diego, CA: Academic Press, pp. 211–46.

Paul, J. H., and M. B. Sullivan. 2005. Marine phage genomics: what have we learned? *Curr. Opin. Biotechnol.* **16**: 299–307.

Prangishvili, D., R. A. Garrett, and E. V. Koonin. 2006a. Evolutionary genomics of archaeal viruses: unique viral genomes in the third domain of life. *Virus Res.* **117**: 52–67.

Prangishvili, D., P. Forterre, and R. A. Garrett. 2006b. Viruses of the Archaea: a unifying view. *Nat. Rev. Microbiol.* **4**: 837–48.

Robb, F. T., and R. T. Hill. 2000. Bacterial viruses and hosts: Influence of culturable state. In R. R. Colwell and D. J. Grimes (eds.), *Nonculturable Microorganisms in the Environment.* Washington DC: ASM Press, pp. 199–208.

Rohwer, F., and R. Edwards. 2002. The phage proteomic tree: a genome-based taxonomy for phage. *J. Bacteriol.* **184**: 4529–35.

Russel, M., and P. Model. 2006. Filamentous phage. In R. L. Calendar and S. T. Abedon (eds.), *The Bacteriophages,* 2nd edn. Oxford: Oxford University Press, pp. 146–60.

Schrader, H. S., J. O. Schrader, J. J. Walker, *et al.* 1997. Effects of host starvation on bacteriophage dynamics. In R. Y. Morita (ed.), *Bacteria in Oligotrophic Environments: Starvation-Survival Lifestyle.* New York, NY: Chapman & Hall, pp. 368–85.

Shilo, M. 1984. *Bdellovibrio* as a predator. In M. J. Klug and C. A. Reddy (eds.), *Current Perspectives in Microbial Ecology.* Washington, DC: ASM Press, pp. 334–9.

Snyder, J. C., J. Spuhler, B. Wiedenheft, F. F. Roberto, T. Douglas, and M. J. Young. 2004. Effects of culturing on the population structure of a hyperthermophilic virus. *Microb. Ecol.* **48**: 561–6.

Sockett, R. E., and C. Lambert. 2004. Bdellovibrio as therapeutic agents: a predatory renaissance? *Nat. Rev. Microbiol.* **2**: 669–75.

Stedman, K. M., D. Prangishvilli, and W. Zillig. 2006. Viruses of Archaea. In R. L. Calendar and S. T. Abedon (eds.), *The Bacteriophages,* 2nd edn. Oxford: Oxford University Press, pp. 499–516.

Steinert, M., U. Hentschel, and J. Hacker. 2002. *Legionella pneumophila*: an aquatic microbe goes astray. *FEMS Microbiol. Rev.* **26**: 149–62.

Stent, G. S. 1963. *Molecular Biology of Bacterial Viruses.* San Francisco, CA: W. H. Freeman.

Sturino, J. M., and T. R. Klaenhammer. 2004. Bacteriophage defense systems and strategies for lactic acid bacteria. *Adv. Appl. Microbiol.* **56**: 331–78.

Sullivan, M. B., J. B. Waterbury, and S. W. Chisholm. 2003. Cyanophages infecting the oceanic cyanobacterium *Prochlorococcus. Nature* **424**: 1047–51.

Summers, W. C. 1991. From culture as organisms to organisms as cell: historical origins of bacterial genetics. *J. Hist. Biol.* **24**: 171–90.

Summers, W. C. 1999. *Felix d'Herelle and the Origins of Molecular Biology*. New Haven, CT: Yale University Press.

Summers, W. C. 2005a. Bacteriophage research: early research. In E. Kutter and A. Sulakvelidze (eds.), *Bacteriophages: Biology and Application*. Boca Raton, FL: CRC Press, pp. 5–28.

Summers, W. C. 2005b. History of phage research and phage therapy. In M. Waldor, D. Friedman, and S. Adhya (eds.), *Phages: Their Role in Bacterial Pathogenesis and Biotechnology*. Washington DC: ASM Press, pp. 3–17.

Summers, W. C. 2006. Phage and the genesis of molecular biology. In R. L. Calendar and S. T. Abedon (eds.), *The Bacteriophages*, 2nd edn. Oxford: Oxford University Press, pp. 3–7.

Sutherland, I. W., K. A. Hughes, L. C. Skillman, and K. Tait. 2004. The interaction of phage and biofilms. *FEMS Microbiol. Lett.* **232**: 1–6.

Suttle, C. A. 1994. The significance of viruses to mortality in aquatic microbial communities. *Microb. Ecol.* **28**: 237–43.

Suttle, C. A. 2000. The ecology, evolutionary and geochemical consequences of viral infection of cyanobacteria and eukaryotic algae, In C. J. Hurst (ed.), *Viral Ecology*. San Diego, CA: Academic Press, pp. 248–86.

Suttle, C. A. 2002. Community structure: viruses. In C. J. Hurst, G. R. Knudson, M. J. McInerney, L. D. Stezenbach, and M. V. Walter (eds.), *Manual of Environmental Microbiology* (2nd edn.). Washington, DC: ASM Press, pp. 364–370.

Takao, Y., K. Mise, K. Nagasaki, T. Okuno, and D. Honda. 2006. Complete nucleotide sequence and genome organization of a single-stranded RNA virus infecting the marine fungoid protist *Schizochytrium* sp. *J. Gen. Virol.* **87**: 723–33.

Tao, L., S. I. Pavlova, and A. O. Kiliç. 2005. Phages and bacterial vaginosis. In M. K. Waldor, D. I. Friedman, and S. L. Adhya (ed.), *Phages: Their Role in Pathogenesis and Biotechnology*. Washington DC: ASM Press, pp. 256–79.

Thingstad, T. F. 2000. Elements of a theory for the mechanisms controlling abundance, diversity, and biogeochemical role of lytic bacterial viruses in aquatic systems. *Limnol. Oceanogr.* **45**: 1320–8.

Trotter, M., O. McAuliffe, M. Callanan, *et al.* 2006. Genome analysis of the obligately lytic bacteriophage 4268 of *Lactococcus lactis* provides insight into its adaptable nature. *Gene* **366**: 189–99.

Twort, F. W. 1915. An investigation on the nature of the ultra-microscopic viruses. *Lancet* **ii**: 1241–3.

Wang, A. L., and C. C. Wang. 1991. Viruses of the protozoa. *Ann. Rev. Microbiol.* **45**: 251–63.

Weinbauer, M. G. 2004. Ecology of prokaryotic viruses. *FEMS Microbiol. Rev.* **28**: 127–81.

Weinbauer, M. G., and F. Rassoulzadegan. 2004. Are viruses driving microbial diversification and diversity? *Environ. Microbiol.* **6**: 1–11.

Weinbauer, M. G., M. Agis, O. Bonilla-Findji, A. Malits, and C. Winter. 2007. Bacteriophage in the environment. In S. Mcgrath and D. van Sinderen (eds.), *Bacteriophage: Genetics and Molecular Biology*. Norfolk: Caister Academic Press, pp. 61–92.

Whitman, W. B., D. C. Coleman, and W. J. Wiebe. 1998. Prokaryotes: the unseen majority. *Proc. Natl. Acad. Sci. U.S.A.* **95**: 6578–83.

Wickner, R. B. 1989. Yeast virology. *FASEB J.* **3**: 2257–65.

Wilhelm, S. W., and L. Poorvin. 2001. Quantification of algal viruses in marine samples. *Meth. Microbiol.* **30**: 53–65.

Wilhelm, S. W., and C. A. Suttle. 1999. Viruses and nutrient cycles in the sea. *BioScience* **49**: 781–8.

Williamson, K. E., K. E. Wommack, and M. Radosevich. 2003. Sampling natural viral communities from soil for culture-independent analyses. *Appl. Environ. Microbiol.* **69**: 6628–33.

Williamson, K. E., M. Radosevich, and K. E. Wommack. 2005. Abundance and diversity of viruses in six Delaware soils. *Appl. Environ. Microbiol.* **71**: 3119–25.

Wommack, K. E., and R. R. Colwell. 2000. Virioplankton: viruses in aquatic ecosystems. *Microbiol. Mol. Biol. Rev.* **64**: 69–114.

Yin, J., K. Duca, V. Lam, *et al.* 2001. Quantifying viral propagation in vitro: toward a method for characterization of complex phenotypes. *Biotechnol. Prog.* **17**: 1156–65.

Part I Phage ecology

CHAPTER 2

Bacteriophages: models for exploring basic principles of ecology

Benjamin Kerr,* Jevin West, and Brendan J. M. Bohannan

2.1 INTRODUCTION

A virus depends intimately upon its host in order to reproduce, which makes the host organism a crucial part of the virus's environment. This basic facet of viral existence means that ecology, the scientific field focusing on how organisms interact with each other and their environment, is particularly relevant to the study of viruses. In this chapter we explore some of the ways in which the principles of ecology apply to viruses that infect bacteria – the bacteriophages (or "phages" for short). In turn, we also discuss how the study of phages and their bacterial hosts has contributed to different subfields of ecology.

Due to their ease of manipulation, large population sizes, short generation times, and wealth of physiological and genetic characterization, laboratory communities of microbial organisms have been popular experimental tools for testing ecological theory (Drake *et al.*, 1996; Jessup *et al.*, 2004). Building upon this foundation of knowledge, the ecological experimentalist can explore whether mechanistic understanding at the organismal level informs an understanding of patterns at the community level (Bohannan and Lenski, 2000a). Further, the initial composition of microbial consortia can be controlled, and thus researchers are able to probe the effects of different community structures on ecological phenomena, such as stability, diversity, and resilience to invasion.

In particular, microcosms of bacteria and phages have been used by many researchers to study the ecology of victim–exploiter interactions (Chao

* Corresponding author

Bacteriophage Ecology: Population Growth, Evolution, and Impact of Bacterial Viruses, ed. Stephen T. Abedon. Published by Cambridge University Press.

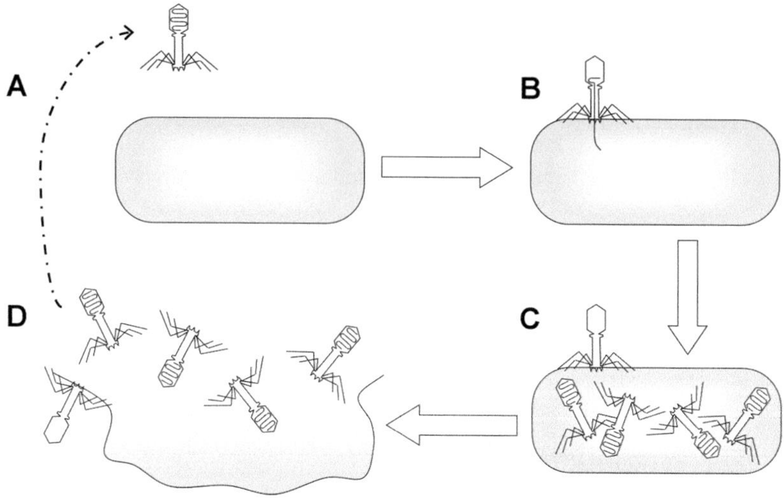

Figure 2.1 The life cycle of a typical lytic bacteriophage. (A) A single infective phage particle is shown with its bacterial host. (B) The phage binds to receptors on the surface of its host (adsorption) and injects its genome into the bacterial cytoplasm. (C) The bacterial cell becomes a phage factory, where the phage genome is copied, phage structural proteins are synthesized, and the genome is packaged. (D) At the end of the latent period (which starts with phage genome injection), the host cell lyses, releasing the progeny phages. A progeny phage particle can then infect a new host, starting the cycle anew (dot-dashed arrow).

et al., 1977; Levin *et al.*, 1977; Lenski and Levin, 1985; Schrag and Mittler, 1996; Bohannan and Lenski, 1997, 1999, 2000b). These relationships are ubiquitous within ecosystems (e.g., prey and their predators, plants and their herbivores, hosts and their pathogens, etc.). In this chapter, we will focus on the obligately lytic (*sensu* Chapter 1) bacteriophages of *Escherichia coli*. The archetypical life cycle is illustrated in Fig. 2.1. Here, we discuss how ecological theory and concepts apply to these phages and how viruses have inspired new conceptual developments within ecology.

2.2 MATHEMATICAL ECOLOGY

Since ecologists deal with complex systems of interacting species, mathematical frameworks have proved extremely useful for both predictive purposes and conceptual understanding. Studies of victim–exploiter interactions have a particularly distinguished history within mathematical ecology, from the continuous-time theory of predators and their prey (Lotka, 1925; Volterra, 1926) to the discrete-time theory of parasites and their hosts (Nicholson and

Bailey, 1935). Such mathematical models have highlighted some basic principles of victim–exploiter interactions, including the fundamental occurrence of oscillations and the factors contributing to community stability.

2.2.1 Lotka–Volterra modeling

One of the simplest mathematical models for understanding the dynamics of predators and their prey was formulated independently by Lotka (1925) and Volterra (1926). This model assumes that prey increase exponentially without predators present and that predators decrease exponentially without prey present. The model also assumes that the growth of the predator population is proportional to the food intake through predation, which depends on the likelihood of predator–prey encounters. If we assume mass action (the rate of interaction between predators and prey being proportional to their product), then the dynamics for predator density (P) and prey density (N) are:

$$\frac{dN}{dt} = \Psi \cdot N - \alpha \cdot N \cdot P \tag{2.1}$$

$$\frac{dP}{dt} = \beta \cdot \alpha \cdot N \cdot P - \delta \cdot P \tag{2.2}$$

with $\boldsymbol{\Psi}$ the growth rate of the prey, α the attack rate of predators, β the predator's efficiency in converting its food to offspring, and δ the death rate of the predator.

There is a single non-trivial equilibrium for this system:

$$(\hat{N}, \hat{P}) = (\delta/(\beta \cdot \alpha), \Psi/\alpha) \tag{2.3}$$

which is neutrally stable (i.e., small perturbations from the equilibrium are neither amplified nor damped). Interestingly, the equilibrial density of the prey depends only on the parameters of the predator and *not* on the growth rate of the prey itself. The full dynamics of this system consist of neutrally stable oscillations, the amplitude of which are determined by the initial densities of predators and prey (May, 1974; Edelstein-Keshet, 1988; Fig. 2.2A). This model has been criticized for being (1) biologically oversimplistic and (2) incapable of explaining sustained oscillations (with neutrally stable cycles, a perturbation of either predator or prey to low density gives rise to cycles regularly visiting low densities, which increases the chance of extinction; see Fig. 2.2A). However, as this model sits on the border between stability and instability, it is a conceptually useful baseline from which to explore stabilizing and destabilizing factors in predator–prey relationships (Edelstein-Keshet, 1988). For instance, if the prey's per capita growth rate decreases as

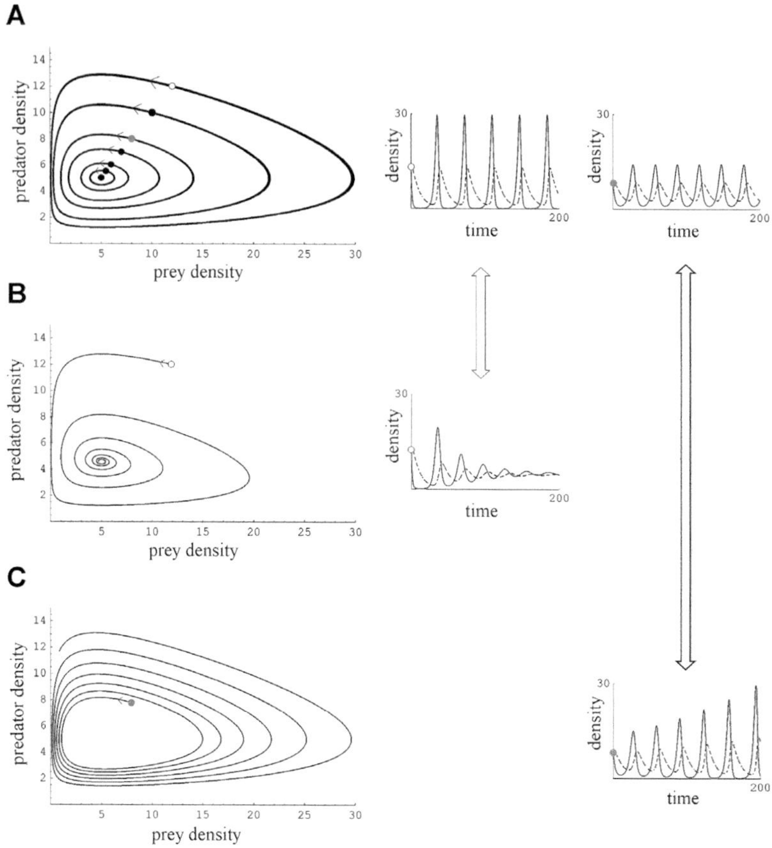

Figure 2.2 Dynamics of a predator–prey community. (A) We use the standard δ Lotka–Volterra model: $dN/dt = \Psi \cdot N - \alpha \cdot N \cdot P$, $dP/dt = \beta \cdot \alpha \cdot N \cdot P - \delta \cdot P$, with $\Psi = 0.5$, $\alpha = 0.1$, $\beta = 0.2$, and $\delta = 0.1$. In the main graph, we show stable limit cycles in a phase plane, where initial densities of the predator and prey determine the amplitude of the cycles (thick lines). The filled circles give the various starting conditions: $N_0 = P_0 = 5$ (the neutral equilibrium), $N_0 = P_0 = 5.5$, $N_0 = P_0 = 6$, $N_0 = P_0 = 7$, $N_0 = P_0 = 8$ (marked with the gray circle), $N_0 = P_0 = 10$, and $N_0 = P_0 = 12$ (marked with the white circle). The small graphs to the right of the phase plane give the time trajectories for the starting conditions $N_0 = P_0 = 8$ (gray circle) and $N_0 = P_0 = 12$ (white circle), where the bacterial density is a solid line and the phage density is a dashed line. (B) We change the model slightly to add negative density dependence for the prey: $dN/dt = \Psi \cdot N(1 - N/K) - \alpha \cdot N \cdot P$, $dP/dt = \beta \cdot \alpha \cdot N \cdot P - \delta \cdot P$, with $\Psi = 0.5$, $\alpha = 0.1$, $\beta = 0.2$, $\delta = 0.1$, and the carrying capacity $K = 60$. Starting with $N_0 = P_0 = 12$ (the white circle), we see that the community exhibits damped oscillations (see also the small graph to the right). Compared to the standard model, negative density dependence stabilizes the predator–prey dynamics (the double-edged arrow points to the graph of the

its density increases (negative density-dependent regulation) then the system is stabilized; on the other hand, a time lag in the processing of the prey can destabilize the system (May, 1972, 1974; Holling, 1973; Fig. 2.2B and C; Chapter 15).

2.2.2 Modeling predator–prey dynamics using phages

Communities of bacteriophages and bacteria serve as ideal test systems of predator–prey theory, and several mathematical models have been tailored to explore the dynamics of these microbial communities (Campbell, 1961; Chao *et al.*, 1977; Levin *et al.*, 1977; Lenski and Levin, 1985; Schrag and Mittler, 1996; Bohannan and Lenski, 1997, 1999, 2000b; Chapter 15). Many of these models make the assumption that the microbes inhabit a "chemostat-like" environment – one which is spatially homogeneous, constant in abiotic variables and continuously supplied with nutrients. In the chemostat, medium containing resources flows into a vessel at a constant rate and depleted medium flows out at the same rate, such that a stable vessel volume is maintained (Chapters 9 and 15). In Section 2.8 we give an example of a continuous-time chemostat model.

As was the case for the simple Lotka–Volterra model, chemostat models predict that equilibrium bacterial density is a function of phage effectiveness only (Levin *et al.*, 1977; Lenski and Levin, 1985; Equation 2.3). Also, these models incorporate explicitly some of the stabilizing and destabilizing factors mentioned above. For instance, bacteria compete for limiting resources and thus there is negative density-dependent regulation in the chemostat. Indeed, the limiting resource concentration in the vessel is often modeled as a dynamic variable, where its rate of change is negatively related to bacterial density (Section 2.8). While such density dependence is a stabilizing force, time lags inherent to this system destabilize dynamics. One fundamental

Figure 2.2 (*continued*)
standard model and the graph of the negative-density-dependence model for the same starting conditions). (C) We now add a time delay: $dN/dt = \Psi \cdot N - \alpha \cdot N \cdot P$, $dP/dt = e^{-\tau \cdot \delta}\beta \cdot \alpha \cdot N' \cdot P' - \delta \cdot P$, with $\Psi = 0.5$, $\alpha = 0.1$, $\beta = 0.2$, $\delta = 0.1$; the time delay (for the predator to convert captured prey items into predator progeny) is $\tau = 0.17$, and the primed variables are evaluated τ time units in the past (i.e., $N' = N(t - \tau)$ and $P' = P(t - \tau)$). Starting with $N_0 = P_0 = 8$ (the gray circle), we see that the community exhibits expanding oscillations (see also the small graph to the right). Compared to the standard Lotka–Volterra model, a time delay destabilizes the predator–prey dynamics (the double-edged arrow points to the graph of the standard model and the graph of the time-lag model for the same starting conditions).

time lag involves the latent period of the phage – the interval between phage "capture" of a bacterial prey (Fig. 2.1B) and the transfer of that capture into phage progeny (Fig. 2.1D; Chapter 15). As a consequence of this lag, phage–bacteria dynamics are generally written as time-delay differential equations (Section 2.8).

If coexistence is predicted within these models, then generally both predator and prey either approach fixed densities (stable fixed point) or both parties cycle in steady fashion (stable limit cycle). However, sometimes these cycles reach densities so low that extinction, due to finite population size, is predicted. In actual chemostats, phages and bacteria tend to coexist (sometimes for long intervals), approaching semi-fixed densities or undergoing regular cycles (Horne, 1970; Paynter and Bungay, 1971; Chao *et al.*, 1977; Levin *et al.*, 1977; Lenski and Levin, 1985; Schrag and Mittler, 1996; Bohannan and Lenski, 1997, 1999, 2000b; Fig. 2.4). Such coexistence can occur even when theory predicts otherwise (Levin *et al.*, 1977; Schrag and Mittler, 1996; Bohannan and Lenski, 1997, 2000b). In such cases, the theoretical framework is likely missing key components, such as spatial, numerical, or physiological refugia from phage attack (Lenski and Levin, 1985; Lenski, 1988; Schrag and Mittler, 1996; Bohannan and Lenski, 2000a). For example, hidden layers of surface growth on the walls of the chemostat may serve as a spatial refuge for phage-sensitive bacteria (Schrag and Mittler, 1996).

2.2.3 More complex phage-based models

The models employed in Section 2.8 (Appendix) consider only a single type of predatory phage and a single vulnerable strain of bacteria. However, there is frequently evolution of new members within laboratory microbial communities, such as fully or partially phage-resistant bacterial mutants and host-range phage mutants. In fact, in nearly all the chemostat studies listed above there was evidence of phage-resistant bacterial mutants arising during the experiment. To probe this added complexity, some researchers have increased the number of dynamic players in their models (Levin *et al.*, 1977; Bohannan and Lenski, 1999, 2000b). For example, Levin *et al.* (1977) constructed a model with an arbitrary number of predatory phage species, prey bacterial species, and resources. They found that, at equilibrium, the number of phage species could not exceed the number of bacterial species; and the number of bacterial species could not exceed the sum of the number of resources and the number of phage species. This "exclusion principle" applies only to equilibrium conditions, and violations of the principle have been described in communities exhibiting periodic and chaotic dynamics (Armstrong and McGehee, 1980; Huisman and Weissing, 1999). Systems with multiple predators and multiple prey are a topic of interest in community ecology, a subject to which we now turn.

2.3 COMMUNITY ECOLOGY

Understanding how interactions among species affect the structure and spatio-temporal dynamics of biological communities is a core objective within community ecology. Many natural communities are a spaghetti-like tangle of interactions. One approach to unraveling this complexity involves the study of stripped-down communities comprising a smaller number of interacting species, so-called "community modules" (Holt *et al.*, 1994; Holt, 1995; Bohannan and Lenski, 2000a). In communities of predators and their prey, the examination of such modules has led to a deeper understanding of the circumstances in which different prey species negatively impact one another through a shared predator, as well as of the conditions under which a predator plays a "keystone" role in maintaining diversity (Holt, 1977, 1995; Holt *et al.*, 1994; Leibold, 1996).

2.3.1 Microbial modules

Microbial microcosms of bacteria and their viruses are well suited for the construction of module communities (Drake *et al.*, 1996; Bohannan and Lenski, 2000a; Jessup *et al.*, 2004). In laboratory settings, the experimenter can easily control the structure of the community by initiating the microcosm with a select set of interacting species. The experimenter can then track the population dynamics and monitor how the community structure changes over time. Such changes can occur when (1) invading species are intentionally added to an established community, (2) new community members evolve *de novo* within the microcosm, and (3) when community players become extinct.

Researchers using bacterium–phage modules have explored a number of topics of interest within community ecology. For instance, the conditions favoring bottom-up (resource-driven) versus top-down (predator-driven) control of prey populations have been investigated with phage and bacteria (Chao *et al.*, 1977; Bohannan and Lenski, 1997, 1999, 2000b). The impact of invading species on community structure and dynamics has been probed using these microbial modules (Levin *et al.*, 1977; Lenski and Levin, 1985;

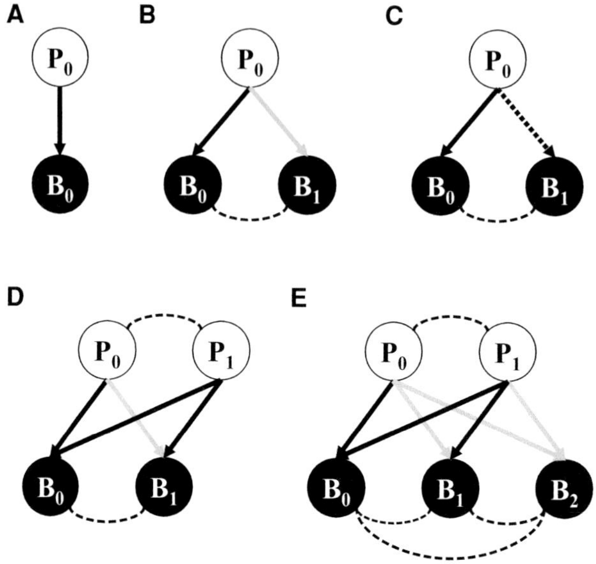

Figure 2.3 Different community modules. In each module, the phage predators are shown as empty circles and the bacterial prey are shown as filled circles. If a phage strain can infect a particular bacterial strain, then a black arrow points from the phage to the host (where a dotted arrow indicates a host *partially* resistant to phage entry). If the phage cannot infect a particular strain of bacterium, then a gray arrow points from the phage to the bacterial strain. Lastly, dashed lines connect strains that are "mutant neighbors" of one another (that is, strains that can mutate to become the attached strain). Shown are (A) the simple single phage strain and single sensitive bacterial strain, (B) a single phage strain, a sensitive bacterial strain, and a fully phage-resistant strain, (C) a single phage strain, a sensitive bacterial strain and a partially resistant strain, (D) two phage strains (an ancestor and a host-range mutant) and two bacterial strains: the phage ancestor can only attack one of the two bacterial strains, whereas the host-range mutant can attack both, and (E) two phage strains and three bacterial strains: exactly the situation in part D, but with an extra bacterial strain that is fully resistant to both phage strains.

Bohannan and Lenski, 1999). The competitive effects that take place between prey species through a common predator have also been explored (Levin *et al.*, 1977; Bohannan and Lenski, 2000b). Several researchers have reported the importance of phenotypic trade-offs in determining community structure (Chao *et al.*, 1977; Levin *et al.*, 1977; Lenski and Levin, 1985; Bohannan and Lenski, 1999). Finally, the effect of resource enrichment on stability and diversity has been studied within these microbial communities (Bohannan and Lenski, 1997, 1999, 2000b). Below we discuss these findings, organizing the experimental communities by their structural complexity (Fig. 2.3).

2.3.2 Communities with phages and sensitive bacteria

Most experimental microcosms are initiated with a single strain of bacteria and a single strain of phage (Fig. 2.3A). For instance, chemostats have been inoculated with *E. coli* and one of the following phage types: λ*vir* (Schrag and Mittler, 1996), T1X (Schrag and Mittler, 1996), T2 (Paynter and Bungay, 1971; Levin *et al.*, 1977; Lenski and Levin, 1985; Bohannan and Lenski, 2000b), T3 (Horne, 1970), T4 (Horne, 1970; Lenski and Levin, 1985; Bohannan and Lenski, 1997, 1999), T5 (Lenski and Levin, 1985), or T7 (Chao *et al.*, 1977; Lenski and Levin, 1985; Forde *et al.*, 2004). In most cases, a phage-resistant strain of bacteria evolved and increased in frequency (see next two sections); thus, a *de novo* increase in complexity occurred. However, in a few cases, some hypotheses about two-member community dynamics could be tested before invasion took place. For instance, Bohannan and Lenski (1997, 1999) found that resource enrichment led to a significant increase in T4 phage density and only a small increase in bacterial density. Furthermore, these authors observed that higher resource concentration destabilizes predator–prey dynamics, an example of the so-called "paradox of enrichment" (Rosenzweig, 1971; Section 2.8; Fig. 2.4A and B; Chapter 10).

2.3.3 Communities with fully resistant bacteria

As mentioned above, even in communities started with only a single sensitive bacterial strain and a single strain of phage, a phage-resistant bacterial strain often evolved and invaded. In some cases, a resistant strain was added to the microcosm and this initially rare species similarly increased in density (Levin *et al.*, 1977; Bohannan and Lenski, 1999). In these cases, the community module had three players, two consumer species (the bacteria) and one predator (the phage) that made a living on one of the consumers (Fig. 2.3B). In the majority of cases, this community maintained all three members (Paynter and Bungay, 1971; Levin *et al.*, 1977; Lenski and Levin, 1985; Schrag and Mittler, 1996; Bohannan and Lenski, 1999).

2.3.3.1 Trade-offs are stabilizing

What allows sustained coexistence? The invasion of the resistant strain occurs because it is invulnerable to the phage. However, why does the resistant bacterial strain not displace the sensitive strain? In most cases, this is due to a trade-off between competitive ability and resistance to phage infection (Bohannan *et al.*, 2002). A competitive cost to phage resistance has been shown for *E. coli* resistant to phage T2 (Levin *et al.*, 1977; Lenski, 1984; Lenski

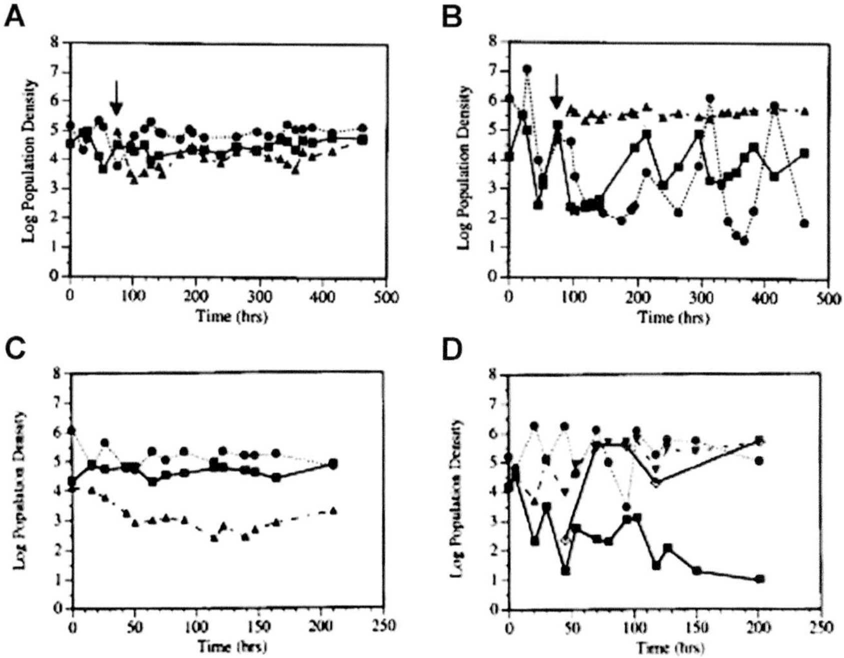

Figure 2.4 Chemostat community dynamics. (A) The dynamics of phage-sensitive *E. coli* B (squares), phage-resistant *E. coli* B (triangles), and phage T4 (circles) in glucose-limited chemostats with a reservoir concentration of 0.1 mg mL^{-1} (low productivity). The arrow indicates when phage-resistant bacteria were added to the chemostat. (B) The dynamics of the sensitive *E. coli*, resistant *E. coli*, and phage T4 in chemostats with a reservoir glucose concentration of 0.5 mg mL^{-1} (high productivity). (C) The dynamics of a sensitive *E. coli* B (squares), partially resistant *E. coli* B (triangles), and phage T2 (circles) in a chemostat with a reservoir glucose concentration of 0.1 mg mL^{-1}. (D) The dynamics of the sensitive *E. coli*, partially resistant *E. coli*, phage T2, and fully resistant *E. coli* (diamonds) in chemostats with a reservoir glucose concentration of 0.5 mg mL^{-1}. (Reprinted with permission from Bohannan and Lenski, 1999 [*Am. Nat.* **153**: 73–82], and Bohannan and Lenski, 2000b [*Am. Nat.* **156**: 329–340]. © 1999 and 2000, University of Chicago.)

and Levin, 1985), phage T4 (Lenski and Levin, 1985; Lenski, 1988; Bohannan and Lenski, 1999), and phage T7 (Chao *et al.*, 1977; Lenski and Levin, 1985). This cost has been shown to be a function of the particular mutation conferring resistance, the genetic background, and the abiotic environment (Lenski, 1988; Bohannan and Lenski, 2000a; Bohannan *et al.*, 2002). Trade-offs between competitive ability and protection from predation are found in other organisms as well (Grover, 1995; Kraaijeveld and Godfray, 1997; Gwynn *et al.*, 2005).

Such trade-offs are predicted by chemostat models to allow coexistence in these three-member modules (Levin *et al.*, 1977). The sensitive bacteria persist due to a higher growth rate than the resistant prey when the latter approach their resource-limited equilibrium, while the resistant bacteria persist due to phage control of the sensitive population. When a trade-off between phage resistance and competitive ability does not exist, the community can lose members. For example, resistance to phage T5 by *E. coli* B comes without a competitive cost in glucose-limited laboratory culture; indeed, this is why T5 resistance is such a popular genetic marker in *E. coli* (Bohannan and Lenski, 2000a). Lenski and Levin (1985) showed that the advent of T5-resistant bacteria in chemostats initiated with phage T5 and T5-sensitive bacteria led to the eventual extinction of both phage and sensitive bacteria.

2.3.3.2 Top-down versus bottom-up control

The presence of a fully resistant bacterial strain is also predicted to shift regulation of prey from top-down (predator control) to bottom-up (resource control). Specifically, resource enrichment is predicted to lead to an increase in the predator density when only a single susceptible prey species is present (Fig. 2.3A), whereas enrichment is predicted to lead only to an increase in the invulnerable prey when both susceptible and invulnerable prey species live with a predator (Leibold, 1989). Bohannan and Lenski (1999) showed that these predictions were qualitatively accurate for a chemostat community of phage T4 and a heterogeneous bacterial population. Furthermore, these authors demonstrated that a version of the "paradox of enrichment" occurred in a subset of their community. Specifically, the phage and sensitive bacteria dynamics were destabilized by increased resource input into the chemostat vessel (whereas the dynamics of the phage-resistant bacterial strain were stabilized with enrichment).

2.3.4 Communities with partially resistant bacteria

Phage T2 can bind to two different receptors on the surface of its bacterial host (Lenski, 1984). This predatory flexibility enables bacteria to evolve partial resistance to T2, in which one of the two receptors has been lost or altered. In communities containing phages, sensitive bacteria, and partially resistant bacteria (Fig. 2.3C), the prey species interact through two indirect routes. First, the bacterial strains compete for resources, a phenomenon termed "exploitative competition" (Holt *et al.*, 1994). Second, each bacterial strain feeds a predator that attacks the other strain, a phenomenon termed "apparent competition" (Holt, 1977; Holt *et al.*, 1994). Levin *et al.* (1977) describe a counterintuitive situation in which the *decrease* in resistance of a partially resistant strain leads to a slight increase in its density and a dramatic decrease in the density of the completely sensitive strain in the presence of a common phage predator. Here the maxim "the enemy of my enemy is my friend" has particular salience. The partially resistant strain gains the upper hand against its superior competitor (the completely sensitive strain) by acting as a phage "carrier." In this scenario, the inferior competitor "delivers" the predator to the more susceptible strain (Levin *et al.*, 1977).

Bohannan and Lenski (2000b) investigated the effect of resource enrichment (i.e., an increase in the input concentration of growth-limiting resource) on the dynamics of these three-member modules. Under low resource input, competition between prey species is expected to influence community patterns; whereas under high resource input, predation is expected to exert a stronger shaping force (Holt *et al.*, 1994; Leibold, 1996). In the absence of phages, bacterial strains partially resistant to phage T2 incur a competitive disadvantage relative to fully sensitive strains (Bohannan and Lenski, 2000b). Bohannan and Lenski predicted that an increase in resource input would shift a community from one where the partially resistant strain was excluded by the sensitive strain to one where the sensitive strain was excluded by the partially resistant strain, with a narrow region of coexistence at intermediate resource input. While they did not observe exclusions, their data did suggest that the sensitive strain (the superior competitor) fared better at low resource input and that the partially resistant strain (the less susceptible) fared better at high resource input (Fig. 2.4C and D).

2.3.5 Communities with phage host-range variants

Even after the evolution of fully resistant bacteria, some bacteriophages can evolve a counter-response, in which mutants can infect the previously resistant bacterial strain. Such host-range mutants have been reported in phage T2 and phage T7 (Chao *et al.*, 1977; Lenski and Levin, 1985). In chemostats inoculated with sensitive *E. coli* B (B_0) and phage T7 ($T7_0$), Chao *et al.* (1977) witnessed the evolution of T7-resistant bacteria (B_1) and then the evolution of phage that could infect B_1 ($T7_1$). In such a four-member community (Fig. 2.3D), prey regulation seemed to be largely top-down, with low densities of bacteria maintained. However, in some of their chemostats, another bacterial strain (B_2) evolved, which was resistant to both $T7_0$ and $T7_1$. In this five-member community (Fig. 2.3E), prey regulation seemed to be primarily bottom-up, with high bacterial densities found.

These four- and five-member communities persist experimentally (Chao *et al.*, 1977), and the importance of phenotypic trade-offs resurfaces (Section 2.3.3.1). Bacterial strains with resistance (B_1 and B_2) are competitively inferior to the fully sensitive strain (B_0). The phage host-range mutant ($T7_1$) is competitively inferior to its ancestor ($T7_0$) in competition for sensitive hosts (B_0). Using a combination of adaptive dynamics and stochastic simulations, Weitz *et al.* (2005) demonstrated that a large number of host-range phage mutants and partially resistant bacterial strains can theoretically coexist. This occurs if the growth costs accompanying partial host resistance to ancestral phages are small relative to the host-range trade-offs in phages (where higher adsorption on one host can result in lower adsorption on a different host). Indeed, the existence of trade-offs may be a general theme in the maintenance of diversity (Tilman, 2000; Bohannan *et al.*, 2002).

The coevolutionary arms races exhibited by phages such as T7 appear lop-sided, with the bacterial host having the last word (Bohannan and Lenski, 2000a). This is because bacteria have the option to alter or lose a phage receptor. While a phage may be able to cope with receptor alterations, the complete loss of a sole binding site is a difficult challenge to answer evolutionarily (although not impossible; see Morona and Henning, 1984). At this point, we should mention that there are other meaningful evolutionary changes that take place in phages besides host-range shifts. Some mutations will alter how the virus uses its bacterial host during its infection; that is, its "foraging strategy," which is the subject of the next section.

2.4 BEHAVIORAL ECOLOGY

All organisms must secure resources in order to develop, survive, and reproduce. Consequently, how organisms find and use resources are critical components of any fitness measure. At least in part, the success of the foraging strategy employed will depend on the organism's environment. In particular, the quality, quantity, distribution, permanence, and heterogeneity of critical resources will influence the optimal foraging strategy; and there has been much theoretical progress towards understanding these relationships (MacArthur and Pianka, 1966; Charnov, 1976; Stephens and Krebs, 1986).

2.4.1 Optimal foraging theory

As a concrete example of foraging-strategy optimization, imagine a hummingbird foraging for nectar in a landscape in which flowers are patchily distributed (Fig. 2.5). Assume that travel from patch to patch takes a certain amount of time, d (dispersal time). Once inside a patch, our hummingbird starts to feed. However, due to a limited amount of nectar, the energetic gain over time shows diminishing returns (the concave function in Fig. 2.5A). How long should the hummingbird stay inside one patch before leaving for the next one? That is, what is its optimal "patch residence time" (or t)? This residence time determines the energetic gain from the patch; specified by the function $g(t)$ in Fig. 2.5A (note, we assume $g(0) = 0$, $g'(t) > 0$, and $g''(t) < 0$). Using the rate of energetic gain (r) as a proxy for fitness, we would like to determine the value of t (call it t^*) that maximizes the quantity $r(t) = g(t)/(d + t)$.

It turns out that $r(t)$ can be visualized graphically: simply draw a line from the filled point on the x-axis a distance d behind the origin to the point $(t, g(t))$; the slope of the resulting line gives $r(t)$. If the residence time is very short or very long (e.g., t_{short} and t_{long}, respectively, in Fig. 2.5A), then the slope is shallow and $r(t)$ is small. The optimal residence time is the value t^* such that the aforementioned line is tangent to the gain function (Fig. 2.5A). This is the marginal value theorem (Charnov, 1976), which states that the optimal residence in a patch is given by the time at which the instantaneous rate of energetic gain (the slope of the tangent to $g(t)$) is equivalent to the long-term rate of energetic gain (the slope of the line connecting the filled point on the x-axis to the point $(t, g(t))$.

We can now ask what would happen to our hummingbird's optimal residence time if its world changes. What happens if the dispersal time between patches becomes longer (compare Fig. 2.5C with Fig. 2.5B) or patch quality increases (compare Fig. 2.5D with Fig. 2.5B)? As shown in Fig. 2.5E and F, a longer dispersal time leads to an increase in optimal residence time, whereas an increase in patch quality leads to a decrease in optimal residence time.

2.4.2 Optimal "phoraging" theory

How might the marginal value theorem apply to bacteriophages? In a very real sense, bacteriophages are foragers of "bacterial resource patches." The time from host lysis to subsequent infection is a dispersal period, D (i.e., the time from Fig. 2.1D to Fig. 2.1B). The dispersal period is inversely proportional to host cell density (Fig. 2.6) and the rate of phage adsorption. The "residence time" of the virus is its latent period, L (i.e., the time from the state shown in Fig. 2.1B to the state shown in Fig. 2.1D). The latent period is generally broken into two periods: (1) the eclipse period, E, which spans

A

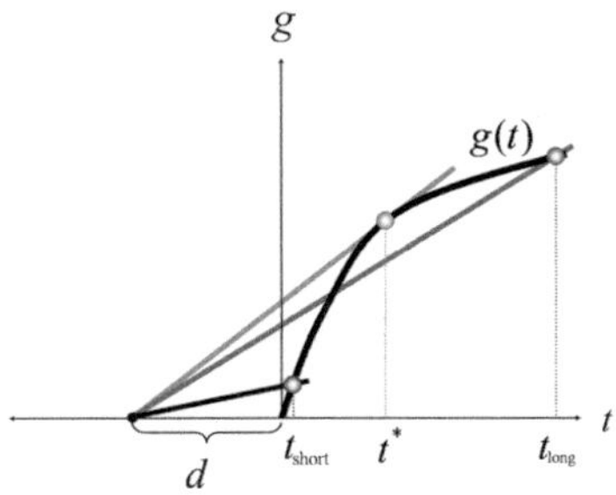

Figure 2.5 Optimal foraging theory. (A) Assume that the time to travel from patch to patch is d, which is why we place a point on the abscissa a distance d to the left of the origin. The thick curve, $g(t)$, gives the energetic gain (on the ordinate) as a function of time spent in the patch (on the positive abscissa). Now, it is simple to graphically describe the long-term average energetic gain, $r(t)$. For a residence time of t_0, simply connect the two points $(t_0, g(t_0))$ and $(-d, 0)$: the slope of this line is $r(t_0) = g(t_0)/(t_0 + d)$. With a concave increasing gain function, we see that short residence times (e.g., t_{short}) or long residence times (e.g., t_{long}) give suboptimal slopes (the shallow black and blue lines, respectively). The maximal slope is obtained for the residence time ($t = t^*$) in which the connecting line is also tangent to the gain curve, in this case the steeper red line. (B) In environment 1, the hummingbird encounters relatively low-productivity patches (with gain function $g_1(t)$) with a relatively short dispersal time between patches (d_1). (C) In environment 2, the hummingbird encounters low-productivity patches (with gain function $g_2(t) = g_1(t)$), but with a longer dispersal time between patches ($d_2 > d_1$). (D) In environment 3, the hummingbird encounters more productive patches ($g_3(t) > g_1(t)$ for all $t > 0$), but with a relatively short dispersal time ($d_3 = d_1$). (E) When we compare optimal residence time in environment 1 (the point of tangency of the purple line, $t_1{}^*$) to the optimal residence time in environment 2 (the point of tangency of the green line, $t_2{}^*$), we see that increasing the dispersal time tends to increase the optimal residence time. (F) When we compare optimal residence time in environment 1 (the point of tangency of the purple line, $t_1{}^*$) to the optimal residence time in environment 3 (the point of tangency of the orange line, $t_3{}^*$), we see that increasing the patch productivity tends to decrease the optimal residence time. *See color plate section.*

the time from initial infection (Fig. 2.1B) to the maturation of infective progeny phage within the host (prior to Fig. 2.1C) and (2) the adult period, A, which lasts from initial progeny maturation to host lysis (Fig. 2.1D). During the adult phase, progeny are produced at a rate, R (see Chapters 3 and 15 for additional discussion of these various phage growth-parameter concepts). Here, we will assume that the number of progeny released at host lysis (the burst size, B) is a linear function of L, namely

$$B(L) = R \cdot (L - E) \quad (2.4)$$

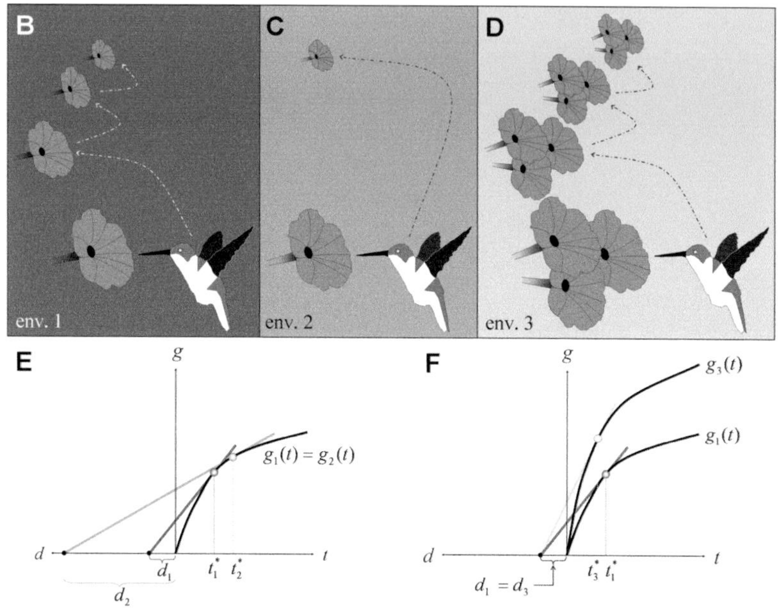

Figure 2.5 (*cont.*)

(Wang *et al.*, 1996; Abedon *et al.*, 2001; Bull *et al.*, 2004). There is empirical support for this assumption of linearity (Hutchinson and Sinsheimer, 1966; Josslin, 1970; Wang *et al.*, 1996; Wang, 2006), although other functions have also been considered (Wang *et al.*, 1996; Abedon *et al.*, 2001).

Equation 2.4 spells out the fundamental trade-off between fecundity and generation time. Each foraging phage particle can only gain more offspring at the expense of a longer generation time (Abedon *et al.*, 2003; Chapter 3). This predicament is similar to the hummingbird's problem, where energetic benefit within a patch is obtained only through costly time investment. Is there an optimal phage solution to this quandary?

When phages are rare in an environment with constant host quantity and quality, an appropriate measure of fitness is the population growth rate:

$$G = [B(L)]^{1/(D+L)} \tag{2.5}$$

where $D + L$ is the phage generation time. The latent period that maximizes G is the optimal latent period, L^*. This optimal latent period will also maximize $\ln(G)$. It can be shown that the optimal latent period is the value L^* such that the instantaneous rate of increase in the log of the burst size and the long-term rate of increase in the log of the burst size are equal. Thus, we

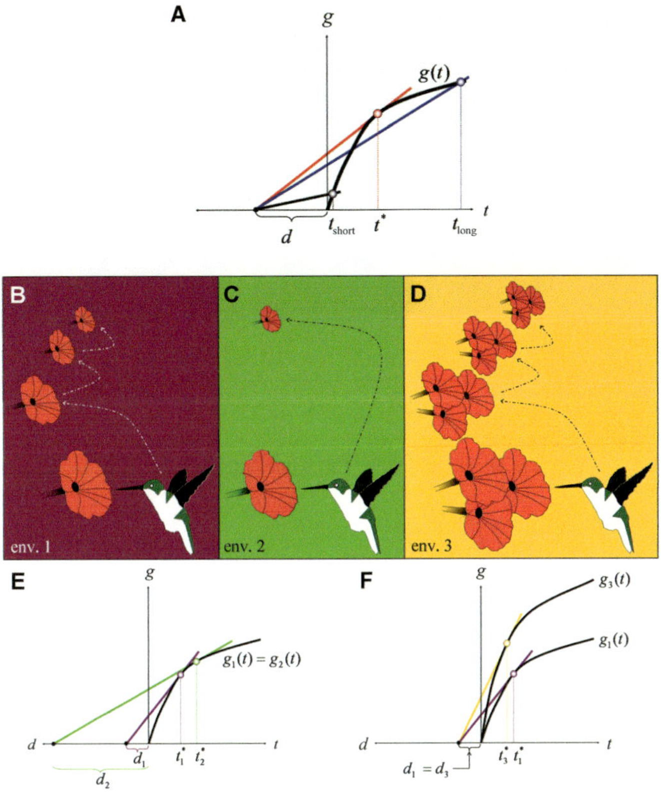

Figure 2.5 Optimal foraging theory. (A) Assume that the time to travel from patch to patch is d, which is why we place a point on the abscissa a distance d to the left of the origin. The thick curve, $g(t)$, gives the energetic gain (on the ordinate) as a function of time spent in the patch (on the positive abscissa). Now, it is simple to graphically describe the long-term average energetic gain, $r(t)$. For a residence time of t_0, simply connect the two points $(t_0, g(t_0))$ and $(-d, 0)$: the slope of this line is $r(t_0) = g(t_0)/(t_0 + d)$. With a concave increasing gain function, we see that short residence times (e.g., t_{short}) or long residence times (e.g., t_{long}) give suboptimal slopes (the shallow black and blue lines, respectively). The maximal slope is obtained for the residence time ($t = t^*$) in which the connecting line is also tangent to the gain curve, in this case the steeper red line. (B) In environment 1, the hummingbird encounters relatively low-productivity patches (with gain function $g_1(t)$) with a relatively short dispersal time between patches (d_1). (C) In environment 2, the hummingbird encounters low-productivity patches (with gain function $g_2(t) = g_1(t)$), but with a longer dispersal time between patches ($d_2 > d_1$). (D) In environment 3, the hummingbird encounters more productive patches ($g_3(t) > g_1(t)$ for all $t > 0$), but with a relatively short dispersal time ($d_3 = d_1$). (E) When we compare optimal residence time in environment 1 (the point of tangency of the purple line, $t_1{}^*$) to the optimal residence time in environment 2 (the point of tangency of the green line, $t_2{}^*$), we see that increasing the dispersal time tends to increase the optimal residence time. (F) When we compare optimal residence time in environment 1 (the point of tangency of the purple line, $t_1{}^*$) to the optimal residence time in environment 3 (the point of tangency of the orange line, $t_3{}^*$), we see that increasing the patch productivity tends to decrease the optimal residence time.

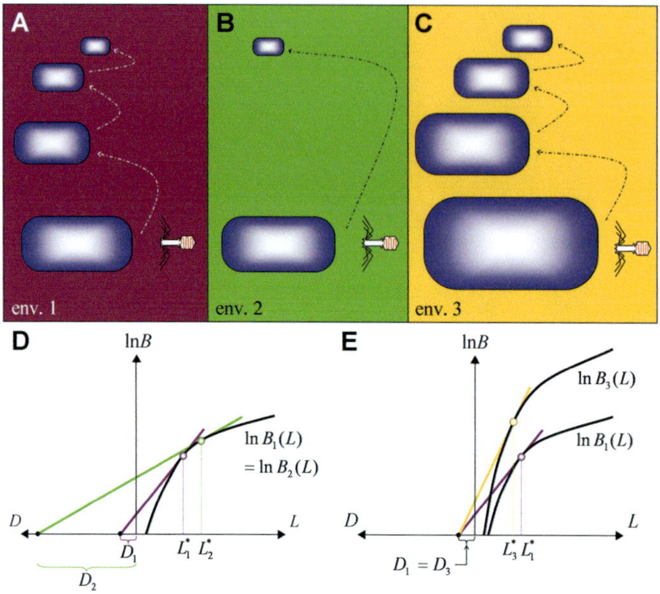

Figure 2.6 Optimal "phoraging" theory. This figure replicates Fig. 2.5 except that the forager's inter-patch dispersal time (d) is replaced by the inter-host dispersal time of the phage (D), the forager's residence time in a patch (t) is replaced by the latent period of a phage in its host (L), and the forager's energetic gain ($g(t)$) is replaced by the log of the phage burst size, $\ln[B(L)]$. As before, we consider (A) environment 1, (B) environment 2, with the same host quality, but a larger inter-host dispersal time than environment 1, and (C) environment 3, with the same inter-host dispersal time, but better host quality than environment 1. As before, we see that (D) increased inter-host travel time (e.g., by decreasing host density) favors a longer optimal latent period and (E) increased host quality favors a shorter optimal latent period. Subscripts in these figures refer to the environment to which the parameter pertains.

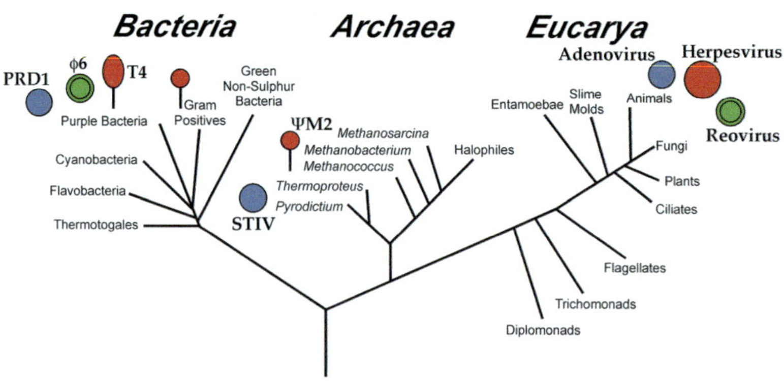

Figure 7.2 Cellular tree of life, with a selection of viruses placed near the hosts they infect. Viruses indicated in the same shade have capsids with shared features (protein fold and aspects of capsid assembly, structure and function) that indicate common ancestry for the genes encoding the capsids in that group. (Modified from Hendrix, 1999, with permission.)

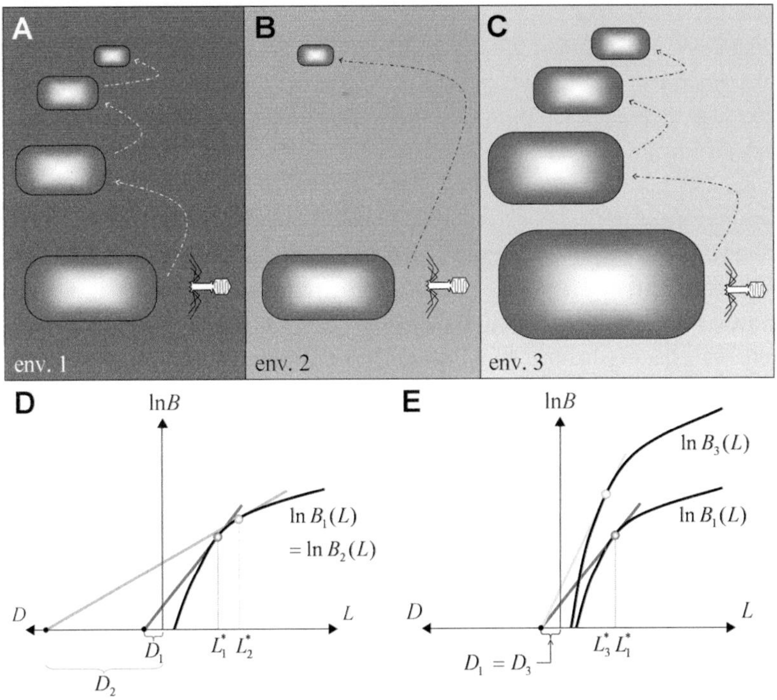

Figure 2.6 Optimal "phoraging" theory. This figure replicates Fig. 2.5 except that the forager's inter-patch dispersal time (d) is replaced by the inter-host dispersal time of the phage (D), the forager's residence time in a patch (t) is replaced by the latent period of a phage in its host (L), and the forager's energetic gain ($g(t)$) is replaced by the log of the phage burst size, $\ln[B(L)]$. As before, we consider (A) environment 1, (B) environment 2, with the same host quality, but a larger inter-host dispersal time than environment 1, and (C) environment 3, with the same inter-host dispersal time, but better host quality than environment 1. As before, we see that (D) increased inter-host travel time (e.g., by decreasing host density) favors a longer optimal latent period and (E) increased host quality favors a shorter optimal latent period. Subscripts in these figures refer to the environment to which the parameter pertains. *See color plate section.*

have a microbial marginal value theorem, where the logarithm of the burst size plays the role of the "energetic gain function," inter-host travel gives the "dispersal time," and latent period is the "residence time." Using a series of phage λ lysis-time mutants, Wang (2006) explored how fitness varied with latent period and demonstrated that an intermediate optimal latent period existed.

In Fig. 2.6, we see that we recover all the results from the hummingbird example. When dispersal period is lengthened (due to a decrease in host quantity or adsorption rate), the optimal latent period is predicted to increase (Fig. 2.6A, B, D). When the rate of maturation for progeny phage increases (due to an increase in host quality), the optimal latent period is predicted to decrease (Fig. 2.6A, C, E). Besides several theoretical investigations into factors affecting optimal latent period (Abedon, 1989; Wang *et al.*, 1996; Abedon *et al.*, 2001; Bull *et al.*, 2004; Bull, 2006; Chapter 3), there have also been two fascinating empirical contributions. In the first study, Abedon *et al.* (2003) competed phage RB69 against a mutant with a short latent period at a variety of host densities. They found that the mutant out-competes the RB69 wild type at high host density, but the reverse occurs at low host density, consistent with the predictions (see Fig. 2.6D). In a second study, Heineman and Bull (2007) used an experimental evolution approach with phage T7. Under high host density, T7 evolved a shorter latent period (close to the predicted optimal latent period). However, these authors did not find the evolution of an optimally longer latent period when T7 was propagated under low host density (the authors attribute this result to a violation of the assumption of linear phage accumulation spelled out by Equation 2.4). Nonetheless, this work is partially consistent with earlier predictions.

2.4.3 Complications on phage optimal foraging

While there has been qualitative empirical agreement with the marginal value theorem, there are some important caveats. First, we have assumed that the dispersal time between hosts, *D*, is constant. Even for constant host density this is unlikely to be true (as there will be variance in waiting times between burst and infection). Using simulation models, Abedon *et al.* (2001) compared the optimal latent period of phages having a constant dispersal time between hosts to the optimal latent period of phages having an exponentially distributed dispersal time between hosts. They demonstrated that optimal latent period, especially at lower bacterial densities, is lower under the exponential model (where phage dispersal times vary). In natural phage populations, host density itself is likely to be variable, which should affect optimal latent period (Bull *et al.*, 2004). Second, we have assumed constant host quality. In natural phage populations, host quality is likely heterogeneous. This is especially true for populations of phage-sensitive and partially resistant bacteria (such as the case of phage T2 described above) and for polyvalent phages (strains that can infect multiple host species). Such heterogeneity in "patch" quality should also affect the optimal latent period (Bull, 2006).

Third, it is important to keep in mind that latent period, adsorption rate, burst size, and eclipse period are dependent not only on the phage, but also on the host and the environment. A given phage genotype can display tremendous phenotypic plasticity. For instance, as the growth rate of the bacterial host increases, the same strain of phage T4 will increase its rate of phage production and burst size and decrease its latent and eclipse periods (Hadas *et al.*, 1997). If a single phage genotype is unable to attain the optimal latent period phenotype in every environment it encounters, then the optimal form of plasticity should depend on both the frequency of different environments and the nature of the genetic constraints (Abedon *et al.*, 2001). See Chapter 3 for discussion of additional constraints on the evolution of latent-period optimality.

Genetic constraints do not usually enter into optimality analysis, where it is assumed that selection will find a genetic path to the optimal strategy. Indeed, latent period in bacteriophages presents a remarkable test case for optimality arguments such as the marginal value theorem (Bull *et al.*, 2004). So far, the empirical results are largely consistent with the predictions of the optimality analysis. And there seem to be some solid genetic reasons for this: (1) it is genetically difficult to transition from a lytic life history to a completely different life history (such as continuous secretion from the host), and (2) mutations in phage holins (the proteins that "time" lysis) can change latent period without concomitant changes to other phage properties (Bull *et al.*, 2004; Wang, 2006). Indeed, the modularity of holins makes the latent period of lytic phages an ideal system for further investigation of predictions of optimal life-history theory.

2.5 METACOMMUNITY ECOLOGY

The models in the last three sections have assumed that interacting species encounter one another in well-mixed conditions. Such an assumption is made primarily for mathematical simplicity and tractability (Tilman and Kareiva, 1997; Dieckmann *et al.*, 2000). However, under natural conditions, this "mass action" assumption (Chapter 15) is surely misplaced. Many biological populations are broken up into subpopulations linked loosely by migration (Hanski and Gaggiotti, 2004; Holyoak *et al.*, 2005), the so-called "metapopulation." If migration is limited, then an individual will have a greater probability of interacting with a member of its own subpopulation than a random individual taken from the entire collection of subpopulations.

2.5.1 Harmony through asynchrony

Does this violation of the mass-action assumption ever matter? It turns out that it can matter significantly. For instance, consider the case in which a subpopulation is reliably extinction-prone through overexploitation of resources. If every individual can access every subpopulation (i.e., migration is unlimited and thus the well-mixed condition is approached), then the population will crash completely. However, if individuals inhabit a metapopulation, even if subpopulations go extinct regularly, persistence at a global scale can be facilitated through asynchrony in extinctions and colonizations (Huffaker, 1958; Holyoak and Lawler, 1996; Ellner *et al.*, 2001; Bonsall *et al.*, 2002; Hanski and Gaggiotti, 2004; Holyoak *et al.*, 2005). That is, resources can be recovering in some places (after the local extinction of the exploiting subpopulation) while the organisms are temporarily thriving in other places (by depleting their local resources). A limited migration rate allows individuals to reach new resources, while avoiding complete synchrony in resource use – and this can allow persistence.

Several researchers have paid particular attention to the behavior of "metacommunities" (Holyoak *et al.*, 2005), a set of interacting species distributed in spatially isolated patches connected by migration of the community members. In a now famous experiment, Huffaker (1958) embedded communities of predatory mites and phytophagous mites (the prey) within a network of oranges. By manipulating the exposed fruit surface and distribution of fruits, Huffaker demonstrated that increased patchiness can promote coexistence of predators and their prey. In essence, the prey is the "resource" for the predator, and limitations in migration (with the resulting asynchrony) can help maintain diversity in the system. This finding has now been confirmed in other laboratory studies as well (Holyoak and Lawler, 1996; Bonsall *et al.*, 2002).

2.5.2 Metacommunities of bacteria and phages

Recently, a similar result was described in bacterium–phage metacommunities (Kerr *et al.*, 2006). These authors embedded phage T4 and *E. coli* B in 96-well microtiter plates and used a high-throughput liquid-handling robot to perform serial transfers of the entire metacommunity. The robot was also used to execute microbial migrations between subpopulations. While the bacteria and phages could not coexist within a single microtiter well (a subpopulation), these authors demonstrated that coexistence was possible at the level of the metacommunity (Fig. 2.7). The precise structure of the metacommunity (e.g., the rate and pattern of migration) turns out to influence evolutionary dynamics within this bacteria–phage system. Such evolutionary dynamics are the subject of the next section.

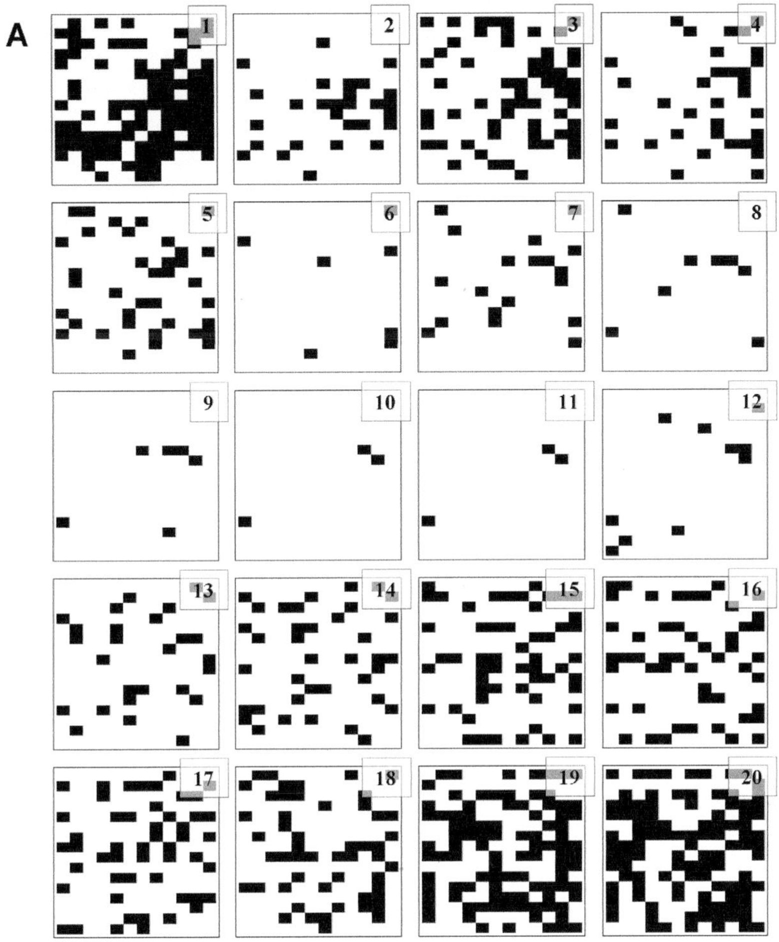

Figure 2.7 A bacterium–phage metacommunity. (A) Each plot gives the spatial distribution of bacteria-filled wells (black squares) in a 192-well metacommunity at different transfers (the numbers in the upper right of each plot) throughout the serial propagation. The white squares are either medium-filled or phage-filled. The migration pattern applied at each transfer (termed "unrestricted" in Kerr *et al.*, 2006) allowed any well to receive microbial migrants, from one of any other well in the metacommunity, with 45% probability. The number of bacteria-filled wells approached low levels during transfers 8 through 12. However, the bacterial population rebounded shortly thereafter. (B) This graph shows the bacterial (squares) and phage (circles) densities corresponding to the spatial distributions shown in part A. Even though bacteria and phages could not coexist within a single well, bacteria and phages are maintained in the metacommunity.

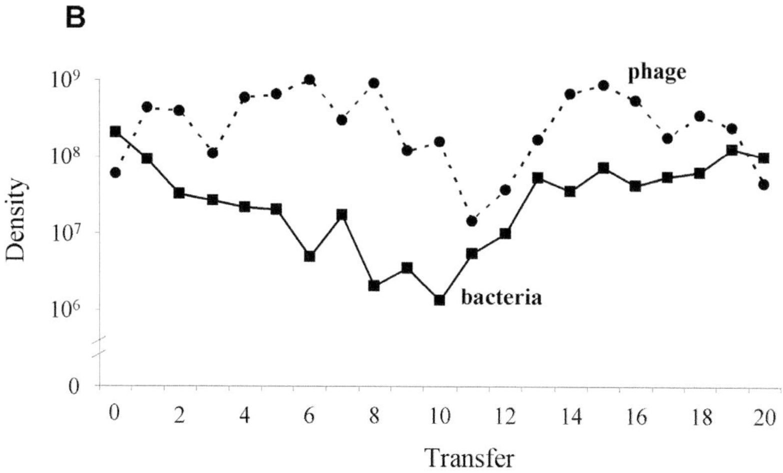

Figure 2.7 (*cont.*)

2.6 EVOLUTIONARY ECOLOGY

How ecological factors shape and are shaped by evolutionary change forms the subject matter of evolutionary ecology. A major challenge within this field is understanding the adaptive significance of the very nature of ecological interactions. Victim–exploiter relationships have an intrinsically antagonistic component (i.e., exploiters damage their victims for personal gain). However, how damaging should an exploiter be? Can exploitative restraint ever be evolutionarily favored? What are the ecological circumstances that select for different ways to use victims? These questions illustrate that ecological interactions (such as those between predators and prey, parasites and hosts, herbivores and plants, etc.) are not static, but have an evolutionary history. It turns out that factors such as migratory range of exploiters and their mode of travel between victims can profoundly influence the nature of these interactions.

2.6.1 Phage productivity versus competitive ability

In the aforementioned metacommunity study of Kerr *et al.* (2006), treatments consisted of different migration patterns within the collection of subpopulations. One treatment restricted migration of microbes to neighboring microtiter wells; whereas in another treatment, migration was unrestricted (i.e., microbes could move from any subpopulation to any other subpopulation with some probability). After several transfers, these authors measured phage productivity (the number of phages produced per parent in a well with host bacteria) and competitive ability (fitness of the phages relative to a marked strain when exposed to shared hosts). They discovered that phages in the restricted treatment evolved high productivity and low competitive ability. On the other hand, phages in the unrestricted treatment evolved low productivity and high competitive ability. In fact, a negative correlation was found between productivity and competitive ability overall, suggesting a trade-off. See Chapter 3 for additional consideration of phage trade-offs between productivity and competitive ability, and Chapter 6 for a more general discussion of trade-offs in the evolution of microbial systems (e.g., in the context of phage host range).

2.6.2 Tragedy of the commons

The results of Kerr *et al.* (2006) suggested the possibility of a 'tragedy of the commons' (Hardin, 1968). This phrase pertains to scenarios in which multiple users exploit the same resource. Although unrestrained users may out-compete restrained users, the rise of unrestrained users may cause the resource (the 'commons') to be overexploited, lowering overall user productivity (the 'tragedy'). Within a subpopulation, resident phages exploit bacterial resources. Prudent use of this resource (due to longer latent period or slower adsorption) leads to higher productivity of the phages (Chapter 3). The 'tragedy of the commons' is that rapacious types lower overall productivity as they displace more restrained competitors. Abedon *et al.* (2003) demonstrated that rapacious variants of phage RB69 can out-compete their prudent ancestors by limiting future access to the host through their own rapid consumption. Further, these authors found that when grown in pure culture, the prudent ancestor was more productive than its rapacious descendant. The findings of Abedon *et al.* (2003) are consistent with the trade-off postulated by Kerr *et al.* (2006) and the potential for a 'tragedy of the commons' (Chapter 3).

2.6.3 Evolution of restraint given spatial structure

Why might different patterns of migration favor different resolutions of this tragedy? Within any subpopulation, rapacious phage mutants always out-compete their prudent ancestors, but these superior competitors are less productive and therefore more likely to go extinct (given periodic dilution). Low productivity is the Achilles heel of the rapacious phage, so its success hinges on gaining sufficient access to fresh hosts. Unrestricted migration ensures that rapacious phages have continual access to hosts. Furthermore, unrestricted migratory patterns increase the probability of mixing phage types, which favors the rapacious competitor. In contrast, restricted migration lowers the likelihood that different phage types mix and lowers the accessibility of fresh hosts. Thus, rapacious phages are more vulnerable to extinction in the restricted treatment, leaving the metacommunity relatively prudent by default (Kerr *et al.*, 2006).

In the 'tragedy of the commons' scenario, the trade-off between competitive ability and productivity is key to understanding the evolution of restraint in one of the aforementioned treatments. Trade-offs are a general theme in evolutionary ecology (Bohannan *et al.*, 2002). For example, trade-offs figure prominently in the literature concerning the evolution of pathogen virulence (Bull *et al.*, 1991; Herre, 1993; Ebert, 1994; Nowak and May, 1994; Lipsitch *et al.*, 1995; Boots and Sasaki, 1999; O'Keefe and Antonovics, 2002; Galvani, 2003; Thrall and Burdon, 2003). In this work, virulence is assumed to covary with pathogen transmission (Herre, 1993; Ebert, 1994; Nowak and May, 1994; Lipsitch *et al.*, 1995; Boots and Sasaki, 1999; Galvani, 2003) or competitive ability (Nowak and May, 1994; Keeling, 2000). Assuming these or similar trade-offs, many researchers have investigated the role of host population structure on the evolution of pathogen strategy (Bull *et al.*, 1991; Herre, 1993; Lipsitch *et al.*, 1995; Boots and Sasaki, 1999; Keeling, 2000; O'Keefe and Antonovics, 2002; Galvani, 2003; Thrall and Burdon, 2003). There are theoretical and experimental results that suggest that as populations become less structured, virulence in pathogens is favored (Bull *et al.*, 1991; Boots and Sasaki, 1999; Keeling, 2000; O'Keefe and Antonovics, 2002; Galvani, 2003). See Chapters 4 and 16 for additional consideration of the phage ecology of spatially structured habitats, and Chapter 3 for an additional consideration of phage virulence.

2.6.4 Evolution of restraint with vertical transmission

Using phage f1 and *E. coli* K12, Bull *et al.* (1991) demonstrated that horizontal transmission of the phage (as would occur in a less structured population) selected for variants that were relatively damaging to their hosts (i.e., highly virulent); whereas vertical transmission (as would occur in a more structured population) selected for avirulent phage. Furthermore, using the same system, Messenger *et al.* (1999) demonstrated that a trade-off existed between virulence and reproductive output. The level of virulence that evolved depended on the ratio of vertical to horizontal transmission, with viral lines evolving higher virulence and higher reproductive output when the ratio was low. The 'tragedy of the commons' illustrates that strategies yielding short-term gain (high viral reproductive output or high viral competitive ability) may be disastrous for long-term success. When hosts are unlimited (i.e., horizontal transmission readily occurs), then maximizing short-term gain can be optimal; whereas when access to fresh hosts is limited (i.e., vertical transmission is more prevalent), then investment in long-term gain pays off and avirulent strategies become more favorable.

2.6.5 Linking evolution and ecology

As the topic of pathogen virulence illustrates, the synergy of evolutionary and ecological approaches can provide useful tools for a deeper understanding of biological systems. This synergy is particularly appropriate for the study of bacteriophages because of their short generation times, large population sizes, heterogeneous environments, and diverse set of biotic interactions. Complementarily, bacteriophage systems form ideal models for exploring the intersection of ecology and evolution: such systems not only demonstrate basic principles of ecology, but enlarge our ecological perspective to include evolutionary dynamics.

For example, the evolution of resistant hosts and host-range phage mutants within experimental microbial communities shows that the study of community structure is enriched by a consideration of *de novo* evolutionary contributions. As another example, optimal life-history theory assumes that evolutionary change will produce the strategy most appropriate for a set of ecological conditions; thus, how phage latent period changes in response to host quantity and quality is an issue well-suited for evolutionary ecology. Similarly, the patterns of migration in metacommunities not only influence the abundances of phages and bacteria, but also their evolutionary trajectories. Overall, research on bacteriophages exemplifies how evolutionary approaches inform various fields within ecology, including community ecology, behavioral ecology, and metapopulation ecology.

2.7 FUTURE DIRECTIONS

In this chapter we have highlighted only a few of the ecological concepts that can be fruitfully explored with, and applied to, phages and bacteria, using studies of coliphages and *E. coli* as examples. These examples demonstrate that ecological concepts can be very useful for increasing our understanding of phages and bacteria, and that in turn phages and bacteria can be very powerful tools for exploring such concepts. Our examples, and indeed most studies of the interactions between phages and bacteria, are of relatively simple communities maintained in homogeneous laboratory environments. The most exciting future directions for such research require the relaxation of these constraints.

Phages and bacteria are ideal for exploring the ecological implications of spatial structure and environmental heterogeneity. Indeed, several recent studies have begun to do exactly this (Brockhurst *et al.*, 2003, 2006; Forde *et al.*, 2004; Kerr *et al.*, 2006; Chapters 3, 4, and 16). Promising areas of focus include the effects of heterogeneous environments on stability, diversity, and evolution within communities consisting of bacteria and phages. In order to explore such topics, microbial community modules could be embedded within metapopulation apparatus (e.g., microtiter plates) or spatially continuous surfaces (e.g., agar-filled Petri dishes; Chapters 4 and 16). Such setups could also be employed to explore the optimal latent period of lytic phages in environments with spatially heterogeneous host density and quality.

Most studies of interactions between phages and bacteria have utilized only one phage type. However, ecological theory predicts that population dynamics, the opportunities for coexistence, and the trajectory of evolution can be very different when multiple predator types, rather than just one, are present (Weitz *et al.*, 2005). It would be interesting to explore the potential for coexistence of phages with different life histories (e.g., obligately lytic and temperate phages) living with a shared pool of hosts. Also, by using phages with different life histories, researchers could explore the evolutionary effects of one phage on another through exploitative competition for hosts. The vast majority of ecological studies of phages and bacteria have focused on obligately lytic phages, i.e., phages that invariably kill their hosts after a relatively short latent period (Chapter 1). However, phages exhibit a range of exploitative lifestyles, including those displayed by temperate phages (which can incorporate into their host's genome for a variable time period, before re-emerging and killing their host; Chapter 5) and filamentous phages (which can release progeny from their hosts without host death). Utilizing the full range of phage lifestyles will permit the application of a wide range of ecological concepts to phages and bacteria, and allow the exploration using laboratory communities of a much longer list of ecological questions. For example, temperate and filamentous phages are closer to traditional parasites than are obligately lytic phages, which are most precisely parasitoids.

Lastly, future research will ideally begin to bridge the gap between ecological studies of phages and bacteria in the laboratory and recent surveys of viruses in natural environments (e.g., Chapters 10 and 11). The study of viruses in nature (the majority of which appear to be bacteriophages) has grown rapidly in recent years, due to methodological developments that permit the quantitative study of viral populations without requiring laboratory culture (Fuhrman, 1999; Chapter 10). Future research could assess the relevance of experimental results with simple phage modules – the topics discussed in this chapter – to the complex webs of natural phage communities.

2.8 APPENDIX

Here we explore the chemostat model of Bohannan and Lenski (1997). They write the following system of differential equations describing the dynamics for the concentration of resource (C), the concentration of the bacterial prey (N), and the concentration of the phage predator (P):

$$dC/dt = (C_0 - C)\cdot\omega - \varepsilon\cdot N\cdot\Psi\cdot C/(K+C) \quad (2.6)$$

$$dN/dt = N\cdot\Psi\cdot C/(K+C) - \alpha\cdot N\cdot P - \omega\cdot N \quad (2.7)$$

$$dP/dt = B\cdot e^{-L\cdot\omega}\alpha\cdot N'\cdot P' - \alpha\cdot N\cdot P - \omega\cdot P \quad (2.8)$$

where C_0 is the concentration of glucose feeding into the chemostat, ω is the flow rate through the chemostat, ϵ is the reciprocal yield of the bacteria, Ψ is the maximal bacterial growth rate, K is the resource concentration at which the bacteria grow at one-half their maximal rate, α is the adsorption rate of phage, B is the burst size, and L is the latent period. All primed variables are evaluated L time units ago, e.g., $N' = N(t-L)$. As a baseline, we use the following values: $C_0 = 0.25\ \mu g\ mL^{-1}$, $\omega = 0.2\ h^{-1}$, $\varepsilon = 2\times 10^{-6}\ \mu g$, $\Psi = 0.7726\ h^{-1}$, $K = 0.0727\ \mu g\ mL^{-1}$, $\alpha = 3\times 10^{-7}\ mL\ h^{-1}$, $B = 80$, and $L = 0.6$ h.

We manipulate the productivity of the system by altering C_0. In Fig. 2.8A we see that the predator–prey cycles increase in amplitude as the system is enriched. We manipulate the negative density dependence of the prey by altering K. In Fig. 2.8B we see that negative density dependence has a stabilizing effect (as K is increased, the amplitude of the cycles diminish). We manipulate the time lag in the system by altering the latent period (L). In Fig. 2.8C, we see that increasing the latent period slightly destabilizes the system. As we discuss in Section 2.4, phage burst size is positively correlated with latent period; we will assume $B = R\cdot(L - E)$. If we perform the same perturbations in latent period as shown in Fig. 2.8C (but allow burst size to

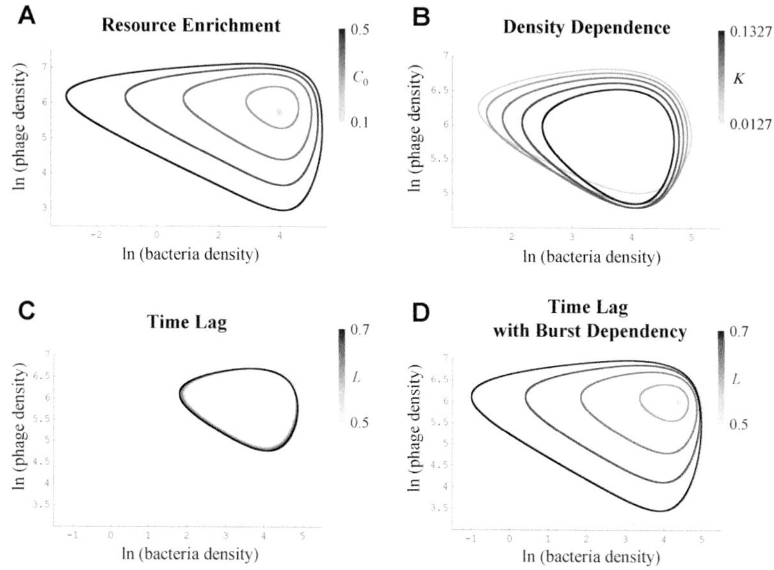

Figure 2.8 Predicted chemostat limit cycles. Unless otherwise noted, parameter values are as presented in Section 2.8. (A) We varied glucose reservoir concentration C_0 from 0.1 μg L^{-1} to 0.5 μg L^{-1}. Resource enrichment destabilizes the system. (B) Letting $C_0 =$ 0.25 μg L^{-1}, $L = 0.6$ h, and $B = 80$, we varied the half-saturation constant K from 0.0127 μg L^{-1} to 0.1327 μg L^{-1}. Negative density dependence (achieved through larger K values) stabilizes the system. (C) Letting $C_0 = 0.25$ μg L^{-1}, $K = 0.0727$ μg L^{-1}, and $B = 80$, we varied the latent period L from 0.5 h to 0.7 h. These increases in time lag slightly destabilize the system. (D) Letting $C_0 = 0.25$ μg L^{-1} and $K = 0.0727$ μg L^{-1}, we varied the latent period L from 0.5 h to 0.7 h, but we let the burst size be a linear function of the latent period (see Section 2.4): $B = R \cdot (L - E)$. We set $E = 0.4333$ h (Abedon *et al.*, 2001) and $R = 480$ h^{-1}: this rate of phage increase was chosen so that the burst size function would intersect the point $(L, B) = (0.6, 80)$. Compared to the case where burst size is held constant (C), an increase in latent period with a simultaneous increase in burst size dramatically destabilizes the system.

change according to the above function with eclipse period, $E = 0.4333$ h, and the rate of progeny production, $R = 480$ h^{-1}) then we see that the system is substantially destabilized by increases in the latent period (Fig. 2.8D).

REFERENCES

Abedon, S. T. 1989. Selection for bacteriophage latent period length by bacterial density: a theoretical examination. *Microb. Ecol.* **18**: 79–88.

Abedon, S. T., T. D. Herschler, and D. Stopar. 2001. Bacteriophage latent-period evolution as a response to resource availability. *Appl. Environ. Microbiol.* **67**: 4233–41.

Abedon, S. T., P. Hyman, and C. Thomas. 2003. Experimental examination of bacteriophage latent-period evolution as a response to bacterial availability. *Appl. Environ. Microbiol.* **69**: 7499–506.

Armstrong, R. A., and R. McGehee. 1980. Competitive-exclusion. *Am. Nat.* **115**: 151–70.

Bohannan, B. J. M., and R. E. Lenski. 1997. Effect of resource enrichment on a chemostat community of bacteria and bacteriophage. *Ecology* **78**: 2303–15.

Bohannan, B. J. M., and R. E. Lenski. 1999. Effect of prey heterogeneity on the response of a model food chain to resource enrichment. *Am. Nat.* **153**: 73–82.

Bohannan, B. J. M., and R. E. Lenski. 2000a. Linking genetic change to community evolution: insights from studies of bacteria and bacteriophage. *Ecol. Lett.* **3**: 362–77.

Bohannan, B. J. M., and R. E. Lenski. 2000b. The relative importance of competition and predation varies with productivity in a model community. *Am. Nat.* **156**: 329–40.

Bohannan, B. J. M., B. Kerr, C. M. Jessup, J. B. Hughes, and G. Sandvik. 2002. Trade-offs and coexistence in microbial microcosms. *Antonie Van Leeuwenhoek* **81**: 107–15.

Bonsall, M. B., D. R. French, and M. P. Hassell. 2002. Metapopulation structures affect persistence of predator–prey interactions. *J. Anim. Ecol.* **71**: 1075–84.

Boots, M., and A. Sasaki. 1999. "Small worlds" and the evolution of virulence: infection occurs locally and at a distance. *Proc. Roy. Soc. London Ser. B Biol. Sci.* **266**: 1933–8.

Brockhurst, M. A., A. D. Morgan, P. B. Rainey, and A. Buckling. 2003. Population mixing accelerates coevolution. *Ecol. Lett.* **6**: 975–9.

Brockhurst, M. A., A. Buckling, and P. B. Rainey. 2006. Spatial heterogeneity and the stability of host–parasite coexistence. *J. Evol. Biol.* **19**: 374–9.

Bull, J. J. 2006. Optimality models of phage life history and parallels in disease evolution. *J. Theor. Biol.* **241**: 928–38.

Bull, J. J., I. J. Molineux, and W. R. Rice. 1991. Selection of benevolence in a host–parasite system. *Evolution* **45**: 875–82.

Bull, J. J., D. W. Pfennig, and I. N. Wang. 2004. Genetic details, optimization and phage life histories. *Trends Ecol. Evol.* **19**: 76–82.

Campbell, A. 1961. Conditions for the existence of bacteriophage. *Evolution* **15**: 153–65.

Chao, L., B. R. Levin, and F. M. Stewart. 1977. Complex community in a simple habitat: experimental study with bacteria and phage. *Ecology* **58**: 369–78.

Charnov, E. L. 1976. Optimal foraging, marginal value theorem. *Theor. Pop. Biol.* **9**: 129–36.

Dieckmann, U., R. Law, and J. A. J. Metz (eds.). 2000. *The Geometry of Ecological Interactions: Simplifying Spatial Complexity*. Cambridge: Cambridge University Press.

Drake, J. A., G. R. Huxel, and C. L. Hewitt. 1996. Microcosms as models for generating and testing community theory. *Ecology* **77**: 670–7.

Ebert, D. 1994. Virulence and local adaptation of a horizontally transmitted parasite. *Science* **265**: 1084–6.

Edelstein-Keshet, L. 1988. *Mathematical Models in Biology*. New York, NY: McGraw-Hill.

Ellner, S. P., E. McCauley, B. E. Kendall, *et al.* 2001. Habitat structure and population persistence in an experimental community. *Nature* **412**: 538–43.

Forde, S. E., J. N. Thompson, and B. J. M. Bohannan. 2004. Adaptation varies through space and time in a coevolving host–parasitoid interaction. *Nature* **431**: 841–4.

Fuhrman, J. A. 1999. Marine viruses and their biogeochemical and ecological effects. *Nature* **399**: 541–8.

Galvani, A. P. 2003. Epidemiology meets evolutionary ecology. *Trends Ecol. Evol.* **18**: 132–9.

Grover, J. P. 1995. Competition, herbivory, and enrichment: nutrient-based models for edible and inedible plants. *Am. Nat.* **145**: 746–74.

Gwynn, D. M., A. Callaghan, J. Gorham, K. F. A. Walters, and M. D. E. Fellowes. 2005. Resistance is costly: trade-offs between immunity, fecundity and survival in the pea aphid. *Proc. Roy. Soc. London Ser. B Biol. Sci.* **272**: 1803–8.

Hadas, H., M. Einav, I. Fishov, and A. Zaritsky. 1997. Bacteriophage T4 development depends on the physiology of its host *Escherichia coli*. *Microbiology-UK* **143**: 179–85.

Hanski, I., and O. E. Gaggiotti (eds.). 2004. *Ecology, Genetics, and Evolution of Metapopulations*. New York, NY: Elsevier Academic Press.

Hardin, G. 1968. The tragedy of the commons. *Science* **162**: 1243–8.

Heineman, R. H., and J. J. Bull. 2007. Testing optimality with experimental evolution: lysis time in a bacteriophage. *Evolution*. **61**: 1695–709.

Herre, E. A. 1993. Population-structure and the evolution of virulence in nematode parasites of fig wasps. *Science* **259**: 1442–5.

Holling, C. S. 1973. Resilience and stability of ecological systems. *Ann. Rev. Ecol. Syst.* **4**: 1–23.

Holt, R. D. 1977. Predation, apparent competition, and structure of prey communities. *Theor. Pop. Biol.* **12**: 197–229.

Holt, R. D. 1995. Community modules. In: *Multitrophic Interaction in Terrestrial Systems*. Oxford: Blackwell, pp. 333–50.

Holt, R. D., J. Grover, and D. Tilman. 1994. Simple rules for interspecific dominance in systems with exploitative and apparent competition. *Am. Nat.* **144**: 741–71.

Holyoak, M., and S. P. Lawler. 1996. Persistence of an extinction-prone predator–prey interaction through metapopulation dynamics. *Ecology* **77**: 1867–79.

Holyoak, M., M. A. Leibold, and R. Holt (eds.). 2005. *Metacommunities: Spatial Dynamics and Ecological Communities*. Chicago, IL: University of Chicago Press.

Horne, M. T. 1970. Coevolution of *Escherichia coli* and bacteriophages in chemostat culture. *Science* **168**: 992–3.

Huffaker, C. B. 1958. Experimental studies on predation: dispersion factors and predator–prey oscillations. *Hilgardia* **27**: 343–83.

Huisman, J., and F. J. Weissing. 1999. Biodiversity of plankton by species oscillations and chaos. *Nature* **402**: 407–10.

Hutchinson, C. A., and R. L. Sinsheimer. 1966. The process of infection with bacteriophage ϕX174. X. Mutations in a ϕX174 lysis gene. *J. Mol. Biol.* **18**: 429–47.

Jessup, C. M., R. Kassen, S. E. Forde, *et al.* 2004. Big questions, small worlds: microbial model systems in ecology. *Trends Ecol. Evol.* **19**: 189–97.

Josslin, R. 1970. The lysis mechanism of phage T4: mutants affecting lysis. *Virology* **40**: 719–26.

Keeling, M. 2000. Evolutionary trade-offs at two time-scales: competition versus persistence. *Proc. Roy. Soc. London Ser. B Biol. Sci.* **267**: 385–91.

Kerr, B., C. Neuhauser, B. J. M. Bohannan, and A. M. Dean. 2006. Local migration promotes competitive restraint in a host–pathogen "tragedy of the commons." *Nature* **442**: 75–8.

Kraaijeveld, A. R., and H. C. J. Godfray. 1997. Trade-off between parasitoid resistance and larval competitive ability in *Drosophila melanogaster*. *Nature* **389**: 278–80.

Leibold, M. A. 1989. Resource edibility and the effects of predators and productivity on the outcome of trophic interactions. *Am. Nat.* **134**: 922–49.

Leibold, M. A. 1996. A graphical model of keystone predators in food webs: trophic regulation of abundance, incidence, and diversity patterns in communities. *Am. Nat.* **147**: 784–812.

Lenski, R. E. 1984. 2-Step resistance by *Escherichia coli* B to bacteriophage T2. *Genetics* **107**: 1–7.

Lenski, R. E. 1988. Experimental studies of pleiotropy and epistasis in *Escherichia coli*.1. Variation in competitive fitness among mutants resistant to virus T4. *Evolution* **42**: 425–32.

Lenski, R. E., and B. R. Levin. 1985. Constraints on the coevolution of bacteria and virulent phage: a model, some experiments, and predictions for natural communities. *Am. Nat.* **125**: 585–602.

Levin, B. R., F. M. Stewart, and L. Chao. 1977. Resource-limited growth, competition, and predation: a model and experimental studies with bacteria and bacteriophage. *Am. Nat.* **111**: 3–24.

Lipsitch, M., E. A. Herre, and M. A. Nowak. 1995. Host population-structure and the evolution of virulence: a law of diminishing returns. *Evolution* **49**: 743–8.

Lotka, A. J. 1925. *Elements of Physical Biology*. Baltimore, MD: Williams & Wilkins.

MacArthur, R. H., and E. R. Pianka. 1966. On optimal use of a patchy environment. *Am. Nat.* **100**: 603–9.

May, R. M. 1972. Limit cycles in predator–prey communities. *Science* **177**: 900–2.

May, R. M. 1974. *Stability and Complexity in Model Ecosystems*, 2nd edn. Princeton, NJ: Princeton University Press.

Messenger, S. L., I. J. Molineux, and J. J. Bull. 1999. Virulence evolution in a virus obeys a trade-off. *Proc. Roy. Soc. London Ser. B Biol. Sci.* **266**: 397–404.

Morona, R., and U. Henning. 1984. Host range mutants of bacteriophage Ox2 can use 2 different outer-membrane proteins of *Escherichia coli* K-12 as receptors. *J. Bacteriol.* 159: 579–82.

Nicholson, A. J., and V. A. Bailey. 1935. The balance of animal populations. *Proc. Zool. Soc. Lond.* **1**: 551–98.

Nowak, M. A., and R. M. May. 1994. Superinfection and the evolution of parasite virulence. *Proc. Roy. Soc. London Ser. B Biol. Sci.* **255**: 81–9.

O'Keefe, K. J., and J. Antonovics. 2002. Playing by different rules: the evolution of virulence in sterilizing pathogens. *Am. Nat.* **159**: 597–605.

Paynter, M. J. B., and H. R. Bungay. 1971. Characterization of virulent bacteriophage infections of *Escherichia coli* in continuous culture. *Science* **172**: 405.

Rosenzweig, M. l. 1971. Paradox of enrichment: destabilization of exploitation ecosystems in ecological time. *Science* **171**: 385–7.

Schrag, S. J., and J. E. Mittler. 1996. Host–parasite coexistence: the role of spatial refuges in stabilizing bacteria–phage interactions. *Am. Nat.* **148**: 348–77.

Stephens, D. W., and J. R. Krebs. 1986. *Foraging Theory*. Princeton, NJ: Princeton University Press.

Thrall, P. H., and J. J. Burdon. 2003. Evolution of virulence in a plant host–pathogen metapopulation. *Science* **299**: 1735–7.

Tilman, D. 2000. Causes, consequences and ethics of biodiversity. *Nature* **405**: 208–11.

Tilman, D., and P. Kareiva (eds.). 1997. *Spatial Ecology: the Role of Space in Population Dynamics and Interspecific Interactions.* Princeton, NJ: Princeton University Press.

Volterra, V. 1926. Variazioni e fluttuazioni del numero d'individui in specie animali conviventi. *Mem. R. Accad. Naz. dei Lincei.* Ser. VI, 2.

Wang, I.-N. 2006. Lysis timing and bacteriophage fitness. *Genetics* **172**: 17–26.

Wang, I.-N., D. E. Dykhuizen, and L. B. Slobodkin. 1996. The evolution of phage lysis timing. *Evol. Ecol.* **10**: 545–58.

Weitz, J. S., H. Hartman, and S. A. Levin. 2005. Coevolutionary arms races between bacteria and bacteriophage. *Proc. Natl. Acad. Sci. U.S.A.* **102**: 9535–40.

CHAPTER 3

Phage population growth: constraints, games, adaptation

Stephen T. Abedon

3.1 INTRODUCTION

Pick up an introductory biology textbook that describes bacteriophages. The presented phage life cycle, often using phage λ as an example, will typically be differentiated into two distinguishable types: the lytic cycle and the lysogenic cycle. This differentiation is real but overly simplistic. First, both the lytic cycle and the lysogenic cycle differ among different phages by numerous molecular details (Calendar and Abedon, 2006). Second, and as typically mentioned, not all phages display a lysogenic cycle (Chapter 5). Third, though atypically mentioned, not all phages display a lytic cycle (Russel and Model, 2006), at least in the sense of productive phage infection followed by a phage-induced bacterial lysis. In this chapter I provide an overview of the ecology of the virion-mediated population growth displayed by obligately lytic phages (*sensu* Chapter 1, Section 1.2.2.5). That is, I explore the ecology of phage adsorption and infection of susceptible bacteria, virion maturation within those bacteria, and then lytic release of phage progeny. I take an evolutionary ecological perspective, considering the impact of phage adaptations on phage population growth (as also does Chapter 2).

3.1.1 Productive phage infection

As described in greater detail in Chapter 1, phage infections may be differentiated into productive versus non-productive (the latter including lysogeny; Chapter 5), with production referring to the intracellular maturation and then release of phage virion particles. Among productive infections

Bacteriophage Ecology: Population Growth, Evolution, and Impact of Bacterial Viruses, ed. Stephen T. Abedon. Published by Cambridge University Press.

are those in which the release of phage progeny is associated with termination of the phage infection (lysis as seen with lytic infections; Fig. 2.1). Alternatively, phage progeny are released by chronically infecting phages into the extracellular environment without loss of the phage-maturation capacity of the original infection (Russel and Model, 2006). In this chapter I consider the ecological implications of phage productive growth, with strong emphasis on phage lytic growth, particularly as it occurs within non-spatially structured (i.e., well-mixed) environments. For consideration of the impact of spatial structure on phage ecology see Chapter 4, for the ecology of phage plaque formation, and Chapter 2, for the ecology of phage metapopulations.

3.1.2 Life-history optimization

The density of bacterial cells available to phages – which constitute the primary phage resource, i.e., their "food" – and the relative survivability of phage virions are crucial to the evolution of phage life-history characteristics, with phage survivability evolvable in its own right. These characteristics, in turn, impact both the rate and end-point productivity of phage population growth (Chapter 15). Phage evolution, as a function of bacterial density and virion survivability, can occur especially through optimization of phage lysis timing (Chapter 2), with rates of lysis ranging from rapid to rare (Section 3.2.1). Bacterial quantity, in turn, is subject to both phage-density-dependent and phage-density-independent factors (Chapters 2, 10, and 15), with phage-density-dependent factors conveniently expressed in terms of the multiplicity of phages that actually adsorb bacteria (Section 3.5). Abedon and LeJeune (2005) and Chapter 14 postulate additional phage control – via expression of phage-encoded bacterial exotoxins – of the characteristics, including density, of uninfected, phage-permissive bacteria that are found within a phage's immediate environment.

In terms of multiplicities of actual phage adsorption, we can describe (I) phage population growth at low phage multiplicities, (II) phage multiplicities during the transition from relatively few bacteria infected to relatively few bacteria not infected, (III) phage multiplicities of approximately greater than one, (IV) the resulting community-wide phage-induced lysis of bacteria (assuming lytic rather than chronic infection) that gives rise to even higher phage multiplicities, and then, finally, (V) phage survival at high multiplicities but low bacterial densities post this lysis. These steps represent a sequence over which virion-mediated phage population growth should typically occur. The basic premise of this chapter is that the phage optimal latent period may vary over the course of this sequence. See Chapter 2 for additional consideration of latent period optimization especially as it occurs during step II.

Whether any of the latter stages of this sequence may be reached, i.e., stages II–V, depends on phage properties that are sometimes cumulatively referred to as a phage's virulence (d'Hérelle, 1922; Smith *et al.*, 1987; Daniels and Wais, 1998; Section 3.3.3), and also on bacterial densities. Alternatively, phages may become locally extinct (Chapters 2, 15, and 17) or may enter into a steady state in which phage numbers never increase sufficiently to wipe out phage-susceptible bacteria (Chapter 2). Phages also can employ means whereby their long-term survival is enhanced such as by complexing with virion-stabilizing environmental materials (Chapter 11), including biofilms (Lacroix-Gueu *et al.*, 2005), or, upon infecting bacteria, by temporarily abandoning virion production via entrance into lysogeny or pseudolysogeny (Chapter 5).

3.2 SELECTION ACTING ON PHAGE GROWTH

Like any creature, a phage must increase its numbers via some mechanism of reproduction, and over time must do so at least as fast as intrinsic or environmental factors can reduce those numbers. As noted, we can consider such growth under at least five circumstances (Fig. 3.1): (I) low multiplicity (Section 3.2.1), (II) multiplicities approaching approximately one (Section 3.2.2), (III) multiplicities exceeding one (Section 3.2.3; Chapter 8), (IV) phage-induced community-wide lysis (3.2.4), and (V) post-lysis phage survival (Section 3.2.5). In all cases I employ the term multiplicity in the sense of multiplicity of adsorption (U_a) rather than multiplicity of infection (U; Section 3.5). By doing so I am implying that I am more concerned with the actual potential of phages to adsorb bacteria rather than with a hypothetical maximum multiplicity given infinitely long incubation (that is, not just a ratio of starting phage to bacterial densities; see Section 3.5).

3.2.1 Selection during low-multiplicity growth (I)

The population growth of obligately lytic phages at lower multiplicities – as achieved by phage exposure to an excess of bacteria – involves multiple rounds of phage adsorption, infection, and lysis (Fig. 3.1, I). Selection during low-multiplicity phage population growth may be considered in terms of exploitative competition, i.e., competition for resources without direct physical confrontation between competitors. During this stage a simple truism holds: All else held constant, those phages that reach bacteria sooner will infect more bacteria than those phages that lag in their bacterial

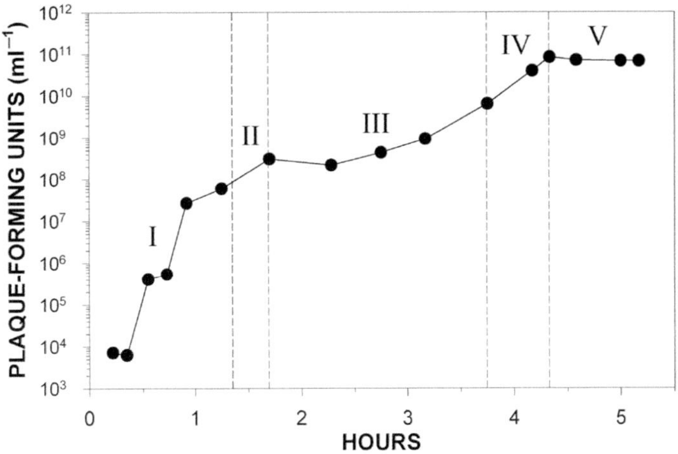

Figure 3.1 Phage T4D population growing on *Escherichia coli* S/6/5 in well-mixed broth. Shown are (I) phage growth at low multiplicities, (II) a transition from low to high phage multiplicities, (III) a period of high phage multiplicities and high cell densities, (IV) population-wide bacterial lysis, and (V) high phage multiplicities but low cell densities. The period indicated by III is greatly extended, compared with other phage systems, due to the display of lysis inhibition by T4 phages. Population-wide phage lysis occurs by an unusual mechanism in phage T4 (Abedon, 1992; Section 3.4.8) but should be relatively rapid in most phage systems, in well-mixed environments, given short phage latent periods. Placement of dashed lines separating different intervals is for illustration purposes rather than indicating exact transitions. Portions of this figure were originally published by Abedon (1992) and are used here with permission from the American Society for Microbiology.

acquisition. The most competitive lysis-timing strategy a phage may employ under these circumstances, however, can vary as a function of bacterial density (Chapter 2).

3.2.1.1 Strategies for effective exploitative competition

There are three basic means by which phages can achieve faster population growth and thereby more quickly exploit their bacterial resource. The first is to display a shorter generation time via latent-period reduction or by displaying a faster rate of adsorption (Fig. 15.2B and D, Chapter 15). The second is to display a greater fecundity, i.e., a larger burst size (Fig. 15.2 A, Chapter 15). The third is to display reduced rates of loss, such as resistance to virion decay or resistance to infection restriction. This resistance to decay or restriction can be considered in terms of the phage *effective* burst size (Section 3.4.2), which is the number of phage particles produced by an infection that succeed in successfully infecting new bacteria. As I will consider in the next section (3.2.1.2), there are trade-offs associated with displaying shorter latent periods that are felt especially in terms of the phage burst size and particularly at lower bacterial densities. It is conceivable that there are also trade-offs associated with displaying faster rather than slower adsorption rates (Daniels and Wais, 1998), though such trade-offs remain at best speculative. Similarly, there presumably are trade-offs, such as reduced per-infection fecundities (de Paepe and Taddei, 2006) or extended generation times, that are associated with displaying adaptations that give rise to greater phage survival.

It is important to note that exploitative competition as a driver towards faster phage population growth should be selected only if there is some sort of time component to phage competition. That is, if phages are competing for the same bacteria within the same environment, then this competition will select for faster exploitation of that environment. If one removes this time component, then exploitation may be better achieved by producing larger overall phage populations within a given environment than by phages more rapidly exploiting individual bacteria. These latter considerations, however, may be relevant particularly as phage multiplicities transition toward past 1.0, which will be more fully considered in Section 3.2.2.

3.2.1.2 Impact of bacterial density on latent-period evolution

The impact of bacterial density on the evolution of phage latent period has been given ongoing consideration within the phage literature (Levin and Lenski, 1983; Abedon, 1989, 1994, 2006; Wang *et al.*, 1996; Abedon *et al.*, 2001, 2003; Bull *et al.*, 2004; Wang, 2006; Bull, 2006; see Chapter 2 for additional consideration of this issue). The bottom-line conclusion is that selection for shorter phage latent periods should accompany higher bacterial densities, whereas evolution should favor relatively longer phage latent periods at lower bacterial densities. The intuitive underpinnings to this conclusion are that bacterial densities impact phage generation times by controlling the duration of the phage extracellular search (Abedon, 1992) for new bacteria. At lower bacterial densities this search takes longer, thereby increasing the overall phage generation time. The longer this "search" or "non-replicative" portion of the phage life cycle, the less of a relative impact a given change in phage latent period can have on the overall phage generation time. At higher bacterial densities the non-replicative portion of the phage life cycle is relatively short, so therefore even small changes in the phage latent period can have a significant impact on the overall phage generation time.

These various points would not necessarily result in evolution favoring different phage latent periods at different bacterial densities except that latent period, in addition to impacting phage generation time, also impacts phage burst size. Thus, longer latent periods at lower bacterial densities can result in a cost that is modest in terms of increasing the phage generation time, but simultaneously in a significant benefit in terms of greater phage fecundity. Alternatively, a similar absolute change in latent period experienced at higher bacterial densities would have a much greater impact on the phage generation time (e.g., one minute subtracted from 20 min. vs. one minute subtracted from 40 min.) such that latent-period reductions at higher bacterial densities can be advantageous even given resulting burst-size costs.

Of interest, the impact of lower bacterial densities on the evolution of phage latent periods is smaller than one might expect based solely on the above intuitive understanding. That is, phage population growth is not driven by the average search time for new bacteria, and therefore not by the average phage generation time. Instead, rates of phage population growth are most affected by those phages that by chance most rapidly find new bacteria and thereby display the shortest generation time within their population. Thus, optimal phage latent periods for phage population growth at lower bacterial densities in fact may be shorter than one might otherwise expect (Abedon *et al.*, 2001; Abedon, 2006).

3.2.2 Selection acting at multiplicities approaching one (II)

Everything changes as the majority of bacteria become phage-infected, and that is because this stage initiates the last round of phage infection within a culture (Fig. 3.1, II). At the end of this stage of phage population growth there no longer exists selection for rapid bacteria procurement, at least within the now phage-saturated environment in question, and this is because there no longer are uninfected bacteria to procure. Selection instead should be for larger phage burst sizes, since the total number of phage progeny produced by a given environment should be approximately equal to the product of the total number of bacteria infected and the phage burst size during this last round of infection. Stating this latter point again, but without seeming to imply group selection (which I'm not): within a phage-saturated environment, reductions in generation time are no longer a means by which an individual phage may increase its fitness, so selective constraints on displaying larger burst sizes are somewhat reduced.

For most phages it is probable that their burst size as phage multiplicities approach and then exceed one is no different from that displayed during previous rounds of replication. Alternatively, for some phages, especially the T-even-like coliphages (Abedon, 2000), this final round of replication can involve a transition to infections displaying greater burst sizes (Abedon, 1990, 1994; Section 3.4.5). Lysogeny, too, represents an adaptation to phage-multiplicity-associated losses of uninfected bacteria (Herskowitz and Banuett, 1984). That is, phage virions are optimized for relatively rapid exploitation of phage-uninfected bacteria whereas both lysogeny and lysis inhibition are strategies that instead allow a longer-term exploitation of the infected bacterium. Lysogeny, like lysis inhibition, may even be viewed explicitly in terms of phage latent periods and burst sizes: a prophage has the potential to produce much larger numbers of phage progeny if induction follows lysogen binary fission – that is, multiple bursts are then possible – versus the single burst that follows the lytic infection of an individual bacterium. The resulting increase in phage progeny produced by a single temperate phage, following the display of lysogeny, comes at a cost in generation time, however, since the necessary multiple rounds of binary fission take time.

Why don't all phages modify their infection strategies as a function of phage multiplicity? There exist a number of possible explanations. First, it may be that many different phages are optimized for effective exploitative competition rather than for a more economical exploitation of their environment (concepts which I will discuss in greater detail in Section 3.3). Alternatively, it may be that many environments do not display the necessary physical limits on phage population growth that could result in a loss of access to new bacteria to infect. In essence, such phages would be permanently competing for new bacteria to quickly infect (stage I). Finally, it might be that bacterial hosts for many phages simply do not exist at high enough densities for phage multiplicities of adsorption to ever approach one. In addition to too-low bacterial densities, phage burst sizes may not be sufficiently high, or rates of virion inactivation could be too high, to allow phage populations to reach sufficient numbers such that bacteria are killed faster than they can reproduce.

3.2.3 Selection acting at multiplicities of greater than one (III)

When phage multiplicities of adsorption are greater than one (Fig. 3.1, III), then there is potential for phage coinfection and superinfection (Chapter 8). These, respectively, are the infection of a single bacterium by two or more phages and the adsorption of a second (or more) phage to an already infected bacterium. The term "secondary" (Doermann, 1948) portrays these later adsorbing but not necessarily infecting phages better than the more commonly employed "superinfecting." In particular, many phages display mechanisms of superinfection exclusion (Abedon, 1994) or immunity (Hershey and Dove, 1983; Roberts and Devoret, 1983) that either prevent phage entrance into an already infected bacterium or prevent phage replication within that bacterium. Thus, coinfecting or (literally) superinfecting phages will contribute productively or reductively to an infection (i.e., produce phage progeny or form a prophage) while a secondary phage may or may not.

A coinfecting phage, given a burst that is otherwise fixed in size, will produce fewer progeny than a phage that has singly infected a bacterium. To avoid displaying the reduced fecundity that comes with coinfection, then either (1) bacterial infections would need to altruistically block subsequent phage adsorption (e.g., as seen with phage T5; Dunn and Duckworth, 1977), (2) phages would need to be smart enough to not adsorb already-infected bacteria, or (3) infected bacteria would need to avoid lysing so as to prevent exposure of their otherwise intracellular phage progeny to already-infected bacteria. Secondary-phage loss of fecundity due to coinfection may not be too great a price to pay for obtaining a bacterium, however, particularly if bacteria are rare. This would be an example of a phage equivalence to the old parable that a bird in the hand (here, an adsorbed bacterium that is already phage-infected) is worth two in the bush (a potentially uninfected but as yet unadsorbed bacterium). This bird-in-the-hand advantage is especially the case if the infection status of an obtained bacterium is predictive of the infection status of neighboring bacteria. See Bull *et al.* (2006) for additional consideration of phage evolutionary ecology when multiplicities exceed one.

3.2.4 Selection acting during population-wide lysis (IV)

If lysis is to be avoided, e.g., so as to avoid exposing phage progeny to superinfection exclusion, then there still will remain a need to reverse the unlysed state so that, ultimately, the protected progeny may disseminate to new hosts. Lysing at an optimal moment during the lysis of the rest of a phage population (Fig. 3.1, IV) could represent a difficult balancing act, however. For instance, significant progeny loss, due to adsorption to already infected bacteria, can occur if lysis occurs somewhat prior to the lysis of the rest of the population of phage-infected bacteria versus somewhat later. It is conceivable, though, that no phage is sophisticated enough to so precisely vary its lysis timing. Indeed, few phages are known which can control the length of their productive latent period in response to environmental stimuli. Exceptional are T-even phages (Section 3.4.5).

3.2.5 Selection acting post population-wide lysis (V)

In the absence of phage-susceptible bacteria to infect (Fig. 3.1, V), one would expect selection to favor adaptations that contribute to virion durability and/or dissemination to new bacteria-containing environments. See

Chapter 14 and Abedon and LeJeune (2005) for speculation as to how exotoxins, such as the phage-encoded Shiga toxin, could contribute to virion dissemination within or between bacteria-containing environments.

3.3 COOPERATION AND DEFECTION

The trade-off between latent period and burst size, as seen especially at higher bacterial densities (Section 3.2.1.2), can be expressed as one between a cooperative behavior, on one the hand, and a "cheating" (or defecting) behavior on the other (e.g., Ackermann and Chao, 2004; Travisano and Velicer, 2004; West *et al.*, 2006). In cooperative behavior, the cooperating group (or, minimally, two cooperating individuals) gains from the cooperative actions of its members, but the cooperative action itself may be costly for individuals to implement. Cheaters, by contrast, avoid the costs associated with displaying a cooperative behavior, but also do not contribute to the cooperation displayed by two or more individuals. Often the cheater may also benefit from the behavior displayed by cooperative individuals.

In this section I will further explore adaptations to phage population growth, framing them in terms of these cooperation versus cheating ideas. I will do so by employing a number of metaphors: expedient actions versus economical ones (Section 3.3.1), the 'tragedy of the commons' (Section 3.3.1.1), and the prisoner's dilemma (PD; Section 3.3.5). I will also describe factors impacting the evolution of degrees of phage virulence (Section 3.3.3) and then consider a complicated game of cooperation and defection displayed by T-even phages during their population growth (Section 3.4). Many of these concepts are also considered in Chapter 2.

3.3.1 Expedient versus economical phage population growth

Within a hypothetical environment lacking in limits, there may exist no brakes upon organismal population growth. In the absence of constraints on population growth, as otherwise would be represented by the concept of carrying capacity, a population would be expected to display an indefinite exponential increase. Maximal Darwinian fitness in such an (admittedly unrealistic) environment would be displayed by maximizing this exponential growth. We probably would describe such organisms as displaying an expediency whereby all organismal qualities except those that contribute to faster population growth are evolutionarily de-emphasized. These are the grasshoppers of fable fame, enjoying an endless summer of immediate (i.e., expedient) resource utilization, while the hardworking ant displays an economy by delaying the exploitation of gathered resource in anticipation of winter's harshness. In this section I consider phage population growth particularly from the perspective of expedient versus economic utilization of common resources (Pfeiffer *et al.*, 2001; Kreft, 2004; MacLean and Gudelj, 2006) and I do so especially by considering phage adaptations that give rise to more rapid phage population growth (expediency) versus greater overall phage yield (economy). See Chapter 2 for a similar discussion, based upon phage adaptation during growth within metacommunities, where the terms "competitive ability" and "productivity" (as well as "rapacious" and "prudent") are substituted for "expediency" and "economy" as used here.

3.3.1.1 The "tragedy of the commons"

It is not at all unusual for unrelated individuals to compete over a common resource. One name for such antagonism is interspecific competition, which can be described as negative–negative dealings between two or more individuals. That is, both parties lose. Parties, however, can lose to different extents. For example, an efficient resource utilizer may find itself at a disadvantage when competing against individuals that are enhancing their growth at the *expense* of efficient resource utilization. That is, a trade-off can exist between population growth rates and resource utilization efficiency.

We can describe the various resources for which organisms can compete as "commons." It is possible to husband these commons so that the common resource is maintained in perpetuity, rather than overly exploited and thereby diminished or destroyed. Such husbanding, however, requires cooperation among those individuals exploiting the common environment. It is thus the "tragedy of the commons" (Hardin, 1968; Chapter 2) that the short-term gain by some individuals (cheaters or defectors) can result in a diminishment of access to the common's resource by all members of a group.

We can describe cheaters as displaying expediency, sacrificing long-term productivity for the sake of short-term gain. Alternatively, the cooperative husbanders of a common resource we can describe as displaying an economy whereby an attempt is made to maximize productivity over the long term rather than just over the short term. It is of interest to ask how it is that long-term interests can come to supersede the much more immediate interests of the short term. I will frame answers to this question specifically from the perspective of phage growth within well-mixed batch cultures.

3.3.1.2 Shorter latent periods versus larger burst sizes

During phage population growth the expediency versus economy trade-offs that can give rise to a "tragedy of the commons" can be framed in terms of a phage population's growth rate versus growth yield. This is a trade-off because both strategies cannot be simultaneously maximized (Kreft, 2004), which in the extreme bars the evolution of that ultimate organism that can display an absolutely minimal generation time and simultaneously an absolutely maximal fecundity. For phages, particularly at higher bacterial densities, we can describe this trade-off as one seen between latent period and burst size, where reduced latent periods can result in the expediency of high rates of phage population growth (Section 3.2.1.2). That is, shorter latent periods allow phages to enter into and then exit infections rapidly, thereby rapidly releasing phage progeny which can then quickly acquire abundantly available uninfected bacteria. This expediency, however, comes at the expense of per-bacterium phage yield. By contrast, a large burst size represents an economic exploitation of each bacterial resource. This trade-off between latent period and burst size also exists at lower bacterial densities, but is of reduced relevance owing to the smaller influence of latent-period length on phage population growth rate when bacterial densities are low (Section 3.2.1.2).

The trade-off between expediency and economy, between growth rate and growth yield, is envisaged for phages most easily in terms of population growth within volume-limited environments, e.g., as in a bacteria-containing flask. An expedient phage, such as one displaying shorter latent periods at higher bacterial densities, will more quickly exploit the bacterial resource, potentially driving that resource to extinction, especially given a lack of either spatial structure (Schrag and Mittler, 1996) or phage mortality (Chao *et al.*, 1977). By contrast, the economic, longer-latent-period phage will less quickly exploit the bacterial resource. Place both phage types within the same environment, and the expedient phage will more quickly acquire bacteria. As environmental limits are approached, this faster acquisition means that the economic phage will be denied similar access to the resource. The net result is selection for the expedient phage given mixed-culture (i.e., non-clonal) phage population growth (see also Chapter 2).

Note that selection for expediency, when that exists, should be greater in the absence of volume limitation than in its presence. This is because the advantage of economy is only seen at the point where the environment has matured, as indicated by phage multiplicities exceeding 1.0 (Fig. 3.1, III). The larger burst size observed with economically longer latent periods will contribute to a greater yield particularly during the final round of bacteria infection within a culture – when the most bacteria are infected. Under these conditions phage population-wide fecundity is approximately equal to the total number of bacteria infected during this final round of phage infection

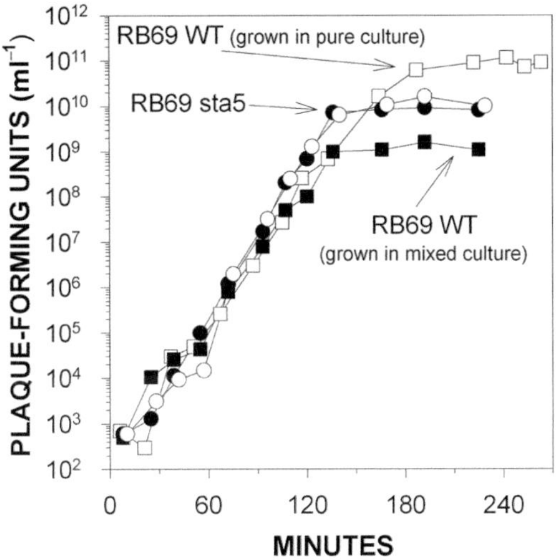

Figure 3.2 Batch competition between expedient and economical phages. RB69 wild type (WT) (■, □) displays a longer-latent-period economy whereas the phage RB69 mutant, sta5 (○, ●), displays a shorter-latent-period expediency. Cultures containing only a single phage type are shown using open symbols. Cultures initially containing a 1 : 1 mixture of WT to sta5 are shown as solid symbols. An *Escherichia coli* CR63 host was employed in this experiment. This figure was originally published by Abedon *et al.* (2003) and is used here with permission from the American Society for Microbiology.

multiplied by the phage burst size. The expedient strategy is to increase the total fraction of bacteria infected, but the efficacy of this strategy is eroded somewhat, given limited environmental volumes, because the burst size of each infected bacterium is reduced with shorter latent periods. The expedient strategy can still provide greater final phage titers, given within-culture competition between expedient and economic phages. However, in the absence of such competition, a pure culture of expedient phages will display a reduced productivity (overall fecundity) as compared to a pure culture of economic phages.

See Fig. 3.2 for experimental illustration of these points. Phage RB69 wild type displays an economic strategy of longer latent period relative to the expedient strategy of the phage RB69 mutant, sta5, which has a shorter latent period. A striking difference among these cultures is seen with the high final yield observed with wild-type phage RB69 when grown in pure culture as compared with the comparatively low final yield observed when phage RB69 wild type is grown in competition with the phage RB69 sta5 mutant.

3.3.1.3 Bacterial population growth versus phage expediency

The trade-off between expediency and economy, as manifest in terms of phage population growth rates versus final per-bacterium phage yield, is not the entire story explaining the great disparity between phage RB69's yield when grown in pure versus mixed culture (Fig. 3.2). An additional consideration derives from the impact of expedient phage growth on the peak bacterium density. That is, when phage populations are growing faster, the bacterial population comes to be fully phage-infected sooner, therefore replicates for a shorter period, and as a consequence consists of fewer bacteria (Figs. 3.3 and 15.3). Thus, in pure culture the economical wild type experiences a larger burst size, infects essentially all of the cells within the culture, *and* infects a larger pool of cells than is seen with either the mixed culture or with the RB69 sta5 pure culture. The latter effect may be equivalent to pathogens allowing their multicellular hosts to grow for longer or to larger sizes, thereby supporting greater pathogen replication (Frank, 1996; Cooper *et al.*, 2002).

3.3.1.4 Periodic dissemination as a selective force

The most obvious means of enforcing cooperation is to physically bind would-be cooperators together, much as those cells making up multicellular organisms are bound together. However, even with multicellularity there must exist mechanisms that inhibit cheating, including clonal relation among organisms/cells, mechanisms of internal policing (such as to destroy cancer cells), and sequestration of germ lines (so as to minimize the long-term payoff for cheating; Michod, 1996; Michod *et al.*, 2003). Microorganisms that tend to invade otherwise empty niches, and do so in very small numbers (e.g., initially as only one individual), may also be able to maintain genes encoding cooperative behaviors, such as genes that restrain the virulence of viral or bacterial pathogens (Frank, 1996; Ebert, 1998; Bergstrom *et al.*, 1999).

One means by which we may attain a similar level of physical binding among would-be phage cooperators is for phage population growth to occur within finite environments that at least temporarily lack competing phages and also which display some degree of cohesiveness such that escape is only periodic (or episodic) rather than continuous. In other words, these would be physically structured environments which in some manner alternate between being closed off and being open (e.g., metapopulation experiments as described in Chapter 2). A test tube or flask during serial transfer experiments represents one such alternatingly closed and then open environment, with pipet insertion or decanting giving rise to the open state. The gut of animals similarly may serve as such an environment, the contents of which are sometimes disseminating into the extra-colonic environment and sometimes not. Physically structured microenvironments, such as might be found in soil or as a consequence of biofilm formation, similarly could give rise to relatively closed chambers within which phage population growth might occur that is temporarily sealed off from the outside world.

Besides physical structure and alternating closed and open states, an additional ingredient potentially necessary for the evolution of a behavior of economy is the occurrence of an environmental maturation prior to environmental opening. This is another way of saying that resource limitations must occur or, in other words, that an environment's carrying capacity must to some extent impose itself upon a population's growth. In the case of phages this means that phage multiplicities must come to exceed approximately one (Section 3.2.3; Fig. 3.1, III) – prior to phage dissemination out of an environment – such that the economy of burst size becomes more relevant to phage fitness, even at higher bacterial densities, than does the expediency of shorter phage latent periods.

3.3.1.5 Founder effects and maintenance of clonality

A dominant take-home message from Fig. 3.2, as well as a significant conclusion within the literature on the evolution of cooperation (e.g., Pfeiffer *et al.*, 2001; Michod *et al.*, 2003; Ackermann and Chao, 2004; Kreft, 2005; West *et al.*, 2006), is that for cooperative behavior to evolve there must be some means by which cheaters are excluded from cooperators. Spatial structure, such that cooperators can congregate away from cheaters, is one important step towards achieving this cooperation. However, also key is the exclusion of cheaters from these potentially safe havens for cooperation. There exist two means by which this cheater exclusion from environments may be accomplished. Stated from a phage perspective, the first is to initiate infections within environments with only one or, at best, a few phage particles. In other words, phages must experience a founder effect, and thereby some reasonable likelihood of within-environment clonal relatedness. Microbiologists obtain a similar result when employing pure-culture technique. Typically this is done by inoculating bacteria or phage stock cultures using isolated colonies or plaques, which, in principle, are initated by only a single cell, phage, or clone. In this way cultures may be propagated without introducing excessive genetic heterogeneity. See Chapter 2 for arguments that reduced mixing of different phage types may also be accomplished by resticting migration within metacommunities.

The second key means of reducing likelihoods of a co-occurrence of cheaters among cooperators is to limit phage population sizes within environments such that mutation to cheater genotypes does not occur with high likelihood (see Chapter 6 for discussion of phage mutation). Thus, even given limits, environments displaying too-large volumes may allow an evolution of cheating prior to completion of the phage exploitation of those environments. Larger environments also may be too difficult to temporarily seal off from phage entrance, or escape, to allow effective selection for economy (i.e., such as may occur given only periodic dissemination out of otherwise closed environments; Section 3.3.1.4).

Thus, for cooperation among phages to evolve during the exploitation of well-mixed environments, it would appear that the following criteria should be met. The environment should be closed, but not too closed, thereby allowing individual phages to enter but not many phages. The environment should allow phage exit only following environment maturation (i.e., somewhat complete phage-mediated bacterial exploitation). Environments in addition should not be so large, nor phage mutation rates so great, that cheating genotypes evolve sufficiently early that significant cheater replication can occur at the expense of cooperator replication. Given these many constraints, evolution should then tend to favor phages that are economical rather than expedient resource exploiters.

3.3.2 Phage virulence

From a bacterium's perspective, death following (obligately) lytic infection may be unavoidable, implying a phage virulence (as measured in terms of reduction in host Darwinian fitness) of essentially 100% per successful phage infection. From a phage's perspective, however, shorter latent periods are more virulent because infection destruction, via lysis, occurs sooner. Alternatively, we can emphasize the multicellularity of the host victims of pathogen virulence, rather than emphasizing the infection of individual cells. An animal's body, for example, is a spatially limited multicellular environment within which pathogen replication takes place (Levin and Antia, 2001). Phages similarly can be described as replicating within spatially limited environments, e.g., laboratory or colonic cultures that contain multiple, phage-susceptible bacteria.

While from a modern perspective it is perhaps a tortured view to consider *phage* "virulence" as occurring against whole cultures, particularly given usage of the term "virulent" as the converse of "temperate" (meaning potentially lysogen-forming; Chapter 1), in fact phages were originally described

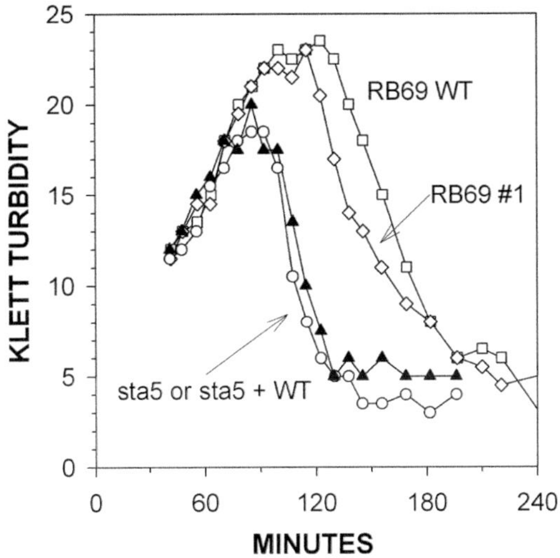

Figure 3.3 Phage virulence against batch cultures of bacteria (*E. coli* CR63). Earlier culture lysis represents greater phage virulence against bacterial populations and culture turbidity was employed in this experiment to assess the ongoing "health" of the bacterial culture. Mutant and wild-type phage RB69 are as indicated, and were added at low densities, i.e., $\sim 10^{3}$ phage per mL, final concentration, versus peak bacterial densities in the range of $\sim 10^{8}$ mL^{-1}. The "sta5 + WT" curve is of a 1 : 1 mixture of WT to sta5. RB69 #1 is a mutant that displays a larger plaque than RB69 WT along with a slightly shorter latent period. This figure was originally published by Abedon *et al.* (2003) and is used here with permission from the American Society for Microbiology.

as entities that lysed bacterial cultures (Twort, 1915; d'Hérelle, 1917), and this was consistent with the more whole-culture emphasis of microbiology at the time (Summers, 1991; Chapter 1). Those phages more capable of lysing these cultures, following growth from lower phage densities, therefore could reasonably be described as more virulent (e.g., Smith *et al.*, 1987; Summers, 2001). Given such a scenario, we can describe phages displaying shorter latent periods within environments possessing higher densities of phage-susceptible bacteria as possessing a growth advantage that is associated with greater virulence towards these multicellular environments, i.e., earlier lysis of whole bacterial cultures (Fig. 3.3). Note, however, that this greater virulence may be short-sighted, particularly if phage transmission to new bacteria-containing environments is a function of peak phage densities (i.e., economy) rather than within-culture phage-population growth rates (i.e., expediency; Fig. 3.2; Section 3.3.1).

3.3.3 Selection for economy during dissemination

High phage death rates during phage transmission between environments, as well as significant phage dilution, could reduce the likelihood of phage transmission to new environments. Note that the marginal value theorem from optimal foraging theory (Charnov, 1976; Stephens and Krebs, 1986), as has been applied to phages elsewhere to derive within-culture optimal latent periods (Wang *et al.*, 1996; Chapter 2), similarly suggests that greater *distances* or *costs* between exploitable environments should select for more complete (such as longer-latent-period-like) exploitation of resources within individual environments. Note also that we can similarly view this between-culture selection from the perspective of later-offspring discounting (Horn, 1978): sooner-produced offspring can be more valuable, but only if they themselves can *quickly* contribute to population growth. However, a quick contribution of offspring to population growth between environments would be the case only if environments were sufficiently close together. As exploitable resource-containing environments become ever closer, then a well-mixed total environment is increasingly approximated (Chapter 2). It is within such a well-mixed total environment that we would expect more-virulent, shorter-latent-period phages to out-compete longer-latent-period phages (Fig. 3.2; Abedon *et al.*, 2001; Kerr *et al.*, 2006).

The trade-off model of pathogen virulence evolution (Ewald, 1994) – which states that pathogen virulence should evolve towards some compromise between selection for more-rapid within-host replication and more-effective between-host transmission – similarly concludes that an evolution of higher virulence is likely particularly when new hosts (i.e., new animals or plants to infect) are readily acquired. We could generalize this prediction. An increase in pathogen virulence might represent an evolutionary response to a relative absence of limits on within-environment (or within-culture) propagation or, similarly, a breakdown in the distinction between within-environment growth and between-environment transmission. It is precisely those pathogens whose within-multicellular-host growth places little constraint on between-multicellular-host transmission that we expect will evolve (or, at least, maintain) more rapid within-host pathogen replication (Ewald, 1994; Ebert and Bull, 2003). See Chapter 2 for consideration of the evolution of phage cooperation versus defection (equivalent to benignness versus virulence, respectively) within spatially and temporally differentiated bacteria-containing environments called metapopulations.

3.3.4 The prisoner's dilemma

Prisoner's dilemmas are games of cooperation and defection (Rapoport and Chammah, 1965; Trivers, 1971; Axelrod, 1984; Dugatkin, 1997; Turner and Chao, 1999, 2003; Chapter 8). For example, a behavior of economical growth (as displayed by phage RB69 wild type in Fig. 3.2) represents a cooperative behavior, whereas the expediency displayed by the RB69 sta5 mutant (also as seen in Fig. 3.2) may be interpreted as a behavior of defection (a. k. a., "cheating"). In all prisoner's dilemmas, over the course of a single round of interaction between two individuals, it is always preferable to defect rather than cooperate. Thus, defecting against a cooperator is preferable to cooperating with a cooperator and, similarly, defecting against a defector is preferable to cooperating when one's opponent defects. Prisoner's dilemmas are dilemmas, however, because the payoff for cooperating with a cooperator is greater than the payoff for the otherwise more rational defecting with a defector. Algebraically, a prisoner's dilemma is thus described as an inequality of payoff values, $T > R > P > S$, that are associated with: (1) unilateral defection (T; one individual defects in the presence of one or more cooperators), (2) mutual cooperation (R; all individuals cooperate), (3) mutual defection (P; all individuals defect), and (4) unilateral cooperation (S; one individual cooperates in the presence of one or more defectors).

Typically one illustrates a prisoner's dilemma as a 2×2 matrix such as that presented in Fig. 3.4. Note that illustrating the interactions between two (or more) individuals as a 2×2 matrix can be helpful towards understanding social behaviors even if the resulting payoffs (e.g., as measured in terms of organism fitness) cannot be arranged so as to strictly conform to a prisoner's dilemma. Thus, in Fig. 3.4 I illustrate behaviors of cooperation and defection using phage RB69 wild type and sta5 as displayed in Fig. 3.2, but otherwise make no claim that those interactions conform to a prisoner's dilemma. Nevertheless, in the following section I will draw upon these ideas of interaction between potentially cooperating or defecting individuals to illustrate the selective forces, and resulting dilemmas, that can be present during phage population growth.

3.4 SCENARIOS FOR PHAGE POPULATION GROWTH

Based on the many considerations outlined above, I will now consider phage population growth from the perspectives of phage adaptation, environmental constraints, and game theory. This will serve as a recap of the

	Cooperate (e.g., RB69 wild type)	**Defect** (e.g., RB69 sta5 mutant)
Cooperate (e.g., economy)	***R**eward* for Mutual Cooperation (e.g., RB69 wild type in pure culture)	***S**ucker's* Payoff (e.g., RB69 wild type in mixed culture)
Defect (e.g., expediency)	***T**emptation* to Defect (e.g., RB69 sta5 in mixed culture)	***P**unishment* for Mutual Defection (e.g., RB69 sta5 in pure culture)

Figure 3.4 Cooperation–defection payoff matrix. Shown are all possible single-round interactions between two players where each player is limited to one behavior, either cooperate or defect. Furthermore, each player is not aware of the other player's behavior until it has chosen its own behavior. Shown are example interactions based upon the economy versus expediency behaviors displayed by phage RB69 wild type and sta5 mutant, respectively, as illustrated in Figs. 3.2 and 3.3. Note that by presenting these examples I am not implying that the interactions between phage RB69 wild type and sta5 mutant strictly conform to a prisoner's dilemma, which would be approximated only if the shown payoff values could be arranged as $T > R > P > S$.

trade-off between expediency and economy which I have emphasized throughout much of this chapter. I will focus on the population growth of T-even phages for the exercise, both because of my familiarity with this system and because of the interesting complexity of their population growth including, in particular, their display of lysis inhibition.

3.4.1 Initiating phage population growth

Upon entrance into a bacteria-containing environment, either physically or as a host-range mutant able to infect a previously inaccessible bacterial population, a phage either adsorbs a bacterial cell or, in a sense, dies trying. This "death" can occur through decay of virion particles (Chapters 5 and 11), through outflow from the environment (Chapter 15), or – though, like outflow, not technically death – via temporary sequestration away from phage-susceptible bacteria (such as via reversible adsorption to non-bacterial materials; Chapter 11). The likelihood of absence of bacterial adsorption by a single phage is a function of the rates of the above processes along with a combination of the phage adsorption constant and bacterial density (Chapter 15). That is, the lower the bacterial density or the smaller the phage adsorption constant – along with the greater the rate of phage loss from the environment – the less likely that a given phage entering a given environment will succeed in acquiring a bacterium and thereby initiating phage population growth. For phages entering environments containing relatively few bacteria, and in low numbers (founder effect), the likelihood of an initiation of phage population growth thus may be low (Abedon, 1989).

3.4.2 Sustaining phage population growth

Phage population growth occurs so long as bacteria are present above a "critical" density. Conditions must be such that the effective phage burst size is greater than one, with effective burst size defined as the average number of phages released per infected bacterium that – assuming an otherwise unchanging environment – go on to successfully infect new bacteria. Factors that can affect this effective burst size include phage genetics, bacterial genetics, and bacterial physiology. These factors together can impact actual burst sizes, while effective burst sizes can also be affected by likelihood of phage-virion decay, likelihood of phage restriction (or abortive infection) given successful adsorption (Chapter 1), and, of course, bacterial density. Thus, all other factors held constant, then only above a certain bacterial density may successful phage population growth be achieved. At this critical density a steady-state phage population size may be attained, and below this density one would expect declines in phage population growth toward extinction (note that the above statements do not contradict the arguments of Kasman *et al.*, 2002, since those authors did not consider phage decay as a balance against phage replication; see instead Chapters 2 and 15 for consideration of phage decline given insufficient bacterial density combined with bacterium-independent mechanisms of virion loss). This idea of a "critical" bacterial density is equivalent to the concept of a critical mass of radioactive material such as uranium-235, with infection-initiating phage virions equivalent to radioactive-decay-initiating neutrons, phage-susceptible bacteria equivalent to uranium-235, and phage population growth equivalent to a radioactive chain reaction.

3.4.3 Exponential population growth

Given an initiation of phage population growth, and a critical density of bacteria, then phage exponential growth can proceed. We expect maximal population growth from those phages that have maximized their adsorption constant, minimized their eclipse period, maximized their rate of intracellular phage-progeny maturation (Chapter 15), and optimized their latent period (Section 3.2.1.2; Chapter 2). Little is known of how phages may evolve shorter eclipse periods or greater rates of phage-progeny maturation. By contrast, latent period may be readily optimized though phage mutation (Abedon *et al.*, 2003; Wang, 2006; Section 3.2.1.2). However, particularly at higher bacterial densities, and given growth within closed environments, a latent period that is optimized for exponential growth (expediency) may not be simultaneously optimized for overall fecundity (economy), especially as the bacterial resource becomes fully phage-infected (Section 3.2.2; Fig. 3.1, II).

3.4.4 Solving the exponential growth dilemma

The above-noted trade-off between rates of phage population growth and overall population fecundity represents a dilemma which is prisoner's-dilemma-like, though not necessarily technically a prisoner's dilemma (see Fig. 3.2). In short, in a population of longer-latent-period phages (cooperators), it pays to display a shorter (especially more optimized) latent period (that is, this defection imparts a reward equal to T). Overall, the cooperators are better off as a pure culture (payoff equals R) while pure cultures of defectors achieve a worse payoff in terms of final phage titers (payoff equals P) than pure cultures of cooperators. Finally, cooperators can lose big if growing in the presence of defectors (Fig. 3.2; payoff equals S). Thus, at the very least, it pays to defect, whether or not one replicates in the presence of cooperators or other defectors. One approach to solving this dilemma is for cooperators to avoid growing in the presence of defectors, as outlined above (Section 3.3.1.5), via some combination of spatial structure, founder effect, and replicative fidelity. These are essentially the same factors that allow cooperation to occur among the cells within multicellular organisms (Section 3.3.1.4). Alternatively, one can solve this dilemma by changing the rules of the game, as I outline in the following section.

3.4.5 Lysis inhibition as a strategy of dilemma avoidance

Lysis inhibition is a phenotype, peculiar to T-even-like phages, that involves changes in phage latent period in response to environmental stimuli (Doermann, 1948). In particular, lysis inhibition is induced by secondary adsorption of phage-infected bacteria, which is likely only once a majority of bacteria within an environment have become phage-infected. With lysis inhibition as an inducible phenotype, T-even-like phages are thus able to first employ relatively short latent periods when uninfected bacteria are abundant and then switch to longer latent periods when uninfected bacteria are no longer readily available. In other words, there is no (or at least less) trade-off in this system between latent period during exponential growth (expediency) and burst size during the final round of phage infection (economy).

3.4.6 Longer latent periods as defection

During exponential bacterial growth in the presence of relatively high densities of bacteria it is those phages displaying the shorter, optimal latent period (despite the smaller burst size) that maximally take advantage of this type of environment in the short run, and which we would describe as defectors (Section 3.4.4). However, just past exponential bacterial growth, once a majority of bacteria are phage-infected, then we can come to describe those phages displaying longer latent periods as the defectors. The latter is a consequence of the display of superinfection exclusion by these phage-infected bacteria (Section 3.2.3; Chapter 8). That is, superinfection exclusion kills the free phages released by other phage infections, and the longer the latent period of a bacterium displaying superinfection exclusion, the more free phages it can kill. Lysis-inhibited bacteria, as well as phages delaying their lysis via a display of lysogeny, thus act as free-phage-killing defectors. Thus, late in cultures, it is those phages displaying shorter latent periods (kinder because superinfection exclusion is expressed for shorter periods!) that are the cooperators.

3.4.7 Positive-feedback selection and other problems

In the face of phages displaying long latent periods, and thereby extensive killing of free phages, an appropriate response is for an infected bacterium to display yet longer latent periods as either a unilateral defection (payoff $= T$) or as part of a mutual defection (payoff $= P$), rather than being a unilateral cooperator (payoff $= S$). This game is not necessarily a prisoner's dilemma, however, since at this point in phage population growth there is not necessarily a cost associated with defection, either unilaterally or mutually (since uninfected bacteria for phage acquisition are likely not be available anyway). That is, *R* (for mutual cooperation, i.e., mutually shorter latent periods) does not necessarily exceed *P*. Selection for longer latent periods thus would not give rise to a counter-selection for some optimized shorter latent period, but to avoid unilateral cooperation, selection instead could favor ever longer latent periods. This longer-latent-period-mediated selection for ever longer latent periods would represent a kind of evolutionary positive feedback, essentially an intraspecific arms race.

Assuming that phages eventually must lyse the cells they infect, a cost to defection may in the end be manifest. Thus, a lysis-inhibiting phage could become faced with a problem whereby induction of lysis is beneficial, but being an early lyser within a given environment is not. There exist perhaps three plausible solutions to this latent-period face-off. One is to lyse and thereby take one's chances vis-à-vis superinfection exclusion. Two is to avoid lysing while retaining some potential to lyse later (especially given an opportunity for bacterial dissemination to some new environment that does not contain phage-infected bacteria). The third is for all infections to lyse at once.

3.4.8 Synchronized lysis-inhibition collapse

T-even-like phages may very well employ all three of these strategies, though only the latter, called synchronized lysis-inhibition collapse, has been studied in any detail (Abedon, 1992). In this phenomenon, the early lysis of some phage-infected bacteria produces phage virions that are able to adsorb unlysed bacteria, resulting in what may be inferred as some sort of destabilization of these unlysed bacteria and a resulting increased likelihood of bacterial lysis (effectively a lysis from without; Abedon, 1994). With more lysis there are more phages released into the environment, resulting in more secondary adsorption, and thereby greater instability in those infected bacteria that remain. The result is a positive-feedback lysis mechanism that is saltatory. That is, population-wide lysis occurs at a rate that is much greater than can be achieved without secondary adsorption giving rise to additional secondary adsorptions.

3.4.9 Conclusion

For T-even-like phages, secondary adsorption induces lysis inhibition, potentially as a means of avoiding the short versus long latent period dilemma of exponential growth within environments containing limits. However, while at lower phage densities it is phages displaying longer latent periods that are the cooperators, at higher phage densities, and due to the expression of superinfection exclusion, it is those phages displaying shorter latent periods that are the cooperators. The resulting cost of lysing sooner than the rest of the phage population creates a dilemma whereby selection favors an avoidance of lysis by phages infecting within lysis-inhibited phage populations. This second dilemma is avoided, however, through a synchronized population-wide phage-induced lysis, one, like lysis inhibition itself, which is induced as a consequence of phage secondary adsorption and which, in this case, induces a lysis from without.

3.5 APPENDIX: PHAGE MULTIPLICITY OF ADSORPTION

For the sake of avoiding confusion, I will here review concepts of phage multiplicity (see Hyman and Abedon, in press, for additional consideration). A very commonly used definition of phage multiplicity (U) is the ratio of phages to bacteria, $U = P/N$. This definition is problematic, however, because both phage population growth and the phage impact on bacteria are dependent on phage adsorption, the rate of which is dependent on absolute phage densities rather than phage densities relative to bacterial densities (Fig. 12.1, Chapter 12). For example, a phage multiplicity of 10 and a cell density of 10 will yield a phage density of 100, whereas the same multiplicity with a cell density of one million (10^6) will yield a phage density of 10^7 (all in arbitrary volume units). Thus, phage multiplicity, in terms of phage ecology, is preferably defined as synonymous with what may be described as a phage multiplicity of adsorption (Abedon, 1990), multiplicity of attachment (Spouge, 1994), or multiplicity of actual infections ("$\text{MOI}_{\text{actual}}$"; Kasman *et al.*, 2002).

As suggested, in this latter concept of multiplicity there is a key phage-density component plus a time component. That is, given the presence of either more phages or longer adsorption times, then the number of phages adsorbed will increase and therefore so will the phage multiplicity of adsorption, U_a (see Fig. 12.1, Chapter 12, for illustration of this increase). In other words, $U_a \approx P \cdot k \cdot N \cdot t / N$ where k is the phage adsorption constant, which is the per-bacterium rate of phage adsorption; N is bacterial density; P is phage density; and $P \cdot k \cdot N \cdot t$ is the total number of phage that have adsorbed over time, t. The formula is presented as an approximation (i.e., using $\approx$ rather than $=$) because it assumes that free phages adsorb with replacement, that is, that P is a constant. When using this formula, then, U_a is reasonably approximated only if $k \cdot N \cdot t$ is small (Fig. 15.1, Chapter 15).

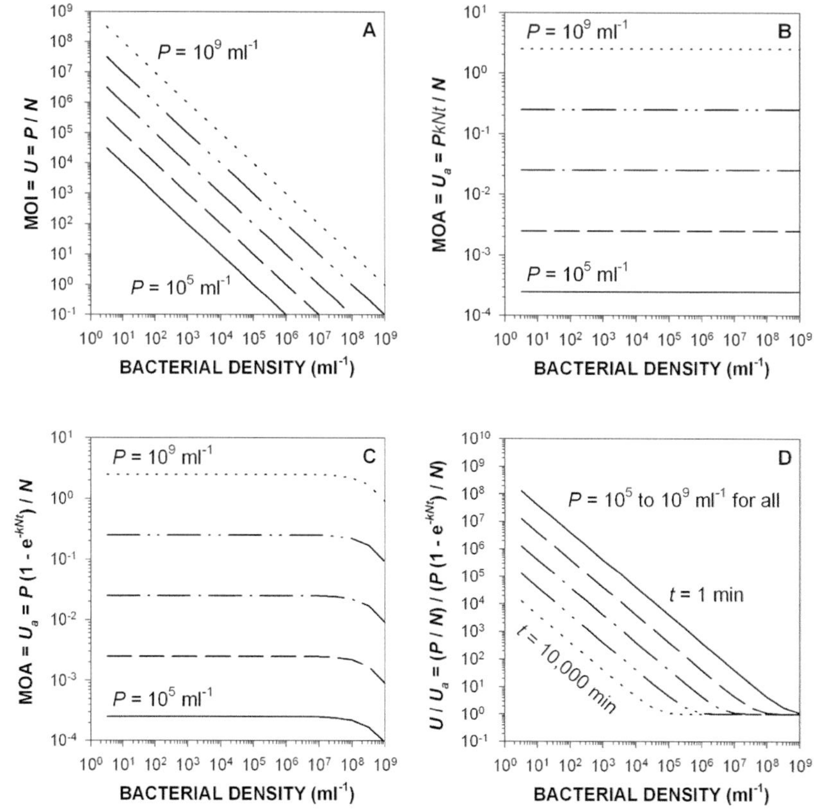

Figure 3.5 Phage multiplicity of infection (MOI, U) versus phage multiplicity of adsorption (MOA, U_a). Multiplicities are calculated as functions of phage density (P, which ranges from $P = 10^5$ to 10^9 by ten-fold increments), bacterial density (N), and adsorption constant (k) $= 2.5 \times 10^{-9}$ mL min^{-1}, after Stent, 1963). Time (t) = 1 min. unless otherwise indicated, and is the interval over which adsorption occurs. (A) Phage multiplicity as typically defined, i.e., MOI $= U = P/N$. (B) A more realistic (but still simplified) definition of the multiplicity of the actual number of phages which have adsorbed bacteria, i.e., MOA $= U_a = P \cdot k \cdot N \cdot t/N$; note the independence of N. (C) A more realistic definition of the actual multiplicity, as based upon assumptions of Poisson distributions of phages adsorbed to bacteria, i.e., MOA $= U_a = P \cdot (1 - e^{-k \cdot N \cdot t})/N$. (D) Ratio of typical to preferred definitions of phage multiplicity: MOI / MOA $= U/U_a = (P/N)/(P \cdot (1 - e^{-k \cdot N \cdot t})/N) = 1 - e^{-k \cdot N \cdot t}$. Note the independence of this ratio of P, the dependence on t (which is set equal to 10, 100, 1000, and 10 000 min. in the various presented curves), and that the ratio of MOI to MOA grows with decreasing bacterial densities, though flattens out given sufficient magnitude of $k \cdot N \cdot t$. The latter occurs because free phage densities decline to zero due to adsorption given sufficient time or bacterial densities (that is, all phages have adsorbed). The latter is seen to increasing extent in panel D with increasing adsorption intervals, t. Note that fractional changes in k would be equivalent in their impact to fractional changes in t.

More correctly, $U_a = P \cdot (1 - e^{-k \cdot N \cdot t})/N < P/N$ where $e^{-k \cdot N \cdot t}$ is the fraction of free phages that remain unadsorbed after an adsorption period of t. The expression reaches a limit as $k \cdot N \cdot t \to \infty$ of $U_a = P/N$. This limit is seen as a flattening of curves in Fig. 3.5D given increasing time of adsorption (t) or increasing bacterial density (N; k is held constant).

Quantitative differences between multiplicity of infection (U) and multiplicity of adsorption (U_a) are illustrated in Fig. 3.5. Figure 3.5A shows the dependence of U on both phage (P) and bacterial (N) densities. Figure 3.5B shows the lack of dependence of U_a on bacterial density for $U_a \approx P \cdot k \cdot N \cdot t/N$. This lack of dependence is corrected at high bacterial densities in Fig. 3.5C because phages there are not assumed to adsorb with replacement (i.e., there $U_a = P \cdot (1 - e^{-k \cdot N \cdot t})/N$). Finally, Fig. 3.5D shows the ratio of U to U_a, indicating that this value can deviate dramatically from 1.0 (i.e., where $U = U_a$) as bacterial densities decline. Bottom line: phage multiplicity of adsorption is dependent on both phage and bacterial densities (see especially Fig. 3.5C), but also on the phage affinity for bacteria (k) as well as the duration of incubation of free phages with bacteria (t). See Chapter 12 (Appendix) for additional discussion of phage multiplicity and adsorption.

ACKNOWLEDGMENTS

Thanks go to Rick Michod, who introduced me to the prisoner's dilemma some 20 years past, and which I first presented as applying to phage T4 lysis inhibition at the Evergreen International Phage Meeting (Olympia, WA) in 1995.

REFERENCES

Abedon, S. T. 1989. Selection for bacteriophage latent period length by bacterial density: a theoretical examination. *Microb. Ecol.* **18**: 79–88.

Abedon, S. T. 1990. Selection for lysis inhibition in bacteriophage. *J. Theor. Biol.* **146**: 501–11.

Abedon, S. T. 1992. Lysis of lysis-inhibited bacteriophage T4-infected cells. *J. Bacteriol.* **174**: 8073–80.

Abedon, S. T. 1994. Lysis and the interaction between free phages and infected cells. In J. D. Karam (ed.), *The Molecular Biology of Bacteriophage T4*. Washington, DC: ASM Press, pp. 397–405.

Abedon, S. T. 2000. The murky origin of Snow White and her T-even dwarfs. *Genetics* **155**: 481–6.

Abedon, S. T. 2006. Phage ecology. In R.L. Calendar and S.T. Abedon (eds.), *The Bacteriophages*, 2nd edn. Oxford: Oxford University Press, pp. 37–46.
Abedon, S. T., and J. T. LeJeune. 2005. Why bacteriophage encode exotoxins and other virulence factors. *Evol. Bioinf. Online* **1**: 97–110.
Abedon, S. T., T. D. Herschler, and D. Stopar. 2001. Bacteriophage latent-period evolution as a response to resource availability. *Appl. Environ. Microbiol.* **67**: 4233–41.
Abedon, S. T., P. Hyman, and C. Thomas. 2003. Experimental examination of bacteriophage latent-period evolution as a response to bacterial availability. *Appl. Environ. Microbiol.* **69**: 7499–506.
Ackermann, M., and L. Chao. 2004. Evolution of cooperation: two for one? *Curr. Biol.* **14**: R73-R74
Axelrod, R. 1984. *The Evolution of Cooperation.* New York, NY: Basic Books.
Bergstrom, C. T., P. McElhany, and L. A. Real. 1999. Transmission bottlenecks as determinants of virulence in rapidly evolving pathogens. *Proc. Natl. Acad. Sci. U.S.A.* **96**: 5095–100.
Bull, J. J. 2006. Optimality models of phage life history and parallels in disease evolution. *J. Theor. Biol.* **241**: 928–38.
Bull, J. J., D. W. Pfening, and I.-W. Wang. 2004. Genetic details, optimization, and phage life histories. *Trends Ecol. Evol.* **19**: 76–82.
Bull, J. J., J. Millstein, J. Orcutt, and H. A. Wichman. 2006. Evolutoinary feedback mediated through population density, illustrated with viruses in chemostats. *Am. Nat.* **167**: E39–51.
Calendar, R. L., and S. T. Abedon (eds.). 2006. *The Bacteriophages*, 2nd edn. Oxford: Oxford University Press.
Chao, L., B. R. Levin, and F. M. Stewart. 1977. A complex community in a simple habitat: an experimental study with bacteria and phage. *Ecology* **58**: 369–78.
Charnov, E. L. 1976. Optimal foraging: the marginal value theorem. *Theor. Pop. Biol.* **9**: 129–36.
Cooper, V. S., M. H. Reiskind, J. A. Miller, *et al.* 2002. Timing of transmission and the evolution of virulence of an insect virus. *Proc. Roy. Soc. Lond. B* **269**: 1161–5.
d'Hérelle, F. 1917. Sur un microbe invisible antagoniste des bacilles dysentériques. *C. R. Acad. Sci. Ser. D* **165**: 373–5. English translation at www.phage.org/bgnws001.htm#submissions.
d'Hérelle, F. 1922. *The Bacteriophage: its Role in Immunity.* Baltimore, MD: Williams and Wilkins/Waverly Press.
Daniels, L. L., and A. C. Wais. 1998. Virulence of phage populations infecting *Halobacterium cutirubrum. FEMS Microbiol. Ecol.* **25**: 129–34.

de Paepe, M., and F. Taddei. 2006. Viruses' life history: towards a mechanistic basis of a trade-off between survival and reproduction among phages. *PLoS Biol.* **4**: e193.

Doermann, A. H. 1948. Lysis and lysis inhibition with *Escherichia coli* bacteriophage. *J. Bacteriol.* **55**: 257–75.

Dugatkin, L. A. 1997. *Cooperation Among Animals: an Evolutionary Perspective.* New York, NY: Oxford University Press.

Dunn, G. B., and D. H. Duckworth. 1977. Inactivation of receptors for bacteriophage T5 during infection of Escherichia coli B. *J. Virol.* **24**: 419–21.

Ebert, D. 1998. Experimental evolution of parasites. *Science* **282**: 1432–5.

Ebert, D., and J. J. Bull. 2003. Challenging the trade-off model for the evolution of virulence: is virulence management feasible? *Trends Microbiol.* **11**: 15–20.

Ewald, P. W. 1994. *Evolution of Infectious Disease.* New York, NY: Oxford University Press.

Frank, S. A. 1996. Models of parasite virulence. *Q. Rev. Biol.* **71**: 37–78.

Hardin, G. 1968. The tragedy of the commons. *Science* **162**: 1243–8.

Hershey, A. D., and W. Dove. 1983. Introduction to Lambda. In R. W. Hendrix, J. W. Roberts, F. W. Stahl, and R. A. Weisberg (eds.), *Lambda II.* Cold Spring Harbor, NY: Cold Spring Harbor Laboratory Press, pp. 3–11.

Herskowitz, I., and F. Banuett. 1984. Interaction of phage, host, and environmental factors in governing the λ lysis–lysogeny decision. In V. L. Chopra, B. C. Joshi, R. P. Sharma, and H.C. Bansal (eds.), *Genetics, New Frontiers: Proceedings of the XV International Congress of Genetics, Vol I.* New Delhi: Oxford and I.B.H., pp. 59–73.

Horn, H. S. 1978. Optimal tactics of reproduction and life-history. In J. R. Krebs and N. B. Davies (eds.), *Behavioural Ecology: an Evolutionary Approach.* Oxford: Oxford University Press, pp. 411–29.

Hyman, P., and S. T. Abedon, in press. Practical methods for determining phage growth parameters. In M. Clokie and A. Kropinski (eds.), *Bacteriophages: Methods and Protocols.* Totowa, NJ: Humana Press.

Kasman, L. M., A. Kasman, C. Westwater, J. Dolan, M. G. Schmidt, and J. S. Norris. 2002. Overcoming the phage replication threshold: a mathematical model with implications for phage therapy. *J. Virol.* **76**: 5557–64.

Kreft, J.-U. 2004. Biofilms promote altruism. *Microbiology* **150**: 2751–60.

Kreft, J.-U. 2005. Conflicts of interest in biofilms. *Biofilms* **1**: 265–76.

Kerr, B., C. Neuhauser, B. J. M. Bohannan, and A. M. Dean. 2006. Local migration promotes competitive restraint in a host–pathogen "tragedy of the commons." *Nature* **442**: 75–8.

Lacroix-Gueu, P., R. Briandet, S. Leveque-Fort, M. N. Bellon-Fontaine, and M. P. Fountaine-Aupart. 2005. In situ measurements of viral particles diffusion inside mucoid biofilms. *C. R. Biol.* **328**: 1065–72.

Levin, B. R., and R. Antia. 2001. Why don't we get sick? The within-host population dynamics of bacterial infections. *Science* **292**: 1112–15.

Levin, B. R., and R. E. Lenski. 1983. Coevolution in bacteria and their viruses and plasmids. In D. J. Futuyama and M. Slatkin (eds.), *Coevolution*. Sunderland, MA: Sinauer Associates, Inc., pp. 99–127.

MacLean, R. C., and I. Gudelj. 2006. Resource competition and social conflict in experimental populations of yeast. *Nature* **441**: 498–501.

Michod, R. E. 1996. Cooperation and conflict in the evolution of individuality. II. Conflict mediation. *Proc. R. Soc. Lond. B Biol. Sci.* **263**: 813–22.

Michod, R. E., A. M. Nedelcu, and D. Roze. 2003. Cooperation and conflict in the evolution of individuality. IV. Conflict mediation and evolvability in *Volvox carteri*. *Bio Systems* **69**: 95–114.

Pfeiffer, T., S. Schuster, and S. Bonhoeffer. 2001. Cooperation and competition in the evolution of ATP-producing pathways. *Science* **292**: 504–7.

Rapoport, A., and A. M. Chammah. 1965. *Prisoner's Dilemma*. Ann Arbor, MI: University of Michigan Press.

Roberts, J., and R. Devoret. 1983. Lysogenic induction. In R. W. Hendrix, J. W. Roberts, F. W. Stahl, and R. A. Weisberg (eds.), *Lambda II*. Cold Spring Harbor, NY: Cold Spring Harbor Laboratory Press, pp. 123–44.

Russel, M., and P. Model. 2006. Filamentous phage. In R. L. Calendar and S. T. Abedon (eds.), *The Bacteriophages*, 2nd edn. Oxford: Oxford University Press, pp. 146–60.

Schrag, S. J., and J. E. Mittler. 1996. Host–parasite coexistence: the role of spatial refuges in stabilizing bacteria–phage interactions. *Am. Nat.* **148**: 348–77.

Smith, H. W., M. B. Huggins, and K. M. Shaw. 1987. Factors influencing the survival and multiplication of bacteriophages in calves and in their environment. *J. Gen. Microbiol.* **133**: 1127–35.

Spouge, J. L. 1994. Viral multiplicity of attachment and its implications for human immunodeficiency virus therapies. *J. Virol.* **68**: 1782–9.

Stent, G. S. 1963. *Molecular Biology of Bacterial Viruses*. San Francisco, CA: W. H. Freeman.

Stephens, D. W., and J. R. Krebs. 1986. *Foraging Theory*. Princeton, NJ: Princeton University Press.

Summers, W. C. 1991. From culture as organisms to organisms as cell: historical origins of bacterial genetics. *J. Hist. Biol.* **24**: 171–90.

Summers, W. C. 2001. Bacteriophage therapy. *Ann. Rev. Microbiol.* **55**: 437–51.

Travisano, M., and G. J. Velicer. 2004. Strategies of microbial cheater control. *Trends Microbiol.* **12**: 72–8.

Trivers, R. 1971. The evolution of reciprocal altruism. *Q. Rev. Biol.* **46**: 35–57.

Turner, P. E., and L. Chao. 1999. Prisoner's dilemma in an RNA virus. *Nature* **398**: 441–3.

Turner, P. E., and L. Chao. 2003. Escape from prisoner's dilemma in RNA phage phi6. *Am. Nat.* **161**: 497–505.

Twort, F. W. 1915. An investigation on the nature of the ultra-microscopic viruses. *Lancet* **ii**: 1241–3.

Wang, I.-N. 2006. Lysis timing and bacteriophage fitness. *Genetics* **172**: 17–26.

Wang, I.-N., D. E. Dykhuizen, and L. B. Slobodkin. 1996. The evolution of phage lysis timing. *Evol. Ecol.* **10**: 545–58.

West, S. A., A. S. Griffin, A. Gardner, and S. P. Diggle. 2006. Social evolution theory for microorganisms. *Nat. Rev. Microbiol.* **4**: 597–607.

CHAPTER 4

Impact of spatial structure on phage population growth

Stephen T. Abedon* and John Yin

4.1 INTRODUCTION

Limitations on bacterial mobility may be described in terms of an environment's spatial structure: the degree to which diffusion, motility, and mixing are hindered or selectively enhanced. A large fraction of bacteria live within environments that possess spatial structure – within biofilms, within soil, or when growing in or on the tissues of plants and animals. This spatial structure also affects phage movement and therefore phage impact on bacterial populations. Gaining an understanding of phage growth within spatially structured environments consequently is pertinent to developing a comprehensive understanding of phage ecology. Here we employ a working assumption that phage population growth in the laboratory in semisolid media (as within phage plaques) represents one approximation of phage population growth within spatially structured environments in the wild. We thus present an introduction to phage plaque formation with the hope that at least some of our discussion may be relevant to future studies of *in situ* phage ecology, such as spatially fine-grained analyses of phage population expansion within biofilms.

See Chapter 16 for methods in modeling phage plaque formation, Abedon and Yin (in press) for a complementary review of phage plaque formation, Chapter 2 for phage growth given spatial structure as introduced in the guise of metapopulations, Chapter 3 for consideration of phage population growth as it occurs within non-spatially structured environments, and Chapter 11 for discussion of the phage ecology of soils.

* Corresponding author

Bacteriophage Ecology: Population Growth, Evolution, and Impact of Bacterial Viruses, ed. Stephen T. Abedon. Published by Cambridge University Press.

4.1.1 Diffusion versus flow

The contrast between plaque growth and phage growth within a well-mixed broth culture provides a useful way of thinking about the conditions that limit phage movement within the laboratory. Plaque growth exists at the limit of slow movement, consisting of local diffusion and progeny dispersal within an agar or gel matrix. Laboratory mixing by stirring or shaking of liquid media (flow), by contrast, constitutes fast movement, and, owing to an absence of spatial impediments or heterogeneity, the affected environment is typically homogeneous. Alternatively, fast movement can be represented by more constrained, well-delimited currents, or may be a consequence of host motility (global or delimited) rather than motion of the medium itself.

In cases where flow is present, it can dominate phage movement. That is, diffusion is comparably negligible because phage progeny from an initial infected cell, in contact with a convective flow, can in principle much more quickly access all susceptible hosts within a single environment (those that are downstream, in contact with the same flow) than they would be able to by diffusion alone. Natural environments typically present complex combinations of two of these cases – slow movement and fast but delimited movement – where phages, depending on either location or temporal considerations, can move by both diffusion and convection, or, instead, by diffusion alone. In this chapter we consider the slow-movement, diffusion-only limit. That is, phage growth within environments lacking in both flow and bacterial motility.

4.2 HOW PLAQUES FORM

Plaques traditionally are defined in terms of their visibility, particularly to the naked eye, which is a function of either phage-induced bacterial lysis or, in the case of chronically infecting phages (Russel and Model, 2006), hindrances to bacteria population growth. Phages may also deplete bacterial densities within more heterogeneous, solid-phase bacterial communities (Section 4.3.2.1.2) such as within biofilms, or over spans too small to be easily visualized (sub-stage 3a of plaque formation, Section 4.2.3.3). Plaques therefore may be more broadly described as localized regions of reduced density of host bacteria rather than in terms of their macroscopic visibility.

Whether or not we can see them, plaques are formed through the net outward diffusion of a population of virions from a central focus, fed by bacterial infection and subsequent phage release. During this movement virions can be lost to bacterial adsorption or otherwise can find themselves temporarily sequestered away from intact bacteria (as toward the center of a plaque). Like phage growth within broth culture (Chapter 3), phage growth within plaques can be considered to occur in a number of stages. We differentiate these stages using terms that may be commonly employed during both broth and plaque growth: (1) the extracellular search, (2) the first adsorption/infection, (3) phage population growth, and (4) termination of phage population growth. To these we also add (5) a post-termination stage, during which additional plaque modification can occur, though without continued phage replication.

4.2.1 Stage 1: the extracellular search

The first stage of phage population growth is a period prior to initial phage attachment to a phage-susceptible bacterium. Though a plaque at this free-phage stage technically has not yet begun to form, nevertheless we will designate this as the first stage of plaque formation, a period of extracellular search by a phage for an uninfected bacterium (Abedon, 1992). Spatial structure will affect this first stage in terms of what is free to move (i.e., phages and bacteria or just phages) and how fast that movement can occur. During plaque growth within a semisolid, agar-based medium, it is likely that bacteria, at a minimum, will display substantially reduced movement relative to within broth.

As with broth growth (Chapter 3), the likelihood of an initial phage adsorption in agar depends on the initial density of bacterial cells. Just as low bacterial densities can greatly delay phage adsorption within broth culture (Abedon, 1989), the combination of low bacterial densities coupled with significant impediments to diffusion can considerably delay the initiation of plaque formation. See Hyman and Abedon (in press) and Chapters 2, 3, 12, and 15 for further discussion of phage dissemination and adsorption to bacteria.

4.2.2 Stage 2: first adsorption/infection

Plaque development proper begins with the "primary adsorption event" (Koch, 1964). The timing of this adsorption depends primarily on initial bacterial densities, rates of phage diffusion, and other aspects of phage adsorption such as likelihood of phage adsorption given encounter with a bacterium (Chapter 15). Alternatively, a plaque may be initiated via bacterium infection that occurs prior to solidification of the soft-agar layer. This infection can occur by chance (i.e., phage and bacterial addition to molten agar initiates the adsorption process, some of which may occur prior to agar pouring), can be forced across the whole population (via a pre-adsorption step), or can occur when sampling a culture that consists of both phages and infected bacteria (e.g., as seen when performing single-step growth experiments; Carlson, 2005; Hyman and Abedon, in press). Together this potential of plaques to be initiated by either a phage particle or by an infected bacterium gives rise to the concept of a plaque-forming unit (PFU), which for the purposes of understanding plaque formation represents a plaque-initiating focus (Stent, 1963).

4.2.3 Stage 3: phage population growth

The next stage of plaque formation consists of phage population growth, which in terms of plaque morphology is a spreading stage during which bacteria are infected and then lysed or otherwise inhibited in terms of their division. Because bacteria are fairly immobilized within a solidified soft agar, while the smaller phages are free to diffuse, presumably the majority of plaque spreading occurs as a consequence of extracellular phage-virion diffusion that is coupled with phage amplification in infected cells. This virion diffusion represents random phage movement, which in the absence of phage reproduction would be unable to sustain a constant rate of spread. Koch (1964) differentiated this stage of plaque development into "the first round or first few rounds of viral multiplication" and "the enlargement phase."

4.2.3.1 Alternating infection and diffusion, plus lawn growth

The spreading (or enlargement) stage of plaque formation consists of expansion of the phage population as mediated by multiple rounds of phage adsorption, infection, and lysis of individual bacteria, plus phage diffusion. Bacterial infection, in other words, feeds plaque formation by increasing phage numbers. It does not, in itself, give rise to virion movement other than over the length of an individual bacterial cell during an individual infection (so long as bacteria are immobile – otherwise phage movement within bacteria could be more extensive than their movement as free phages). Plaque development thus consists of two alternating events, plaque-size increase via phage diffusion and phage-pool increase that is a consequence of phage infection of bacteria. Plaques thus enlarge as a consequence of simultaneous parallel processes of virion diffusion and virion propagation, expanding outward from a central focus (Yin, 1991). This representation of plaque expansion, however, assumes the host cells are homogeneously distributed in space, an assumption that can be relaxed (Chapter 16).

Plaque formation during phage population growth is dynamic for reasons that extend beyond the phage infection–diffusion process. For instance, bacterial lawns typically are initiated using log-phase bacteria, which themselves are free to multiply within the soft-agar overlay. Thus, as with most phage-bacterial systems, bacterial density increases over time during lawn growth and therefore during plaque development. Bacteria also may vary physiologically over time – minimally a bacterial lawn will pass through the various stages of the standard bacterial growth curve, particularly as they pass from log (exponential) phase into stationary phase. In addition, for lawns initiated with bacterial overnights (stationary-phase bacteria), bacteria will pass from stationary phase to lag phase to log phase before returning to stationary phase upon lawn maturation.

4.2.3.2 Constancy in spread rate

Despite these changes in bacterial density and physiology, from observation of previous studies as well as his own experiments, Koch (1964) reviewed evidence that plaque diameter increases linearly as a function of time during this third stage of plaque formation (see also Kaplan *et al.*, 1981; Yin, 1991). Because at least some of this growth occurs prior to the point that a plaque becomes visible as a hole in a mature or maturing bacterial lawn (sub-stage 3a of plaque-size increase; see Section 4.2.3.3) at least some of the growth in plaque diameter must be inferred when employing standard means of plaque visualization. For phage T7, by contrast, the increase in plaque radius occurs at a faster rate while the bacterial lawn is still growing in density, then slows (but does not stop) to an again constant rate once the bacterial lawn enters into stationary phase (Lee and Yin, 1996b).

4.2.3.3 Three sub-stages of plaque-size increase

From these observations we suggest that stage 3 of plaque formation, representing both phage population growth and plaque-size increase, may be differentiated for practical reasons into two or three sub-stages. The first, stage 3a, takes place prior to lawn visibility. This stage has been little studied, both because rates of virion spread or bacterial lysis are not yet measurable during this stage as a function of clearing of a visible bacterial lawn (the standard basis of plaque-size analysis) and because diameters inherently are much smaller during this early stage than they are later in plaque formation. Note that this stage 3a of plaque formation may be equivalent to Koch's (1964) description of a stage 2 of plaque formation consisting of "the first round or *first few rounds* of viral multiplication" (emphasis ours).

The pre-visibility stage (3a) may be followed by a second, much-better-characterized stage (3b) during which plaque growth is visible against an otherwise turbid bacterial lawn. During stage 3b plaques increase in radius at a constant rate as a function of time (Section 4.2.3.2), though not necessarily at the same rate as during stage 3a. For phages such as T7 (Yin, 1991) and others (Robb and Hill, 2000) whose plaques continue to grow, even after the bacterial lawn has moved into stationary phase, stage 3b of plaque-size increase may be followed by a post-lawn-maturation stage (3c) of plaque growth. Stage 3c, if present, also can occur at a constant rate, though again that rate may be different from that observed during the previous stages of plaque growth (Lee and Yin, 1996b). Owing to an inability to replicate on stationary-phase bacteria, for most phages stage 3c does not occur.

4.2.3.4 Factors affecting rate of spread

In Chapter 16 (this volume), in Abedon and Yin (in press), and in Abedon and Culler (2007) a number of models of plaque formation are reviewed. What these models all have in common is their emphasis on stage 3 of plaque formation, particularly sub-stages 3b or 3c. More precisely, they are models that predict rates of plaque diameter increase, which we expect to occur at a more or less constant rate (Section 4.2.3.2).

4.2.3.4.1 Predicting plaque front velocities

Based on published parameters for phage T7 plaques expanding on lawns of *Escherichia coli* (Yin and McCaskill, 1992), one can estimate rates of plaque expansion using Koch's heuristic model (Koch, 1964; Chapter 16), or using a mechanistic model that assumes adsorption of the phages is a fast equilibrated process (Yin and McCaskill, 1992; Chapter 16). Alternatively, estimations may be made via numerical solution of the mechanistic model that is not restricted to a limiting case (You and Yin, 1999). These models yield similar magnitudes of their velocities of radial expansion of 4 mm h^{-1}, 3 mm h^{-1}, and 1 mm h^{-1}, respectively. Experimentally observed velocities are significantly lower, at 0.2 mm h^{-1}.

To address the mismatch, Fort and colleagues incorporated time delays into their model for growing plaques (Fort and Méndez, 2002). They reasoned that the production of phages in each infected cell occurs only after a delay, the period of time between the initial adsorption of the phage to the host and the initial production of phage progeny. By explicitly accounting for this delay, they developed a time-delay model that yielded velocities of 0.2 mm h^{-1}, consistent with experimental observations (see Weld *et al.*,

2004, for consideration of similar issues during phage broth growth). This work has been extended to provide approximate analytical solutions for the spread velocity in terms of biophysical parameters (Ortega-Cejas *et al.*, 2004). Although the models of Fort and coworkers need as input an additional parameter, the time-delay, this parameter can be readily estimated from standard one-step growth experiments (Carlson, 2005; Hyman and Abedon, in press). On the other hand, so far this model has been tested against only a single experimental determination of plaque formation, that of Yin and McCaskill (1992), so its robustness is not known. See Section 4.5 (Appendix) for discussion of the means of incorporation of this time dependence.

4.2.3.4.2 Sensitivity analysis

To better understand mechanistically how phage infections spread, model building can help (Chapter 16 and Section 4.2.3.4.1), but ultimately models rely on parameters. Because phage characteristics in liquid media are more easily quantified, most parameters for phage growth in models of plaque formation are obtained from liquid-culture experiments. This is unfortunate because culture conditions in broth, including especially bacterial physiology, are not necessarily equivalent to those in solid media. Since the bacteria physiological state can impact phage growth parameters (e.g., Chapter 5 and You *et al.*, 2002), a reasonable hypothesis is that the phage growth parameters used in models of plaque formation are incorrect. For example, burst sizes during plaque growth may be smaller than observed under ideal conditions in broth (Mayr-Harting, 1958; Koch, 1964). Deviations from broth determinations may be especially large in the Yin and McCaskill (1992) experiments since their bacterial lawn was in stationary phase, not the log phase at which their phage growth parameters are typically determined (e.g., Chapter 5). If bacterial physiologies are less conducive to phage growth within lawns than within well-mixed (and aerated) liquid media, or when bacteria are in stationary phase, then it is likely that employing broth-derived growth parameters would lead to an overestimation of plaque growth rates (front velocity).

4.2.4 Stage 4: termination of phage population growth

For plaque growth Koch (1964) described a "final phase, in which viral multiplication ceases." This end point of plaque growth can also differ from that within broth. For those phages that cannot productively infect stationary-phase bacteria – which likely is the majority of phage types – phage population growth can continue only so long as sufficient quantities of non-stationary-phase, phage-susceptible bacteria are present. For a culture of a given size and containing a given number of bacteria, it is more likely, if spatial structure is present, that the stationary phase will be reached while a substantial portion of the phage-susceptible bacterial population remains uninfected. In other words, starting with otherwise similar numbers of bacteria, phages, and culture volumes, broth-culture clearing is more likely than confluent lysis of bacterial lawns. Whether or not the changes in bacterial physiology that we readily observe in the laboratory are relevant to spatially structured phage growth in the wild, however, is a matter for future research.

4.2.5 Stage 5: post-termination plaque modification

Even after phage replication ceases, a plaque can continue to change, in terms of both phage distribution and the appearance (or properties) of the bacterial lawn in the immediate vicinity of the plaque.

4.2.5.1 Continued virion spreading

Within natural environments, we should expect at least some phage diffusion out of "plaques." In terms of biofilms, this movement either is out of the bacteria-containing matrix, is within the bacteria-containing matrix but into regions containing bacteria that are not phage-adsorption permissive, or is also within the matrix but into regions containing adsorption-permissive but infection-non-permissive host bacteria. In other words, whether in biofilm or some other spatially structured bacteria-containing environment, those regions containing actively metabolizing, phage-permissive bacteria should serve as phage-virion sources while all other regions should serve instead as phage-virion sinks.

Within the laboratory, in phage plaques, similar forces should be at work. Thus, phage virions are expected to diffuse into regions lacking in permissive bacteria, the most obvious of such regions being within a plaque's clearing where phage-permissive bacteria are no longer present. Such diffusion implies, even following lawn maturation, that plaques may be "frozen in time" only with regard to phage multiplication (stage 3 as described above) but not with regard to within-plaque virion mixing. It is also possible that, after lawn maturation and thereby past the time where lawn bacteria are capable of supporting productive phage infections, virions may be able to continue to spread beyond a plaque's visible boundary.

Along with this continued diffusion, for phages that are capable of adsorbing to stationary-phase bacteria (which are an example of the adsorption-permissive but infection-non-permissive host bacteria suggested above), free phages may continue to be lost to stationary-phase bacteria at a plaque's zone of infection, the region near a plaque's periphery where the majority of bacterial adsorption takes place (Chapter 16). Depending on the specifics of the phage–bacterium–medium combination being observed, these stationary-phase adsorptions may or may not be able to initiate infections given subsequent dilution of plaques, and their zone of infection, into fresh media. Thus, after lawn maturation, phage virions may continue to diffuse and therefore mix within plaques, they may be lost to adsorption to stationary-phase bacteria following entry into a plaque's zone of infection, and, to the degree that bacteria-free space is available within the mature lawn, phages may even diffuse into the plaque periphery, which we define here as phage-containing but otherwise uninfected bacterial lawn (Chapter 16).

4.2.5.2 Continued enzyme diffusion

Phages can release relatively low-molecular-weight factors, such as unincorporated capsomere, that can enzymatically affect the morphology of bacterial lawns. As smaller-than-phage entities, the potential of these factors for diffusion is greater than that of phages, and thereby these factors can extend the reach of a plaque beyond the clearing resulting from phage-induced bacterial lysis. As a consequence, plaque morphology can change even after lawn maturation, with the lawn modified well beyond the reach of phage virions (Adams, 1959; Stent, 1963; Stirm, 1994; see also references cited in Sutherland *et al.*, 2004). Thus, the fourth stage of plaque formation is defined by a termination in phage production rather than by a termination in diffusion-mediated plaque dynamics or even a halting of a plaque's morphological development.

4.3 PLAQUE SIZE AND MORPHOLOGY

Plaque size and morphology depend on the phage strain, the bacterial host, and plating conditions. In the resulting variance in plaque appearance we can distinguish between quantitative differences (particularly in terms of plaque size, i.e., diameter) and qualitative differences (such as differences in plaque turbidity). In this section we begin with the former but emphasize the latter. We discuss these variations in plaque formation as a primer on future interpretation of macroscopic differences among phages replicating within naturally occurring spatially structured environments.

4.3.1 Quantitative differences between plaques

Plaque formation is a dynamic process. Not surprisingly, a number of factors can affect plaque development and, ultimately, plaque size. Key to understanding plaque size is that plaque formation basically represents a race between maturation of the bacterial lawn and spreading of the developing plaque. If the spreading wave of bacterial lysis is fast relative to the maturation of the bacterial lawn (e.g., McConnell and Wright, 1975) or if spreading continues past that maturation (Yin, 1991), then a plaque will be larger. However, if spreading is slow relative to lawn maturation, with productive phage infections ceasing as bacteria enter stationary phase, then the plaque will be smaller. Specific factors that can affect plaque size in the laboratory are discussed by Abedon and Yin (2007). Chapter 16 considers the theory of the impact especially of phage growth parameters on plaque size. See also Section 4.2.3.4 of this chapter.

Plaque size does not necessarily correlate with per-infection productivity (see Abedon and Yin, in press, for a more detailed discussion of this claim). Phages with smaller burst sizes, for instance, could display larger plaques if those smaller burst sizes are a consequence of shorter latent periods (Koch, 1964; Yin and McCaskill, 1992). Consistently, phage RB69 mutants have been isolated which simultaneously display larger plaques, shorter latent periods, and smaller burst sizes (Abedon *et al.*, 2003).

4.3.2 Qualitative differences within and among plaques

Plaques can differ not only in size but also in their general appearance. Here we consider factors that can affect plaque qualitative appearance, focusing on the effects of mutations in phage functions.

4.3.2.1 Plaque-morphology mutations

Phage plaque-morphology mutants played important roles in the development of the field of molecular genetics, particularly for early recombination experiments including Benzer's genetic fine-structure studies (1955; 1961). It is telling, in fact, that in Stent's 1963 monograph the various plaque-morphology mutants monopolize the entire first page of his table (8.1) summarizing the then-known plaque and non-plaque phage mutant types. In this section we explore the phenotypic underpinnings of various plaque-morphology mutations, with the caveat that we have made no attempt to exhaustively review known variation on plaque morphologies.

4.3.2.1.1 Temperate phage clear mutants

It was well before the application of phages to problems of molecular genetics that the first phage plaque-morphology mutant was described. This mutant is of particular historical importance because it apparently represents the first description of a phage mutant phenotype:

> the first clear-cut demonstration of the occurrence of phage mutants was provided only in 1936. In that year Burnet and Lush [118 as cited by Stent] reported that the staphylococcal phage C, which produces plaques containing a dense central growth of phage resistant bacteria . . . sports a stable hereditary mutant C', which produces clear plaques without any central growth. Burnet and Lush showed that this difference in plaque morphology reflected the loss by the C' mutant phage of the capacity possessed by the normal C phage to induce phage resistance in the phage-sensitive staphylococci. (Stent, 1963, p. 176)

Clear- or "*c*"-type phage mutants often are a consequence of a temperate phage losing its ability to effect lysogeny (see Chapters 1 and 5 for discussion of these latter topics). Phages that normally produce plaques that are turbid, containing microcolonies of lysogenic bacteria, thus come to produce clear plaques.

Note though that Stent (1963) confusingly also employs the symbol "*c*" to describe the morphologies of plaques formed by cofactor-requiring phages.

4.3.2.1.2 Host-range mutants visualized using mixed indicator

A second example of variation in plaque morphology is achieved via a "technique of mixed indicators" (Stent, 1963) whereby a phage host-range mutant is able to lyse a mixture of bacterial indicators while its wild-type parental phage can lyse only one of two indicators. That is, a bacterium that displays resistance to the original, wild-type phage may be selected by mixing bacteria with a large excess of phages. The selecting phages are unable to form plaques on these bacterial mutants. By plating an excess of phages using the above-selected resistant bacteria as indicator, some phage host-range mutants may be selected that are able to form plaques on the otherwise resistant bacterium. Certain host-range (*h*-type) mutants, ones capable of infecting both host types (bacteria that are resistant to the parental phage type as well as those that are permissive), will display relatively clear plaques when plated using the mixed indicators whereas the wild-type parental phages will display easily distinguished plaques that are turbid owing to the presence of resistant bacteria throughout the plaque (Stent, 1963).

Note though that Baylor *et al.* (1957) suggested the possibility of phage evolution during mixed-indicator growth. They therefore warned against using mixed-indicator plaques to initiate phage stocks, though presumably the posited dual-host selection could be of interest in terms of phage experimental evolution (Chapters 6 and 9). Of related interest, especially to concerns of phage ecology, is the supposition that phage growth within naturally spatially structured environments occurs within heterogeneous bacterial communities in which a fraction – or perhaps the majority – of bacteria are not susceptible to a specific phage. See Chapter 10 for additional consideration of the impact of phage growth within mixed-phage-susceptibility bacterial communities.

4.3.2.1.3 Minute plaques

Another early example of phage plaque-morphology mutants were the so-called minute (*m*-type) plaques (Stent, 1963). A number of factors could give rise to these smaller-than-wild-type plaques, though presumably the most likely are latent-period extensions, even if only moderate extensions, or dramatic reductions in phage burst size (e.g., to 10; Stent, 1963; Carlson and Miller, 1994). One can imagine that a number of metabolism-disrupting phage mutations could result in *both* extended latent periods and reduced burst sizes, thereby giving rise to minute plaques. Amla (1981), for example, reports on a minute mutant of cyanophage AS-1 which displays a reduced burst size, lengthened eclipse period, and lengthened latent period, plus a reduced rate of adsorption.

4.3.2.1.4 Rapid-lysis mutants

Quite famous, and centrally important to the development of molecular genetics, were the T-even-phage rapid-lysis (*r*-type) mutants (Stent, 1963; Abedon, 1990, 1994). These phages fail to display an inducible lysis inhibition, which is a latent-period extension that is coupled in wild-type T-even phages, such as phage T4 (Abedon, 2000), to dramatic burst-size increases. Both the latent-period extension and burst-size increase are induced when the adsorbing phage multiplicity to a given bacterium exceeds one (i.e., when adsorption by one phage is followed, after a sufficient period, by adsorption by a second or additional phage; Chapter 3). Lysis inhibition, by increasing average phage latent periods during plaque development, results in production of smaller plaques than those produced by rapid-lysis mutants.

4.3.2.1.5 Plaques with greater turbidity

Phage mutants that display plaques that are more turbid than wild type have also been described, and have even been mapped to a number of distinct loci (Adams, 1959). With so-called "*ht*"-type plaques the excessive turbidity is speculated to result from slow phage adsorption (Stent, 1963). For more on plaque turbidity, see Section 4.3.2.2.

4.3.2.1.6 Additional variation in plaque morphology

There exist yet additional plaque-morphology mutants. Phages can vary in the amount of soluble "lysins" they produce (e.g., as against glycocalyx) and therefore in the size of halos that can form around plaques (Section 4.2.5.2). There also exist phage mutants that display sectored plaques (reviewed in Abedon, 1994), also known as star or "*s*"-type plaques (Stent, 1963), with sectors representing faster-spreading mutants of what otherwise presumably are more slowly growing or spreading phages (for more on such overgrowth, see Yin, 1993; Lee and Yin, 1996a; Wei and Krone, 2005). A related plaque morphology variant is the so-called mottled plaque (Adams, 1959; Stent, 1963), which is a plaque initiated by two or more phages that together display more than one plaque-morphology phenotype, such as rapid lysis versus lysis inhibition (Section 4.3.2.1.4). Mottled plaques essentially are mosaics where some regions are dominated by one parental phage (and therefore one phenotype) while other regions are dominated by the second parental phage (and therefore by the second phenotype). Mottled plaques can also be formed by individual phages, though this occurs owing to a heterozygosity in some phages that results from genomic terminal redundancy (Stent, 1963).

4.3.2.2 Within-plaque inhomogeneity

Plaque formation also results in inhomogeneities across a plaque's breadth. Plaques, for example, differ in terms of their turbidity, ranging from quite clear with sharp edges to quite turbid with poorly defined edges; the typical plaque, as produced by a non-temperate phage, is a hybrid of these two extremes, displaying a more or less clear center with increasing turbidity toward the plaque periphery. This can be seen in Fig. 16.1, where a turbid area – "II" in that figure – surrounds a central "zone of clearing," indicated with "I."

Turbidity presumably occurs as a consequence of the retention of intact bacteria within the plaque clearing. Very little effort, however, has been made towards the study of inhomogeneities within plaques. Our relative ignorance is unfortunate given that plaques are the best model we have for phage growth within spatially structured environments and, as noted in the introduction, a great deal of bacterial ecology, including phage–bacterial interaction, likely occurs within spatially structured environments (e.g., Chapter 11).

Here we consider three possible explanations for why plaques often display a "bull's eye" morphology – that is, greater turbidity toward plaque peripheries.

4.3.2.2.1 Mutation to phage resistance

Because the density of lawn bacteria increases over the course of plaque formation, the density of bacteria present during the earlier stages of plaque formation can be substantially lower than the density found, at a plaque's periphery, toward the end of plaque formation. In addition, bacterial physiology likely is changing over time. Any process causing intact bacteria to be retained within a plaque that can be affected by bacterial density or bacterial physiology should therefore vary in its impact going from a plaque's center, which forms earlier during plaque formation, toward its periphery, which forms later. An example is bacterial mutation to phage resistance. If mutation rates, on a per-bacterium basis, remain constant over the course of plaque development, then greater numbers of resistant bacterial clones should arise wherever higher densities of bacteria are present at the point of phage exposure.

Since the density of bacterial lawns increases as lawns age, and therefore as plaques spread outward, presumably more resistant bacteria are retained toward the periphery of plaques relative to their centers. These resistant bacteria can replicate, forming relatively large microcolonies given a relative lack of bacterial competition (Kaplan *et al.*, 1981), as may be experienced within the otherwise bacteria-free environment formed in the interior of a plaque. The result should be greater turbidity away from a plaque's center, consisting of microcolonies of resistant bacteria surrounding a bull's eye of less turbidity toward the plaque's center.

4.3.2.2.2 Microcolony resistance to phage penetration

Lawns of phage-permissive bacteria consist of bacterial microcolonies initiated by individual bacteria, most of which are founded by bacteria that are sensitive, at least in log phase, to phage infection and lysis. Should those microcolonies become physically more resistant to phage penetration over time, then this could give rise to an increased turbidity toward a plaque's periphery. That is, when first reached by phages, microcolonies at a plaque's periphery will have been growing for longer than microcolonies that are first reached earlier in lawn development. See Yin (1991) for photographs of plaque development that seem to suggest the occurrence of delays in microcolony elimination.

We envisage three mechanisms that could contribute to greater retention of microcolonies toward a plaque's periphery: First, the delay between phage contact and phage-induced complete lysis of a given microcolony could be longer for larger microcolonies, where larger microcolonies occur later during a lawn's development. Second, microcolony centers may enter stationary phase, and thereby become resistant to phage infection and/or phage-induced lysis (e.g., pseudolysogeny; Chapter 5). The existence of bacteria that are at least temporarily refractory to phage-induced lysis could result in their retention, at least temporarily, well within a plaque's periphery. Third, it is conceivable that the physiology of microcolonies found nearer to a plaque's periphery is under greater influence from the intact bacterial lawn found outside these plaques than is the case for microcolonies found closer to a plaque's interior. The more peripheral microcolonies consequently may be more likely to enter stationary phase earlier, either wholly or in part, relative to microcolonies found further toward a plaque's center.

4.3.2.2.3 Infection period extension

An additional reason for bull's-eye formation is a special case, though we mention it because of its long and important history in the study of plaque morphology: the impact of T-even phage lysis inhibition. In lysis inhibition, phage multiplicities that are greater than one give rise to phage infections with greatly extended latent periods. Plaques produced by phages that can display lysis inhibition show a significant turbidity toward their periphery, and this turbidity is thought to consist of lysis-inhibited bacteria. It is conceivable that for many of these bacteria lysis never occurs since changes in lawn physiology could affect infection physiology such that lysis is delayed indefinitely. Alternatively, it may be that phages that display lysis inhibition are especially ill-equipped to penetrate large bacterial microcolonies. Long infections at the edges of microcolonies may not lyse fast enough to allow phage penetration to the interior of microcolonies before stationary phase there, or within the lawn as a whole, is reached. See also Chapter 5 for consideration of how pseudolysogeny might similarly contribute to increased plaque turbidity.

4.3.2.2.4 Spatial distributions of phages in plaques

We have reason to expect that all but the tiniest plaques will be inhomogeneous with respect to phage density. In particular, phage density at the center of plaques does not necessarily equal that toward the periphery, where either higher bacterial densities later during plaque formation (Kaplan *et al.*,

1981) or longer periods of virion inactivation toward the center (Yin, 1991) give rise to higher phage numbers toward plaque peripheries. Alternatively, given non-productive phage adsorption to stationary-phase bacteria found at the periphery of plaques (Section 4.2.5.1), we might expect a depletion of free phages (reduced phage density) at a plaque's edges.

4.3.2.2.5 Selective inhomogeneity

Selection acting on phages likely also varies across a plaque's diameter (Abedon, 2006). For instance, toward a plaque's periphery there should be selection for more rapid exponential growth, e.g., for shorter phage latent periods when host densities are high (Section 4.3.1) because this could allow so-located phages more rapid access to uninfected bacteria. For similar reasons, during plaque enlargement there should be selection for more rapid virion diffusion. Alternatively, towards a plaque's center, once densities of uninfected bacteria have been depleted, there should exist selection for greater burst sizes even at the expense of longer latent periods, such as selection for the T-even phage lysis-inhibition phenotype (Section 4.3.2.2.3). Finally, we can envisage a selection for a continuation of phage growth even as the bacterial indicator enters into stationary phase (such as is seen with phage T7; Yin, 1991).

4.4 CONCLUSION

We end this chapter with a reiteration that phage population growth within a soft-agar overlay, as a plaque, may represent a plausible model for phage population growth within more natural spatially structured environments, such as biofilms. Determining the actual correspondence between within-plaque and within-biofilm dynamics of phage growth, however, will require detailed experimental analysis of both processes, including of the impact of both bacterial heterogeneity and convection in the latter. Furthermore, as biofilms are implicated in infectious disease (Lacroix-Gueu *et al.*, 2005), and further can serve as a means of effecting phage resistance to bacteriophages (Lacqua *et al.*, 2006), a better understanding of phage infection and population dynamics within spatially structured environments should allow for greater refinement of phage-based bacterial eradication strategies (Chapter 17).

4.5 APPENDIX: ROLE OF TIME DEPENDENCE

The incorporation of a latent-period delay into models of plaque formation (Section 4.2.3.4.1) is one means of dealing with a common difficulty in modeling rates of plaque-size increase (also known as spread or front velocity). The difficulty, so far as model derivation is concerned, is a requirement that velocity measures be presented in units of distance over time. We describe methods for achieving these units as follows.

All mechanistic models for the rate of plaque expansion share a dependence of spread velocity on the square root of a diffusion coefficient (D) divided by time (t): $\sqrt{D/t}$. This dependence is what one would expect from dimensional arguments. Briefly, since the observable spread velocity (c) has dimensions of length (d for *d*istance) divided by time, i.e., $c = d/t$, the rates of the microscopic processes that influence c must collectively combine to produce a quantity that matches the d/t dimensions of that velocity. In all plaque-expansion models to date, from Koch through Fort, units or dimensions of length are introduced through the diffusion coefficient or diffusivity of the phage, which has units of length squared as divided by time, $D = d^2/t$, typically of magnitude 10^{-7} cm^2 s^{-1} (or $\sim 10^{-5}$ cm^2 min^{-1}) for a phage-sized particle in water.

In order to recover the unit of length (rather than d^2), we take the square root of the phage diffusivity: $\sqrt{D}$. Of course, taking the square root of the diffusivity yields dimensions of $\sqrt{d^2/t} = d/\sqrt{t}$, which requires further contributions from time-scale (or reaction rate) constants to attain the final dimensions of velocity. These time constants ultimately contribute units of $1/\sqrt{t}$ to the velocity since $\left(1/\sqrt{t}\right) \cdot \left(d/\sqrt{t}\right) = d/t = c$. Typically $1/\sqrt{t}$ is found in contributions from the time scale (or 1/rate) for adsorption of the phage to its host cell, the delay or latent time required for the infected host to produce initial phage progeny (Chapters 2 and 15), and the time scale between the initial release of phages to the time when the total yield of phages has occurred. The recent models by Fort and coworkers (Fort and Méndez, 2002; Ortega-Cejas *et al.*, 2004) have highlighted a potential advantage from explicitly accounting for the contribution from the delay or latent time in models of plaque expansion. In the future, as we expand our understanding of how host cells and microcolonies distribute spatially in both laboratory and natural systems, we anticipate that length scales associated with microcolony sizes or between-colony distances may well also influence the macroscopic rates of infection spread (see Chapter 16 for discussion of one means by which incorporation of microcolonies into plaque models may be achieved).

ACKNOWLEDGMENTS

Thank you to Stephen Krone, who read and commented on this chapter. During the writing of this chapter, J.Y. has been supported by the National Science Foundation (USA) and the National Institutes of Health (USA). S.T.A.

would like to thank Donivan Sphar, who, working in his laboratory as an undergraduate for six weeks in the summer of 2004, inspired much thinking on the biology of phage plaques.

REFERENCES

Abedon, S. T. 1989. Selection for bacteriophage latent period length by bacterial density: a theoretical examination. *Microb. Ecol.* **18**: 79–88.

Abedon, S. T. 1990. Selection for lysis inhibition in bacteriophage. *J. Theor. Biol.* **146**: 501–11.

Abedon, S. T. 1992. Lysis of lysis-inhibited bacteriophage T4-infected cells. *J. Bacteriol.* **174**: 8073–80.

Abedon, S. T. 1994. Lysis and the interaction between free phages and infected cells. In J. D. Karam (ed.), *The Molecular Biology of Bacteriophage T4.* Washington, DC: ASM Press, pp. 397–405.

Abedon, S. T. 2000. The murky origin of Snow White and her T-even dwarfs. *Genetics* **155**: 481–486.

Abedon, S. T. 2006. Phage ecology. In R. Calendar and S.T. Abedon (eds.), *The Bacteriophages.* Oxford: Oxford University Press, pp. 37–46.

Abedon, S. T., and R. R. Culler. 2007. Bacteriophage evolution given spatial constraint. *J. Theor. Biol.* **248**: 111–19.

Abedon, S. T., and J. Yin, in press. Bacteriophage plaques: theory and analysis. In M. Clokie and A. Kropinski (eds.), *Bacteriophages: Methods and Protocols.* Totowa, NJ: Humana Press.

Abedon, S. T., P. Hyman, and C. Thomas. 2003. Experimental examination of bacteriophage latent-period evolution as a response to bacterial availability. *Appl. Environ. Microbiol.* **69**: 7499–506.

Adams, M. H. 1959. *Bacteriophages.* New York, NY: Interscience.

Amla, D. V. 1981. Isolation of characteristics of minute plaque forming mutant of cyanophage AS-1. *Biochem. Physiol. Pflanz.* **176**: 83–9.

Baylor, M. B., D. D. Hurst, S. L. Allen, and E. T. Bertani. 1957. The frequency and distribution of loci affecting host-range in the coliphage T2H. *Genetics* **42**: 104–20.

Benzer, S. 1955. Fine structure of a genetic region in bacteriophage. *Proc. Natl. Acad. Sci. U.S.A.* **41**: 344–54.

Benzer, S. 1961. On the topography of the genetic fine structure. *Proc. Natl. Acad. Sci. U.S.A.* **47**: 403–15.

Carlson, K. 2005. Working with bacteriophages: common techniques and methodological approaches. In E. Kutter and A. Sulakvelidze (eds.), *Bacteriophages: Biology and Application.* Boca Raton, FL: CRC Press, pp. 437–94.

Carlson, K., and E. S. Miller. 1994. Enumerating phage. In J. D. Karam (ed.), *Molecular Biology of Bacteriophage T4*. Washington, DC: ASM Press, pp. 427–9.

Fort, J., and V. Méndez. 2002. Time-delayed spread of viruses in growing plaques. *Phys. Rev. Lett.* **89**: 178101

Hyman, P., and S. T. Abedon, in press. Practical methods for determining phage growth parameters. In M. Clokie and A. Kropinski (eds.), *Bacteriophages: Methods and Protocols*. Totowa, NJ: Humana Press.

Kaplan, D. A., L. Naumovski, B. Rothschild, and R. J. Collier. 1981. Appendix: a model of plaque formation. *Gene* **13**: 221–5.

Koch, A. L. 1964. The growth of viral plaques during the enlargement phase. *J. Theor. Biol.* **6**: 413–31.

Lacqua, A., O. Wanner, T. Colangelo, M. G. Martinotti, and P. Landini. 2006. Emergence of biofilm-forming subpopulations upon exposure of *Escherichia coli* to environmental bacteriophages. *Appl. Environ. Microbiol.* **72**: 956–59.

Lacroix-Gueu, P., R. Briandet, S. Leveque-Fort, M. N. Bellon-Fontaine, and M. P. Fountaine-Aupart. 2005. In situ measurements of viral particles diffusion inside mucoid biofilms. *C. R. Biol.* **328**: 1065–72.

Lee, Y., and J. Yin. 1996a. Detection of evolving viruses. *Nat. Biotech.* **14**: 491–3.

Lee, Y., and J. Yin. 1996b. Imaging the propagation of viruses. *Biotech. Bioeng.* **52**: 438–42.

Mayr-Harting, A. 1958. Die Entwicklung von Phagenloechern und der mechanismus der Phagenwirkung in festen Naehrboeden. *Zbl. f. Bakt. Paras. Infek. u. Hyg.* **171**: 380–92.

McConnell, M., and A. Wright. 1975. An anaerobic technique for increasing bacteriophage plaque size. *Virology* **65**: 588–90.

Ortega-Cejas, V., J. Fort, V. Méndez, and D. Campos. 2004. Approximate solution to the speed of spreading viruses. *Phys. Rev. E* **69**: 031909–1–4.

Robb, F. T., and R. T. Hill. 2000. Bacterial viruses and hosts: influence of culturable state. In R. R. Colwell and D. J. Grimes (eds.), *Nonculturable Microorganisms in the Environment*. Washington, DC: ASM Press, pp. 199–208.

Russel, M., and P. Model. 2006. Filamentous bacteriophages. In R. Calendar and S. T. Abedon (eds.), *The Bacteriophages*. Oxford: Oxford University Press, pp. 145–60.

Stent, G. S. 1963. *Molecular Biology of Bacterial Viruses*. San Francisco, CA: W. H. Freeman.

Stirm, S. 1994. Examination of repeating units of bacterial exopolysaccharides. In J. N. BeMiller, D. J. Manners, and R. J. Sturgeon (eds.), *Methods in Carbohydrate Chemistry*. New York, NY: Wiley, pp. 143–54.

Sutherland, I. W., K. A. Hughes, L. C. Skillman, and K. Tait. 2004. The interaction of phage and biofilms. *FEMS Microbiol. Lett.* **232**: 1–6.

Wei, W., and S. M. Krone. 2005. Spatial invasion by a mutant pathogen. *J. Theor. Biol.* **236**: 335–48.

Weld, R. J., C. Butts, and J. A. Heinemann. 2004. Models of phage growth and their applicability to phage therapy. *J. Theor. Biol.* **227**: 1–11.

Yin, J. 1991. A quantifiable phenotype of viral propagation. *Biochem. Biophys. Res. Commun.* **174**: 1009–14.

Yin, J. 1993. Evolution of bacteriophage T7 in a growing plaque. *J. Bacteriol.* **175**: 1272–7.

Yin, J., and J. S. McCaskill. 1992. Replication of viruses in a growing plaque: A reaction-diffusion model. *Biophys. J.* **61**: 1540–9.

You, L., and J. Yin. 1999. Amplification and spread of viruses in a growing plaque. *J. Theor. Biol.* **200**: 365–73.

You, L., P. F. Suthers, and J. Yin. 2002. Effects of *Escherichia coli* physiology on growth of phage T7 in vivo and in silico. *J. Bacteriol.* **184**: 1888–94.

CHAPTER 5

Contribution of lysogeny, pseudolysogeny, and starvation to phage ecology

Robert V. Miller* and Martin J. Day

5.1 INTRODUCTION

How do bacteriophages exist in the hostile environments that their bacterial hosts inhabit? In most environments, from the desert to the mammalian gut, bacteria live for most of their existence in a starved state (Koch, 1971; Morita, 1997) where energy, carbon, and other resources are in scarce supply. Under such conditions we know that the latency period for phage infection lengthens, that the burst size is greatly reduced (Kokjohn *et al.*, 1991), and that the half-life of virion infectivity (rate of decay) is short (Miller, 2006); yet total counts of virus-like particles present in environmental samples are high. Clearly bacteriophages have evolved strategies for surviving under these unfavorable conditions. As survival-enhancement strategies, many biological entities, from bears to bacteria, have evolved dormant states. During phage infection we recognize analogous dormant states as lysogeny and as pseudolysogeny. In this chapter we explore several aspects of the ecological consequences of these "reductive" infections.

In addition to the material presented here, we direct the reader to additional reviews considering lysogeny, pseudolysogeny, and phage infection of starved bacteria: Barksdale and Arden (1974), Ackermann and DuBow (1987), Schrader *et al.* (1997a), Robb and Hill (2000), and Miller and Ripp (2002). Related issues, especially of phage contribution to bacterial genotype and phenotype, are also considered in Chapters 11 and 14.

* Corresponding author

5.1.1 Phage lifestyles

Bacteriophages can potentially exhibit three different lifestyles, distinguishable as functions of phage genetics, host genetics, and host physiological status. These lifestyles are described as productive, lysogenic, and pseudolysogenic. Productive infection, in turn, may be distinguished into lytic versus chronic, though only the former (lytic infection) will be emphasized here. For an expanded view of the various phage lifestyles, see Table 1.1 (Chapter 1).

Phage productive infection is characterized by bacterial adsorption that is rapidly followed by replication of the phage genome, synthesis of viral components, and release of mature phage particles such as through cell lysis (Fig. 2.1, Chapter 2). Lysogeny is the indefinite persistence of a temperate phage genome, as a mostly quiescent prophage, in the host cell. It is latent and so its presence does not promote cell death or the production of phage particles. Pseudolysogeny, on the other hand, is the unstable presence of a bacteriophage genome in a host bacterium, where that genome fails to replicate either as a productive infection or in step with cell division as a lysogenic infection (see Table 1.1, Chapter 1, for issues concerning use of the term pseudolysogeny). Traditionally only double-stranded (ds) DNA phages have been believed capable of forming lysogens; however, this is now known not to be the case, as the temperate *Vibrio cholerae* phage CTXΦ (Chapter 14) has been shown to have a single-stranded (ss) DNA genome (Val *et al.*, 2005). To date the only report of a lysogeny-like state in RNA bacteriophages is the establishment of a "carrier-state" in which the genomes of dsRNA phages phi-6 and phi-8 have been shown to replicate as stable episomes with the expression of certain, but not all, of the phage-encoded genes (Sun *et al.*, 2004).

5.1.1.1 Lysogeny

Lysogeny, by definition, is one consequence of an infection of a bacterium by a temperate bacteriophage (Miller, 2004), where the phage downregulates its gene expression to establish a quasi-stable, long-term relationship between the phage (now a prophage) and its host. Prophage replication is coordinated with host-genome replication and progeny prophages are distributed to each daughter cell at division. The molecular and physiological basis for lysogeny by bacteriophage λ in *Escherichia coli* has been highly studied and delineated (Ptashne, 2004). Evidence of the importance of lysogeny in bacterial evolution can be found in the bacterial genome, with whole-genome sequencing indicating the common prevalence of prophages, including vestigial prophages (the remains of bacteriophage genomes). Many bacteria, in fact, have multiple prophages present in their genomes – *E. coli* O157:H7 strain Sakai, for example, contains parts of 18 prophage genomes. This amounts to 16% of its total genomic content (Canchaya *et al.*, 2003).

Prophage residence often elicits new phenotypes in the host (lysogenic conversion; Section 5.5.1) that increase host and, presumably, prophage fitness. Frye and colleagues (2005) suggest that successful niche occupation by bacteria can occur through the lysogenic conversion that results from temperate-phage infection of bacterial strains; that the flexibility and mobility of phenotypes delivered by temperate phages thus contributes to the success of the bacterial species. Edlin *et al.* (1975) suggest, further, that lysogeny influences community composition by allowing survival and/or dominance by lysogens. These new phenotypes may even open new ecological niches to the lysogen by providing it with new metabolic functions or virulence factors (Mahenthiralingam, 2004; Lindell *et al.*, 2005; Chapter 14). The adaptation of lysogeny thus modifies the typical predator–prey interaction between phage and bacterium to a uniquely symbiotic relationship (for additional consideration of lysogenic conversion, see also Chapters 11 and 14). The phage–bacterium symbiotic relationship represented by lysogeny also provides safety for the prophage from environmental insults that otherwise could result in phage-virion decay (Section 5.2).

5.1.1.2 Pseudolysogeny

As compared to lysogeny, especially for phage λ, a similarly detailed mechanistic understanding of pseudolysogeny has yet to be developed (Miller and Ripp, 2002). Pseudolysogeny describes a phage–host interaction in which the phage, upon infecting its host, elicits an unstable, non-productive response (Baess, 1971). Neither lysogeny nor productive growth is initiated. However, the viral genome is maintained in the host cell for potentially extended periods of time. Miller and Ripp (2002) christened this unstable phage genome a "preprophage" in keeping with the nomenclature of lysogeny. They hypothesized that the pseudolysogenic response occurs because the host cell is so dramatically starved that it cannot provide the phage with the necessary energy and substrates required for either a productive or lysogenic response.

Cellular starvation in environmental ecosystems is commonplace (Koch, 1971, 1979; Morita, 1997), though in many cases these periods of famine are interspersed with periods of relative feast. Following the acquisition of such a nutrient source by the pseudolysogen (that is, the preprophage-infected bacterium), the preprophage is supplied with sufficient energy to initiate virion formation (lytic or productive response) or establish a stable true-lysogenic (or reductive) relationship. As virions are rapidly inactivated in many hostile environments (Miller, 2006; Section 5.2.2), the overall result of pseudolysogeny is to protect and thereby extend the viral genome's lifetime under conditions where bacteria may be more readily acquired than productively infected. This delay between phage adsorption and productive infection may explain at least in part the numerical "largeness" of environmental phage communities (Section 5.3).

5.2 IMPACT OF ENVIRONMENTS ON PHAGES

In this section we focus on the environmental impact on phages by means of influencing bacterial physiology and thereby phage infection dynamics. As a consequence, many environmental conditions limit production of high viral titers. This impact is especially important given that, for example, phages in aquatic environments are unstable and therefore have short infective half-lives (Saye *et al.*, 1987; Saye and Miller, 1989; Wommack and Colwell, 2000) – decay rates of 5–30% per hour are not uncommon (Miller, 2006) and occur even though several factors in the environment, such as attachment to suspended particulates, may stabilize virions to some extent (Ripp and Miller, 1995; Chapter 11). Thus, environments can impact phages in terms of both titers produced (phage replication) and titers retained (phage decay).

5.2.1 Impact of nutrients on phage replication

The nutritional and physiological states of the bacterial host are paramount in determining the consequences of phage infection (Farrah, 1987; Moebus, 1987; Williams *et al.*, 1987; Miller and Ripp, 2002; Ptashne, 2004; Miller, 2006). They can affect an infecting virus in four ways. First, nutritional status can impact phage receptor presence. For instance, *Escherichia coli* grown on medium containing a carbon source other than maltose will be resistant to phage λ infection because they will lack maltose permease, which otherwise would serve as the phage receptor (the inducible *mal*B gene product; Ptashne, 2004). Second, as reviewed immediately below (Sections 5.2.2 and 5.2.3), nutritional status can impact phage growth parameters such as burst size and latent period. Third, when ATP levels are higher than levels of cyclic-AMP, as in a physiologically energy-rich phage-sensitive bacterium, then the lytic–lysogenic molecular switch can be biased toward lytic infection (i.e., as seen with the virulent response of phage λ), while the opposite physiological state will set the switch towards a lysogenic response (Ptashne, 2004; Section 5.3.1; see also Section 5.4.2). Lastly, phage genes can impact rates of lysogen replication, especially under starvation conditions (Section 5.2.4). In other words, the bacterial nutritional state can profoundly impact the way a temperate phage replicates.

5.2.2 Impact of host starvation on phage lytic growth

Lenski (1988) demonstrated that maximum yields and shortest latency periods for phage infection are achieved when the host is growing under optimal conditions (not often achieved in the environment). Several studies have explored the influence of environmentally significant host generation times on the efficiency of phage production (Kokjohn *et al.*, 1991; Proctor *et al.*, 1993; Guixa-Boixareu *et al.*, 1996; Moebus 1996a, 1996b; Schrader *et al.*, 1997b; Middelboe, 2000). These studies have revealed that latency periods are lengthened and burst sizes of progeny phage particles reduced when compared to laboratory-grown hosts (see also discussion in Chapter 10). Kokjohn *et al.* (1991) demonstrated that the *Pseudomonas aeruginosa* phage F116 had a latency period of 240 minutes in starved hosts while the period from infection to lysis was only 100–110 minutes in host grown under optimal conditions in the laboratory. The average burst size of this phage in starved hosts was four phage particles per infected starving *P. aeruginosa* cell, while 27 particles were produced in a well-fed host cell.

The same authors (Kokjohn *et al.*, 1991) demonstrated that the burst size of UT1, a *P. aeruginosa* phage isolated from the eutrophic Fort Laudon Lake near Knoxville, Tennessee (Ogunseitan *et al.*, 1990), had a latency period of 70–80 minutes with a burst size of 65 in exponentially growing hosts, but the latency period was greater than 110 minutes, with only six particles produced, in a starved host cell. Schrader *et al.* (1997b) observed similar changes with various *P. aeruginosa* and *E. coli* phages. For instance, the burst size of *P. aeruginosa* bacteriophage was reduced ten-fold in starved cells compared to exponentially growing hosts.

5.2.3 Impact of inorganic factors on phage lytic growth

Limitations in other nutritional and environmental factors have also been shown to affect phage production. Wilson *et al.* (1996) reported that nitrate and phosphate depletion reduced the burst size and titer of bacteriophages even though phosphate limitation did not affect adsorption rate of the cyanophage S-PM2. The reduction in titer was due to the large fraction of cells that did not produce any virions. Wilson *et al.* (1998) reported similar finding in a phosphate-limited mesocosm. Addition of inorganic nutrients stimulated phage production in natural subtropical estuarine ecosystems (Hewson *et al.*,

2001) and in both planktonic and algenate-associated *Pseudomonas fluorescens* infected with bacteriophage ϕR2f (Smith *et al.*, 1996).

Various investigators have hypothesized that the greater effect of phosphate over nitrate is due to the higher ratio of nucleic acid to protein in viruses than in other types of organisms (Bratbak *et al.*, 1993a). In this context, it is interesting to note that inspection of the genomic sequence of the marine virus Roseophage SIO1 suggests that this phage has acquired the gene for a host-derived protein that is induced during phosphate limitation (Rohwer *et al.*, 2000). Likewise, phosphate-stress genes have been identified in the complete genome of KVP40, a T4-like phage infecting a broad range of *Vibrio* spp. (Miller *et al.*, 2003), and Sullivan *et al.* (2005) identified two T4-like cyanophages that, in addition to host photosynthetic genes (*psbA* and *hliP*), contained phosphate-inducible genes (*phoH* and *pstS*). These authors believe that these genes are likely involved in adaptation to phosphate stress, which is often a limiting factor in marine environments.

5.2.4 Prophages and nutrient impact on host growth

The finding that the relative reproductive rate (fitness) of an *E. coli* lysogen is raised by the presence of a temperate prophage (Lin *et al.*, 1977) under some environmental conditions indicates an additional complexity in the bacterium–prophage symbiosis. The prophage enhances the growth rate of the lysogen during aerobic growth in glucose-limited chemostats and retards the rate during anaerobic growth. Widely different temperate phages such as λ, P1, P2, and Mu increase the reproductive rate of *E. coli* lysogens in limiting glucose (Lin *et al.*, 1977). Thus, these phages have two reproductive strategies. They adopt a lysogenic life cycle when the host is in a poor physiological condition and convert to a productive life cycle (which is lytic for most temperate phages) under more physiologically favorable ones.

5.2.5 The paradox of viral numbers

The 1990s were characterized by a new interest in phage ecology and the application of new techniques to the study and enumeration of bacteriophages in the environment (Suttle, 1993; Miller, 1998b). These studies made it obvious that earlier estimates of phage numbers in the environment were wrong and that phage numbers – especially total counts in aquatic environments – were quite high, with particle concentrations often exceeding 10^8/mL (Miller and Sayler, 1992; Chapter 10; see also Chapter 11). Today it is well established that bacteriophages play an important role in the ecology of natural ecosystems and that phage-like particles are present in high numbers in most environments (Wommack and Colwell, 2000; Miller, 2006).

We are now left with a dilemma. How are large numbers of bacteriophages maintained in environments that seem hostile to phage development and propagation? How do phages survive long droughts in host energy supply if their virions are unstable? Even though most estimates of phage numbers in the environment do not distinguish between infective and non-infective phage-like particles, the existence of reservoirs of phages in natural environments must be entertained to explain their survival and high numbers. These reservoirs have been created by the implementation of strategies developed by bacteriophages over evolutionary time to raise their endurance and increase their potential for survival. They are based on the adoption of alternate, if quiescent, lifestyles: lysogeny and pseudolysogeny.

5.3 LYSOGENY: STRATEGY FOR LONG-TERM SURVIVAL

5.3.1 The lytic–lysogenic decision

Two lifestyles for the icosahedral-tailed phages have long been recognized. The ecology of the first, productive or lytic infection, is reviewed in detail in Chapters 2, 3, and 4. In addition, many phages (and perhaps all phages when existing under certain environmental conditions) have the ability to initiate a reductive or temperate infection (lysogeny or pseudolysogeny; Miller, 2004). The choice between productive and reductive lifestyles is determined to a great extent by the metabolic health of the host at the time of infection (Ptashne, 2004). When a phage encounters a host that is unlikely to provide the essential building blocks and energy needed for development of progeny virions, a series of molecular events is initiated that leads to the repression of the expression of a majority of phage genes and the establishment of the viral genome (termed a prophage) in the host cell. Anthropomorphically, we can think of this as the decision by the phage to wait out hard times in hopes that the "good times will roll again" at some time in the future. In other words, the phage is betting that the host cell at some point will become metabolically better able to produce a reasonable (that is, a larger) burst of progeny virions.

A bacterium containing a prophage is described as a lysogen. When a lysogen replicates, its chromosome divides and the prophage is also replicated and then partitioned between the daughter cells. Various environmental conditions can lead to the termination of lysogeny (Miller 2000, 2004; Ptashne, 2004). These are often conditions that damage DNA and induce the SOS repair systems of the host (Miller, 2000; Ptashne, 2004). This is not to say that the prophage itself is DNA damaged, but instead that the host is experiencing insults that, perhaps, could impact the host's survival and, by extension, the survival of the prophage. The termination of lysogeny, through induction, serves to sever the link between prophage and bacterial survival, presumably resulting in increased odds of survival of the phage genome.

5.3.2 Lysogeny consequences

In the laboratory, lysogenic strains can be recognized by several characteristics (Miller, 2004; Ptashne, 2004). These include the presence of virions in liquid growth cultures due to spontaneous termination of lysogeny in a minority of cells each generation. This termination of lysogeny is followed by a productive cycle that produces virions and, for lytic phages, ends with the lysis of the host, releasing the phage particles into the culture medium. Lysogenized bacteria are also immune to superinfection by the virions of the same immunity type as the resident prophage (Chapter 8). In the lysogen, a soluble repressor that binds to regulatory operators in the phage genome represses the expression of the prophage. The presence of this repressor in the cytoplasm of the lysogen leads to immediate repression of expression of any sensitive infecting viral chromosome due to binding of the repressor protein to the operator sites of the superinfecting genome (Ptashne, 2004). In addition, a prophage can contain genes that are expressed during reductive growth that alter the phenotype of the host. This phenomenon, known as lysogenic conversion, can provide the host with new phenotypic characteristics (Section 5.5.1; Chapter 14).

5.3.3 Lysogeny in aquatic environments

Since 1990, many studies of lysogeny *in situ* have been carried out in aquatic environments (Miller, 2006). Aquatic environments have many characteristics that do not favor phage production while still containing high numbers of phage-like particles. These environments are populated by slow-growing bacteria at relatively low concentrations of 10^5–10^6/mL (Bratbak *et al.*, 1990, 1993b; Thingstad *et al.*, 1993). The bacteria find themselves in a situation where nutrients are only available on a sporadic basis and growth is likewise sporadic. Bacteria in these, as well as terrestrial environments, often live in biofilms (Chapters 4 and 11). Several reports have indicated that biofilms are excellent environments for the development of lysogeny under carbon-limiting conditions (Corbin *et al.*, 2001; Sutherland *et al.*, 2004).

Many temperate bacteriophages and cyanophages have been isolated from aquatic environments (Baross *et al.*, 1978; Rambler and Margulis, 1979; Muramatsu and Matsumoto, 1991; Ripp *et al.*, 1994; Mitra *et al.*, 1995; Pajni *et al.* 1995; Wilson *et al.*, 1996; Jiang *et al.*, 1998). Ackerman and DuBow (1987) have estimated that between 21% and 60% of environmental bacteria are lysogenic. If anything, recent evidence suggests that this number is low. In a study of lysogeny during a bloom of cyanobacteria in a pristine fjord in British Columbia, Canada, Ortmann *et al.* (2002) found that 80% of heterotrophic bacteria and 0.6% of cyanobacteria contained prophages inducible with mitomycin C. Miller *et al.* (1992) conducted a study in Fort Loudon Lake in Tennessee that showed more than 80% of the *P. aeruginosa* isolates from the lake contained bacteriophage sequences in their genomes.

Most commonly, mitomycin C induction of prophages (Jiang and Paul, 1998; Ortmann *et al.*, 2002; Lisle and Priscu, 2004), ultraviolet (UV) light induction (McKay and Baldwin, 1973; Rambler and Margulis, 1979; Miller, 2000), and *in situ* hybridization (Ogunseitan *et al.*, 1990, 1992; Miller *et al.*, 1992) have been used to identify lysogens in nature. Each of these methods has drawbacks. Not all prophage are inducible with mitomycin C or UV light, so an underestimate is produced. Likewise, DNA probing depends on the isolation of a representation of viral genomic material from the ecosystem to prepare the probe (Miller *et al.*, 1992; Ashelford *et al.*, 2003). The methods currently available to us are thus doomed to miss many lysogens.

The frequency of lysogeny has been investigated in several aquatic environments where the amount of available carbon and energy varied (Weinbauer and Suttle, 1996, 1999; Paul *et al.*, 1997; Jiang and Paul, 1998). In each of these studies, higher frequencies of lysogeny were noted in the more oligotrophic portions of the environment, where carbon and energy were most limited. In the Gulf of Mexico, Weinbauer and Suttle (1996, 1999) found that 2–11% of bacteria contained inducible prophages in oligotrophic offshore waters, compared to 1–2% in coastal waters. Gradients of lytic cyanophages were observed by Sullivan *et al.* (2003), with high levels in coastal waters and low levels in oligotrophic open-ocean waters. These authors speculate that differences in population diversity, growth rates, and the incidence of lysogeny underlay these trends. Finally, Lisle and Priscu (2004) found that the frequency of lysogeny was high in the carbon-limited lakes of the McMurdo dry valleys of Antarctica.

5.3.4 Lysogen induction in aquatic environments

McDaniel *et al.* (2002) followed induction of lysogenic *Synechococcus* spp. in Tampa Bay, Florida. They found seasonal variation in the rates of prophage induction and a presumably associated variation in the frequency of lysogens and the titer of virions in these waters. Induction was highest in the late winter months and was consistent with a pattern of lysogeny maintenance during times of low host availability, resource limitation, and other adverse environmental conditions. Lysogeny was fostered during times when it was most advantageous to ensure viral survival. The authors speculate that the limited number of lysogens and higher titers of virions observed during the summer months might be due to greater exposure to UV light and/or the higher water temperatures. These environmental factors have been shown to increase induction rates of many prophages, leading to bacterial lysis and virion production. These data seem consistent with an earlier study in Tampa Bay that indicated a seasonal switch from "dormant" lysogens (those that were not capable of induction by mitomycin C) in the winter to inducibility in the warmer months (Cochran and Paul, 1998).

Miller (2001) carried out a study in the Gulf of Mexico in which lysogenic and non-lysogenic *Vibrio parahemolyticus* were incubated *in situ* in marine microcosms. Half of the microcosms were exposed to solar radiation containing solar UV light and half were incubated under conditions where they were protected from UV exposure. Those exposed to solar UV irradiation showed a diel induction of the lysogens resulting in an increase in the titer of virions and reduction in the number of lysogenic bacteria in the microcosm. These results were not observed in the microcosms that were not exposed to UV light. This study lends credence to the hypothesis of McDaniel *et al.* (2002) that solar UV can greatly affect the frequency of lysogens in natural populations of aquatic microorganisms. See Chapter 10 for additional consideration of phages as found in aquatic habitats.

5.3.5 Lysogeny in terrestrial environments

Studies in terrestrial ecosystems have also shown correlations between host-cell density, nutrient availability, and the frequency of temperate phages. Ashelford *et al.* (1999) studied the seasonal variation of temperate and lytic *Serratia* phages found in a sugar beet rhizosphere. They found high levels of virions of temperate phages during periods of high host-cell density and elevated metabolic activity early in the growing season, but titers were much lower when cell densities were low due to nutrients dwindling late in the growing season. Consistent with the findings of McDaniel *et al.* (2002) in the aquatic ecosystem, these data are suggestive of the productive outcome of infection of bacteria by temperate phages during times of plenty in the rhizosphere microbial community, and the choice of lysogeny establishment (reductive lifestyle) when environmental factors become limiting. See Chapter 11 for additional consideration of phages as found in terrestrial habitats.

5.3.6 Feast–famine impact on lysogen prevalence

The role of nutrient availability in the induction of lysogens in aquatic environments was studied by Miller and Ripp (1998). They prepared microcosms of starved *P. aeruginosa* and on a sporadic schedule added limiting concentrations of yeast extract as a carbon and energy source. The microcosms contained two genetically identifiable strains of *P. aeruginosa*. The first strain (Strain I) was lysogenic for bacteriophage F116 (Miller *et al.*, 1974). The second (Strain II) was not lysogenic but was sensitive to infection by F116. Initially the microcosms consisted of *Pseudomonas* Minimal Medium (Miller and Ku, 1978) containing a very low level of nutrient (yeast extract at 10^{-5}% [wt/vol]). The microcosms were inoculated and maintained for several days to ensure that the host bacteria were starved for nutrients. Each week, a spike of nutrient consisting of yeast extract (10^{-5}%, final concentration) was added to the culture. These conditions were designed to simulate the feast–famine cycles so often encountered by microorganisms in the natural environment (Koch, 1971). Within the 24–48 hours following each nutrient spike, an induction of Strain I lysogens occurred, resulting in a rise in the relative number of phage particles. At the same time an increase in the percentage of Strain II cells exhibiting the characteristics of lysogeny was seen. Nutrient spikes continued on a weekly basis up to 110 days. Each was followed by increases in the relative numbers of virions and percent lysogeny of Strain II. No increases were observed in control microcosms that were not spiked with nutrient. These data illustrate that nutrient availability dramatically affects phage–host interactions.

Some exceptions have been noted. Burroughs *et al.* (2000) observed little correlation of gross nutritional status with burst size and frequency of lysogeny establishment in a study of streptomycetes and their phages in a soil microcosm. They felt that physical processes within the environment such as nutrient diffusion were more important effectors of the ecosystem's dynamics. Their findings illustrate that consideration of nutritional concentrations are not sufficient in predicting an ecosystem's ability to support viral growth and phage–host interaction. Instead, determination of micro-scale nutrient availability to the host must be considered, particularly in terrestrial ecosystems.

5.3.7 Ecological relevance of lysogeny

John Mittler (1996) has modeled the evolution of the genetic switch used by temperate bacteriophages in the decision-making process of lytic growth versus the establishment of lysogeny. Using a chemostat model, he found that in variable environments those bacteriophages that choose lysogeny more often are favored over those with low probability of establishing lysogeny. He found that the probability of survival also increased when lysogens had higher rates of induction as environmental factors became favorable. The opposite characteristics were found to be important for phage survival in constant environments. Mittler's findings are just those that would be expected if lysogeny is an evolutionary adaptation for survival in changing environments. Consistent with this hypothesis, Mittler's simulations suggest that in both constant and variable environments, phages that evolved moderately low rates of lysogeny establishment and induction are able to "hedge the bets." They can deal with changes in environmental factors (particularly host and nutrient availability) in variable environments without sacrificing their ability to compete in environments where conditions are stable.

The establishment of lysogeny requires that some energy and resources be available to the infecting phage genome (Miller, 2004). In highly oligotrophic environments, bacteria may not possess even these minimal requirements necessary for the establishment of lysogeny. What then is the fate of viruses that infect such hosts? Evidence is mounting for a transient state of host–viral interaction (Section 5.4), referred to as "pseudolysogeny," which was alluded to by Twort (1915) and first fully described in an environmental setting by Romig and Brodetsky (1961).

5.4 PSEUDOLYSOGENY: STRATEGY FOR MEDIUM-TERM SURVIVAL

In 1992, Ogunseitan *et al.* studied the frequency of lysogeny in Fort Loudon Lake, Knoxville, Tennessee. As mentioned earlier, it had been discovered that approximately 80% of *P. aeruginosa* isolates from that lake contained phage-specific DNA homologous to nucleic acid probes available to these investigators. Even so, only 1–7% of the isolates contained inducible prophages or displayed the characteristics commonly associated with true lysogeny (Section 5.3.2). This led these investigators, like many before them, to search for evidence of an alternative relationship, that is, for evidence of pseudolysogeny.

5.4.1 Defining pseudolysogeny

The state of pseudolysogeny was first imagined by Twort (1915). Although discussed in André Lwoff's famous review of lysogeny (1953), the phenomenon was virtually ignored until Romig and Brodetsky (1961) clearly

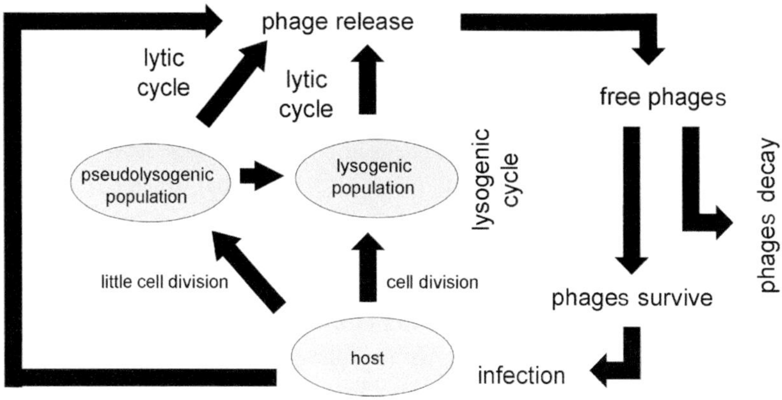

Figure 5.1 The progressive relationships between host cells and temperate and virulent bacteriophages.

described it in exploring the relationship between *Bacillus* spp. and their viruses in desert soils, although they did not use the term specifically. Thus the exact origin of the term is difficult to discover. Even then, pseudolysogeny's potential importance to phage ecology has been recognized only in the recent past. Operationally, the term describes a phage–host interaction in which the nucleic acid of the phage (preprophage), following infection, neither establishes a long-term stable relationship (lysogeny) nor produces virions (which in practice is a lytic response though it could broadly be described as a productive response). The phage nucleic acid simply resides within the cell in a non-active state. Unlike true lysogeny, the preprophage does not replicate and is not segregated equally into all progeny cells (Baess, 1971; Ackerman and DuBow, 1987; Ripp and Miller, 1997, 1998).

The interrelationships among the phage lytic lifestyle, lysogeny, and pseudolysogeny in an environmental habitat are illustrated in Figure 5.1. Both temperate phages and those that express a lytic phenotype under laboratory conditions often exhibit pseudolysogenic characteristics in environmental settings (Ogunseitan *et al.*, 1990; Miller and Ripp, 2002). Ripp and Miller (1997) hypothesized that under environmental conditions where cells are highly starved (Koch, 1971; Morita, 1997) insufficient energy is available in the host cell to initiate either of these typical laboratory-defined responses. However, when nutrients become transiently available to the host cell and sufficient energy becomes available for phage metabolism to proceed, the preprophage initiates either establishment of true lysogeny or expression of the lytic cycle.

Phages capable of establishing pseudolysogeny typically show highly turbid plaques when initially isolated on host bacteria grown under environmental conditions (Ripp and Miller, 1997, 1998; Miller and Ripp, 2002). Upon subculture on hosts grown on rich media, under ideal laboratory conditions, plaque turbidity is highly reduced, often exhibiting a "clear" phenotype (see Chapter 4 for more on plaque turbidity). Thus, pseudolysogeny provides obligately lytic phages (i.e., phages which cannot display lysogeny; Chapter 1) with an alternative strategy, other than reproductive lysis, upon infection of a bacterium found in a poor physiological state. In doing this, their host survives the infection, and may propagate slowly to form a small pool of pseudolysogenized progeny. However, the preprophage does not propagate and is segregated into only one cell in the microclone. As more optimal growth conditions occur and the host cell's physiology returns to normal, then virion production can be initiated.

As indicated above, the major difference between lysogeny and pseudolysogeny is the lack of phage-genome replication. Thus during pseudolysogeny, but not lysogeny, there is a unilateral inheritance of the phage by just one of the daughter cells at cell division. When studied *in situ* or under environmental conditions in the laboratory, pseudolysogens typically do not spontaneously release phage virions, as is seen with true lysogenic cultures grown under ideal conditions. Even so, molecular analysis such as colony hybridization will demonstrate the presence of the preprophage sequences (Ripp and Miller, 1997). Lack of superinfection immunity and superinfection exclusion (Chapter 8) are also characteristic of pseudolysogens (Wommack and Colwell, 2000; Williamson *et al.*, 2001; Miller and Ripp, 2002).

5.4.2 Examples of pseudolysogeny

Pseudolysogeny has been observed in many species of bacteria, which we review below. These observations demonstrate an essential characteristic of pseudolysogeny: regulation by environmental conditions that restrict host growth and survival (Ripp and Miller, 1997; Wommack and Colwell, 2000). Nevertheless, the molecular and physiological basis for pseudolysogeny is not yet understood.

5.4.2.1 Pseudolysogeny with *Pseudomonas* hosts

Ripp and Miller (1997, 1998) carried out several studies with the *P. aeruginosa* phages UT1 (Ogunseitan *et al.*, 1990) and F116 (Miller *et al.*, 1974). The first study established the ability of long-term pseudolysogens of F116 to resolve into active virion production or stable, true lysogenic relationships with their hosts. Even after 43 days of starvation, a nutrient spike to the starved host allowed the F116 preprophage to elicit the production of virions, thereby indicating both the maintenance of a pseudolysogenic state and its resolution into productive infection (Ripp and Miller, 1998).

A second set of experiments utilized continuous culture conditions for *P. aeruginosa* (Replicon *et al.*, 1995) to study the dynamics of the pseudolysogenic relationship in very slow-growing populations (14-hour generation times; Ripp and Miller, 1998). Chemostat samples were analyzed for the numbers of true lysogens (i.e., those isolates that were capable of spontaneous release of mature virions) and for pseudolysogens (i.e., isolates that did not produce virions but demonstrated the presence of phage F116 genomes by colony hybridization). Throughout the extended life of the chemostat, a highly variable portion (18–83%) of the population was shown to be pseudolysogenic. When the progeny of individual pseudolysogenic isolates from the chemostat were analyzed, only a small and variable fraction of the daughter cells contained the phage genome. Ripp and Miller (1998) hypothesized that due to the severely starved state of the host, not enough energy was available to the infecting phage genome to establish stable lysogeny and the preprophage was not replicated in synchrony with the host genome. As in abortive transduction (Arber, 1994), only one of the two daughter cells from each cell division acquired the viral genome.

Similar results were obtained when microcosms of *P. aeruginosa* F116 pseudolysogens were incubated *in situ* in a semi-oligotrophic freshwater lake (Ripp and Miller, 1998). While microcosms incubated at the bottom of the lake produced no phage-releasing cells (true lysogens), about 20% of the starved population exhibited characteristics of pseudolysogeny. In a later publication, Miller and Ripp (2002) speculated that the lack of mature phages in these bottom-incubated microcosms (surface microcosms did produce some inducible lysogens) was due to the lack of solar UV light exposure. Such exposure has been shown to be important in the induction of many environmental lysogens including F116 (Kokjohn and Miller, 1985; Suttle and Chen, 1992; Kidambi *et al.*, 1996; Miller, 2001; see also Section 5.3.4).

Similar experiments were carried out by Miller and Ripp (2002) using the obligately lytic phage UT1 (Ogunseitan *et al.*, 1990). They demonstrated that an environmental equilibrium between starved host cells and obligately lytic phages can be established such that the host is not eradicated but instead serves as a long-term safe haven for the viral genome. When environmental conditions improve (nutrients become available), phage particles can be produced. These investigators found that the frequency of pseudolysogeny establishment was inversely proportional to the availability of nutrients.

Tuomi *et al.* (1995) saw a correlation between an increase in the ratio of phage particles per bacterial cell and an increased availability of carbon and energy in a seawater microbial community. Although the authors did not investigate further, these data suggest the possibility that additional nutrients increase levels of activation of lysogens or pseudolysogens leading to the production of phage virions. The data are consistent with the findings of Ripp and Miller (1997) that addition of an energy source to a freshwater microcosm allowed 15-day-starved pseudolysogens of *P. aeruginosa* to lyse, releasing phage particles. It has been well established in environmental studies that nutrient status influences the decision between lytic and lysogenic growth upon primary infection (Wilson and Mann, 1997). Thus, the data of Tuomi *et al.* (1995) extended the observation of a close link between nutrient concentration and the maintenance and termination of lysogeny and pseudolysogeny in salt-water environments.

5.4.2.2 Additional examples of pseudolysogeny

The *Halobacterium salinarium*–phage Hs1 system illustrates the essential characteristics of pseudolysogeny (Torsvik and Dundas, 1980; Reiter *et al.*, 1988). Cultures of *H. salinarium* are often characterized by sporadic lysis, with the release of phage Hs1. However, stable lysogenic clones of the survivors cannot be subcultured, indicating pseudolysogenic infection. The salt concentration of the medium is crucial to the outcome of infection of *H. salinarium* with Hs1. This halophile will not survive at concentrations of NaCl lower than 17.5% [wt/vol]. At this salt concentration phage infection is invariably productive. However, as the concentration of salt approaches 30%, a concentration that does not favor host growth, the majority of infections lead to pseudolysogens. In other words, pseudolysogeny is favored under conditions where it is likely to increase the potential for phage survival (Torsvik and Dundas, 1980).

Pseudolysogeny of *Azotobacter vinelandii* by phage A21 was studied by Thompson *et al.* (1980a, 1980b). They also observed that the unstable pseudolysogenic state could be converted to a stable one of true lysogeny. This observation is consistent with the idea that pseudolysogens do indeed represent preprophages in the sense that pseudolysogens of temperate phages possess some potential to form lysogens instead of productive infections.

Los *et al.* (2003) studied pseudolysogeny in *E. coli* infected by the obligately lytic phage T4. They used carbon-limited chemostats and found that this phage can form pseudolysogens not only when growth is completely inhibited but also in very slow-growing host cells. The *rI* gene product of this phage was demonstrated to play an important role in the establishment of pseudolysogeny. This gene has previously been shown to be indispensable for lysis inhibition in this phage (Paddison *et al.*, 1998; Tran *et al.*, 2005; see Chapter 3 for further discussion of lysis inhibition). Thus, the transient state of pseudolysogeny in both lytic and temperate phages may be activated by regulation of *rI*-like genes. Clearly additional research is needed to fully understand the molecular genetics of pseudolysogeny.

Besides Romig and Brodetsky's (1961) observations of soil isolates of *B. subtilis*, Bramucci *et al.* (1977) identified a pseudolysogenic phage in a soil isolate of *Bacillus pumilus*. Several species of soil mycobacteria have demonstrated characteristics of pseudolysogeny as well (Baess, 1971; Grange, 1975; Grange and Bird, 1975). Pseudolysogens have also been identified in the mycoplasma species *Acholeplasma laidlawii* (Roger, 1983). Phages isolated from soil and animal droppings have been shown to exist in a state of pseudolysogeny in *Myxococcus virescens* and *M. fulvus* (Brown *et al.*, 1976), and pseudolysogenic states have been identified in soil *Streptomyces* spp. (Marsh and Wellington, 1992).

Aquatic systems have also yielded information on the occurrence of pseudolysogeny. Ogunseitan *et al.* (1990, 1992), as noted above, studied lysogeny and pseudolysogeny among lake *Pseudomonas* spp. Following careful scrutiny of numerous marine phage–host relationships, Moebus (1996b, 1997) concluded that pseudolysogeny is a common phenomenon in marine ecosystems. Williamson *et al.* (2001) determined that isolates of the marine bacterium *Listonella pelagia* were unstable during long-term storage at –80° C and exhibited various characteristics of pseudolysogens.

Pseudolysogeny has also been identified in human- and animal-associated as well as pathogenic bacterial species. Drozhevkina *et al.* (1984) inspected over 1000 isolates of *Vibrio cholerae* strains from ponds, sewage, and fecal samples for the frequency of lysogens and pseudolysogens. Since then, the cholera toxin associated with human disease has been found to be associated with the temperate phage CTXΦ (Davis *et al.*, 2000; Faruque *et al.*, 2000; Mukhopadhyay *et al.*, 2001). In their study, Drozhevkina *et al.* (1984) found that approximately 1% of the strains were pseudolysogens. Thus, pseudolysogeny must be considered when developing epidemiological models for this disease. Pseudolysogens of group A streptococci have also been discovered to be associated with toxigenic conversion in that species (Nida and Ferretti, 1982), and a similar phage-carrier state (pseudolysogeny) is observed among *Bacteroides fragilis* strains using DNA probes of phages isolated from sewage (Booth *et al.*, 1979).

Yet additional microorganisms have been shown to display characteristics of pseudolysogeny. Archaebacterial isolates have been studied (Reiter *et al.*, 1988), and pseudolysogens identified in several environmental strains. Wall *et al.* (1975) similarly identified "phage carriers" that they believed to be pseudolysogens among strains from a collection of the photosynthetic bacterium *Rhodopseudomonas capsulata*.

5.4.3 Ecological relevance of pseudolysogeny

Although described by Romig and Brodetsky (1961) in environmental (soil) isolates of *Bacillus subtilis*, the environmental importance of pseudolysogeny has only been seriously considered in the past 15 years (Miller and Ripp, 2002). The pseudolysogenic state allows increased longevity of the preprophage genome in hostile environments. In this state, the viral DNA has the potential of enduring for extended periods of starvation of a host. In this way, labile phage virions are not required to distinguish between starved and not-starved bacteria, infecting only the former, but instead employ a mechanism of phenotypic plasticity following adsorption: they initiate infection but subsequently employ a productive, pseudolysogenic, or lysogenic infection strategy depending on phage type and host physiology. Through evolutionary modifications, the phage thus has adapted itself for survival under harsh environmental conditions (Roszak and Colwell, 1987).

Evidence is rapidly accumulating that pseudolysogeny is spread widely among divergent environments and a wide variety of bacterial hosts (Section 5.4.2). Wommack and Colwell (2000) suggest that pseudolysogeny affords "phage populations a means of quickly reacting to environmental changes." They hypothesize that influxes of small amounts of nutrients into a nutrient-limited marine ecosystem can stimulate both bacterial production and bacterial mortality through lytic activation of preprophages.

5.5 PHAGE-INDUCED HOST ALTERATION

The evolution of lysogeny has allowed the creation of a symbiotic relationship between phage and host. Not only is the phage genome provided with a safe haven from an antagonistic environment, but the host's fitness to deal with this same intimidating environment can be increased (Lin *et al.*, 1977). Through lysogeny, the host's phenotype is altered by the addition or loss of various characteristics due to expression of genes that are part of the prophage's genome. In some cases this phenomenon may be referred to as various forms of specialized transduction (Breitbart *et al.*, 2005), where bacterial genes are carried by lysogen-forming temperate phages (Chapter 11). Furthermore, some temperate phages have novel genes which when expressed confer new phenotypes on the host cell. When phenotypes are so gained this is termed lysogenic conversion. In addition, bacterial genes can be moved between hosts by viruses, a phenomenon referred to as generalized transduction (Miller, 1998a, 2001, 2004; Chapter 11).

5.5.1 Lysogenic conversion

While supplying protection to the host from superinfection is its most fundamental expression (Ptashne, 2004), the manifestation of lysogenic conversion includes a multitude of characteristics. In different species of phages and bacteria, these changes may include alteration in receptors that protect the bacterium from infection by other bacteriophages (Saye and Miller, 1989); acquisition of components of metabolic systems (Cavenagh and Miller, 1986; Ptashne, 2004) including a portion of the photosynthetic complex in cyanobacteria (Mann *et al.*, 2003; Lindell *et al.*, 2004, 2005); and acquisition of an ability to produce bacteriocins (Ivanovics *et al.*, 1976), antibiotics (Martinez-Molina and Olivares, 1979), or pili (Karaolis *et al.*, 1999). One subcategory of lysogenic-conversion genes encodes toxins that can enhance the virulence of the bacterial host (Chapter 14). Another new trait conferred by many of the larger genome phages is a restriction-modification system (Rocha *et al.*, 2001).

5.5.2 "Lytic" conversion

Bacteriophages sometimes carry host-like genes that are expressed during lytic infection. In temperate phages, these functions may also be expressed during lysogeny. Here we present these phage-expressed functions as examples of "lytic" conversion to mirror the better-appreciated "lysogenic" conversion effected by prophages (Section 5.5.1). Both phenomena, in general, may be described as examples of "phage" conversion (Barksdale and Arden, 1974). The classical case is that of the diphtheria toxin genes that are carried by corynebacteriophage β (Chapter 14) and expressed during both lytic and lysogenic growth (Holmes and Barksdale, 1970).

An additional and interesting example of this phenomenon of "lytic" conversion has recently come to light. Several cyanophages carry genes for polypeptides (D1 and D2) that are found in the core of the photosystem II (PSII) reaction center (Mann *et al.*, 2003; Bailey *et al.*, 2004; Lindell *et al.*, 2004, 2005; Clokie *et al.*, 2006). The *Synechococcus* cyanophage S-PM2 contains the genes *psbA* and *psbD*, which code for D1 and D2. These proteins are very labile and turnover rapidly in the PSII. The phage genes are expressed during lytic infection and are believed to be important in maintaining the photocenter activity during lytic development of this T4-like phage (Mann *et al.*, 2003;

Bailey *et al.*, 2004; Clokie *et al.*, 2006; Lindell *et al.*, 2005). This phage infects not only *Synechococcus* spp. but also *Prochlorococcus* spp. (Bailey *et al.*, 2004). Other phages of *Synechococcus* and *Prochlorococcus* also contain these genes.

Cyanophage examples of "lytic" conversion are not limited to genes actively involved in photosynthesis. Instead, some *Prochlorococcus* phages contain cyanobacterial analogs of the *hli* (high-light-inducible protein, HLIP), *petE* (plastocyanin) and *petF* (ferredoxin) (Lindell *et al.*, 2004). These proteins are important to the photosynthetic process and protection from photobleaching. Thus, phage genes can not only augment otherwise host-mediated processes, but can overtly protect them as well.

Lindell *et al.* (2004) suggested that horizontal gene transfer (HGT) among cyanobacteria mediated by these various cyanophages might play an important role in ensuring the fitness of these organisms in their environmental setting. Zeidner *et al.* (2005) examined uncultured environmental cyanophage and prophage populations of *Synechococcus*. Their study revealed strong evidence supporting the hypothesis that these genes are undergoing transductional HGT among *Synechococcus* and *Prochlorococcus* populations. By carrying genes that enhance bacterial function during lytic phage infection, these phages also serve as a genetic pool and therefore may play important roles in the evolution of their host (Miller, 2004; Zeidner *et al.*, 2005).

5.6 CONCLUSIONS

Bacteriophages are important in natural ecosystems. They regulate the size of bacterial populations (Wommack and Colwell, 2000; Ashelford *et al.*, 2003; Chapters 2 and 10). They also increase bacterial diversity by regulating horizontal gene transfer through virus-mediated transduction (Miller 1998a, 2004; Weinbauer and Rassoulzadegan, 2004; Chapter 11) and by lysogenic conversion (Weinbauer and Rassoulzadegan, 2004; Section 5.5.1). Viruses certainly have increased the diversity of their hosts (Chapter 10) and have likely contributed to bacterial speciation (Miller, 1998a, 2004; Miller and Day, 2004; Weinbauer and Rassoulzadegan, 2004; Bull *et al.*, 2006).

Within this context, lysogeny and pseudolysogeny have important environmental and ecological consequences, increasing the likelihood of the survival of both the phage and its host. As it is clear that no more that 1% of the bacterial "species" on earth can be cultured in the laboratory, it is equally evident that our knowledge of the vast diversity of bacteriophage "species" that influence environmental microbial communities is highly limited. Metagenomics and other modern techniques of molecular microbial ecology need to be applied to the future exploration of the lysogenic, lytic, and pseudolysogenic interactions between bacteriophages and their hosts. Such techniques are critical to the advancement of our understanding of the mechanisms of bacterial diversity and evolution.

ACKNOWLEDGMENTS

This work was supported in part by the Microbial Observatories program of the National Science Foundation (MCB-0132097).

REFERENCES

Ackerman, H. W., and M. S. DuBow. 1987. *Viruses of Prokaryotes: General Properties of Bacteriophages.* Baca Raton, FL: CRC Press.

Arber, W. 1994. Bacteriophage transduction. In R. G. Webster and A. Granoff (eds.), *Encyclopedia of Virology, Vol. 1.* San Diego, CA: Academic Press, pp. 107–13.

Ashelford, K. E., J. C. Fry, M. J. Bailey, A. R. Jeffries, and M. J. Day. 1999. Characterization of six bacteriophages of *Serratia liquefaciens* CP6 isolated from the sugar beet phytosphere. *Appl. Environ. Microbiol.* **65**: 1959–65.

Ashelford, K. E., M. J. Day, and J. C. Fry. 2003. Elevated abundance of bacteriophage infecting bacteria in soil. *Appl. Environ. Microbiol.* **69**: 285–9.

Baess, I. 1971. Report on a pseudolysogenic mycobacterium and a review of the Literature concerning pseudolysogeny. *Acta Path. Microbiol. Scand.* **79**: 428–34.

Bailey, S., M. R. Clokie, A. Millard, and N.H. Mann. 2004. Cyanophage infection and photoinhibition in marine cyanobacteria. *Res. Microbiol.* **155**: 720–25.

Barksdale, L., and S. B. Arden. 1974. Persisting bacteriophage infections, lysogeny, and phage conversions. *Annu. Rev. Microbiol.* **28**: 265–99.

Baross, J. A., J. Liston, and R. Y. Morita. 1978. Incidence of *Vibrio parahaemolyticus* bacteriophages and other *Vibrio* bacteriophages in marine samples. *Appl. Environ. Microbiol.* **36**: 492–9.

Booth, S. J., R. L. van Tassel, J. Johnson, and T. D. Wilkins. 1979. Bacteriophages of *Bacteroides. Rev. Infect. Dis.* **1**: 325–36.

Bramucci, M. G., K. M. Keggins, and P.S. Lovett. 1977. Bacteriophage conversion of spore negative mutants to spore-positive in *Bacillus pumilus. J. Virol.* **22**: 194–202.

Bratbak, G., M. Heldal, S. Norland, and T. F. Thingstad. 1990. Viruses as partners in spring bloom microbial trophodynamics. *Appl. Environ. Microbiol.* **56**: 1400–5.

Bratbak, G., M. Heldal, A. Naess, and T. Poeggen. 1993a. Viral impact on microbial communities. In Guerrero, R., and Pedros-Alio, C. (eds.), *Trends in Microbial Ecology*. Barcelona: Spanish Society for Microbiology, pp. 299–302.

Bratbak, G., J. K. Egge, and M. Heldal. 1993b. Viral mortality of the marine alga *Emiliania huxleyi* (Haptophycaea) and termination of algal blooms. *Mar. Ecol. Prog. Ser.* **93**: 39–48.

Breitbart, M., F. Rohwer, and S. T. Abedon. 2005. Phage ecology and bacterial pathogenesis. In M. K. Waldor, D. I. Friedman, and S. L. Adhya (eds.), *Phages: Their Role in Bacterial Pathogenesis and Biotechnology*. Washington DC: ASM Press, pp. 66–91.

Brown, N. L., R. P. Burchard, D. W. Morris, J. H. Parish, N. D. Stow, and C. Tsopanakis. 1976. Phage and defective phage of strains of *Myxococcus*. *Arch. Microbiol.* **108**: 271–9.

Bull, J. J., J. Millstein, J. Orcutt, and H. A. Wichman. 2006. Evolutionary feedback mediated through population density, illustrated with viruses in chemostats. *Am. Nat.* **167**: E39–E51.

Burroughs, N. J., P. Marsh, and E. M. H. Wellington. 2000. Mathematical analysis of growth and interaction dynamics of Streptomycetes and a bacteriophage in soil. *Appl. Environ. Microbiol.* **66**: 3868–77.

Canchaya, C, C. Proux, G. Fournous, A. Bruttin, and H. Brüssow. 2003. Prophage genomics. *Microbiol. Mol. Biol. Rev.* **67**: 238–76.

Cavenagh, M. M., and R. V. Miller. 1986. Specialized transduction of *Pseudomonas aeruginosa* PAO by bacteriophage D3. *J. Bacteriol.* **165**: 448–52.

Clokie, M. R., H. Shan, S. Bailey, *et al.* 2006. Transcription of "photosynthetic" T4-type phage during infection of a marine cyanobacterium. *Environ. Microbiol.* **8**: 827–35.

Cochran, P. K., and J. H. Paul. 1998. Seasonal abundance of lysogenic bacteria in a subtropical estuary. *Appl. Environ. Microbiol.* **64**: 2308–12.

Corbin, B. D., R. J. C. McLean, and G. M. Aron. 2001. Bacteriophage T4 multiplication in a glucose-limited *Escherichia coli* biofilm. *Can. J. Microbiol.* **47**: 680–4.

Davis, B. M., K. E. Moyer, F. Boyd, and M. K. Waldor. 2000. CTX prophages in classical biotype *Vibrio cholerae*: functional phage genes but dysfunctional phage genomes. *J. Bacteriol.* **182**: 6992–8.

Drozhevkina, M. S., T. I. Kharitonova, L. G. Voronezhskaia, and V. K. Kirdeev. 1984. Lysogeny studies of *Vibrio cholerae* NAG. *Zh. Mikrobiol. Epidemiol. Immunobiol.* Dec (12): 50–4.

Edlin, G., L. Lin, and R. Kudrna. 1975. λ-lysogens of *E. coli* reproduce more rapidly than non-lysogens. *Nature* **255**: 735–7.

Farrah, S. R. 1987. Ecology of phage in freshwater environments. In S. M. Goyal, C. P. Gerba, and G. Bitton (eds.), *Phage Ecology*. New York, NY: Wiley, pp. 125–36.

Faruque, S. M., Asadulghani, M. M. Rahman, M. K. Waldor, and D. A. Sack. 2000. Sunlight-induced propagation of the lysogenic phage encoding cholera toxin. *Infect. Immun.* **68**: 4795–801.

Frye, J. G., S. Porwollik, F. Blackmer, P. Cheng, and M. McClelland. 2005. Host gene expression changes and DNA amplification during temperate phage induction. *J. Bacteriol.* **187**: 1485–92.

Grange, J. M. 1975. Pseudolysogeny in *Mycobacterium diernhoferi* ATCC19341. *J. Gen. Microbiol.* **89**: 387–91.

Grange, J. M., and R. G. Bird. 1975. The nature and incidence of lysogeny in *Mycobacterium fortuitum*. *J. Med. Microbiol.* **8**: 215–23.

Guixa-Boixareu, N., J. I. Calderón-Paz, J. Heldal, G. Bratbak, and C. Pedrnós-Alió. 1996. Viral lysis and bacterivory as prokaryotic loss factors along a salinity gradient. *Aquat. Microb. Ecol.* **11**: 215–27.

Hewson, I., J. M. O'Neill, J. A. Fuhrman, and W. C. Dennison 2001. Virus-like particle distribution and abundance in sediments and overmaying waters along eutrophication gradients in two subtropical estuaries. *Limnol. Oceanogr.* **47**: 1734–46.

Holmes, R. K., and L. Barksdale. 1970. Comparative studies with *tox*$^+$ and *tox*$^-$ corynebacteriophages. *J. Virol.* **5**: 783–94.

Ivanovics, G., V. Gaal, E. Nagy, B. Pragai, and M. Simon, Jr. 1976. Studies on negacinogeny in *Bacillus cerus*. II. *Bacillus cereus* isolates characterized by prophage-controlled production of megacin A (Phsopholipase A). *Acta Microbiol. Acad. Sci. Hung.* **23**: 283–91.

Jiang, S. C., and J. H. Paul. 1998. Significance of lysogeny in the marine environment: studies with isolates and a model of lysogenic phage production. *Microb. Ecol.* **35**: 235–43.

Jiang, S. C., C. A. Kellogg, and J. H. Paul. 1998. Characterization of marine temperate phage-host systems isolated from Mamala Bay, Oahu, Hawaii. *Appl. Environ. Microbiol.* 64: 535–42.

Karaolis, D. K. R., S. Somara, D. R. Maneval, Jr, J. A. Johnson, and J. B. Kaper. 1999.A bacteriophage encoding a pathogenicity island, a type-IV pilus and a phage receptor in cholera bacteria. *Nature* **199**: 375–9.

Kidambi, S. P., M. G. Booth, T. A. Kokjohn, and R. V. Miller. 1996. *recA*-dependence of the response *of Pseudomonas aeruginosa* to UVA and UVB irradiation. *Microbiology* **142**: 1033–40.

Koch, A. L. 1971. The adaptive responses of *Escherichia coli* to a feast and famine existence. *Adv. Microb. Physiol.* **6**: 147–217.

Koch, A. L. 1979. Microbial growth in low concentrations of nutrients. In M. Shilo (ed.), *Strategies in Microbial Life in Extreme Environments.* Dahlem Konferenzen–1978. Weinheim: Verlag Chemie, pp. 261–79.

Kokjohn, T. A., and R. V. Miller. 1985. Molecular cloning and characterization of the recA gene of *Pseudomonas aeruginosa. J. Bacteriol.* **163**: 568–72.

Kokjohn, T. A., G. S. Sayler, and R. V. Miller. 1991. Attachment and replication of *Pseudomonas aeruginosa* bacteriophages under conditions simulating aquatic environments. *J. Gen. Microbiol.* **137**: 661–6.

Lenski, R. E. 1988. Dynamics of interactions between bacteria and virulent bacteriophage. *Adv. Microb. Ecol.* **10**: 99–108.

Lin, L., R. Bitner, and G. Edlin.1977. Increased reproductive fitness of *Escherichia coli* lambda lysogens. *J. Virol.* **21**: 554–9.

Lindell, D., M. B. Sullivan, Z. I. Johnson, A. C. Tolonen, F. Rohwer, and S. W. Chisholm. 2004. Transfer of photosynthesis genes to and from *Prochlorococcus* viruses. *Proc. Natl. Acad. Sci. U.S.A.* **101**: 11013–18.

Lindell, D, J. D. Jaffe, Z. I. Johnson, G. M. Church, and S. W. Chisholm. 2005. Photosynthesis genes in marine viruses yield proteins during host infection. *Nature* **438**: 86–9.

Lisle, J. T., and J. C. Priscu. 2004. The occurrence of lysogenic bacteria and microbial aggregates in the lakes of the McMurdo dry valleys, Antarctica. *Microb. Ecol.* **47**: 427–39.

Los, M., G. Wegrzyn, and P. Neubauer. 2003. A role for bacteriophage T4*rI* gene function in the control of phage development during pseudolysogeny and in slowly growing host cells. *Res. Microbiol.* **154**: 547–52.

Lwoff, A. 1953. Lysogeny. *Bacteriol. Rev.* **17**: 269–337.

Mahenthiralingam, E. 2004. Gene associations in bacterial pathogenesis: pathogenicity islands, and genomic deletions. In R. V. Miller and M. J. Day (eds.), *Microbial Evolution: Gene Establishment, Survival, and Exchange.* Washington, DC: ASM Press, pp. 249–74.

Mann, N. H., A. Cook, A. Millard, S. Bailey, and M. Clokie. 2003. Bacterial photosynthesis genes in a virus. *Nature* **424**: 741.

Marsh, P., and E. M. H. Wellington. 1992. Interactions between actinophage and their streptomycete hosts in soil and the fate of phage borne genes. In M. J. Gauthier (ed.), *Gene Transfer and Environment.* Berlin: Springer-Verlag, pp. 135–42.

Martinez-Molina, E., and J. Olivares. 1979. Antibiotic production by *Pseudomonas reptilivora* as a phage conversion. *Can. J. Microbiol.* **25**: 1108–10.

McDaniel L., L. A. Houchin, S. J. Williamson, and J. H. Paul. 2002. Lysogeny in marine *Synechococcus. Nature* **415**: 496.

McKay, L. L., and K. A. Baldwin. 1973. Induction of prophage in *Streptococcus lactis* C2 by Ultraviolet irradiation. *Appl. Microbiol.* **25**: 682–4.

Middelboe, M. 2000. Bacterial growth rate and marine virus-host dynamics. *Microb. Ecol.* **40**: 114–24.

Miller, E. S., J. F. Heidelberg, J. A. Eisen, *et al.* 2003. Complete genome sequence of the broad-host-range vibriophage KVP40: comparative genomics of a T4-related bacteriophage. *J. Bacteriol.* **185**: 5220–33.

Miller, R. V. 1998a. Bacterial gene swapping in nature. *Sci. Am.* **278**: 66–71.

Miller, R. V. 1998b. Methods for enumeration and characterization of bacteriophages from environmental samples. In Burlage, R. (ed.), *Techniques in Microbial Ecology*. Oxford: Oxford University Press, pp. 218–35.

Miller, R. V. 2000. *recA:* the gene and its protein product. In S. Luria (ed.), *Encyclopedia of Microbiology*, 2nd edn, Vol. 4. San Diego, CA: Academic Press, pp. 43–54.

Miller, R. V. 2001. Environmental bacteriophage–host interactions: factors contributing to natural transduction. *Antonie van Leeuwenhoek* **79**: 141–7.

Miller, R. V. 2004. Bacteriophage-mediated transduction: an engine for change and evolution. In R. V. Miller and M. J. Day (eds.), *Microbial Evolution: Gene Establishment, Survival, and Exchange.* Washington, DC: ASM Press, pp. 144–57.

Miller, R. V. 2006. Marine phages. In R. Calendar and S. T. Abedon (eds.), *The Bacteriophages.* Oxford: Oxford University Press, pp. 535–44.

Miller, R. V. and M. J. Day. 2004. Horizontal gene transfer and the real world. In R. V. Miller and M. J. Day (eds.), *Microbial Evolution: Gene Establishment, Survival, and Exchange.* Washington, DC: ASM Press, pp. 173–7.

Miller, R. V., and C.-M.C. Ku. 1978. Characterization of *Pseudomonas aeruginosa* mutants deficient in the establishment of lysogeny. *J. Bacteriol.* **134**: 875–83.

Miller R. V., and S. Ripp. 1998. The importance of pseudolysogeny to *in situ* bacteriophage-host interactions. In M. Syvanen and C. I. Kado (eds.), *Horizontal Gene Transfer.* London: Chapman & Hall, pp. 179–91.

Miller, R. V., and S. A. Ripp. 2002. Pseudolysogeny: a bacteriophage strategy for increasing longevity *in situ*. In M. Syvanen and C. I. Kado (eds.), *Horizontal Gene Transfer*, 2nd edn. San Diego, CA: Academic Press, pp. 81–91.

Miller, R. V., and G. S. Sayler. 1992. Bacteriophage–host interactions in aquatic systems. In E. M.Wellington and S. J. D. Van Elsa (eds.), *Genetic Interactions Among Microorganisms in the Natural Environment.* Oxford: Pergamon Press, pp. 176–93.

Miller, R. V., J. M. Pemberton, and K. E. Richards. 1974. F116, D3, and G101: temperate bacteriophages of *Pseudomonas aeruginosa*. *Virology* **59**: 566–9.

Miller, R. V., S. Ripp, J. Replicon, O. A. Ogunseitan, and T. A. Kokjohn. 1992Virus-mediated gene transfer in freshwater environments. In Gauthier, M. J. (ed.), *Gene Transfers and Environment*. Berlin: Springer-Verlag, pp. 50–62.

Mitra, S. N., S. Kar, R. K. Ghosh, S. Pajni, and A. Ghosh. 1995. Presence of lysogenic phage in the outbreak strains of *Vibrio cholerae* O139. *J. Med. Microbiol.* **42**: 399–403.

Mittler, J. E. 1996. Evolution of the genetic switch in temperate bacteriophage. I. Basic theory. *J. Theor. Biol.* **179**: 161–72.

Moebus, K. 1987. Ecology of marine bacteriophages. In S. M. Goyal, C. P. Gerba, and G. Bitton (eds.), *Phage Ecology*. New York, NY: Wiley, pp. 136–56.

Moebus, K. 1996a. Marine bacteriophage reproduction under nutrient-limited growth of host bacteria. I. Investigations with six phages. *Mar. Ecol. Prog. Ser.* **144**: 1–12.

Moebus, K. 1996b. Marine bacteriophage reproduction under nutrient-limited growth ob host bacteria II. Investigations with phage-host system [H3;H3/1]. *Mar. Ecol. Prog. Ser.* **144**: 13–22.

Moebus, K. 1997. Investigations of the marine lysogenic bacterium H24. 2. Development of pseudolysogeny in nutrient rich broth. *Mar. Ecol. Prog. Ser.* **148**: 229–40.

Morita, R. Y. 1997. *Bacteria in Oligotrophic Environments: Starvation–Survival Lifestyle*. London: Chapman & Hall.

Mukhopadhyay, A. K., S. Chakraborty, Y. Takeda, G. B. Nair, and D. E. Berg. 2001. Characterization of VPI pathogenicity island and CTXϕ prophage in environmental strains of *Vibrio cholerae*.*J. Bacteriol.* **183**: 4737–46.

Muramatsu, K., and H. Matsumoto. 1991. Two generalized tranducting phages in *Vibrio parahaemolyticus* and *Vibrio alginolyticus*. *Microbiol. Immunol.* **35**: 1073–84.

Nida, S. K., and J. J. Ferretti. 1982. Phage influence on the synthesis of extracellular toxins in group A streptococci. *Infect. Immun.* **36**: 745–50.

Ogunseitan, O. A., G. S. Sayler, and R. V. Miller. 1990. Dynamic interactions of *Pseudomonas aeruginosa* and bacteriophages in lake water. *Microb. Ecol.* **19**: 171–85.

Ogunseitan, O. A., G. S. Sayler, and R.V. Miller. 1992. Application of DNA probes to the analysis of bacteriophage distribution patterns in the environment. *Appl. Environ. Microbiol.* **58**: 2046–52.

Ortmann, A. C., J. E. Lawrence, and C. A. Suttle. 2002. Lysogeny and lytic viral production during a bloom of cyanobacterium *Synechococcus* spp. *Microb. Ecol.* **43**: 225–31.

Paddison, P., S. T. Abedon, H. K. Dressman, *et al.* 1998. The roles of the bacteriophage T4*r* genes in lysis inhibition and fine structure genetics: a new perspective. *Genetics* **148**: 1539–50.

Pajni, S., N. R. Chowdhury, A. Ghosh, S. Kar, and R. K. Ghosh. 1995. Characterization of phage phi O139, a *Vibrio cholerae* O139 temperate bacteriophage with cohesive DNA termini. *FEMS Microbiol. Lett.* **131**: 69–74.

Paul, J. H., J. B. Rose, S. C. Jiang, P. London, X. Xhou, and C. Kellogg. 1997. Coliphage and indigenous phage in Mamala Bay, Oahu, Hawaii. *Appl. Environ. Microbiol.* **63**: 133–8.

Proctor, L. M., A. Okubo, and J. A. Fuhrman. 1993. Calibrating estimates of phage-induced mortality in marine bacteria: ultrastructural studies of marine bacteriophage development from one-step growth experiments. *Microb. Ecol.* **25**: 161–82.

Ptashne, M. 2004. *A Genetic Switch*, 3rd edn. Cold Spring Harbor, NY: Cold Spring Harbor Laboratories Press.

Rambler, M. and L. Margulis. 1979. An ultraviolet light induced bacteriophage in *Beneckea gazogenes*. *Orig. Life* **9**: 235–40.

Reiter, W. D., W. Zillig, and P. Palm. 1988. Archaebacterial viruses. *Adv. Virus Res.* **34**: 143–88.

Replicon, J., A. Frankfater, and R. V. Miller. 1995. A continuous culture model to examine factors that affect transduction among *Pseudomonas aeruginosa* strains in freshwater environments. *Appl. Environ. Microbiol.* **61**: 3359–66.

Ripp, S., and R. V. Miller. 1995. Effects of suspended particulates on the frequency of transduction among *Pseudomonas aeruginosa* in a freshwater environment. *Appl. Environ. Microbiol.* **61**: 1214–19.

Ripp, S., and R. V. Miller. 1997. The role of pseudolysogeny in bacteriophage–host interactions in a natural freshwater environment. *Microbiology* **143**: 2065–2070.

Ripp, S., and R. V. Miller. 1998. Dynamics of pseudolysogenic response in slowly growing cells of *Pseudomonas aeruginosa*. *Microbiology* **144**: 2225–32.

Ripp, S., O. A. Ogunseitan, and R. V. Miller. 1994. Transduction of a freshwater microbial community by a new *Pseudomonas aeruginosa* generalized transducing phage, UT1. *Mol. Ecol.* **3**: 121–6.

Robb, F. T., and R. T. Hill. 2000. Bacterial viruses and hosts: influence of culturable state. In R. R. Colwell and D. J. Grimes (eds.), *Nonculturable Microorganisms in the Environment*. Washington, DC: ASM Press, pp. 199–208.

Rocha E. P. C., A. Danchin, and A. Viari. 2001. The evolutionary role of restriction and modification systems as revealed by comparative genome analysis. *Genome Res.* **11**: 946–58.

Roger, A. 1983. Instability of host–virus association in *Acholeplasma laidlawiii* infected by a mycoplasma virus of the Gourlay group L1. *Zentralbl. Bakteriol. Mikrobiol. Hyg. (A)* **254**: 139–45.

Rohwer, F., A. Segall, G. Steward, *et al.* 2000. The complete genomic sequence of marine phage Roseophage SIO1shares homology with nonmarine phages. *Limnol. Oceanogr.* **45**: 408–418.

Romig, W. R., and A. M. Brodetsky. 1961. Isolation and preliminary characterization of bacteriophages of *Bacillus subtilis*. *J. Bacteriol.* **82**: 135–41.

Roszak, D. B., and R. R. Colwell. 1987. Survival strategies of bacteria in the natural environment. *Microbiol. Rev.* **51**: 365–79.

Saye, D. J., and R. V. Miller. 1989. The aquatic environment: consideration of horizontal gene transmission in a diversified habitat. In Levy, S. B., and Miller, R. V. (eds.), *Gene Transfer in the Environment*. New York, NY: McGraw-Hill, pp. 223–59.

Saye, D. J., O. Ogunseitan, G. S. Sayler, and R. V. Miller. 1987. Potential for transduction of plasmids in a natural freshwater environment: effect of plasmid donor concentration and a natural microbial community on transduction in *Pseudomonas aeruginosa*. *Appl. Environ. Microbiol.* **53**: 987–95.

Schrader, H. S., J. O. Schrader, J. J. Walker, *et al.* 1997a. Effects of host starvation on bacteriophage dynamics. In R. Y. Morita (ed.), *Bacteria in Oligotrophic Environments: Starvation–Survival Lifestyle*. New York, NY: Chapman & Hall, pp. 368–85.

Schrader, J. S., J. O. Schrader, J. J. Walker, T. A. Wolf, K. W. Nickerson, and T. A. Kokjohn. 1997b. Bacteriophage infection and multiplication occur *in Pseudomonas aeruginosa* starved for 5 years. Can.J. Microbiol. **43**: 1157–1163.

Smith, E., A. C. Wolters, H. Lee, J. T. Trevors, and J. D. van Elsas. 1996. Interactions between a genetically marked *Pseudomonas fluorescens* strain and bacteriophage ϕR2f in soil: effects of nutrients, alginate encapsulation, and the wheat rhizosphere. *Microb. Ecol.* **31**: 125–40.

Sullivan, M. B., J. B. Waterbyrn, and S. W. Chisholm. 2003. Cyanophages infecting the oceanic cyanobacterium *Prochlorococcus*. *Nature* **424**: 1947–51.

Sullivan, M. B., M. L. Coleman, P. Weigele, R. Rohwer, and S. W. Chisholm. 2005. Three *Prochlorococcus* cyanophage genomes: signature features and ecological interpretations. *PLoS Biol.* **3**: e144.

Sun, Y., X. Qiao, and L. Mindich. 2004. Construction of carrier state viruses with partial genomes of the segmented dsRNA bacteriophages. *Virology* **319**: 274–9.

Sutherland, I. W., K. A. Hughes, L. C. Skillwan, and K. Tait. 2004. The interaction of phage and biofilms. *FEMS Microbiol. Lett.* **232**: 1–6.

Suttle, C. A. 1993. Enumeration and isolation of viruses. In Kemp, P. F., Sherr, B. F., Sherr, E. B., and Cole, J. J. (eds.), *Handbook of Methods in Aquatic Microbial Ecology*. Boca Raton, FL: Lewis, pp. 121–34.

Suttle, C. A., and F. Chen. 1992. Mechanisms and rates of decay of marine viruses in seawater. *Appl. Environ. Microbiol.* **58**: 3721–9.

Thingstad, T. F., M. Heldal, G. Bratbak, and I. Dundas. 1993. Are viruses important partners in pelagic food webs? *Trends Ecol. Evol.* **8**: 209–12.

Thompson, B. J., E. Domingo, and R. C. Warner. 1980a. Pseudolysogeny of *Azotobacter* phages. *Virology* **102**: 267–77.

Thompson, B. J., M. S. Wagner, E. Domingo, and E. C. Warner. 1980b. Pseudolysogenic conversion of *Azotobacter vinelandii* by phage A21and the formation of a stably converted form. *Virology* **102**: 278–85.

Torsvik, T. and I. D. Dundas. 1980. Persisting phage infection in *Halobacterium salinarium* str. 1. *J. Gen. Virol.* **47**: 29–36.

Tran, T. A. T., D. K. Struck, and R. Young. 2005. Periplasmic domains define holin-antiholin interactions in T4 lysis inhibition. *J. Bacteriol.* **187**: 6631–40.

Tuomi, P., K. M. Fagerbakke, G. Bratbak, and M. Heldal. 1995. Nutritional enrichment of a microbial community: the effects on activity, elemental composition, and community structure and virus production. *FEMS Microbiol. Ecol.* **16**: 123–34.

Twort, F. W. 1915. An investigation on the nature of ultramicroscopic viruses. *Lancet* **ii**: 1241–3.

Val, M.-E., M. Bouvier, J Campos, *et al.* 2005. The single-stranded genome of phage CTX is the form used for integration into the genome of *Vibrio cholerae*. *Mol. Cell* **19**: 559–66.

Wall, J. D., P. F. Weaver, and H. Gest. 1975. Gene transfer agents, bacteriophages, and bacteriocins of *Rhodopseudomonas capsulata*. *Arch. Microbiol.* **105**: 217–24.

Weinbauer, M. G., and F. Rassoulzadegan. 2004. Are viruses driving microbial diversification and diversity? *Environ. Microbiol.* **6**: 1–11.

Weinbauer, M. G., and C. A. Suttle. 1996. Potential significance of lysogeny to bacteriophage production and bacterial mortality in coastal waters of the Gulf of Mexico. *Appl. Environ. Microbiol.* **62**: 4374–80.

Weinbauer, M. G., and C. A. Suttle. 1999. Lysogeny and prophage induction in coastal and offshore bacterial communities. *Aquatic Microb. Ecol.* **18**: 217–25.

Williams, S., A. Mortimer, and L. Manchester. 1987. Ecology of soil bacteriophages. In S. M. Goyal, C. P. Gerba, and G. Bitton (eds.), *Phage Ecology*. New York, NY: Wiley, pp. 136–56.

Williamson, S. J., M. R. McLaughlin, and J. H. Paul. 2001. Interaction of the ΦHSIC virus with its host: lysogeny or pseudolysogeny? *Appl. Environ. Microbiol.* **67**: 1682–8.

Wilson, W. H., and N. H. Mann. 1997. Lysogenic and lytic viral production in marine microbial communities. *Aquat. Microb. Ecol.* **13**: 95–100.

Wilson, W. H., N. G. Carr, and N. H. Mann. 1996. The effect of phosphate status on the kinetics of cyanophage infection in the oceanic cyanobacterium *Synechococcus* sp. WH7803. *J. Phycol.* **32**: 506–16.

Wilson, W. H., S. Turner, and N. Mann. 1998. Population dynamics of phytoplankton and viruses in a phosphate-limited mesocosm and their effect on DMSP and DMS production. *Estuar. Coast. Shelf Sci.* **46**: 49–59.

Wommack, K. E., and R. R. Colwell. 2000. Virioplankton: viruses in aquatic ecosystems. *Microbiol.Mol. Biol. Rev.* **64**: 69–114.

Zeidner, G., J. P. Bielawski, M. Shmoish, D. J. Scanlan, G. Sabehi, and O. Beja. 2005. Potential photosynthesis gene recombination between *Prochlorococcus* and *Synechococcus* via viral intermediates. *Environ. Microbiol.* **7**: 1505–13.

Part II Phage evolutionary biology

CHAPTER 6

Phage evolutionary biology

Siobain Duffy and Paul E. Turner*

6.1 INTRODUCTION

Evolutionary biology and microbiology, with the ushering in of the molecular revolution, developed a tenuous relationship (Woese, 1994). Further isolating these disciplines, once unified university biology departments split in two, with organismal versus molecular emphases. Because phage biologists were pioneers in molecular biology, they were placed on the molecular side of the divide along with the rest of microbiology, whereas evolutionary biologists, with their less reductionist approaches to biology, were grouped with researchers in zoology, botany, paleontology, and ecology (Rouch, 1997). Consequently, microbiologists were physically isolated from model organism researchers such as *Drosophila* evolutionary geneticists, and intellectually removed from discoveries such as rapid ecological radiations of wild populations and theories explaining biodiversity and speciation. Evolutionary biologists, in turn, were isolated from microbial experiments that bore on evolution, even though some historically significant discoveries in evolutionary biology used microbes, especially bacteriophages. Despite these past rifts, there exists a newfound appreciation for the power of using microbes to explore evolution and ecology: as molecular researchers had long realized, there are profound advantages to employing small, relatively simple, and rapidly replicating organisms as models for deciphering universal biological truths. Microbiologists, too, in this age of genomics, are increasingly aware of the crucial importance of ecology and evolution in their research. Within this new research climate, bacteriophages – the most numerous organisms on our

* Corresponding author

Bacteriophage Ecology: Population Growth, Evolution, and Impact of Bacterial Viruses, ed. Stephen T. Abedon. Published by Cambridge University Press.

planet (Wommack and Colwell, 2000; Hendrix, 2002) – have come to be recognized as crucial to our understanding of past, present, and future processes of evolution. This chapter furthers this ongoing synthesis by introducing the principles of evolutionary biology with examples from the phage literature, and by summarizing the current state of phage evolutionary biology.

This chapter also introduces the second emphasis of this monograph (Part II: Phage evolutionary biology), which consists of Chapters 6, 7, 8, and 9. Chapter 7 takes an opposite tack from that employed in this chapter, harnessing our understanding of phage genomics and applying this knowledge to elucidating principles of phage evolution. Chapters 8 and 9 then return to the theme of employing phages as model systems in evolutionary biology: Chapter 8 considers phage–phage interactions at or within individual bacteria while Chapter 9 provides an overview of experimental adaptation from the perspective of phages. See also Chapter 1 for additional discussion of phage evolutionary biology.

6.2 WHAT IS EVOLUTIONARY BIOLOGY?

Evolutionary biology is an incredibly broad field that bridges all other areas of biology, including phage biology. When public health microbiologists show that phages bearing toxin genes can turn benign bacteria into virulent pathogens, they demonstrate the evolutionary importance of horizontal gene transfer and sex in prokaryotes (Brüssow *et al.*, 2004; Chapters 11 and 14). When structural phage biologists note that the shape of the RNA-dependent RNA polymerase in the dsRNA bacteriophage φ6 is similar to that of the mammalian hepatitis C virus – even though the amino acid sequences show no homology – they note either remarkable convergent evolution or the robustness of polymerase shape through common descent (Butcher *et al.*, 2001; Chapter 7).

The decades of effort by molecular biologists that determined the sequences, functions, and structures of phage enzymes involved in DNA synthesis, repair, and protein production were useful to biologists outside the phage community, as both tools and concepts, because all biological entities on our planet are related through evolution. As Jacques Monod famously quipped about the preferred host used in much of this classic phage research, "What is true for *E. coli* is true for the elephant, only more so" (Monod and Jacob, 1961). Consequently, all biologists owe a debt to the phage experiments that revealed basic molecular underpinnings of the whole of biology, such as evidence that nucleic acid (not protein) is the hereditary material (Hershey and Chase, 1952) and elucidation of the general nature of the triplet genetic code underlying protein translation (Crick *et al.*, 1961). Even James Watson and Francis Crick, co-discoverers of the structure of DNA (1953), respectively were or were to become phage biologists.

6.2.1 Microevolution and population genetics

Evolution is change in the genetic makeup of a population that transcends the lifespan of single individuals (Futuyma, 1998). Evolution encompasses two levels of such change – change within a species (microevolution) and change among species, including the origin of new species (macroevolution). While nearly all of this chapter will deal with the processes and outcomes of microevolution, speciation is considered in Section 6.6.

As "population change over time" defines evolution, it is useful to discuss evolution in terms of changing genotype frequencies: population genetics. There are five mechanisms that account for changes in genotype frequencies within populations: mutation, recombination, natural selection, drift, and migration (Hartl and Clark, 1997). This chapter will detail the first four of these mechanisms and highlight the contributions of phage research to our understanding of these forces. The effects of migration on phage evolution (as opposed to the evolutionary ecology of source and sink habitats; Dennehy *et al.*, 2006) have been poorly explored in phages, and will not be discussed (see, however, the discussion of phage metapopulation ecology in Chapter 2 and Morgan *et al.*, 2005). After summarizing the canonical forces involved in population genetics, we discuss several more complicated evolutionary concepts, such as genotype-by-environment interactions (trade-offs, Section 6.5.1) and gene interactions (epistasis, Section 6.5.2).

6.3 CREATION OF VARIATION

The two mechanisms which can create genotypic variation are mutation (Section 6.3.1) and recombination (Section 6.3.2). Mutation changes an existing allele – a particular gene variant – into a different allele by changing its nucleotide sequence. Recombination usually changes which alleles exist in combination rather than immediately altering allele frequencies in a population. These new combinations of alleles are part of the variation among individuals in the population. Other evolutionary forces serve to change the frequencies of these variants over time (Section 6.4).

6.3.1 Mutation

It is simple to understand how mutation can change allele frequencies in a population. Mutation – through polymerase error, natural enzymatic

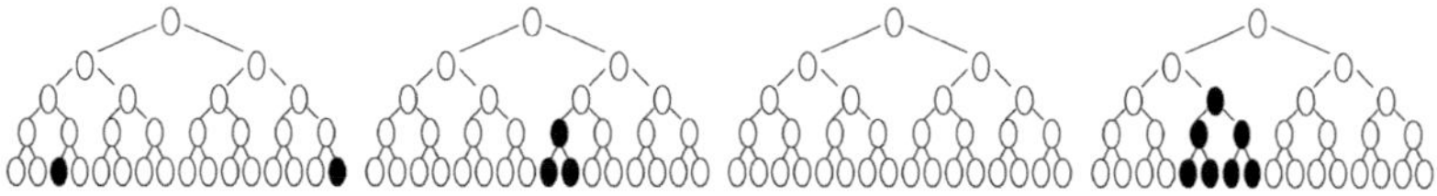

Figure 6.1 Jackpot distribution of mutations, after Luria and Delbrück (1943). The open ovals represent phage-susceptible bacteria; the filled ovals are phage-resistant bacteria. As the four populations grow (in the absence of phages), resistance mutations arise spontaneously and the trait is passed along to direct descendants (daughter cells), causing high variance in the frequency of resistant bacteria across populations (left to right these are 2, 2, 0, and 4 resistant bacteria).

processes, or environmental exposure to ultraviolet (UV) light or other mutagenic agents – results in a gene that differs in nucleotide sequence from its parent copy. In the early 1900s, geneticists realized that mutants could be created using X-rays and with newly discovered mutagenic chemicals. However, it was not clear how the mutations arose. Were mutations induced in an individual in response to an environmental challenge (that is, other than exposure to a mutagen)? Alternatively, were mutations continually produced, regardless of their usefulness in a given environment?

Luria and Delbrück (1943) were the first to demonstrate that mutations were random, and not induced as an adaptive response within individuals. Their research measured the frequency of *Escherichia coli* resistance to the bacteriophage T1 (then called α). If mutations to resistance were induced by contact with the phage, then roughly equivalent proportions of replicate cultures of phage-sensitive *E. coli* should form resistant colonies on agar plates containing an excess of phages. However, if mutations to T1-resistance occurred spontaneously and randomly within a growing culture prior to exposure to phage T1 on the plate, then there should be highly variable numbers of resistant colonies. Luria and Delbrück (1943) found the latter pattern, and deduced that mutation to phage resistance occurred late in some cultures, but very early on in others (see Fig. 6.1 for a schematic of the "jackpot" distributions of mutations).

Phages are still on the cutting edge of mutational research. Genomic mutation rates are unknown for most organisms, but phages have provided several such estimates (mostly due to work by John Drake and colleagues). Table 6.1 lists currently available mutation rates in phages, including two RNA bacteriophages, Qβ and ϕ6, which, due to their replication using lower-fidelity polymerases, have higher mutation rates than DNA phages (Drake, 1993). Investigating the mutation rate in different portions of a phage genome will help determine if the distribution of mutations in genomes is truly random, or if phages instead possess mutational "hot spots." Finally,

Table 6.1 *Mutation rates measured in bacteriophages.*

Phage	Genome	Mutation rate	Reference
Qβ	ssRNA	1.5×10^{-3}, 3.5×10^{-4}	Drake, 1993; Batschelet *et al.*, 1976
φ6	dsRNA	2.7×10^{-6}	Chao *et al.*, 2002
φX174	ssDNA	7.4×10^{-6}	Raney *et al.*, 2004
M13	ssDNA	7.2×10^{-7}	Drake, 1991
λ	dsDNA	7.7×10^{-8}	Drake, 1991
T2	dsDNA	2.7×10^{-8}	Drake, 1991
T4	dsDNA	2.0×10^{-8}	Drake, 1991

Mutation rates are compiled from studies determining the per nucleotide, per generation mutation rate. Omitted are the many studies on phage mutation frequency.

molecular biology and experimental evolution in phages has been invaluable in determining the interactions of mutations with different environments (trade-offs, Section 6.5.1; also Chapter 2) and with other mutations (epistasis, Section 6.5.2).

6.3.2 Recombination

Recombination is an important force in the evolution of both prokaryotes and eukaryotes, and was first experimentally demonstrated to consist of DNA-strand breaking and joining in phage λ (e.g., Meselson and Weigle, 1961; Campbell, 1962; Gellert, 1967). Recombination shuffles genetic material, changing allele combinations to ones that may or may not respond better to natural selection. The importance of recombination in phage evolution is covered in Chapter 7, while the role of recombination in evolution experiments is covered in Chapter 9. Here we describe the two basic forms of recombination, and how they pertain to phages.

Homologous recombination, as occurs among DNAs possessing highly similar sequences, is the most common consequence of molecular recombination. It can mix, within individual genomes, which alleles of gene A are physically joined with which alleles of gene B. Homologous recombination typically will occur between genomically similar phages upon coinfection. Coinfection can occur between two newly infecting phage genomes (Chapter 8), between a newly infecting phage and a resident prophage (Lawrence *et al.*, 2002; Chapter 7), or even among resident prophages (or their progeny) within polylysogenic bacteria (Ohnishi *et al.*, 2001).

Heterologous recombination occurs when gene exchange is asymmetrical. For example, templates can be switched between portions of a single phage's genome or between genomes of coinfecting phages, which yields gene (or gene-portion) duplication, insertion, or deletion (Mindich, 1995). Bacterial genes that integrate into phage genomes via illegitimate recombination are another important example of heterologous recombination (Chapter 7). The duplicate copies or regions of genes created by heterologous recombination can mutate and diverge from one another. This can allow one gene copy to take on a new adaptive function, while a functioning copy of the old gene is retained in the phage genome. The redundant copy, however, can also simply be inactivated into a pseudogene by mutation, or deleted from the genome entirely. These processes can lead to closely related phages with different genome sizes (e.g., Bollback and Huelsenbeck, 2001).

6.4 ACTION ON VARIATION

Genotypic variation in a population can underlie differences between individual phenotypes. If some of these phenotypes are fitter than others, then natural selection (Section 6.4.1) can increase the frequency of the alleles carried by the individuals of high fitness. In contrast, drift (Section 6.4.2) does not take the phenotypic effects of alleles into account. Drift explains frequency changes of alleles that have no effect on fitness (neutral alleles), and the change in the frequencies of all alleles when natural selection is not a strong force in a population. Both of these mechanisms act on heritable variation, that is, genetic variation caused by mutation or recombination which can be passed on to offspring. Both mechanisms can also lead to an allele becoming "fixed" (present in all or nearly all individuals) in a population.

6.4.1 Natural selection

Natural selection is the name given to the idea that individuals which are better adapted to the current environment than other members of a population, due to some genetically heritable phenotypic trait (i.e., individuals which are genetically fitter), will have more, and more successful, offspring than individuals that are less fit. Their offspring will, in turn, have greater reproductive success (more children will survive to have more children), and, absent bad luck (Section 6.4.2), the fitter genotypes consequently will increase in frequency over generations. Since this process of natural selection is key to understanding the evolution of organismal adaptation, a fundamental consideration in understanding the evolution of life (Darwin, 1859), we explicitly summarize the three criteria necessary for natural selection to occur within a population:

(1) heritable variation in a phenotypic trait
(2) variation in reproductive success (survival and fecundity)
(3) correlation between variation in the phenotypic trait and variation in reproductive success

The simplest example of natural selection acting within a phage system would be inoculating a bacterial culture with two otherwise identical obligately lytic phages, where one had a burst size of 100 and the other of 110. All else being equal, the phage with the larger burst size should be the majority genotype after a given period of time. In this very simple example, the more generations of phage growth that occur, the bigger numerical advantage – in both absolute and relative terms – for the phage with the larger burst size (see Fig. 15.2 A of Chapter 15). As introduced above, we would describe this more successful phage as fitter than the less successful, in this case smaller-burst-size, phage. This "fitness" is often defined as the relative number of grandchildren an individual produces, but more generally fitness is relative reproductive success: the differential ability for individuals within a population to survive and pass their genes to viable offspring, which in turn survive and reproduce (Futuyma, 1998).

Early phage researchers conducted countless selection experiments, adapting phages to new hosts, to resistant hosts, and to varying environmental conditions. Modern phage evolutionary biologists similarly have researched some selective conditions in great depth, adapting phages to different hosts (e.g., Crill *et al.*, 2000), different temperatures (e.g., Bull *et al.*, 2000; Poon and Chao, 2004; Knies *et al.*, 2006), different levels of coinfection (e.g., Turner and Chao, 1999), unnatural amino acids (Bacher *et al.*, 2003), and the presence of antisense RNA (Bull *et al.* 1998). Other selection experiments have allowed phages to recover fitness from very low values, imposed through population bottlenecking (Burch and Chao, 1999) or by gene deletions (Heineman *et al.*, 2005). Selection experiments are discussed further in Chapter 9 (see also Chapter 1).

6.4.1.1 Measuring fitness

To quantify fitness differences between bacteriophage genotypes, researchers typically measure the absolute or relative numbers of offspring after a given time period in a given environment. In the simplest example below, the fitness of phage X is the number of phage produced over a given period of time, t (X_t), given a starting quantity of phage at time 0 (X_0). Fitness is abbreviated as W (Futuyma, 1998).

$$W = \frac{X_t}{X_0} \tag{6.1}$$

To compare the fitness of phage X to another genotype, for example a common competitor, phage Y:

$$\text{Relative}\, W = \frac{W_X}{W_Y} \tag{6.2}$$

The selective advantage of one genotype over another is called the selective coefficient, s, and is often given as log(fitness).

There are many ways to measure the fitness of a phage. Published examples include absolute phage population growth rate in broth (e.g., Bull *et al.*, 2004), total phage production per plaque (Poon and Chao, 2005), phage population growth rate relative to a marked competitor phage (e.g., Duffy *et al.*, 2006), and plaque size (Burch and Chao, 1999; Schuppli *et al.*, 2000). Still others measure parameters that are only part of the phage life cycle, such as average burst size or attachment rate, which they use as surrogates for more comprehensive fitness measures (Pepin *et al.*, 2006). Fitness can be measured using culture-based methods (usually involving plating), or genotypically with techniques such as real-time qPCR (Edelman and Barletta, 2003), sequencing, and northern (or southern) blots (de la Peña *et al.*, 2000). There are advantages and disadvantages to all of these methods, though replicable methods that are robust to day-to-day environmental changes in the laboratory are preferred (Chapter 9 contains additional thoughts on fitness measurements).

6.4.1.2 Consequences of selection

Natural selection can only "see" the overall fitness of an individual (the phenotype), and is the process of more fit phenotypes leaving more descendants, thereby increasing the frequency in the population of whatever alleles these fit individuals carry. Because fitter individuals usually carry beneficial alleles, selection tends to increase the frequency of beneficial alleles. The classic example of natural selection is the "selective sweep," where a newly created genotype (created via mutation or recombination) replaces a previously dominant genotype within a population (Atwood *et al.*, 1951).

A sweep occurs when a beneficial genotype unequivocally has a more fit phenotype than the other phenotypes present (for an example in phage φX174, see Wichman *et al.*, 1999). The greater the fitness advantage (selective coefficient), the faster the genotype will increase in frequency. However, when two genotypes have equally fit phenotypes, selection cannot differentiate between them, and neither genotype can sweep through the population. This "clonal interference" can cause two or more genotypes (the "clones") to co-occur in a population over time (that is, each "interferes" with the other becoming fixed), until one genotype happens to fix stochastically.

An example of clonal interference in phages would be a population that consists of two different but equally fit phage clones that have mutated to overcome the phage resistance of their common host. Each clone has a random chance of becoming the fixed clone; the probability is proportional to the clone's existing frequency in the population. However, it is likely that more mutations will arise in one or both clonal backgrounds prior to this random fixation. If a genotype with a higher fitness arises, it will selectively sweep through the population, purging the population of the previously clonally interfering strains (for more information, see Chapter 9). Interested readers are encouraged to check the recent literature for experimental evolution research on the coliphages φX174 and MS2 (e.g. Bollback and Huelsenbeck, 2007).

If a population is adapting to a fixed environment – such as phages evolved in the laboratory on a naive strain of bacteria – then it is predicted that the first mutation fixed should be the most beneficial (greatest increase in s) of the beneficial mutations which the phage population happens to find by spontaneous mutation. The next mutation should then be the most fit of those sampled (happened upon by spontaneous mutation) in the genetic background of the genotype that had swept to fixation (Orr, 2005; see also Chapter 9). It is important to note that not all possible beneficial mutations will be fixed. A few of the reasons include: not all may be spontaneously found in the population; they may not be as beneficial as another mutation present; they may be clonally interfered with; or they may be lost by pure chance when there are few copies of the mutation in a population (i.e., soon after it has arisen). Even mutations possessing the greatest possible benefit could either fail to arise in a population or, if they do arise, could by chance go extinct prior to becoming fixed.

6.4.1.3 Neutrality and selection against deleterious alleles

Many, if not most, mutations are deleterious (Futuyma, 1998). Of the remainder, the majority appear to be neutral – they neither increase nor decrease organismal fitness. An example of a neutral mutation is a synonymous substitution in a protein-coding gene where the protein consequently produced is identical, and there is no codon bias influencing the speed with which the protein is translated. The remaining very small fraction of mutations, which are neither deleterious nor neutral, are beneficial in a given environment.

Deleterious mutations and alleles are selected against, causing their genotypes to leave fewer descendants over time. The speed with which a deleterious mutation disappears from the gene pool is proportional to the magnitude of its selection coefficient; that is, as for beneficial mutations (Section 6.4.1.2), selection acts most efficiently on harmful alleles of largest effect. As neutral mutations do not influence an individual's fitness, selection does not influence the allelic frequencies of these mutations.

Because experimental evolution of microbes (see Chapter 9) generally involves very large population sizes, natural selection overwhelms the effects of drift, and adaptation usually occurs through fixation of mutations that are beneficial in the treatment environment. However, mutations that are neutral or nearly neutral in the current environment are constantly being generated, and some might be adaptive in a different environment. The example of bacterial mutation to phage-resistance (Section 6.3.1) demonstrates this "exaptation" – a trait that is not adaptive in the current environment but which inadvertently anticipates a need in a future environment (Fig. 6.1). An exapted allele will increase in frequency if the population is placed in a new environment where it is beneficial. An additional example of an exaptation is a phage carrying a spontaneous mutation that can attach to an altered host receptor. The altered receptor is not yet present but subsequently evolves in the bacterial population as a means of phage resistance, only to be exploited by the exapted phage.

6.4.1.4 Levels of selection

This chapter has covered how natural selection acts on individual phenotypes to change allele frequencies in a population. But selection can act on other levels as well. For example, selection can act at the level of the gene, instead of at the level of organismal phenotype. A selfish gene promotes its own fitness by increasing its allele frequency at the encoding locus, other than by means of natural selection acting on the organism as a whole, or by replicating to create multiple copies of itself within the genome. By definition, these actions of a selfish gene are to the detriment of the overall individual organism, thus creating a conflict between selection acting at the level of the gene and selection at the level of the organism.

Conflicts between levels of selection arise within even relatively simple biological systems, such as bacteria and phages (Burt and Trivers, 2006). One level of conflict is that between a temperate phage and its host bacterium. Until the phage becomes lytic again, an integrated phage and its host share a common interest in bacterial ability to undergo rapid cell division (Chapters 5 and 14). The association allows efficient vertical transmission of the phage inside the host genome, and the prophage may simultaneously enhance bacterial fitness, for instance, by providing the host with immunity against superinfecting phages (Chapter 8).

As the host bacterium and lysogenizing phage have ultimately different fitness goals (more bacterial offspring versus more phage progeny), there can exist a conflict between the prophage and the bacterial genes within the host genome. Therefore, it is in the host's interest not to carry the potentially lethal parasite, and it should inactivate the prophage if possible (Lawrence *et al.*, 2001). Selection on phage fitness similarly will be acting in a different way than selection on bacterial fitness, and there may be separate selection acting on the fitness of individual genes. The conflicting selection pressures acting on the entire prophage versus genes within a prophage have been poorly articulated in the literature, but see Chapter 14, and also the analysis of Abedon and LeJeune (2005) on the selective pressures on a phage-encoded bacterial virulence factor.

Another example of differing levels of selection is a temperate phage that is parasitized by a transposon. The transposable element could move from the prophage and reproduce itself elsewhere in the host genome. The transposon and the alleles within it can thus improve their fitness by increasing transposon copy number, but this may harm the fitness of the bacterial host (e.g., through new insertions that inactivate genes), in turn potentially decreasing fitness of the lysogenized phage. Phage ϕR73, which appears to be a P4-like satellite phage parasitized by a retron (Ec73), is a real-life example of this complex situation (Inouye *et al.*, 1991). The retron is capable of independent replication and integration, but its impact on the ecology and evolution of phage ϕR73 and *E. coli* has not been investigated. This system (Ec73, ϕR73, and their *E. coli* host) is even more complicated, as P4-like phages are not capable of independent replication and require a genetically distinct helper phage such as phage P2 (Dehó and Ghisotti, 2006). The levels of selection conflicts arising between coinfecting phages are explored in Chapter 8.

6.4.2 Genetic drift

Drift was first described by Sewall Wright in his attempts to explain how populations may not always achieve their maximum fitness in a fixed environment (Hartl and Clark, 1997). He envisioned an adapting population on a landscape, a geometric plane representing genotype space, with peaks and valleys of fitness corresponding to the phenotype associated with each of those genotypes. In this adaptive landscape, selection would bring the population to the top of the closest peak, even if that were not the highest peak (most fit genotype) on the landscape. If the environment changes, then the location of the optimal peak may shift, perhaps allowing the population to ascend a higher peak via natural selection. For instance, if a phage is well adapted to a bacterium, but the bacterium becomes resistant to the phage, then the adaptive peak will have shifted from its previous location. The best phage host-range genotype currently in the population would be that closest to the top of this shifted adaptive peak, and therefore the most favored (most fit) genotype. This metaphor, known as the shifting balance theory of Wright's adaptive landscape, is a useful heuristic to explain how environmental and adaptive change can together cause populations to explore different fitness optima through time.

Environmental change is not the only way that populations can shift to new fitness optima. Wright was also the first to conceptualize a way for populations to meander (drift) down an adaptive peak, when drift becomes a more potent force than natural selection. Most easily understood as inbreeding depression, drift is strongest when population sizes are too small for natural selection to effectively alter allele frequencies. A small population may have too little genetic variation on which natural selection can act. Alternatively, there may be plenty of genetic variation, correlating with variation in reproductive fitness, but the population may continue to experience intermittent small population sizes ("bottlenecks"). The sampling error of individuals that survive these bottlenecks can overwhelm changes in genetic frequency due to selection. Bottlenecks thus can disrupt the correlation between a trait and reproductive success, and thereby weaken the effects of natural selection.

In the absence of natural selection, all mutations are neutral – even if they could be deleterious or beneficial in a different selective environment. Consequently, independent of selection all alleles have a probability of becoming fixed in the population that is equal to the frequency of the allele in its population. In the case of largely haploid organisms such as phages and bacteria, the probability of a new mutation (copy number $= 1$) *drifting* to fixation thus is $1/N$, the frequency of a new mutation in a population of size N.

The time needed to fix the mutation through drift is proportional to N_e, the effective population size (Hartl and Clark, 1997). Most microbial evolution experiments involve serial transfer, where a propagated population fluctuates in size between a maximum value and the bottleneck (minimum population size) imposed by the researcher. As N_e equals the harmonic mean of population size over the course of these fluctuations, not the arithmetic mean, then even infrequent small bottleneck sizes disproportionately lower N_e. That is, for means of population size,

$$N_e = \text{harmonic mean} \leq \text{arithmetic mean} \quad (6.3)$$

Represented mathematically this is

$$N_e = n/(1/N_1 + 1/N_2 + \Lambda + 1/N_n) \leq (N_1 + N_2 + \Lambda + N_n)/n \quad (6.4)$$

where n is the number of measurements of population size, N_x is the population size as measured at time or generation x, and $x = 1, 2, 3, \ldots, n$. From Equation 6.4 we can illustrate the impact of smaller numbers on harmonic means versus arithmetic means: for example, for the list 1, 5, 5, 5 the harmonic mean is 2.5 while the arithmetic mean is 4. Small population bottlenecks consequently have a disproportionate influence on effective population sizes, resulting in strong effects of drift relative to those of natural selection.

6.4.2.1 Muller's ratchet

Because sexual (genetic exchange) mechanisms are very common in biological populations, evolutionary biologists have continually asked the question (e.g., Chapters 8 and 9): what is the adaptive significance of sex? One classic hypothesis is that sex evolved to combat 'Muller's ratchet', the buildup of deleterious mutations in small populations (Muller, 1964). In asexual populations of small size there is a tendency for slightly deleterious mutations to accumulate; mutation-free individuals become rare and they can be easily lost due to drift. Asexual populations can only reconstitute mutation-free individuals through reversions or compensatory mutations, which are less likely the smaller the population. Hence the term Muller's ratchet, because of the similarity to a ratchet tool that only clicks forward and cannot move in reverse. In contrast, sex can stop or slow down Muller's ratchet because it allows two parents to exchange their genetic material and produce offspring that contain fewer harmful mutations.

Muller's ratchet was first demonstrated empirically with the RNA bacteriophage φ6 (Chao, 1990). As phage plaques can be initiated with a single virus particle, plaque-to-plaque serial transfers of phages can be used to force an evolving population through bottlenecks consisting of a single individual ($N = 1$). Chao (1990) passaged 20 replicate lineages of φ6 through 40 such bottlenecks, and observed that all 20 populations evolved to be less fit than the wild type ancestor (6–71% drop in fitness). Clearly, drift rather than natural selection drove the fitness changes. This result was also shown using another RNA phage, MS2 (de la Peña *et al.*, 2000). Chao (1990) noted that asexual organisms can avoid Muller's ratchet if they acquire compensatory mutations (discussed in detail in Section 6.5.2.1) sufficiently often. Other phage studies have examined hypotheses for the evolutionary advantage of sex in populations of large size, such as the ability for sex to bring together multiple deleterious alleles so that selection can more efficiently purge deleterious alleles from a population (Malmberg, 1977; Froissart *et al.*, 2004).

6.5 MUTATIONS AND THEIR ENVIRONMENT

Mutations do not exist in a vacuum; they arise in a genome containing other loci, and are expressed differently depending on the environment. Mutations therefore can interact with other loci within a genotype, and the same mutation can be beneficial in one environment and deleterious in another. These complex relationships are explored below.

6.5.1 Trade-offs

Trade-offs occur when an increase in fitness in one aspect of an organism's phenotype correlates with a decrease in fitness in another aspect of an organism's phenotype. Classic examples of evolutionary trade-offs occur in life-history traits such as age of first reproduction and the number of offspring produced, analogous to phage latent period and burst size (Chapters 2 and 3). Briefly, a phage can improve its reproduction in two ways: by increasing its burst size to produce many progeny per round of replication, or by shortening its latent period to more rapidly initiate the next cycle of replication. There is an assumed trade-off underlying the two life-history strategies: if a phage maximizes its burst size, it must lengthen its latent period to allow sufficient time to make a large number of progeny. If a phage minimizes its latent period (approaching the lower limit of the eclipse period), then it cannot produce as large a burst.

The genetic basis for trade-offs has been a fruitful area in phage evolution research. There are two genetic mechanisms for trade-offs, antagonistic pleiotropy, where the mutation conferring the benefit in one dimension of the phenotype necessarily is deleterious in another, and accumulation of conditionally deleterious alleles (Elena and Lenski, 2003). In the second, neutral or nearly neutral mutations accumulate in genotypes in one environment, either by drift or by hitchhiking with a beneficial mutation in a selective sweep. These neutral mutations can have deleterious effects in another environment, and therefore the phage shows the same trade-off between improved fitness

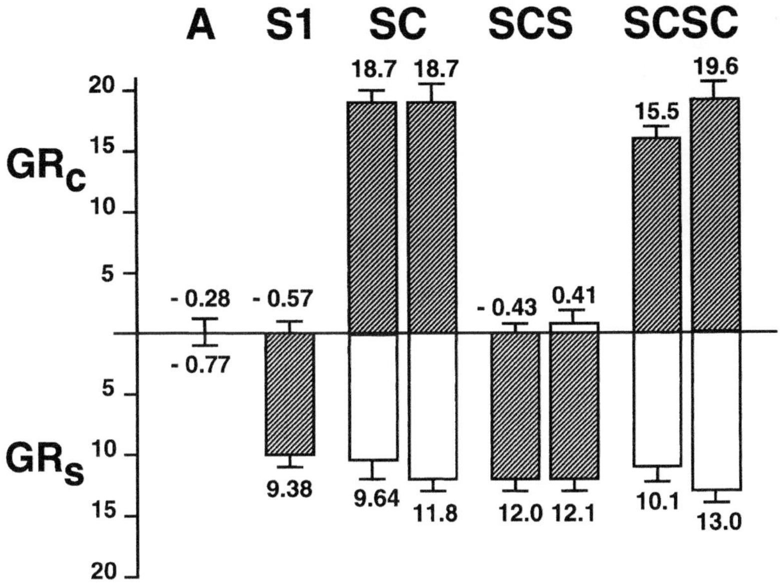

Figure 6.2 Fitness ($\log_2$ growth rate, "GR") of two populations of phage ϕX174 on *E. coli* (C) and *S. enterica* (S) over the course of a chemostat evolution experiment where phages evolved on a single bacterial host at a time. The first adaptation to *S. enterica* (S1) did not cause a further loss in fitness on *E. coli* relative to the ancestor (A). Similarly, when S1 was evolved on *E. coli*, fitness increased on *E. coli* without a loss of fitness on *S. enterica* (SC), indicating no trade-off in adaptive performance on *E. coli*. However, trade-offs were seen when the populations were evolved again on *S. enterica* (SCS). (Reprinted with permission from Crill *et al.*, 2000 [*Genetics* **154**: 27–37], (c) 2000, Genetics Society of America.)

in the selected environment and decreased fitness in an unselected environment. However, different mutations are responsible for the different fitness effects in the two environments.

Short-term experimental evolution studies in phages have revealed that most but not all mutations affecting host range (Duffy *et al.*, 2006; Ferris *et al.*, 2007) and virion-to-host attachment rate (Pepin *et al.*, 2006) show antagonistic pleiotropy. An example of fitness changes due to an unknown number of mutations in phage ϕX174 is shown in Fig. 6.2. Mutations that increased ϕX174 fitness on *E. coli* did not cause a reduction in fitness on *Salmonella enterica*, but ϕX174 adaptation to *S. enterica* showed a fitness trade-off on *E. coli*. After adapting further to *E. coli*, both populations regained fitness on *E. coli* without losing fitness on *S. enterica*. It is therefore possible that the mutation conferring the trade-off was not antagonistically pleiotropic, but instead was a hitchhiking, conditionally deleterious mutation, which was lost by both populations during further adaptation to *E. coli*. Alternatively, the mutation might have been antagonistically pleiotropic, and the fitness loss on *E. coli* may have been compensated by additional mutational change (Section 6.5.2.1). Sequencing data alone could not rule out either explanation, but direct molecular tests of one reversion mutation indicated that it was antagonistically pleiotropic for attachment rate, a component of fitness (Crill *et al.*, 2000).

No phage studies have yet confirmed the evolution of conditionally deleterious mutations, but with further analysis of some very long-term evolution experiments (Wichman *et al.*, 2005), their prevalence in phages may be determined. Small DNA phages like φX174 offer the best opportunity to test the effects of single mutations in multiple environments, as evolved mutations can be placed into the ancestral genetic background with site-directed mutagenesis (Crill *et al.*, 2000; Pepin *et al.*, 2006). This combination of molecular tools and rapid evolution also makes phages ideal for studying another kind of interaction between mutations, epistasis.

6.5.2 Epistasis

Understanding mutation and adaptation would be easier if biology followed the simplifying assumptions necessarily used in theoretical models that attempt to mathematically describe the complexities of the real world (e.g., see Chapters 2, 10, 15, 16, and 17). It would be trivial to understand mutational effects if they were absolutely beneficial or deleterious in all environments, for example, but there exist obvious and strong genotype-by-environment interactions (Section 6.5.1). In the same vein, it would be convenient if mutations had the same magnitude of effect (in a fixed environment) in all genetic backgrounds, but this is not the case (Weinreich *et al.*, 2005). The contingent fitness or phenotypic effect of one mutation on another mutation is called epistasis.

Epistasis can cause a population to be trapped on a smaller peak in an adaptive landscape, preventing it from scaling a larger peak: this results from the number of mutational changes necessary for drift to move the population away from the current peak, which can be higher given gene-to-gene phenotypic interactions (Poon and Chao, 2005). Epistasis can also help reduce the fitness cost of mutational load within an individual genotype. Mutations to drug resistance in the human immunodeficiency virus (HIV) can exhibit "antagonistic" epistasis (also described as "diminishing-returns" epistasis) where two harmful mutations are less deleterious in combination than expected by their individual effects. This result explains the unexpectedly

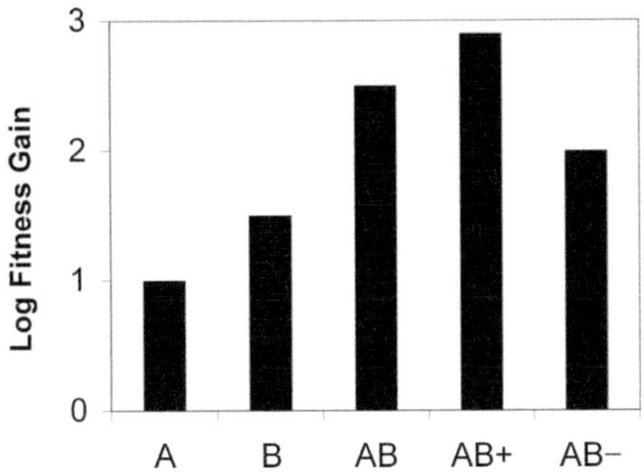

Figure 6.3 A schematic of epistatic interactions between two beneficial mutations. Mutation A has a fitness effect of 1 unit, mutation B has a fitness effect of 1.5 units. If these mutations act independently, then a genotype with both A and B mutations has a fitness improvement of 2.5 (AB). If the mutations interact synergistically through positive epistasis (AB+), then the fitness benefit is greater than 2.5. If the mutations interact antagonistically (AB–, meaning diminishing returns or, among beneficial mutations, negative epistasis), then their combined effect is less than 2.5.

high fitness of multi-drug-resistant HIV strains that hamper efforts to control the AIDS pandemic (Bonhöffer *et al.*, 2004).

A hypothetical example of epistasis is shown in Fig. 6.3. If two beneficial mutations interact synergistically ("synergistic" epistasis; Fig. 6.3, AB+), then their effects are greater together than one would expect from their separate effects (which in the absence of epistasis would be exactly additive on a selection coefficient, or log(fitness) scale, i.e., Fig. 6.3, AB). If two deleterious mutations interact synergistically then their combined fitness effect is worse than expected. The opposite of synergistic epistasis is antagonistic epistasis, where two beneficial mutations are not as beneficial in combination (Fig. 6.3, AB–), or two deleterious mutations are not as deleterious in combination. The reader should be warned, however, that different papers on epistasis use different terms for these various concepts of epistasis.

Phages offer many advantages for the study of epistasis. In addition to the well-established biology and molecular tools (such as site-specific mutagenesis) of a phage like ϕX174, the small genome sizes of some phages allow frequent whole-genome sequencing to assess the number of mutations under study). Antagonistic (diminishing-returns) epistasis has been demonstrated for beneficial mutations in phage ϕX174 (Bull *et al.*, 2000), for deleterious mutations in phage ϕ6 (Burch and Chao, 2004), and *in silico* for phage T7 (You and Yin, 2002); phage T7 has experimentally been shown to exhibit epistasis as well (Bull *et al.*, 2003). Synergistic epistasis has been observed between certain deleterious mutations in phage ϕ6 (Froissart *et al.*, 2004), among beneficial mutations in Qβ (Schuppli *et al.*, 2000), and in *Pseudomonas fluorescens* when coevolving with a bacteriophage (Buckling *et al.*, 2006). Lack of epistatic interactions between deleterious mutations has been found in phage MS2 (de la Peña *et al.*, 2000).

6.5.2.1 Compensatory mutations

A classic example of epistasis is compensatory mutation. Compensatory mutations are beneficial in genotypes that have a specific deleterious mutation, but are neutral or even deleterious in more fit genotypes. They are thought to be more likely to arise and fix than a reversion of the deleterious mutation itself, though reversion mutations have been observed in phage experimental evolution studies (Crill *et al.*, 2000). In the phage molecular genetics literature, compensatory mutations may instead be described as *pseudo* reversions (e.g., Nelson *et al.*, 1982; Klovins *et al.*, 1997). Compensatory mutations can arise within the same gene or in other parts of the genome. A compensatory mutation arising within a damaged allele can be understood intuitively, because it allows the protein to stabilize itself. A study in ϕX174 showed an equal number of intra- and extragenic compensatory mutations (within and external to the mutation-containing gene, respectively; Poon and Chao, 2005), and experiments involving gene deletion can only compensate for the deleterious mutation with extragenic changes (e. g., Nelson *et al.*, 1982; Heineman *et al.*, 2005).

Compensatory mutations could also be selected to stabilize single-stranded phage genomes to make up for a deleterious mutation affecting genomic secondary structure. For example, when most of a non-coding stem-loop structure was deleted from phage MS2, the phage compensated either with two insertion mutations, which re-created the structure, or by evolving a further deletion, which sacrificed the remaining stem-loop structure but reconstructed the ribosome binding sequence (Olsthoorn and van Duin, 1996; see also the follow-up study by Klovins *et al.*, 1997, and further work in MS2 by Licis and van Duin, 2006). A similar example of intergenomic compensation (really coevolution of co-replicating genetic entities) has been found in phage Qβ, which repeatedly mutated its 3' sequence to get rid of secondary structure to compensate for the deletion of a host factor which otherwise disrupts phage internal base pairing (Schuppli *et al.*, 1997).

The fitness effects of compensatory mutations can be small, or large enough to restore fitness to the level of the mutation-free genotype (Burch and Chao, 1999), but the confounding problem is that it is difficult to distinguish between a compensatory mutation and a universally beneficial mutation arising in a mutation-bearing genotype. Again, the molecular tools available for phages provide experimentalists with the means to test whether beneficial mutations are compensatory, and therefore conditional on the presence of the deleterious mutation. Poon and Chao (2005), for example, tested five presumptive compensatory mutations and found that indeed they were either neutral or had a reduced fitness in the deleterious-mutation-free genotype, thereby proving their epistatic effect. Future studies following this pattern of testing mutations in the ancestral background would enhance our understanding of the frequency and importance of compensatory mutation.

6.5.3 Evolvability and robustness

Two other fertile areas of research concern the interaction between a new mutation and the rest of the genome: evolvability and robustness. Evolvability is the ability of an organism to gain fitness in a given environment or, more specifically, the potential to acquire new beneficial mutations or traits (Kirschner and Gerhart, 1998). Robustness, by contrast, describes the constancy of phenotype (including fitness) in the face of genomic mutations (genetic robustness) or changing environments (environmental robustness; de Visser *et al.*, 2003; Montville *et al.*, 2005). Evolvability fosters the rapid ability to change, innovate, and adapt, though in principle it also is susceptible to the much more frequent deleterious mutations that arise and affect phenotype. Genetically robust organisms suppress the effects of deleterious mutations, and presumably also suppress the more rare beneficial mutations. These mutations – potentially beneficial or deleterious in less robust organisms – fix by drift since they have no phenotypic effect (in the more robust organism) upon which natural selection can act. The relationship between evolvability and robustness is still under investigation (Kirschner and Gerhart, 1998; Kitano, 2004). As there are advantages and disadvantages to both strategies (high evolvability versus high robustness), it is an open question how genetically robust a genome, or any portion of a genome, should be. We can speculate, however, that greater genetic robustness would be beneficial given exposure to more, potentially disruptive, mutations whereas greater evolvability could be beneficial, at least in the short term, given exposure to a novel environment to which an organism is poorly adapted.

Experimental work suggests that phages are quite genetically robust. Mutagenesis has revealed the robustness of the phage λ regulatory circuitry (Little *et al.*, 1999). It was recently shown that phage genotypes with more frequent gene exchange ("sexually" evolved) are less robust than those evolved asexually (Montville *et al.*, 2005; Chapter 8). Simulation studies have also indicated that phage T7's growth rate is robust to mutational change (You and Yin, 2006). Phages nonetheless are capable of rapid adaptation: they display significant evolvability. Experimental evolution with phage ϕ6 showed that the limitation on evolvability is the one-step mutational neighborhood (i.e., what mutations are easily acquired), and perhaps the epistatic effects (Section 6.5.2) of those mutations (Burch and Chao, 2000). The relative importance of evolvability and robustness is a question of interest in all organisms, but it is perhaps most easily studied in rapidly evolving systems of small genome size, such as bacteriophages.

6.6 DIVERGENCE OF PHAGE POPULATIONS

In the laboratory, phages have rapid generation times and large population sizes, which together promote rapid evolution (Malmberg, 1977; Studier, 1980). If there are multiple adaptive solutions to a common environmental challenge, then replicate populations independently evolving in the same habitat can diverge from one another. RNA phages additionally can evolve rapidly due to their characteristically high mutation rates (Table 6.1). This outcome can also be matched in DNA phages by using mutagenic chemicals such as nitrosoguanidine (e.g., Bull *et al.*, 1993). Phages therefore are powerful systems for studying how drift or divergent selection can cause populations to become very different from one another.

6.6.1 Inadvertent divergence

Standard pure-culture technique attempts not only to minimize contamination but also to minimize strain evolution, but these efforts are not always successful, resulting in divergence among laboratory-propagated strains. One example of this inadvertent divergence was that observed in Max Delbrück's lab-derived phage T7 stocks, which were subsequently maintained in different laboratories. Studier (1979, 1980) used restriction fragment analysis to show that half of the phage T7 stocks contained deletions or point mutations compared to the canonical Delbrück T7, and differed from this original genotype in phenotypic traits, such as plaquing behavior. This population divergence, which occurs readily in the lab, can lead to phage "speciation," which we consider in the next section.

6.6.2 Phage "speciation"

Rapid generation time and ease of conducting laboratory selection experiments are appealing features for studying population divergence. If population divergence leads to barriers against gene flow, then populations may become reproductively isolated, even if they are not spatially separated. Distinct populations with little or no gene flow can be described as separate species, and thus the evolution of reproductive isolation is the link between microevolution (change within a population) and macroevolution (divergence between reproductively isolated populations; Futuyma, 1998).

Phages, unfortunately, do not conform to any species concept developed for diploid eukaryotes. There are, however, identifiable units of phages (Lawrence *et al.*, 2002). Genetic divergence among phages gives rise to detectable barriers to hybridization (e.g., phage T2 and T4; Gary *et al.*, 1998), and phage biologists are researching the mechanisms which create and reinforce those barriers (Silander *et al.*, 2005). While a common criterion for defining species boundaries is a virtual absence of gene flow across those boundaries (Kirkpatrick and Ravigné, 2002), some genetic exchange between apparent phage "species" still occurs (Lawrence *et al.*, 2002; Rokyta *et al.*, 2006). In fact, over large time scales there appear to be few insurmountable barriers to gene flow among dsDNA phages as a whole, although gene flow predominantly occurs between phages which infect the same host (Hendrix *et al.*, 1999). Barriers to gene flow can arise either from coinfection (Sections 6.6.3 and 6.6.4) or from recombination incompatibilities (Section 6.6.5).

6.6.3 Host shifting

In the laboratory, bacteriophage host range can readily expand and contract in response to exposure to different host strains, and presumably this expansion and contraction in phage host range happens often in nature as well. Host-range alteration has obvious ecological benefits, either in expanding the resource (i.e., host) niche or in optimizing phage exploitation of select host types (specialization; Levins, 1968). As a phage can only exchange genetic material with another phage during coinfection, changes in host range can reduce gene flow between phages, resulting in divergence between the now partially reproductively isolated phage populations (Gary *et al.*, 1998). Phage host-shifting can thus result in two phage strains no longer sharing any of the same hosts (Duffy *et al.*, 2007), and subsequent genetic divergence between those phage strains can ultimately lead to the formation of new phage "species."

Phage host-shifting usually requires changes in the proteins a phage uses for host attachment, and therefore genes for host-recognition proteins often experience much faster rates of evolution than other genes. This has been experimentally observed in many phages – e.g., T-even (Montag *et al.*,

1987; Hashemolhosseini *et al.*, 1994), ϕX174 (Crill *et al.*, 2000) – and is seen between natural phage isolates. For example, in geographically distinct tectivirus isolates from around the world, over 37% of the differences were in the host-attachment protein (called P2 in PRD1, the type phage of the group), which comprises only 12% of the protein-coding region in the genome (statistic based on our analysis of Saren *et al.*, 2005).

6.6.4 Phage–host coevolution can speed divergence

Bacteria, as noted, can mutate to phage resistance via loss or change of phage receptors, but bacteria also can acquire post-adsorption resistance mechanisms (Chapter 1). As a phage coevolves to maintain infection of the hosts in its host range in the face of host evolution toward phage resistance, it should be less able to respond to the selection pressures of another host evolving different resistance mechanisms (Kawecki, 1998). Consequently, phages may specialize in a relatively narrow range of hosts and become caught in an arms race where the bacterium continually evolves resistance to the phage and the phage continually evolves to surmount each new mechanism of bacterial resistance. This is a microbial example of the "red queen hypothesis" (Van Valen, 1973), the concept of continually evolving to maintain the same niche, which recalls the Red Queen in Lewis Carroll's *Through the Looking Glass* (1872), who stated "It takes all the running you can do, to keep in the same place." Coevolution can therefore speed population divergence and could be the primary selective pressure driving phage specialization and fission of a single population gene pool into separate components. See Chapters 2 and 10 for consideration of phage–host coevolution from more ecological perspectives.

6.6.5 Recombination and barriers to gene flow

Mutations elsewhere in the phage genome can also determine the amount of gene flow that can occur between coinfecting phages. For example, sequence homology in surrounding genes has been shown to determine the rate of recombination between phage T2 and T4, whereas more divergent sequences greatly inhibited T4's acquisition of a T2 gene (Gary *et al.*, 1998). In another example, epistatic interactions among mutations accumulated in divergent populations of phage T7 passaged on the same *E. coli* host caused recombinant progeny to be of lower fitness (Bull *et al.*, 2003). This hybrid incompatibility (reduced viability or fecundity of recombinant offspring) generally is theorized to lead to reinforcement of barriers to gene flow (Hartl and Clark, 1997) and therefore to speciation. This would correspond to a "postzygotic" barrier to gene flow, unlike the other mechanisms of reproductive isolation mentioned above (Futuyma, 1998).

6.7 FUTURE DIRECTIONS

In this age of whole-genome sequencing and proteomics, biologists are poised to connect genotype and phenotype in very absolute terms. There are few organisms as streamlined as bacteriophages, which can have vanishingly small genomes (some less than 4 kb), and which often are solely composed of nucleic acid and protein. The expression and structure of those proteins are often well understood due to the rigorous molecular work conducted with a few bacteriophages of interest. As a harbinger of these trends, the first genomes ever sequenced were those of phages MS2 (Fiers *et al.*, 1976) and ϕX174 (Sanger *et al.*, 1977), both of which now serve as important model organisms in evolutionary research (Wichman *et al.*, 1999, 2005; Rokyta *et al.*, 2005; Bollback and Huelsenbeck, 2007).

Evolutionary research with bacteriophages extends this elegantly simple relationship between genotype and phenotype through time and over entire adaptive landscapes. It seems likely that the first adaptive landscape which could be mapped would be that of a bacteriophage of small genome size grown in a controlled laboratory environment (Rokyta *et al.*, 2005; see Chapter 9). Whether adaptation to one portion of a niche causes a trade-off in the rest of the niche is being investigated in both discrete environments such as different bacterial hosts and over continuous environments such as temperature (Bull *et al.*, 2000; Holder and Bull, 2001; Knies *et al.*, 2006). Many such studies with phage other than ϕX174 and its relatives are forthcoming in the literature, including RNA phages MS2 and ϕ6.

Laboratory coevolution of phages and their hosts is another current trend in the field. Chao *et al.* (1977) and Lenski and Levin (1985) studied the first rounds of the arms race between phage T7 and *E. coli* B, but were limited by *E. coli* B's effective ability to escape the phage through certain mutations. More recent studies (Buckling and Rainey, 2002; Morgan *et al.*, 2005), as well as work yet to be published on marine phage–host coevolution, have had more success – and this should prove productive both for phage biologists and for testing coevolutionary theory.

6.8 CONCLUDING REMARKS

It should be explicitly noted that while this chapter has focused solely on bacteriophage evolutionary biology, there are many analogous results that have come from research on viruses of eukaryotes. In addition, as bacteria are haploid, typically asexual microbes, their experimental evolution results are often informative to the study of phage experimental evolution. One of the hallmarks of evolutionary biology is the synthesis of work from different taxonomic groups, and experiments in other microbes should not be overlooked in phage papers (e.g., Chapter 9 has many examples drawn from eukaryotic virus and bacterial work).

The only major caveat in phage evolutionary biology is that, at its best, it requires researchers who understand the field of evolutionary biology *and* who understand the physiology and genetics of their bacteriophage and its host bacterium. While many evolutionary biologists are drawn to phage research because of the organism's simplicity, too often the complexities of the phage are ignored. Phage research was immeasurably advanced by physicists turned novice biologists, but it is precisely the subsequent sixty years of detailed research into phage biochemistry and structural biology that should attract evolutionary biologists to these organisms (e.g., Chapters 1 and 7). This is an emerging field of biological research, and there are still myriad questions, as well as many opportunities both to explore the evolution of bacteriophages themselves and to apply what is learned to the problems of evolutionary biology in general.

ACKNOWLEDGMENTS

S.D. was supported by a predoctoral fellowship from the Howard Hughes Medical Institute. This work was supported by the Woodrow Wilson Foundation and the US National Science Foundation grant DEB-04—52163 to P.E.T.

REFERENCES

Abedon, S. T., and J. T. LeJeune. 2005. Why bacteriophage encode exotoxins and other virulence factors. *Evol. Bioinf. Online* **1**: 97–110.

Atwood, K. C., L. K. Schneider, and F. J. Ryan. 1951. Periodic selection in Escherichia coli. *Proc. Natl. Acad. Sci. U.S.A.* **37**: 146–55.

Bacher, J. M., J. J. Bull, and A. D. Ellington. 2003. Evolution of phage with chemically ambiguous proteomes. *BMC Evol. Biol.* **3**: 24.

Batschelet, E., E. Domingo, and C. Weissmann. 1976. The proportion of revertant and mutant phage in a growing population, as a function of mutation and growth rate. *Gene* **1**: 27–32.

Bollback, J. P., and J. P. Huelsenbeck. 2001. Phylogeny, genome evolution and host specificity of single-stranded RNA bacteriophage (family Leviviridae). *J. Mol. Evol.* **52**: 117–28.

Bollback, J. P., and J. P. Huelsenbeck. 2007. Clonal interference is alleviated by high mutation rates in large populations. *Mol. Biol. Evol.* **24**: 1397–406.

Bonhöffer, S., C. Chappey, N. T. Parkin, J. M. Whitcomb, and C. J. Petropoulos. 2004. Evidence for positive epistasis in HIV-1. *Science* 306: 1547–50.

Brüssow, H., C. Canchaya, and W.-D. Hardt. 2004. Phages and the evolution of bacterial pathogens: from genomic rearrangements to lysogenic conversion. *Microbiol. Mol. Biol. Rev.* **68**: 560–602.

Buckling, A., and P. B. Rainey. 2002. Antagonistic coevolution between a bacterium and a bacteriophage. *Proc. Biol. Sci.* **269**: 931–6.

Buckling, A., Y. Wei, R. C. Massey, M. A. Brockhurst, and M. E. Hochberg. 2006. Antagonistic coevoltuion with parasites increases the cost of host deleterious mutations. *Proc. Biol. Sci.* **273**: 45–9.

Bull, J. J., C. W. Cunningham, I. J. Molineux, M. R. Badgett and D. M. Hillis. 1993. Experimental molecular evolution of bacteriophage T7. *Evolution* **47**: 993–1007.

Bull, J. J., A. Jacobson, M. R. Badgett, and I. J. Molineux. 1998. Viral escape from antisense RNA. *Mol. Microbiol.* **28**: 835–46.

Bull, J. J., M. R. Badgett, and H. A. Wichman. 2000. Big-benefit mutations in a bacteriophage inhibited with heat. *Mol. Biol. Evol.* **17**: 942–50.

Bull, J. J., M. R. Badgett, D. Rokyta, and I. J. Molineux. 2003. Experimental evolution yields hundreds of mutations in a functional viral genome. *J. Mol. Evol.* **57**: 241–8.

Bull, J. J., M. R. Badgett, R. Springman, and I. J. Molineux. 2004. Genome properties and the limits of adaptation in bacteriophages. *Evolution* **58**: 692–701.

Burch, C. L., and L. Chao. 1999. Evolution by small steps and rugged landscapes in the RNA virus φ6. *Genetics* **151**: 921–7.

Burch, C. L., and L. Chao. 2000. Evolvability of an RNA virus is determined by its mutational neighborhood. *Nature* **406**: 625–8.

Burch, C. L., and L. Chao. 2004. Epistasis and its relationships to canalization in the RNA virus φ6. *Genetics* **167**: 559–67.

Burt, A., and R. Trivers. 2006. *Genes in Conflict: the Biology of Selfish Genetic Elements.* Cambridge, MA: Belknap Press.

Butcher, S. J., J. M. Grimes, E. V. Makeyev, D. H. Bamford, and D. L. Stuart. 2001. A mechanism for initiating RNA-dependent RNA polymerization. *Nature* **410**: 235–40.

Campbell, A. M. 1962. Episomes. *Adv. Genet.* **11**: 101–45.

Carroll, L. 1872. *Through the Looking Glass, and What Alice Found There.* London: Macmillan. www.online-literature. com/carroll/lookingglass.

Chao, L. 1990. Fitness of RNA virus decreased by Muller's ratchet. *Nature* **348**: 454–5.

Chao, L., B. R. Levin, and F. M. Stewart. 1977. A complex community in a simple habitat: an experimental study with bacteria and phage. *Ecology* **58**: 369–78.

Chao, L., C. U. Rang, and L. E. Wong. 2002. Distribution of spontaneous mutants and inferences about the replication mode of the RNA bacteriophage ϕ6. *J. Virol.* **76**: 3276–81.

Crick, F. H. C., L. Barnett, S. Brenner, and R. J. Watts-Tobin. 1961. General nature of the genetic code for proteins. *Nature* **192**: 1227–32.

Crill, W. D., H. A. Wichman, and J. J. Bull. 2000. Evolutionary reversals during viral adaptation to alternating hosts. *Genetics* **154**: 27–37.

Darwin, C. 1859. *On the Origin of Species by Means of Natural Selection, or the Preservation of Favoured Races in the Struggle for Life.* London: John Murray.

Dehó, G. and D. Ghisotti. 2006. The satellite phage P4. In R. L. Calendar and S. T. Abedon (eds.), *The Bacteriophages*, 2nd edn. Oxford: Oxford University Press, pp. 391–408.

de la Peña, M., S. F. Elena, and A. Moya. 2000. Effect of deleterious mutation-accumulation on the fitness of RNA bacteriophage MS2. *Evolution* **54**: 686–91.

Dennehy, J. J., N. A. Friedenberg, R. D. Holt, and P. E. Turner. 2006. Viral ecology and the maintenance of novel host use. *Am. Nat.* **167**: 429–39.

de Visser, J. A. G. M., J. Hermisson, G. P. Wagner, *et al.* 2003. Perspective: evolution and detection of genetic robustness. *Evolution* 57: 1959–72.

Drake, J. W. 1991. A constant rate of spontaneous mutation in DNA-based microbes. *Proc. Natl. Acad. Sci. U.S.A.* **88**: 7160–4.

Drake, J. W. 1993. Rates of spontaneous mutation among RNA viruses. *Proc. Natl. Acad. Sci. U.S.A.* **90**: 4171–5.

Duffy, S., P. E. Turner, and C. L. Burch. 2006. Pleiotropic costs of niche expansion in the dsRNA bacteriophage ϕ6. *Genetics* **172**: 751–7.

Duffy, S., C. L. Burch, and P. E. Turner. 2007. Evolution of host specificity drives reproductive isolation among RNA viruses. *Evolution* **61**: 2614–22.

Edelman, D. C., and J. Barletta. 2003. Real-time PCR provides improved detection and titer determination of bacteriophage. *Biotechniques* **35**: 368–75.

Elena, S. F., and R. E. Lenski. 2003. Evolution experiments with microorganisms: the dynamics and genetic bases of adaptation. *Nat. Rev. Genet.* **4**: 457–69.

Ferris, M. T., P. Joyce and C. L. Burch. 2007. High frequency of mutations that expand the host range of an RNA virus. *Genetics* **176**: 1013–22.

Fiers, W., R. Contrera, F. Duerinck, *et al.* 1976. Complete nucleotide sequence of bacteriophage MS2 RNA: primary and secondary structure of the replicase gene. *Nature* **260**: 500–7.

Froissart, R., C. O. Wilke, R. Montville, S. K. Remold, L. Chao, and P. E. Turner. 2004. Co-infection weakens selection against epistatic mutations in RNA viruses. *Genetics* **168**: 9–19.

Futuyma, D. F. 1998. *Evolutionary Biology*. Sutherland, MA: Sinauer Associates.

Gary, T. P., N. E. Colowick, and G. Mosig. 1998. A species barrier between bacteriophages T2and T4: exclusion, join-copy and join-cut-copy recombination and mutagenesis in the dCTPase genes. *Genetics* **148**: 1461–73.

Gellert, M. 1967. Formation of covalent circles of lambda DNA by *E. coli* extracts. *Proc. Natl. Acad. Sci. U.S.A.* **57**: 148–55.

Hartl, D. L., and A. G. Clark. 1997. *Principles of Population Genetics*. Sutherland, MA: Sinauer Associates.

Hashemolhosseini, S., D. Montag, L. Krämer, and U. Henning. 1994b. Determinants of receptor specificity of coliphages of the T4 family: a chaperone alters the host range. *J.Mol. Biol.* **241**: 524–33.

Heineman, R. H., I. J. Molineux, and J. J. Bull. 2005. Evolutionary robustness of an optimal phenotype: re-evolution of lysis in a bacteriophage deleted for its lysin gene. *J. Mol. Evol.* **61**: 181–91.

Hendrix, R. W. 2002. Bacteriophages: evolution of the majority. *Theor. Popul. Biol.* **61**: 471–80.

Hendrix, R. W., M. C. Smith, R. N. Burns, M. E. Ford, and G.F. Hatfull. 1999. Evolutionary relationships among diverse bacteriophages and prophages: all the world's a phage. Proc. Natl. Acad. Sci. U.S.A. **96**: 2192–2197.

Hershey, A. D., and M. Chase. 1952. Independent functions of viral protein and nucleic acid in growth of bacteriophage. *J. Gen. Physiol.* **36**: 39–56.

Holder, K. K., and J. J. Bull. 2001. Profiles of adaptation in two similar viruses. *Genetics* **159**: 1393–404.

Inouye, S., M. G. Sunshine, E. W. Six, and M. Inouye. 1991. Retronphage phi R73: an *E. coli* phage that contains a retroelement and integrates into a tRNA gene. *Science* **252**: 969–71.

Kawecki, T. J. 1998. Red queen meets Santa Rosalia: arms races and the evolution of host specialization in organisms with parasitic lifestyles. *Am. Nat.* **152**: 635–51.

Kirkpatrick, M., and V. Ravigné. 2002. Speciation by natural and sexual selection: models and experiments. *Am. Nat.* **159**: S22–35.

Kirschner, M., and J. Gerhart. 1998. Evolvability. *Proc. Natl. Acad. Sci. U. S. A.* **95**: 8420–7.

Kitano, H. 2004. Biological robustness. *Nat. Rev. Genet.* **5**: 826–37.

Klovins, J., N. A. Tsareva, M. H. de Smit, V. Berzins, and D. Van. 1997. Rapid evolution of translational control mechanisms in RNA genomes. *J. Mol. Biol.* **265**: 372–84.

Knies, J. L., R. Izem, K. L. Supler.J. G. Kingsolver, and C. L. Burch. 2006. The genetic basis of thermal reaction norm evolution in lab and natural phage populations. *PLoS Biol.* **4**: e201.

Lawrence, J. G., R. W. Handrix, and S. Casjens. 2001. Where are the pseudogenes in bacterial genomes? *Trends Microbiol.* **9**: 535–40.

Lawrence, J. G., G. F. Hatfull, and R.W. Hendrix. 2002. Imbroglios of viral taxonomy: genetic exchange and failings of phenetic approaches. *J. Bacteriol.* **184**: 4891–905.

Lenski, R. E., and B. R. Levin. 1985. Constraints on the coevolution of bacteria and virulent phage: a model, some experiments and predictions for natural communities. *Am. Nat.* **125**: 585–602.

Levins, R. 1968. *Evolution in Changing Environments: Some Theoretical Explorations.* Princeton, NJ: Princeton University Press.

Licis, N., and J. van Duin. 2006. Structural constraints and mutational bias in the evolutionary restoration of a severe deletion in RNA phage MS2.*J. Mol. Evol.* **63**: 314–29.

Little, J. W., D. P. Shepley, and D.W. Wert. 1999. Robustness of a gene regulatory circuit. *EMBO J.* **18**: 4299–307.

Luria, S. E., and M. Delbrück. 1943. Mutations of bacteria from virus sensitivity to virus resistance. *Genetics* **28**: 491–511.

Malmberg, R. L. 1977. The evolution of epistasis and the advantage of recombination in populations of bacteriophage T4. *Genetics* **86**: 607–21.

Meselson, M., and J. J. Weigle. 1961. Chromosome breakage accompanying genetic recombination in bacteriophage. *Proc. Natl. Acad. Sci. U.S.A.* **47**: 857–68.

Mindich, L. 1995. Heterologous recombination in the segmented dsRNA genomes of bacteriophage ϕ6. *Semin. Virol.* **6**: 75–83.

Monod, J., and F. Jacob. 1961. General conclusions: teleonomic mechanisms in cellular metabolism, growth and differentiation. Cold Spring Harbor Symposium. *Quant. Biol.* **26**: 389–401.

Montag, D., I. Riede, M.-L. Eschbach, M. Degen, and U. Henning. 1987. Receptor-recognizing proteins of T-even type bacteriophages. *J. Mol. Biol.* **196**: 165–74.

Montville, R., R. Froissart, S. K. Remold, O. Tenaillon, and P. E. Turner. 2005. Evolution of mutational robustness in an RNA virus. *PLoS Biol.* **3**: e381.

Morgan, A. D., S. Gandon, and A. Buckling. 2005. The effect of migration on local adaptation in a coevolving host-parasite system. *Nature* **437**: 253–6.

Muller H. J. 1964. The relation of recombination to mutational advance. *Mutat. Res.* **1**: 2–9.

Nelson, M. A., M. Ericson, L. Gold, and J.F. Pulitzer. 1982. The isolation and characterization of TabR bacteria: Hosts that restrict bacteriophage T4rII mutants.*Mol. Gen. Genet.* **188**: 60–8.

Ohnishi, M., K. Kurokawa, and T. Hayashi. 2001. Diversification of *Escherichia coli* genomes: are bacteriophages the major contributors? *Trends Microbiol.* **9**: 481–5.

Olsthoorn, R. C. L., and J. van Duin. 1996. Evolutionary reconstruction of a hairpin deleted from the genome of an RNA virus. *Proc. Natl. Acad. Sci. U.S.A.* **93**: 12256–61.

Orr, H. A. 2005. Theories of adaptation: what they do and don't say. *Genetica* **123**: 3–13.

Pepin, K. M., M. A. Samuel, and H. A. Wichman. 2006. Variable pleiotropic effects from mutations at the same locus hamper prediction of fitness from a fitness component. *Genetics* **172**: 2047–56.

Poon, A., and L. Chao. 2004. Drift increases the advantage of sex in RNA bacteriophage ϕ6. *Genetics* **166**: 19–24.

Poon, A., and L. Chao. 2005. The rate of compensatory mutation in the DNA bacteriophage ϕX174. *Genetics* **170**: 989–99.

Raney, J. L., R. R. Delongchamp, and C. R. Valentine. 2004. Spontaneous mutant frequency and mutation spectrum for gene A of ϕX174 grown in *E. coli*. *Environ. Mol. Mutagen.* **44**: 119–27.

Rokyta, D. R., P. Joyce, B. Caudle, and H. A. Wichman. 2005. An empirical test of the mutational landscape model of adaptation using a single-stranded DNA virus. *Nat. Genet.* **37**: 441–4.

Rokyta, D. R., C. L. Burch, S. B. Caudle, and H.A. Wichman. 2006. Horizontal gene transfer and the evolution of microvid coliphage genomes. *J. Bacteriol.* **188**: 1134–42.

Rouch, W. 1997. Biology departments restructure. *Science* **275**: 1556–8.

Sanger F., G. M. Air, B. G. Barrell, *et al.* M. Slocombe, and M. Smith. 1977. Nucleotide sequence of bacteriophage ϕX174 DNA. *Nature* **265**: 687–95.

Saren, A.-M., J. J. Ravantti, S. D. Benson, *et al.* 2005. A snapshot of viral evolution from genome analysis of the *Tectiviridae* family. *J. Mol. Biol.* **350**: 427–40.

Schuppli, D., G. Miranda, H. C. T. Tsui, M. E. Winkler, J. M. Sogo, and H. Weber. 1997. Altered 3'-terminal RNA structure in phage Qβ adapted to host factor-less *Escherichia coli*. *Proc. Natl. Acad. Sci. U.S.A.* **94**: 10239–42.

Schuppli, D., J. Georgijevic, and H. Weber. 2000. Synergism of mutations in bacteriophage Qβ RNA affecting host factor dependence of Qβ replicase. *J. Mol. Biol.* **295**: 149–54.

Silander, O. K., D. M. Weinreich, K. M. Wright, *et al.* 2005. Widespread genetic exchange among terrestrial bacteriophages. *Proc. Natl. Acad. Sci U.S.A.* **102**: 19009–14.

Studier, F. W. 1979. Relationships among different strains of T7 and among T7-related bacteriophages. *Virology* **95**: 70–84.

Studier, F. W. 1980. The last of the T phages. In N.H. Horowitz and E. Hutchings, Jr. (eds.), *Genes, Cells, and Behavior: a View of Biology Fifty Years Later*. San Francisco, CA: W. H. Freeman and Co., pp. 72–8.

Turner, P. E., and L. Chao. 1999. Prisoner's dilemma in an RNA virus. *Nature* **398**: 441–3.

Van Valen, L. 1973. A new evolutionary law. *Evol. Theor.* **1**: 1–30.

Watson, J. D., and F. H. C. Crick. 1953. Molecular structure of nucleic acids: a structure for deoxyribose nucleic acid. *Nature* **171**: 737–8.

Weinreich, D. M., R. A. Watson, and L. Chao. 2005. Perspective: sign epistasis and genetic constraint on evolutionary trajectories. *Evolution* **59**: 1165–74.

Wichman, H. A., M. R. Badgett, L.A. Scott, C. M. Boulianne, and J. J. Bull. 1999. Different trajectories of parallel evolution during viral adaptation. *Science* **285**: 422–4.

Wichman, H. A., J. Wichman, and J. J. Bull. 2005. Adaptive molecular evolution for 13,000 phage generations: a possible arms race. *Genetics* **170**: 12–31.

Woese, C. 1994. There must be a prokaryote somewhere: microbiology's search for itself. *Microbiol. Rev.* **58**: 1–9.

Wommack, K. E., and R. R. Colwell. 2000. Virioplankton: viruses in aquatic ecosystems. *Microbiol. Mol. Biol. Rev.* **64**: 69–114.

You, L., and J. Yin. 2002. Dependence of epistasis on enviroment and mutation severity as revealed by in silico mutagenesis of phage T7. *Genetics* **160**: 1273–81.

You, L., and J. Yin. 2006. Evolutionary design on a budget: robustness and optimality of bacteriophage T7. *Syst. Biol.* **153**: 46–52.

CHAPTER 7
Phage evolution

Roger W. Hendrix

7.1 INTRODUCTION

Bacteriophages have a long history as objects of biological study. They were discovered about 90 years ago and have engaged biologists ever since, initially for their potential in combating human disease through phage therapy. Later, phages served as arguably the most important model systems in the development of the discipline of molecular biology and the associated explosion of knowledge about the detailed workings of genes and cells. Yet it is only in very recent years that the study of phage evolution has attracted the attention of more than a handful of individuals. The primary reasons for the current increased interest in phage evolution, I would suggest, are two: discovery, over the past 20 years, of astonishingly high phage population numbers in the natural environment, and improved, low-cost methods of phage genotypic analysis, especially DNA sequencing. In this chapter I discuss the abundance and diversity of the global phage population, with an emphasis on what we are learning from comparative genomic studies about the mechanisms by which it has evolved to its current state.

Chapter 6 provides an introduction to basic evolutionary mechanisms of phage evolution. See Hendrix (2003), Casjens (2005), and Brüssow and Desiere (2006) for additional reviews of phage evolution from the perspective of genomic studies. See also Chapter 8 for complementary consideration of phage evolution as it occurs in the case of multiple phage infection of individual bacteria.

Bacteriophage Ecology: Population Growth, Evolution, and Impact of Bacterial Viruses, ed. Stephen T. Abedon. Published by Cambridge University Press.

7.1.1 Phage environmental abundance

Direct counts of phage particles in environmental samples (Suttle, 2005) have shown us that the number of tailed phages on Earth is literally astronomical. If 10^{31} tailed phages – a current estimate of their global population – were laid end to end they would reach out into space about 200 million light-years (see also Chapters 1, 5, 10, and 11). We have also learned from environmental studies that the turnover time of this population, at least in marine environments, is surprisingly short, probably on the order of a few days (Suttle and Chen, 1992). To replace this population on such a short time scale, it has been estimated, would require $\sim 10^{24}$ productive infections per second.

These numbers are vastly greater than anyone had imagined before the measurements were made, and they have a wonderfully liberating effect on how we can think about the mechanisms of phage evolution. For example, non-homologous recombination, which occurs in laboratory genetic crosses at frequencies on the order of $\sim 10^{-9}$ or less per generation, was seen by most people who thought about it at all as too infrequent to have a major role in phage evolution. Yet now that we know there are $\sim 10^{24}$ opportunities for non-homologous recombination per second, it becomes clear that this mechanism is perfectly capable of producing enough non-homologous recombinants to have a significant role in phage evolution, even if we imagine that, say, only one in 10^{10} such recombinants is functional and able to survive the scrutiny of natural selection.

7.1.2 Phage diversity

The second change that has brought phage evolution studies to their current prominence is a technological one, the development of cheap, high-throughput methods of DNA sequencing. The sequence of a single phage genome, as for example the phage λ sequence published in 1982 (Sanger *et al.*, 1982), can be quite informative about the biology of the phage, but it is only when we can compare sequences from multiple genomes that we can recognize and interpret the signals in the sequence that are, in essence, fossils of past evolutionary events.

Most of our understanding of the diversity of phage genomes, as well as phage environmental abundance, pertains primarily to the tailed phages, members of the order *Caudovirales* of the ICTV classification (Fauquet *et al.*, 2005), which have linear dsDNA genomes and include the familiar *Escherichia coli* phages λ and T4, among many others. There are, of course, several other types of bacteriophages and archaeal viruses known, often categorized according to the nucleic acid type of their chromosome. These include ssRNA phages (*Leviviridae*), dsRNA phages with segmented chromosomes (*Cystoviridae*), phages with circular ssDNA (*Microviridae, Inoviridae*) and circular dsDNA (*Corticoviridae, Plasmaviridae, Fuselloviridae, Guttaviridae*), and other phages with linear dsDNA (*Tectiviridae, Lipothrixviridae, Rudiviridae, Salterprovirus*; see also Chapter 1). We know by far the most about the evolution of the tailed phages, however, and I will spend most of this chapter discussing them, touching on some of the others briefly at the end. The other groups all have smaller genomes than the tailed phages and, in the cases where we know something about their evolution, they appear to evolve by mechanisms similar to the tailed phages, but to do so less flamboyantly.

7.1.3 Biases in phage sampling

The tailed phages are generally thought to constitute the great majority of all phages, based on the fact that about 96% of the 5000+ phage isolates reported in the literature are tailed phages (Ackermann, 2006). This view is supported qualitatively by direct observation of aquatic samples, where the majority of objects identified as virus particles are tailed phages (Suttle, 2005). However, it is good to bear in mind that there are almost certainly significant biases in our sampling of the environmental phage populations. Tailed phages, for example, may be overrepresented because they often make more dramatic plaques than the other phages and therefore beckon more alluringly to the scientist looking to isolate a new phage. A related point comes from *Bacillus* phage G, the largest known phage at a genome size of 500 kbp (unpublished result); the phage G virion is so big that it does not diffuse well through agar, so plaques are too small to be seen unless unusually low concentrations of top agar are used in plaque assays (Serwer *et al.*, 2007). This must mean that there has been a systematic bias against isolating exceptionally large phages for the entire time we have been isolating phages using a plaque assay (see Chapters 4 and 16 for additional consideration of phage plaque formation).

In addition, the distinctive shapes of tailed phages are easier to identify unambiguously as phages in electron micrographs of environmental samples. Furthermore, even among the tailed phages there is clearly a very strong sampling bias toward phages that grow on a small number of intensively studied bacterial types; additional possible phage isolation biases, with regard to broth-culture growth rates and host-range breadth, are discussed in Chapter 1. Though these various sampling biases seem unlikely to have had a significant impact on our understanding of genetic mechanisms of evolution in phages, it is likely they have an adverse effect on our understanding of the genetic diversity and the genetic structure of phage populations.

7.1.4 Nature of phage genomes

A striking characteristic of phage genomes is their efficient use of genetic real estate. Typically 90–95% of the DNA of a tailed phage is protein-coding sequence. Phages encode some of the largest proteins known (e.g., the 7312 aa putative tail fiber protein of cyanophage P-SSM4; Sullivan *et al.*, 2005), but curiously the average gene size of most phages is about 200 codons, about two-thirds the size of the average gene of their hosts (Hatfull *et al.*, 2006). Some phages encode a few tRNAs, and there is a small number of examples of phages encoding a tmRNA (a small RNA with characteristics of both tRNA and mRNA). There must certainly be other functional small RNAs encoded by phages, but the only example known to date is the "packaging RNA" of *Bacillus* phage ϕ29 (Guo *et al.*, 1991) and its close relatives. Much of the non-coding parts of the sequence can be assigned to regulatory elements like promoters, transcription terminators, operators, etc.

Genes tend to be arranged in co-transcribed groups. Such groups of genes are often very tightly packed, in many cases with an overlap of the initiation codon of one gene with the termination codon of the upstream gene. In some genomes all the genes are transcribed in one direction, but in others there are multiple transcription units distributed between the two directions of transcription. The size of tailed phage genomes varies from ~19 kbp, which appears to be near the minimum size needed to encode the virion structural proteins plus a few regulatory proteins, up to the ~500 kbp of phage G and possibly beyond.

7.2 GENOME COMPARISONS OF dsDNA TAILED PHAGES

With only rare exceptions, our knowledge of the mechanisms of phage evolution comes not from examination of individual genomes (or individual virions, or individual life cycles), but from a comparison of two or more phages related to some degree through common ancestry. It is the differences revealed by such comparisons, whether they be the result of point mutations, DNA insertions, duplications or deletions, or other sorts of rearrangements of the DNA in the genome, that give critical insight into past genetic events that have led to new versions of phage.

As will become clear in the discussion below, the prevalence of horizontal exchange of genetic material among phages makes the concept of a common ancestor for any two phages an elusive one. That is, two phages are typically mosaic with respect to each other, so a given gene may or may not have been present in a putative common ancestor inferred from other genes in the genome. Thus at the level of the phage as a whole, there is in general no single common ancestor with respect to another phage. However, once we understand this, the complexity introduced into the phages' history by horizontal exchange enriches rather than confuses our understanding of the evolutionary process.

7.2.1 Evolution by recombination: mosaicism

DNA recombination is a major source of genetic diversity in phage populations (Hendrix, 2002). It is convenient to discuss recombination events as being either homologous recombination, in which two identical sequences recombine in register to reproduce the starting sequence, or non-homologous recombination, in which two different sequences are joined together to make a "novel joint." In reality, the degree of similarity between two sequences that are going to recombine is not all or none but instead can fall anywhere on a continuum from undetectable to 100% identity.

7.2.1.1 Non-homologous recombination

When whole genome sequences of genetically similar tailed phages were first compared it became clear that their sequences are mosaic with respect to each other (Simon *et al.*, 1971). In a pairwise comparison of genomes, the two sequences might match at a high level of identity for a gene's length or more, and then suddenly stop matching at all, or match with a very different level of identity (Fig. 7.1). Such an observation implies that there has been a non-homologous recombination event in the ancestry of one of the phages being compared, which has created a novel joint in the recombinant sequence – a juxtaposition of two sequences that were not adjacent before. Note that in cases where two adjacent gene sequences match the two homologous genes in a comparison phage at only moderately different levels, then different apparent mutation rates in the two genes may be the explanation for the difference rather than near-homologous recombination. Examples of this, which can be distinguished by bioinformatic analysis, are found, but do not enter into the analysis of recombinants discussed here.

Examination of the positions of novel joints in recombinant sequences makes it apparent that they are not distributed randomly over the genome. Rather, they are predominantly found at or very near to gene boundaries (Juhala *et al.*, 2000). There are, however, examples of recombination events occurring within protein coding regions, and other examples of groups of genes that are apparently never split apart by recombination. Genes with

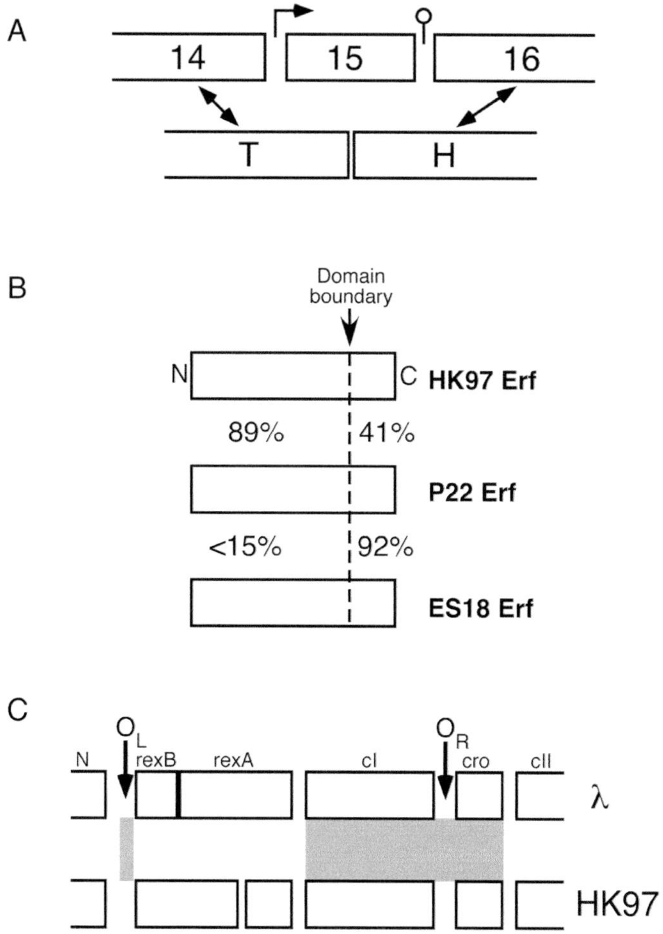

Figure 7.1 Selected regions from four lambdoid phages, showing "fossils" of horizontal gene exchange made apparent by genome comparisons. (A) A moron (gene *15* plus the flanking transcription promoter and terminator) inserted between two tail genes in phage HK97. Double arrows show functional correspondence between HK97 and λ genes. (B) Recombination protein Erf of phages HK97, P22, and ES18, with evidence for non-homologous recombination within the coding region at the position corresponding to the domain boundary of the protein. Percentages indicate amino acid identity between the protein domains above and below the number. (C) Immunity region of phages λ and HK97, with evidence for recombination at gene boundaries. The shaded areas indicate >95% nucleotide sequence identity; unshaded areas are not detectably related at the nucleotide level. O_L and O_R are binding sites on the DNA that are shared by the CI and Cro repressors.

exchanges within the coding region, at least in cases where we have some knowledge of the encoded protein, generally fall into two groups: genes encoding proteins of multiple domains, in which case the recombination is found at the domain boundary, and tail fiber genes. In both of these cases it is believed that the gene has tolerated the recombination event because the structural properties of the encoded protein mean that its function is not disrupted by the resulting reassortment of protein parts. In the case of the multi-domain proteins, this is presumed due to the absence of intimate interactions between the domains that would be disturbed by combining domains from two lineages. In the case of tail fibers the explanation is similar. Tail fibers have an essentially one-dimensional fold along the length of the fiber structure, which means that there are no important interactions between amino acids that are distant in the primary sequence that would be expected to be disrupted by joining parts of two separately evolved precursor fibers.

These observations argue against models in which non-homologous recombination is initially restricted to certain positions in the genome, like gene boundaries. Instead, they favor the view that such recombination occurs indiscriminately across the genome, but that most recombinants are rapidly removed from the population by natural selection because they compromise function. Such a process is a classical Darwinian scenario in which diversity is generated in the population – in this case by non-homologous recombination – and then all but the fittest recombinants are eliminated by natural selection.

7.2.1.2 Homologous recombination

There is also certainly homologous recombination, occurring between phage sequences at many orders of magnitude higher frequency than non-homologous recombination. Homologous recombination does not leave a signature in the sequence at the site of recombination, as does non-homologous recombination. However, novel sequence joints that have been created earlier by non-homologous recombination in the sequences flanking the site of homologous recombination can be reassorted. In this way, homologous recombination provides a mechanism for novel joints to be grouped with each other in novel combinations and to be moved around in the population. Finally, there are also likely situations in which short regions of partial sequence identity ("micro-homologies") bias the selection of recombination pairings away from complete randomness and in favor of pairings between such regions of weak similarity (Baker *et al.*, 1991).

7.2.1.3 Coevolving gene groups

Some groups of functionally related genes, such as head genes or tail genes, appear never to be reassorted by recombination. The probable explanation for this is that the encoded proteins interact intimately and have coevolved with each other to carry out their joint function, with the result that genes in the separately evolved homologous groups are not functionally interchangeable between the groups (see also Chapter 14). We presume, and there is some supporting evidence in phage sequences (Juhala *et al.*, 2000), that recombination *per se* is not excluded from these regions of the genome; in other words, selection against non-functional recombinants occurs at the level of function of the recombinant phage rather than at the level of recombinant formation.

In phages with smaller genomes, such as λ and its relatives, the head genes remain together over evolutionary time, with the occasional exception of a non-homologous substitution of the head maturation protease (Liu and Mushegian, 2004). Tail genes in these phages also tend to remain together, though less absolutely so than in the case of the head genes. With minor exceptions the rest of the genes in these phages exchange horizontally as individual genes.

A somewhat different picture is seen in a group of phages with larger genomes, typified by the well-studied coliphage T4 (Desplats and Krisch, 2003). These "T4-type" phages share a large number of genes, recognizable as homologous by sequence similarity and encompassing head and tail genes, DNA replication genes, and nucleotide metabolism genes. Remarkably, in light of what is seen in the smaller phages, this entire group of "core" genes shows almost no evidence for horizontal exchange within the group (Filee *et al.*, 2006). In marked contrast, the genes outside the "core" group, which in some of these phages can constitute more than half the genome, appear to be the product of prolific and promiscuous horizontal exchange through non-homologous recombination. Furthermore, the non-core genes of these phages are extremely diverse and largely non-homologous between phages. The origin of this Janus-like character of the T4-type phage genomes – one group of stable, sober, and functionally important genes and another group of genes with non-essential or unknown functions that jump into and out of genomes with manic abandon – is not well established, but it may be useful to think of it as the result of a variation on the process described above for phages with smaller genomes. That is, promiscuous non-homologous recombination generates pervasive variation in genome composition, but any recombinants that disrupt the essential core genes, or their essential interactions with each other, are stringently counterselected.

7.2.2 Evolution by point mutation

Phage genes, of course, accumulate point mutations over time, as in any other organism (Chapter 6). Pairwise comparison between phages of homologous genes and their encoded proteins yields everything from 100% nucleotide sequence identity to protein sequences with only barely detectable sequence similarity (<15% amino acid identity). When the nature of point mutations has been examined in comparisons of closely similar sequences, mutations that are synonymous in the amino acids encoded are substantially more frequent than non-synonymous mutations, implying that the genes are under purifying selection (Weigele *et al.*, 2007; see also Chapter 6). The fact that gene sequences and the associated protein sequences have diverged to the point of barely recognizable similarity suggests that they may have in some cases diverged past that point, and there are a few examples in which proteins are known to be homologous (i.e., to have common ancestry) even though there is no remaining evidence of homology in the primary sequences, as for example in the major capsid protein sequences discussed below.

The high degree of sequence divergence seen in phage genes across the population suggests that the population is ancient. However, the fact that there is no accurate understanding of how fast the mutational clock runs for phages means that it is not possible to get an accurate estimate from sequence divergence data of the time scale of phage evolution. Note, however, that data from structural studies of capsids, described below, argue that phages resembling contemporary phages have likely been evolving for at least 3.5 billion years.

7.2.3 Additions to the genome and genome rearrangements

There are a number of mechanisms by which sequences have been added to phage genomes, and these can be discerned from examination of the genomes. The simplest is the direct addition of mobile genetic elements including transposons, introns, and inteins (the latter, essentially, are protein introns). While all of these are seen, they are not abundant, and it therefore seems unlikely that they are a major source of new functionality for phage genomes. However, a second class of apparently mobile elements, the morons ("units of more DNA"), may have a much more important role in adding novel functions to existing genomes (Lawrence *et al.*, 2001). Morons are typically single genes, usually flanked by a transcription promoter and terminator, which in favorable cases can be recognized as evolutionarily recent additions to a phage genome because they have a different nucleotide composition than flanking sequence and because they are situated between two genes that are adjacent in a comparison phage.

Based on examples for which a function is known or inferred for the moron gene, some morons are thought to confer a selective advantage on the phage indirectly, by providing a benefit to the host cell and thereby selection against loss of the prophage (see, for example, Chapters 5 and 14). Addition of morons to phage genomes may also provide novel functions that benefit lytic growth of the phage, and it has been suggested that this may be a mechanism by which a substantial fraction of phage genes have entered the genomes (Hendrix *et al.*, 2000). The biochemical mechanism by which morons insert is not known; no sequences reminiscent of a conventional transposon are evident, and insertion may be either the result of "random" non-homologous recombination or the product of a more directed mechanism, as yet undefined.

Gene duplication and divergence provides in principle another mechanism for acquiring new genes, but clear evidence for this is lacking in phages with smaller genomes (<100 kbp). However, a good case can be made for such a mechanism in the instance of the capsid genes of phage T4 (170 kbp genome) and its close relatives. T4's major capsid protein, gp23, constitutes the 155 hexons that form the bulk of the capsid, while gp24, encoded by the adjacent gene and distantly but unequivocally related to gp23 in sequence, makes the pentons on eleven icosahedral vertices of the capsid. To date, tailed phage capsids with separate proteins for pentons and hexons are found only in phages with prolate capsids (heads elongated "top" to "bottom"), so it is tempting to think that duplication and divergence of a single ancestral capsid protein gene may have been associated with a transition from an isometric to a prolate capsid. For some of the unusually large phage genomes that have become available recently, including phage G (500 kbp), we can see families of homologous proteins within one genome, none of which have identified functions (unpublished observations). It remains to be seen whether the apparent correlation between genome size and presence of gene duplications will hold up as further genome sequences are determined.

There is also good evidence for exchange of analogous but not homologous genes between phages. There are clear examples of such exchanges among cell wall hydrolases (the transglycosylase of phage λ sits in a homologous surrounding context to the lysozyme of phage P22) and among capsid maturation proteases (Liu and Mushegian, 2004). Such exchanges in the ancestry of phages can also involve larger blocks of genes, and these can potentially lead to increases (or decreases) in the total number of functions encoded by genes in a phage genome. Thus phage SfV has head genes in the phage HK97 head-gene family, but instead of the simple non-contractile tail of HK97, SfV has a contractile tail with genes in the sequence family of the phage Mu tail genes.

7.2.4 Inventing new genes

As mentioned above, phages make very efficient use of their genomic DNA, with more than 90% of the sequence devoted to protein-coding genes (Section 7.1.4). In a well-studied phage like λ, a majority of the genes are identified as essential for lytic or lysogenic growth, and several more can be associated with biochemical functions that are non-essential but presumed or demonstrated to be useful in certain hosts or ecological situations. After all these, there still remain a few apparent genes for which no plausible function can be assigned. More tellingly, comparisons with other genomes show that some of these genes are intragenic mosaics or truncations of genes found in other genomes (Juhala *et al.*, 2000), raising the possibility that they may in fact provide no useful function for the phage.

One might expect that such putatively useless genes would have long ago been lost in the stringently selective environment where phages earn their living. However, it may be true that a phage like λ can accommodate some "useless" DNA without penalty as a result of the constraints on genome size imposed by the DNA packaging process. That is, λ packages DNA between two specific sequences (*cos* sites) located one genome's length apart on the genomic concatemer produced by replication (Feiss and Becker, 1983). When the length of the genome is changed by insertion or deletion, the efficiency of packaging drops off quite steeply (Feiss and Siegele, 1979). As a result, there is a selection for a particular length of genome independent of any particular selective benefit due to biochemical functions provided by its constituent genes; the "useless" gene(s) may provide a selective benefit to the phage simply by filling out the genome to the optimal size for packaging. Assuming this is correct, it would seem to afford the evolutionary process the possibility of "inventing" new genes with novel functions. That is, as ongoing non-homologous recombination mixes fragments of existing genes in novel combinations, there is, for these regions of "useless" genes, no selective pressure to eliminate the multitude of non-functional intermediates on the way to producing, by chance, a fortuitous combination of gene parts that would provide a novel selective benefit.

7.3 POPULATION STRUCTURE AND METAGENOMIC STUDIES

The sparse and biased sampling of the global tailed phage population, as represented in complete phage genome sequences, makes it very difficult to get a clear picture of the genetic structure of the phage population as a whole. On the one hand, no two independently isolated phages to date have been found to have identical genome sequences. Furthermore, while in some cases two such phages may be very similar in sequence (say, 95% identical in nucleotide sequence), a more typical finding is that less than half of the genes in a newly sequenced phage match any sequence in the databases, even when examined at the level of amino acid sequences. Thus, the sequence diversity in the phage population seems quite high, judging from sequenced genomes. On the other hand, it seems clear from an examination of available phage genomes that they fall into a discrete number of types. In the relatively well-studied and well-sequenced phages of *E. coli* and closely related hosts we usually think of the λ-like, T4-like, and T7-like phages, among a few others, as prominent groups of phage types, based on shared sequence, genome organization and size, lifestyle, and other features. Other groups of phages, say the mycobacteriophages, also show some clustering into types, but these do not correlate with the types defined for the coliphages (Hatfull *et al.*, 2006). Furthermore, as more genome sequences are determined, the diversity within each of these types increases, and in some cases the phages on the fringes of one group begin to resemble the phages on the fringes of another group. In sum, it is clear that the diversity in the phage population is not uniformly distributed across phages but rather clusters to some extent in phage types – but how many such types there are, how they relate to each other evolutionarily, and how distinct they are from each other remains largely unknown.

A more comprehensive assessment of genetic diversity in the phage population comes from metagenomic analyses in which all of the phages from a particular environmental source are collected, and their DNA is extracted and sequenced in bulk. Because the phages are not required to grow on a laboratory host, there is in principle little discrimination against any particular segment of the population, and the resulting sequences should be much more representative of the diversity of the entire population. Such analyses have been carried out on a number of environmental samples, principally ocean water but also including some more earthy sources (Breitbart *et al.*, 2003; Zhang *et al.*, 2006). The metagenomic analyses confirm the impression from the genomic studies that the genetic diversity in the phage population is very high, but allow a more quantitative characterization of that diversity. The conclusion from these studies, in fact, is that there is more genetic diversity in the tailed phage population than in any other compartment of the biosphere (Edwards and Rohwer, 2005). Metagenomic analysis of phage DNA has recently been extended with the aid of new sequencing technology to analyze ~1.8 Mb of sequence from four widely separated marine sites (Angly *et al.*, 2006). This gives a view not only of the diversity of sequences, which is in accord with earlier metagenomic analyses, but also a first look at the geographical distribution of sequences. The analysis found considerable overlap in the populations of sequences at the four sites, but also good evidence for site-specific differences in the abundances of particular sequences. These differences were best explained by assuming that essentially the same types of viruses were found at all four sites but that the relative abundances of those virus types varied among the sites.

The metagenomic approach is a powerful one, but it suffers from the drawback that it does not reveal how genes are associated in individual genomes, and so cannot tell which gene combinations have withstood the challenges of natural selection. Even if enough sequence data were available to reconstruct complete genome sequences from the metagenomic data, the mosaic nature of phage genomes means that it would not be possible to certify that a given genome sequence assembly represented an actual phage genome in the environment rather than an "*in silico* recombinant." Given the current limitations of sequencing technology, an understanding of the genetic structure of the global phage population is therefore best served by a combination of the genomic and metagenomic approaches.

7.4 OTHER PHAGES AND DEEP PHYLOGENY

A few phylogenetic analyses have been carried out on phages outside the tailed phage group, principally the phages with circular ssDNA chromosomes (see Chapters 6 and 9 for examples of experimental evolution employing such phages, especially coliphage ϕX174, and Chapter 8 for consideration of phylogenetic analysis of dsRNA phages). Although there is clear evidence for horizontal exchange of sequences in both families of phages with this type of chromosome, it appears less prominent than what is seen in the tailed phages. In the most extensive study to date, Rokyta *et al.* (2006) isolated 42 new examples of members of the *Microviridae* family on an *E. coli* host. Most of the divergence among these phages, though not all, could be explained by a conventional scheme of point mutations inherited vertically. However, when these *E. coli* phages are compared to members of the *Microviridae* growing on hosts such as *Chlamydia*, it is clear that there have been some insertions and deletions of genes in the ancestors of these phages. There appear not to be any comparable analyses on other families of bacterial or archaeal viruses, and for most of the families there are not enough individual isolates in hand to allow such an analysis (see, however, Chapter 8).

As described above, there has been such extensive divergence in the sequences of homologous proteins in the tailed phages that there is no possibility of tracing the ancestral relationships of, for example, all the known major capsid proteins of tailed phages based on sequence alone. There is even less possibility of using sequences to detect ancestral relationships that may exist between phages and other virus groups. However, recent work on the structure and assembly of virions has begun to give convincing evidence for deep phylogenetic connections among not only all the tailed phages' capsids but among viruses that infect hosts from all three domains of life. For example, structural data argue that the major capsid proteins of four tailed phages (HK97, T4, P22, and ϕ29), all with no detectable mutual sequence similarity, have the same polypeptide fold (Wikoff *et al.*, 2000; Jiang *et al.*, 2003; Fokine *et al.*, 2005; Morais *et al.*, 2005). Strikingly, the same fold appears to be in the major capsid protein of herpesvirus (Baker *et al.*, 2005), and these observations argue for common ancestry not only for the capsid proteins of all the tailed phages but of the herpesviruses as well. Similar deep connections can be made for two other, different groups of viruses. Thus, phage PRD1 (*Tectiviridae*), animal virus adenovirus, algal virus PBCV1 (*Phycodnaviridae*), and archaeal virus STIV all share a "double jelly-roll" fold of their capsid hexon proteins (Benson *et al.*, 2004; Khayat *et al.*, 2005). Yet another fold and an unusual double-shelled capsid are shared by phage ϕ6 (*Cystoviridae*) and the reoviruses of plants and animals (Huiskonen *et al.*, 2006; see Chapter 8 for additional consideration of phage ϕ6). Another possible set of ancestral connections, somewhat less compelling because of the lesser complexity of the fold, is implied by the single jelly-roll capsid protein fold of phage ϕX174 (*Microviridae*) and many small plant and animal viruses (Shepherd *et al.*, 2006). This fold is also found in the outer capsid shell of the Reovirus group (above) and as part of the double jelly-roll fold of the Adenovirus group (above), suggesting more complex ancestral interconnections.

The most straightforward (but not the only) interpretation of these several observations is that there were already at least three lineages of viruses resembling contemporary viruses infecting the cells that preceded the division of life into Bacteria, Archaea, and Eukarya. Members of each of those lineages have subsequently coevolved with their hosts following division of cellular life into three domains and up to the present (Fig. 7.2). An important caveat, however, is that the lineages implied by these arguments are based solely on structural features of the capsid proteins: consequently the inferred lineages are, strictly speaking, lineages of the capsid structures and not of the viruses as a whole. In fact, some other features of these viruses,

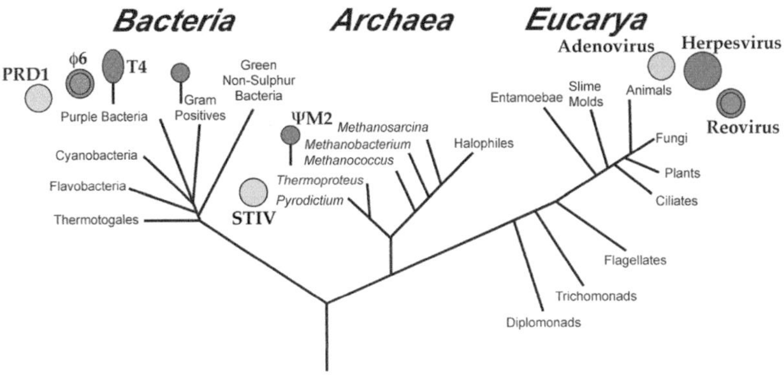

Figure 7.2 Cellular tree of life, with a selection of viruses placed near the hosts they infect. Viruses indicated in the same shade have capsids with shared features (protein fold and aspects of capsid assembly, structure and function) that indicate common ancestry for the genes encoding the capsids in that group. (Modified from Hendrix, 1999, with permission.) *See color plate section.*

such as mode of genome replication, distribute somewhat differently across the virus types than do the capsid protein folds. It seems that viruses considered at these deep phylogenetic levels are genetic mosaics with respect to each other, just as the tailed phages are so clearly mosaic within their group.

7.5 CONCLUSION

Evidence from genomic sequencing argues strongly that the most important creative force in tailed phage evolution is non-homologous recombination, which creates novel joints in the genomic sequence and in the process creates novel associations of genes in a genome, rearrangements of gene organization in a genome, insertions of new genes, and deletions. Non-homologous recombination works in conjunction with genetic change through point mutation and homologous recombination to produce an extravagance of genetic diversity in the population. In classical Darwinian fashion, the population is subject to stringent natural selection, producing the population of phages we observe. The extreme diversity in homologous sequences argues that the tailed phages are ancient, and recent structural data point to a concrete if still tentative conclusion that these phages originated prior to the divergence of cellular life into the three contemporary domains some 3.5 billion years ago. They have been evolving and shaping the ecology and evolution of their hosts ever since.

ACKNOWLEDGMENTS

Work in the author's laboratory on phage evolution is supported by NIH grant GM51975 and NSF grant EF-0333112. I am grateful to the many colleagues who have educated me on this topic, especially Dennis Bamford, Allan Campbell, Sherwood Casjens, Graham Hatfull, Jeffrey Lawrence, Maggie Smith, and Ry Young.

REFERENCES

Ackermann, H.-W. 2006. Classification of bacteriophages. In R. L. Calendar and S. T. Abedon (eds.), *The Bacteriophages*, 2nd edn. Oxford: Oxford University Press, pp. 8–16.

Angly, F. E., B. Felts, M. Breitbart, *et al.* 2006. The marine viromes of four oceanic regions. *PLoS Biol.* **4**: e368.

Baker, J., R. Limberger, S. J. Schneider, and A. Campbell. 1991. Recombination and modular exchange in the genesis of new lambdoid phages. *New Biol.* **3**: 297–308.

Baker, M. L., W. Jiang, F. J. Rixon, and W. Chiu. 2005. Common ancestry of herpesviruses and tailed DNA bacteriophages. *J. Virol.* **79**: 14967–70.

Benson, S. D., J. K. Bamford, D. H. Bamford, and R. M. Burnett. 2004. Does common architecture reveal a viral lineage spanning all three domains of life? *Mol. Cell.* **16**: 673–85.

Breitbart, M., I. Hewson, B. Felts, J. M. Mahaffy, J. Nulton, P. Salamon, and F. Rohwer. 2003. Metagenomic analyses of an uncultured viral community from human feces. *J. Bacteriol.* **185**: 6220–3.

Brüssow, H., and F. Desiere. 2006. Evolution of tailed phages: insights from comparative phage genomics, p. 26–36. In R. L. Calendar and S. T. Abedon (eds.), *The Bacteriophages*, 2nd edn. Oxford: Oxford University Press, pp. 26–36.

Casjens, S. R. 2005. Comparative genomics and evolution of the tailed-bacteriophages. *Curr. Opin. Microbiol.* **8**: 451–8.

Desplats, C., and H. M. Krisch. 2003. The diversity and evolution of the T4-type bacteriophages. *Res. Microbiol.* **154**: 259–67.

Edwards, R. A., and F. Rohwer. 2005. Viral metagenomics. *Nat. Rev. Microbiol.* **3**: 504–10.

Fauquet, C., M. A. Mayo, J. Maniloff, U. Desselberger, and L. A. Ball. (eds.). 2005. *Virus Taxonomy: Classification and Nomenclature of Viruses.* London: Elsevier/Academic Press.

Feiss, M., and A. Becker. 1983. DNA packaging and cutting, pp. 305–330. In R. W. Hendrix, J. W. Roberts, F. W. Stahl, and R. A. Weisberg (eds.), *Lambda II*. Cold Spring Harbor, NY: Cold Spring Harbor Laboratory Press, pp. 303–30.

Feiss, M., and D. A. Siegele. 1979. Packaging of the bacteriophage lambda chromosome: dependence of cos cleavage on chromosome length. *Virology* **92**: 190–200.

Filee, J., E. Bapteste, E. Susko, and H. M. Krisch. 2006. A selective barrier to horizontal gene transfer in the T4-type bacteriophages that has preserved a core genome with the viral replication and structural genes. *Mol. Biol. Evol.* **23**: 1688–96.

Fokine, A., P. G. Leiman, M. M. Shneider, *et al.* 2005. Structural and functional similarities between the capsid proteins of bacteriophages T4 and HK97 point to a common ancestry. *Proc. Natl. Acad. Sci. U.S.A.* **102**: 7163–8.

Guo, P. X., B. S. Rajagopal, D. Anderson, S. Erickson, and C. S. Lee. 1991. sRNA of phage ϕ29 of *Bacillus subtilis* mediates DNA packaging of ϕ29 proheads assembled in *Escherichia coli*. *Virology* **185**: 395–400.

Hatfull, G. F., M. L. Pedulla, D. Jacobs-Sera, *et al.* 2006. Exploring the mycobacteriophage metaproteome: phage genomics as an educational platform. *PLoS Genet.* **2**: e92.

Hendrix, R. W. 1999. Evolution: the long evolutionary reach of viruses. *Curr. Biol.* **9**: R914–7.

Hendrix, R. W. 2002. Bacteriophages: evolution of the majority. *Theor. Popul. Biol.* **61**: 471–80.

Hendrix, R. W. 2003. Bacteriophage genomics. *Curr. Opin. Microbiol.* **6**: 506–11.

Hendrix, R. W., J. G. Lawrence, G. F. Hatfull, and S. Casjens. 2000. The origins and ongoing evolution of viruses. *Trends Microbiol.* **8**: 504–8.

Huiskonen, J. T., F. de Haas, D. Bubeck, D. H. Bamford, S. D. Fuller, and S. J. Butcher. 2006. Structure of the bacteriophage phi6nucleocapsid suggests a mechanism for sequential RNA packaging. *Structure* **14**: 1039–48.

Jiang, W., Z. Li, Z. Zhang, M. L. Baker, P. E. Prevelige, Jr., and W. Chiu. 2003. Coat protein fold and maturation transition of bacteriophage P22 seen at subnanometer resolutions. *Nat. Struct. Biol.* **10**: 131–5.

Juhala, R. J., M. E. Ford, R. L. Duda, A. Youlton, G. F. Hatfull, and R. W. Hendrix. 2000. Genomic sequences of bacteriophages HK97 and HK022: pervasive genetic mosaicism in the lambdoid bacteriophages. *J. Mol. Biol.* **299**: 27–51.

Khayat, R., L. Tang, E. T. Larson, C. M. Lawrence, M. Young, and J. E. Johnson. 2005. Structure of an archaeal virus capsid protein reveals a common ancestry to eukaryotic and bacterial viruses. *Proc. Natl. Acad. Sci. U.S.A.* **102**: 18944–9.

Lawrence, J. G., R. W. Hendrix, and S. Casjens. 2001. Where are the pseudogenes in bacterial genomes? *Trends Microbiol.* **9**: 535–40.

Liu, J., and A. Mushegian. 2004. Displacements of prohead protease genes in the late operons of double-stranded-DNA bacteriophages. *J. Bacteriol.* **186**: 4369–75.

Morais, M. C., K. H. Choi, J. S. Koti, P. R. Chipman, D. L. Anderson, and M. G. Rossmann. 2005. Conservation of the capsid structure in tailed dsDNA bacteriophages: the pseudoatomic structure of phi29. *Mol. Cell.* **18**: 149–59.

Rokyta, D. R., C. L. Burch, S. B. Caudle, and H. A. Wichman. 2006. Horizontal gene transfer and the evolution of microvirid coliphage genomes. *J. Bacteriol.* **188**: 1134–42.

Sanger, F., A. R. Coulson, G. F. Hong, D. F. Hill, and G. B. Petersen. 1982. Nucleotide sequence of bacteriophage lambda DNA. *J. Mol. Biol.* **162**: 729–73.

Serwer, P., S. J. Hayes, J. A. Thomas, and S. C. Hardies. 2007. Propagating the missing bacteriophages: a large bacteriophage in a new class. *Virol J.* **4**: 21–5.

Shepherd, C. M., I. A. Borelli, G. Lander, *et al.* 2006. VIPERdb: a relational database for structural virology. *Nucleic Acids Res.* **34**: D386–9.

Simon, M. N., R. W. Davis, and N. Davidson. 1971. Heteroduplexes of DNA molecules of lambdoid phages: physical mapping of their base sequence relationships by electron microscopy. In A. D. Hershey (ed.), *The Bacteriophage Lambda.* Cold Spring Harbor, NY: Cold Spring Harbor Laboratory, pp. 313–28.

Sullivan, M. B., M. L. Coleman, P. Weigele, F. Rohwer, and S. W. Chisholm. 2005. Three *Prochlorococcus* cyanophage genomes: signature features and ecological interpretations. *PLoS Biol.* **3**: e144.

Suttle, C. A. 2005. Viruses in the sea. *Nature* **437**: 356–61.

Suttle, C. A., and F. Chen. 1992. Mechanisms and rates of decay of marine viruses in seawater. *Appl. Environ. Microbiol.* **58**: 3721–9.

Weigele, P. R., W. H. Pope, M. L. Pedulla, *et al.* 2007. Genomic and structural analysis of Syn9, a cyanophage infecting marine *Prochlorococcus* and *Synechococcus. Environ. Microbiol.* **9**: 1675–95.

Wikoff, W. R., L. Liljas, R. L. Duda, H. Tsuruta, R. W. Hendrix, and J. E. Johnson. 2000. Topologically linked protein rings in the bacteriophage HK97 capsid. *Science* **289**: 2129–33.

Zhang, T., M. Breitbart, W. H. Lee, *et al.* 2006. RNA viral community in human feces: prevalence of plant pathogenic viruses. *PLoS Biol.* **4**: e3.

CHAPTER 8

Evolutionary ecology of multiple phage adsorption and infection

Paul E. Turner* and Siobain Duffy

8.1 INTRODUCTION

Multiple adsorption of phages to individual bacteria can lead to coinfection, which can occur with reasonable likelihood given sufficiently high phage densities. As there can be an incredible diversity of phages that infect the same host (e.g., the coliphages), coinfection between unrelated phages is also possible. This chapter begins with a discussion of phenomena associated with multiple phage adsorption, including coinfection as well as superinfection exclusion and immunity. Subsequently we consider the evolutionary ecological effects specifically of coinfection: genetic exchange via recombination, reassortment, complementation between phages, phenotypic mixing, and within-host competition. Finally, we highlight research with the RNA phage $\phi6$ to illustrate the long-term effects of coinfection on the evolution of phage genes, genomes, individual phages, phage populations, and communities.

8.1.1 Evolutionary ecology of phage–phage interaction

Evolutionary ecology sits at the interface of ecology and evolutionary biology. In considering how organisms have evolved to become adapted to their environments, it examines the selective pressures imposed by the environment and the evolutionary response to these pressures (Pianka, 1999). From a purely evolutionary standpoint, a lytic phage's main goal is to successfully harness the host cell's machinery to make progeny. These offspring determine the growth capacity (fitness) of the virus relative to other genetic

* Corresponding author

Bacteriophage Ecology: Population Growth, Evolution, and Impact of Bacterial Viruses, ed. Stephen T. Abedon. Published by Cambridge University Press.

variants in the population. This genetic variability provides raw material for evolution by natural selection and, over time, a phage population becomes dominated by whichever variants are most successful at producing offspring in the particular selective environment (Chapters 6 and 9).

The study of coinfection of an individual bacterium by two or more phages dates back at least to Delbrück and Luria's analysis of "interference between two bacterial viruses acting upon the same host" (1942). Study of coadsorption (rather than coinfection) can be traced at least to Doermann (1948) and his analysis of the T-even phage lysis inhibition phenomenon (Chapter 3). Further analyses of coinfection considered multiplicity reactivation, a form of DNA repair requiring coinfection (Luria, 1947), and recombination of genetic markers between coinfecting phages (Delbrück and Bailey, 1946). Evolutionarily, the demonstration of phage recombination was fundamentally important because it showed that multi-phage infection could generate novel genetic variation, the same key phenomenon observed in other laboratory genetic models such as *Drosophila*, fungi, and *Escherichia coli*.

This chapter highlights the evolutionary ecological interactions that occur when multiple phages adsorb individual cells. The resulting virus–virus interactions may enhance or suppress the fitness of an individual infecting phage. We argue that the evolutionary ecology of multi-phage infections is inherently interesting because it can influence the course of phage evolution at multiple levels of biological organization: gene, genome, individual, population, community, and perhaps even ecosystem. See Chapter 3 and Abedon (1994) for additional consideration of the biology and evolutionary ecology of multiple phage adsorption.

8.1.2 Multiple adsorption and infection terminology

The adsorption of two or more phages to a single bacterium creates complicated biological interactions that may be further obscured by ambiguities in terminology. Therefore, we begin by defining our terms (in part as from Abedon, 1994). Adsorption (more strictly, irreversible adsorption) is the attachment of a phage to a bacterium that results in virion inactivation but not necessarily phage infection. The first phage to adsorb a bacterium may be described as a primary phage (the event is a primary adsorption). Subsequently adsorbing phages are said to be secondary phages (resulting from secondary adsorptions).

Infection is initiated when a phage genome enters the cytoplasm of a bacterium. Two phages which have infected a single bacterial cytoplasm can be said to have coinfected. Coinfection can also occur between a phage and an already infecting prophage. Superinfection is a consequence of a delay in infection by a second phage of an already infected bacterium. Note that this definition of superinfection is narrower than that generated by the common tendency among researchers to equate secondary adsorption with superinfection. Instead, all bacteria that have been successfully superinfected can be said to have been coinfected, but not all bacteria that have been successfully secondarily adsorbed can be said to be coinfected.

Manipulation of phages and bacteria to create conditions of high multiplicity (Chapter 3) does not necessarily translate into coinfection, or subsequent development of both phages if coinfection is achieved. Superinfection exclusion is a means by which a primary phage prevents a secondary phage from infecting. For example, the temperate phage Tuc2009 of *Lactococcus lactis* can block DNA injection from several phage species (McGrath *et al.*, 2002). Alternatively, some phages achieve superinfection exclusion by altering host receptors, thereby blocking secondary adsorption altogether. Other phages can express an immunity, which is a block not to phage entry into the bacterial cytoplasm but instead to subsequent steps of phage infection; the mechanisms involved in maintaining lysogeny particularly have the effect of inhibiting lytic or lysogenic infection by similar phages (Susskind and Botstein, 1980).

We define superinfection immunity as blocks on the intracellular reproduction of superinfecting phages, via the expression of repressor proteins other than those involved in the maintenance by prophages of the lysogenic state (Chapter 5; the latter is typically described simply as immunity, without the superinfection qualifier). The *Salmonella* temperate phage P22, for example, encodes two superinfection immunity proteins, one of which (sieA) prevents homologous and heterologous infection, the second of which (sieB) is solely active against heterologous phages (Susskind *et al.*, 1974). Superinfection immunity mechanisms can be very specific against competing phages, e.g., the common temperate coliphage P2 produces a protein whose sole known function is to interfere with phage T4 DNA replication (Mosig *et al.*, 1997).

8.2 PROXIMATE EFFECTS OF COINFECTION

Coinfection holds immediate consequences for the genotypic and phenotypic makeup of the resulting viral progeny, as well as the relative fitness of the coinfecting parent viruses. Four of the possible consequences are genetic exchange (sex), phenotypic mixing, complementation, and intracellular competition (Fig. 8.1). It is important to note that these proximate mechanisms do not operate in isolation during coinfection. Determining

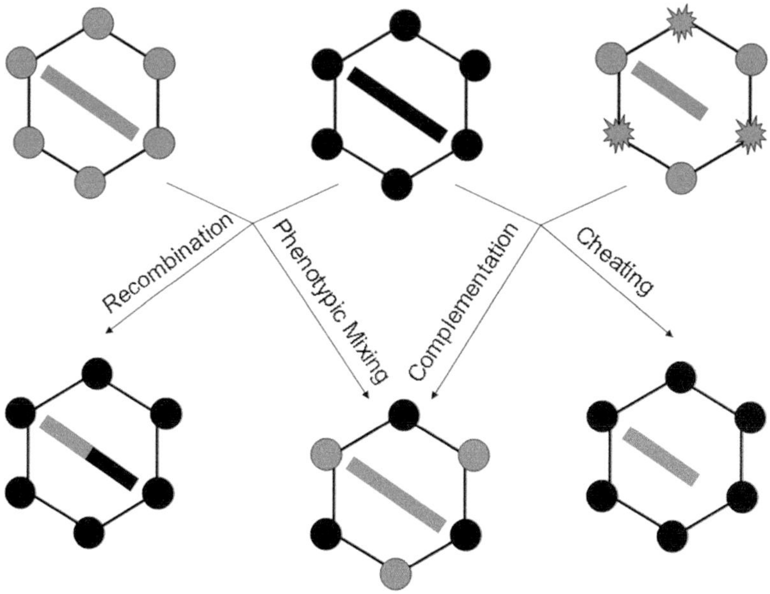

Figure 8.1 Proximate mechanisms occurring as a result of virus interactions within the same cell. The depicted phage particles differ by genotype (black versus gray bars), capsid proteins (black versus gray circles; spiked circles are defective capsid proteins), and genome size (ordinary versus shortened, where the shortened genomes are defective due to deletions). Recombination creates hybrid genotypes bearing a portion of genetic information from two or more of the coinfecting parent viruses. Phenotypic mixing can lead to the incorporation of one parent's genome into a capsid containing proteins from multiple parents. Complementation allows one virus that is relatively deficient for a protein or other product to use that which a coinfecting genotype has provided in the intracellular resource pool. Cheating allows an expectedly low fitness genotype, such as a defective deletion mutant, to gain a replication advantage during coinfection.

the relative importance of the proximate mechanisms operating within the cell is one of the biggest challenges in studying the evolutionary ecology of multi-phage infections. Not surprisingly, few studies have demonstrated the relative strengths of these mechanisms (see Froissart *et al.*, 2004, for a rare example).

8.2.1 Sex

Classic experiments with bacteria and phages established molecular biology as an independent field. A key component of many early studies was exchange of genetic material between coinfecting phages (e.g., Delbrück and

Bailey, 1946; Luria, 1947; Hershey and Rotman, 1949). Many phages with DNA and RNA genomes can experience sex via recombination of homologous genetic material. RNA viruses with segmented genomes, including the phage φ6 (family *Cystoviridae*), can experience a different form of genetic mixis, known as reassortment. When segmented phages coinfect the same cell, their RNA chromosomes are simultaneously replicated in the cytosol, allowing formation of hybrid progeny that contain different segments from the coinfecting parent genotypes. Some reassorting viruses, including influenza-A virus and the *Cystoviridae*, can also undergo heterologous recombination through template switching; however, φ6 and its relatives undergo reassortment much more readily than they do recombination (Steinhauer and Skehel, 2002; Silander *et al.*, 2005; Mindich, 2006). Recombination or reassortment provides variability (new combinations of alleles at different loci) upon which natural selection can act (Chapter 6).

8.2.2 Phenotypic mixing

Phenotypic mixing is an epigenetic (inherited, but not from the phage's own genes) consequence of coinfection. Protein products from all coinfecting phages freely circulate in the host cell's cytoplasm, and related proteins can assemble into chimeric capsids. A single genome, which could be entirely from one parent (no recombination), is then packaged into a capsid which contains proteins from more than one phage or from a different phage entirely. This outcome would result in a virus with at least some phenotypic traits other than those encoded by the genome inside the phage particle. Phenotypic mixing was discovered by Delbrück (Delbrück and Bailey, 1946) and Hershey (Hershey and Rotman, 1949), and can be distinguished from recombination by testing if the shared phenotypic traits breed true. In the case of mixed infections of phages T2 and T4, which have distinct host ranges, at least 15% of T2 genotypes have only the T4 host range (not a combined T2 and T4 host range), and at least 45% of the T4 genotypes have only the T2 host range (Streisinger, 1956). There are additional progeny (at least 16% of T2 genotypes and 12% of T4 genotypes) with a hybrid host range, indicating that the phages have tail fibers from both parental phages (Streisinger, 1956).

8.2.3 Complementation

In the parlance of Mendelian genetics, single infections are strictly haploid whereas coinfections are polyploid. Polyploidy allows for genetic dominance (masking of one gene's effects by another's). With phages this dominance is observed when deleterious mutations are complemented in trans by unmutated proteins of a coinfecting virus (some phenotypic mixing may be viewed as a form of codominance). Complementation occurs because during infection phage gene products, such as replication enzymes and structural proteins, diffuse within the cell to create a resource pool that is commonly available to all coinfecting genotypes (Nee and Maynard Smith, 1990; Turner and Chao, 1999, 2003). As a consequence, inferior and genetically recessive phage genotypes can potentially benefit from genetically dominant superior products circulating in the pool.

Complementation can result in increased or decreased yield of the parental viruses (Mahy, 2001). Extreme examples are seen upon coinfection between an ordinary virus and a defective virus that lacks one or more protein-coding functions. If a defective virus interferes with the reproduction of a coinfecting ordinary virus then it is termed a DI (defective interfering) particle. A DI particle is "defective" because it fatally lacks at least one essential gene (and therefore complementation will increase its fitness to greater than zero), and "interferes" by reducing the fitness of a coinfecting virus through using its gene products (Horiuchi, 1983; Roux *et al.*, 1991; Bull *et al.*, 2006). It is possible for different DI particles to complement each other (García-Arriaza *et al.*, 2004).

Evolutionary biologists often focus on the negative connotations of complementation, especially the masking of inferior genes by superior ones that can weaken the ability for purifying selection to remove deleterious alleles from a population (Froissart *et al.*, 2004); retention of DI particles within virus populations is a good example of this consequence. DI particles have been identified in phages (filamentous phages, reviewed in Horiuchi, 1983; their evolution in phage ϕX174 is considered in Bull *et al.*, 2006). In Section 8.3.3.3 we describe the evolution of less extreme cheating involving full-length genotypes of phage ϕ6.

8.2.4 Intracellular competition

Intracellular competition between coinfecting phages within a single host can be fundamentally similar to the competition by phages for whole hosts, or the general ecological competition for resources occurring elsewhere in the biological world. Here the currency of interest is usually relative fitness: the number of offspring one competitor can create per unit time relative to others in the same habitat. Upon coinfection, phages compete for ribosomes and tRNA pools, for polymerases, and for completed phage protein products such as assembled nucleocapsids. However, if infections are strictly clonal, then the genotype that produces more viral offspring per unit time is expected to win out in competition, and to eventually displace all other genotypes (see Chapters 2, 3, 6, 15, and 16). A phage optimizing the codon bias of its genome to match its host's tRNA pool, for example, should help increase its speed of protein production, and therefore the number of complete virions.

When coinfection is common, phages can adapt to better control the intracellular resource pool. A naive assumption is that the virus that multiplies the fastest during clonal infections would also win a contest between coinfecting viruses. However, our work and that of others suggests that viruses can specialize in intracellular "cheating" to gain superiority during within-cell competition (Horiuchi, 1983; Roux *et al.*, 1991; Turner and Chao, 1999). Here, a virus does not have to be a particularly good replicator during clonal infection to competitively win during coinfection. Rather, a phage could evolve traits that ensure its progeny are overrepresented in the next generation (relative fitness advantage), such as biased entry of its genetic material into capsids provided by other coinfecting genotypes (thereby exploiting mechanisms such as complementation and phenotypic mixing). DI particles, for example, tend to have a replication advantage inside the cell because they have shortened, incomplete genomes.

8.3 ULTIMATE EFFECTS OF COINFECTION

Biological systems are organized at many different levels, and this section describes the longer-term, ultimate effects of phage coinfection on these various levels. Recognized levels of biological organization include genome, individual, population, and community. Phage coinfection can influence phage evolution on all these levels. Due to both space limitations and familiarity, we bias our discussion of the ultimate effects of phage coinfection towards work employing phage ϕ6. Phage ϕ6 is a three-segmented dsRNA phage that infects plant-pathogenic *Pseudomonas* species (Duffy *et al.*, 2006) and which has been extensively used in experimental ecology and evolution.

8.3.1 Phage ϕ6 and coinfection

Many of the proximate mechanisms associated with coinfection – complementation, sex, phenotypic mixing, intracellular competition (Section 8.2) – seem to operate in phage ϕ6. Certain phages can prevent additional viruses from entering or replicating within the cell, thus affecting the opportunity for these mechanisms to operate. Phage ϕ6 is one of the few viruses where the average number of virus particles that can successfully coinfect a single cell (of its usual host, *Pseudomonas syringae* pv. *phaseolicola*) has been enumerated (Olkkonen and Bamford, 1989; Turner *et al.*, 1999): only one, two, or three phages successfully enter, even following adsorption of an average of 40 virion particles.

8.3.2 Ultimate consequences of clonality (asexuality)

For viruses to exchange genes they must coinfect an individual cell. If two viruses are genotypically identical, then this coinfection will not result in the generation of new combinations of viral genes. Thus, a viral population initiated with a single virus particle can reasonably be described as clonal even if coinfection among clones can occur. Such a description, however, ignores viral mutations that can accumulate during population growth, and coinfection among these novel variants would in fact constitute a form of sexuality.

Owing to the nature of viral propagation in the laboratory (Chapter 9), it can be inconvenient to completely eliminate coinfection. Consequently, a common means to minimize the impact of coinfection on viral evolution is to initiate viral populations using a single individual and to otherwise limit the extent of viral population growth. By limiting phage propagation to populations initiated with only a small number of individuals, one introduces a population bottleneck, thereby effecting genetic drift (Chapter 6). Unlike natural selection, drift gives rise to random changes in the allelic frequencies within populations, including especially the loss of beneficial variants (fixing of deleterious alleles), and can have severe effects on population fitness. Consistently, Chao (1990) found that repeated phage $\phi 6$ bottlenecking through single plaques caused lower-fitness individuals to come to dominate laboratory populations (Muller's ratchet; see Chapter 6).

8.3.3 Ultimate consequences of coinfection (sexuality)

Because sex can be defined as any exchange of genetic material between individuals (Section 8.2.1), mechanisms such as recombination and plasmid-mediated conjugation cause many phages and bacteria to be just as "sexy" as eukaryotes that experience obligate sexual reproduction (Turner, 2003; Silander *et al.*, 2005). In this section, contrasting that immediately above (Section 8.3.2), we discuss consequences of aclonal ("sexual") viral replication, i.e., that involving coinfection, especially between genotypically heterogeneous individuals.

8.3.3.1 Preventing Muller's ratchet

Chao *et al.* (1997) showed that sex via segment reassortment can reverse the decline in average fitness of individuals in the population. That is, coinfection permits the genetic exchange whereby mutated parent genotypes can produce hybrid offspring bearing fewer or no mutations. These results suggested that viral sex might have evolved, in part, to combat Muller's ratchet (Chao *et al.*, 1992; Chapter 6). The study also hinted that the ability to exchange genetic material is an individual adaptive trait possessed by a virus, because it can allow a sexual genotype to produce progeny with greater fitness (fewer deleterious mutations) than the parent genotype. We note that a segmented virus such as phage ϕ6, which displays limited molecular recombination, can only experience rescue from Muller's ratchet in this manner if deleterious mutations are not confined or largely restricted to one segment. Had deleterious mutations been largely restricted to a single segment, then it would have decreased the likelihood that Chao *et al.* (1992) would have observed a benefit to sexual reassortment in their study.

8.3.3.2 Selection for intracellular replication strategies

The cell provides a forum for coinfecting genotypes to exchange genetic material (Section 8.2.1), and this may be beneficial at least some of the time (Chao *et al.*, 1997). But competition for resources is a more universal ecological interaction, and intracellular competition can occur between individual coinfecting genotypes whether or not they can or do undergo sex. Since relative ability to exploit a common resource is often the phenotypic gauge for determining the fitness of phages, a phage is more fit if it is more effective at harnessing the cellular machinery and available proteins to make offspring. With coinfection, however, the exploitable resource in fact may consist of more than just the infected bacterium; that is, the coinfecting phage might also be exploitable.

If intracellular competition rests solely on the ability of each phage to exploit the host cell, then one would expect that through time the phage population would experience selection to favor the relatively fitter variant, whether in clonal infection or during coinfection. In this case, we would expect selection to favor the phage which could replicate its genome, produce more phage protein, and package its genome into its nucleocapsid more quickly. However, if the coinfecting phage is also exploitable, then selection can favor that exploitation, and can do so potentially at the expense of exploitation of the cell. For example, with frequent coinfection, a phage may evolve to replicate its genome more quickly, at the expense of protein production, and thereby competitively displace the more efficiently protein-producing competitor genome from packaging into the capsids it so diligently created.

Lewontin (1970) suggested that adaptation to intrahost competition may occur through novel virus traits that detract from the ability of the virus to exploit its host. Turner and Chao (1998) used phage ϕ6 to examine adaptation of viruses to *P. syringae* pv. *phaseolicola* in the presence and absence of intracellular competition. They observed that the lineages that evolved for hundreds of generations under frequent coinfection – and hence with intracellular competition – acquired traits that were beneficial when coinfection was allowed, but traded off with growth performance in an environment where phages infected cells singly (Turner and Chao, 1998, 1999, 2003; Dennehy and Turner, 2004; see also a similar example in foot-and-mouth disease virus, Sevilla *et al.*, 1998). Thus, the derived coinfecting phages were similar to DI particles because they experienced a strong selective advantage when coinfecting with wild-type ϕ6; however, these phages were not completely defective because they were still able to replicate independently and still possessed all of the ϕ6 genes.

The contextual fitness of the evolved phages showed that the net effect of coinfection was negative in the experiment, because the struggle to compete intracellularly led to genetic trade-offs for the viruses in other habitats. One way to view this result is in terms of an organism adapting to a new host that trades off with its fitness in the old host habitat, more generally referred to as the evolution of ecological specialization (Futuyma and Moreno, 1988). That is, the phages became adaptively specialized to a new habitat consisting of bacteria containing wild-type ϕ6, while losing performance ability in the old environment (just bacteria). Another way to view this same adaptation is in terms of game theory, as the evolution of cheating.

8.3.3.3 Coinfection and game theory

Virus variants coinfecting the same cell do not necessarily have exclusive access to their own protein products; rather, the resource pool may be freely accessible to any virus within the cell, creating a potential conflict over using these products equally (Nee and Maynard Smith, 1990; Nee, 2000; Brown, 2001; Turner and Chao, 2003; Section 8.3.3.2). On the one hand, viruses could donate the same number of resources to the pool as they need to make offspring, an action analogous to behavioral cooperation in social organisms. On the other hand, viruses could specialize in parasitizing the pool, and benefit from not equally contributing to resource production. The latter strategy is akin to behavioral cheating, and is exemplified by DI particles (Sections

8.2.3 and 8.2.4). For analogous strategies, as they occur in terms of phage between-infection competition, see Chapters 2 and 3.

Turner and Chao (1999) described how the evolution of cheating over the intracellular resource pool explained earlier results when the phage was evolved in the presence and absence of frequent coinfection (Turner and Chao, 1998). The individual viruses in the coinfection treatment (multiplicity = 5) retained a full set of genes but were greatly advantaged only in mixed infections, suggesting that they evolved traits to selfishly sequester products that other coinfecting viruses contributed to the resource pool. But the drawback to this evolved strategy was that mutations which led to cheating (or loci on the same segment closely linked to that mutation; Duffy *et al.*, unpublished) were inferior when the viruses competed in alternative environments where coinfection was uncommon (Dennehy and Turner, 2004). This result was consistent with the decline in population fitness due to takeover by individual cheaters, termed the prisoner's dilemma in evolutionary game theory (Maynard Smith, 1982; Turner and Chao, 1999; Chapter 3; Fig. 3.4). An additional phage ϕ6 study showed that cheater genotypes can be prevented from taking over the population, but can instead be "tolerated" as a minority subpopulation when the coinfecting helpers are very productive in contributing products to the resource pool (Turner and Chao, 2003).

8.3.3.4 Phages, sex, and adaptation

The importance of genetic mixis in viruses is well known from studies involving viruses that infect eukaryotes, especially those of economic and health importance (Steinhauer and Skehel, 2002; Varma and Malathi, 2003; Heeney *et al.*, 2006). It is rarely asked, however, if the homologous recombination or reassortment that are products of mixis are similarly crucial in the natural adaptation of phages: whether phage sexuality is favored over asexuality because it brings together good combinations of alleles in large populations of closely related phages, thus accelerating the pace of adaptation via natural selection (Fisher, 1930; Muller, 1932; see also Felsenstein, 1974). It is conceivable that genetic exchange among closely related phages could play an important role in phage evolution. Even if coinfection happens rarely, the novel genetic variants produced may spearhead key changes in the phage population, akin to the antigenic shifts occurring in influenza virus (Steinhauer and Skehel, 2002). In the parlance of evolutionary biology, "a little sex" may often be enough to provide the variation needed for major transitions in the adaptive success of a lineage. Only two studies have used phages to examine whether sex speeds the rate of adaptation in large populations (Malmberg, 1977; Turner and Chao, 1998), indicating a need for further work in this area to elucidate the general significance (if any) of sex in phage adaptation.

8.3.3.5 Phages, sex, and purifying selection

Purifying selection is the other general process historically used to argue for the adaptive significance of sex in selecting for sexual lifestyles over asexual ones (Felsenstein, 1974). Here sex is expected to provide a benefit in promoting selection's efficiency at removing deleterious mutations from populations of large size. In such populations, selection will cause harmful mutations to decrease in frequency over time because the genotypes harboring them are of low fitness relative to others in the population. Sex has the ability to combine these alleles into the same genetic background, allowing selection to be more efficient. Synergistic epistasis between deleterious alleles makes combinations of these alleles in the same genetic background more harmful for fitness, relative to their fitness effects when measured alone (Chapter 6). Theoretical work states that if synergistic epistasis exists between deleterious alleles, then selection should be especially efficient at purging deleterious alleles, while the remaining hybrid progeny, having fewer deleterious alleles, will be highly favored. Sex therefore should be advantaged (Eshel and Feldman, 1970; Kondrashov, 1993).

Froissart *et al.* (2004) examined the impact of coinfection on purifying selection using large populations of phage $\phi6$ founded by a mixture of four genotypes: the wild type plus three mutants that differed by a single unique deleterious allele on one of each of the three $\phi6$ segments. These three mutations showed synergistic epistasis such that the double and triple mutants generated through reassortment were of very low fitness. When propagated clonally, selection should cause the asexual population to approach the known fitness optimum consisting of fixation of the wild type. In contrast, coinfecting populations might allow the optimum to be approached more rapidly because sex creates double and triple mutants through reassortment, which, as a consequence of synergistic epistasis, more efficiently eliminates the deleterious alleles from the population. In this way, synergistic epistasis was expected to cause the sexual (coinfecting) populations to increase in fitness faster than their asexual counterparts.

Froissart *et al.* (2004) recognized that complementation (Section 8.2.3) could also operate in populations of phage $\phi6$, whereby coinfection allows viruses of low fitness to have their deleterious alleles masked in trans by other genotypes bearing superior alleles. Their experiment consequently determined which of two intracellular forces, synergistic epistasis versus complementation, was more important in the evolution of phages experiencing multiple infections. If sexuality were more efficient than asexuality in removing harmful alleles from the populations, then the deleterious mutations would disappear faster from the coinfecting lineages. However, if complementation was the more potent force, this masking effect would weaken selection against the harmful alleles during coinfection, causing them to disappear faster in the non-coinfecting lineages. Results showed that effects of complementation overwhelmed the expected advantage of sex in terms of purifying selection.

8.4 COINFECTION BETWEEN HETEROLOGOUS PHAGES

When phage populations exist at high densities, as follows exponential growth within environments containing relatively high bacterial densities, then coinfection of a single cell by closely related phages can occur (Chapter 3). Typically, however, many species of phage can infect the same bacterial strain. It is expected, therefore, that phages experience coinfection with phages of different species at least occasionally, especially given phage infection of a bacterial lysogen (Lawrence *et al.*, 2002; Chapter 7). These events could lead to exchange of genetic material between very different phages. More generally, the scenarios would allow for interspecific competition between phages within the same cell. Concepts used to describe the consequences of these interspecific interactions include mutual exclusion, partial exclusion, and the depressor effect (Abedon, 1994). The evolutionary pressure created by this interspecific phage competition is also seen in heterologous superinfection exclusion and immunity (Section 8.1.2), and in the phenotypic mixing between different phage types (Section 8.2.2). Multiple phage species must sometimes enter the same cell and compete for resources in nature, but this is still a largely untapped area of phage ecological research.

8.4.1 Long-term hyperparasitism (P4-like phages)

One well-understood interspecific phage relationship is the P2–P4 interaction. P2 is a ~33 kbp temperate coliphage, which is somewhat unusual as its prophage is not induced by host SOS response (Nilsson and Ljungquist, 2006). P2, and P2-like prophages, appear to be widespread among *E. coli* isolates (>25% of the ECOR strains; Nilsson and Ljungquist, 2006). P4 is an incomplete (lacking genes for a head, tail, and proteins involved in lysis), temperate, satellite phage of about 11.5 kbp (Dehó and Ghisotti, 2006). As it is incomplete, P4 cannot complete a lytic infection cycle on its own; it can only form plaques on host cells which are already lysogens containing a P2 or certain other prophages. P4 does not appear to be related to or derived from P2, as a DI particle would be. Yet P4 can create infectious phage particles when infecting a P2 lysogen, either by lytic coinfection, by its lytic infection inducing a P2 prophage, or by P2 infecting a host which already contains a P4 prophage (or P4 in a plasmid state; see Dehó and Ghisotti, 2006). Once both phages are in the same cell, and producing transcripts, then P4 co-opts P2 proteins to form smaller capsid heads than P2 would, into which it packages only the P4 genome.

Phage P4 thus has evolved a hyperparasitic lifestyle, parasitizing a phage that must in turn parasitize a bacterial host. It is not the only phage to do so: a number of satellite phages have subsequently been discovered, apparently related to P4 (Dehó and Ghisotti, 2006). The relationship, which is more complicated than simply competing for host ribosomes or available protein products, represents another consequence of coevolution among bacteriophages. While many phages have evolved resistance mechanisms to block further phage infection, there is as yet no evidence that phage P2 has evolved to defend itself against parasitization by phage P4. It may be that P2 instead gains some benefit from increased horizontal transfer when P4 induces P2 lytic infection, and that the interaction reflects mutualism more than parasitism. Alternatively, P2 simply may not encounter P4 enough in the wild for selection to favor the evolution of countermeasures. Future research is needed to tease apart the complex evolutionary ecology of the interactions between these viruses.

8.4.2 Phage biogeography

The gene exchange that can occur during coinfection has important consequences for natural phage populations. Sharing a common gene pool defines a population of organisms (Hartl and Clark, 1997). If phage populations experience barriers to coinfection – either by geographic distance or by evolved molecular mechanisms – then the gene pools of different phages can diverge to the extent that separate phage species develop (Chapter 6).

Silander *et al.* (2005) examined phage biogeography by isolating 25 wild cystoviruses from California and Connecticut that were capable of infecting the usual laboratory host of ϕ6. Sequencing portions of each of the three genomic segments showed frequent reassortment, as demonstrated by low linkage disequilibrium (Chapter 14). In fact, within these *Cystoviridae,* reassortment was as important as mutation in generating genetic variation. Lineages of phage ϕ6 evolved on a novel host, however, can rapidly lose ability to infect their original host, resulting in reduced opportunity for reassortment with the parental phage lineage (Duffy *et al.*, 2007). The ensuing reproductive isolation could represent a phage equivalent to the biological species concept of eukaryotes (Mayr, 1942).

8.5 GENOME AND GENE-LEVEL EFFECTS OF COINFECTION

The previous sections examined the effects of coinfection on phage evolution at individual and higher levels. Here we take a more reductionist view. We will consider how the evolutionary ecology of phage interactions can shape the architecture of the phage genome, especially its ability to maintain a constant phenotype when new mutations are added. In addition, we will consider how the evolution of selfish cheating phages may be explained by selfish actions at the level of their genes.

8.5.1 Coinfection weakens selection for robust genomes

When Darwin (1859) proposed the interplay between variation and natural selection as the driving force of evolution, he had no idea what material produced that variation. While we now know that nucleic acid is the genetic material, and how mutation and recombination create variation, evolutionary biologists are still struggling to elucidate the details of the interplay between variation and selection.

An organism is a collection of genes (the genotype) that dictate physical characteristics (the phenotype); the environment saves or discards a genotype based on its phenotypic performance, or fitness. If the population of genotypes becomes well adapted to the environment, then any mutation is likely to reduce individual performance. Thus, populations at equilibrium in their environment should experience selection for mechanisms that preserve phenotypic expression despite introduced mutations. This phenomenon is called mutational (or genetic) robustness (Chapter 6). A genome is robust if it experiences mutations but maintains a constant phenotype. A brittle genome, on the other hand, is relatively more susceptible to phenotypic change due to mutations. Although the concept of robustness is widely accepted by evolutionary biologists (de Visser *et al.*, 2003), there is little empirical evidence that robustness arises from selection rather than via chance alone. Support for the idea mostly comes from mathematical theory (Wagner *et al.*, 1997) and from studies of "digital organisms" – computer programs that self-replicate, mutate, and evolve (Wilke *et al.*, 2001). It has been notoriously difficult to demonstrate a link between robustness and selection in the laboratory.

Montville *et al.* (2005) provided experimental evidence for adaptive robustness using populations of phage ϕ6. Complementation allows a phage ϕ6 genotype to overcome its own mutational deficiencies by using the proteins produced by more-fit viruses coinfecting the same host cell (Froissart *et al.*, 2004). Complementation is thus a built-in mechanism for robustness, because it serves to buffer mutational effects. By this logic, complementation (and hence, coinfection) should weaken selection for a phage ϕ6 genome to maintain robustness, because the trait is an inherent feature of the coinfection environment. Montville *et al.* (2005) therefore predicted that infection mode – clonality versus coinfection – should impact evolution of robustness in phage ϕ6; in particular, less robust phage genomes should be selected under coinfection.

To test this idea, the authors used phage ϕ6 lineages (Turner and Chao, 1998) evolved for hundreds of generations in the presence or absence of coinfection. They predicted that selection for robustness should be relaxed in the coinfecting phages, because complementation provides a built-in buffering mechanism. Clones were isolated at random from each treatment population and used to generate a set of lineages that were subjected to mutation accumulation via bottlenecking. To examine the effect of input mutations on phenotypic fitness (W, growth rate on the host bacteria; see Chapter 6), they measured pre- and post-bottleneck fitness of a lineage; the difference, $\Delta \log W$, indicated the lineage's sensitivity to phenotypic effects of added mutations. If the lineages founded by clones historically evolved under coinfection were less robust (more brittle) then they would show greater variance in $\Delta \log W$, owing to weak selection for maintaining a robust genome (Fig. 8.2). The data supported this prediction, arguing that selection for mutational robustness is stronger in the absence of virus coinfection (Montville *et al.*, 2005; Fig. 8.2), and suggesting that an additional cost of coinfection in phage ϕ6 is evolution of relatively brittle genomes. Future work is needed to determine the molecular mechanisms for robustness in this phage, as well as to explore the link between coinfection and adaptive robustness in other virus systems.

8.5.2 Sex and the selfish gene

Richard Dawkins popularized the "selfish gene hypothesis," which posits that genes are the fundamental units of selection as well as of inheritance (Dawkins, 1976). All higher-level functions of organisms, which Dawkins calls vehicles, are due to and for the benefit of genes acting selfishly on their own behalf. Selfish genes offer another way of understanding how cheating evolves as a result of frequent phage coinfection.

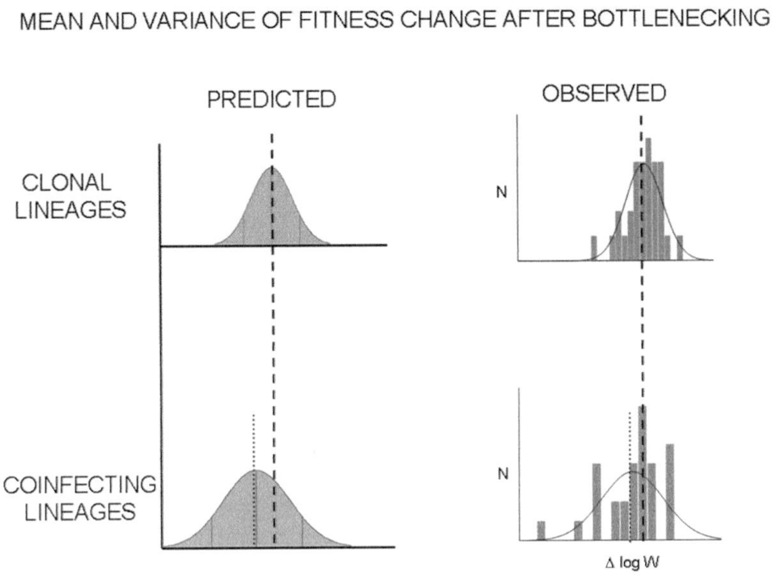

Figure 8.2 Coinfecting phages can experience complementation that masks mutational effects. Montville *et al.* (2005) predicted that this built-in robustness mechanism would cause phages evolved under coinfection to experience weakened selection for individual robustness. Hence, lineages historically evolved under coinfection should show greater change in fitness (Δ log$W$) due to mutation accumulation, relative to clonally evolved lineages. This prediction was supported using experiments with phage ϕ6. Predicted and observed distributions reflect the variance in Δ log$W$ values, and the dashed and dotted lines represent means for the distributions.

If coinfection of the same host cell is common, then a virus can evolve to allocate more energy to replication and come to rely on other viruses to produce the required proteins (Chao *et al.*, 2000). It is relatively easy to understand the cheating of DI particles with shortened genomes. DI particles cannot produce their share of proteins, as they lack some portion of the protein-coding genes, and simultaneously they have smaller genomes, which can replicate faster than complete genomes (for the details of the best-understood DI particles, those of vesicular stomatitis virus, see Roux *et al.*, 1991). They can replicate more genomes per unit time than a complete virus can, and, as long as the packaging region is unaffected, will competitively displace complete genomes from packaging into capsids. But the evolution of cheating in phage ϕ6 is conceptually problematic. The cheaters maintain a full-length and functional genome; thus there is no intuitive genome replication advantage of the cheater genotype over the ordinary ϕ6 (Turner and Chao, 1999). The quandary is: how can full-length cheaters evolve if the phage has a segmented genome and cheating is selected only under frequent coinfection where reassortment is possible?

Dawkins' hypothesis may provide the answer to this paradox. If cheating in phage ϕ6 is a complex trait that requires action of genes located on separate RNA segments, then reassortment can easily blur the distinction between cheater and non-cheater. However, if the cheating trait is reducible to the action of a single segment, then the prisoner's dilemma result of Turner and Chao (1999) might really be due to a selfish RNA segment ensuring that it is overrepresented in the next generation. In certain monosegmented RNA viruses, DI particles gain a replicative advantage through gene duplications or other mutations involved in encapsidation (Roux *et al.*, 1991). Phage ϕ6 packages its three segments into the procapsid in a highly specific and sequential fashion (Mindich, 2006). Therefore, a cheating segment need not contain a gene that efficiently makes capsid proteins. It simply has to ensure that it is preferentially packaged. The exact molecular mechanism for cheating in phage ϕ6, nevertheless, has yet to be determined (Turner *et al.*, unpublished).

8.6 CLOSING REMARKS

In nature, viruses can and do occasionally experience intracellular interactions with other viruses of similar and dissimilar genetic backgrounds and evolutionary histories. Virologists and phage biologists have long recognized this possibility, but empirical research has rarely reflected the complications of coinfection. Phage–phage interactions, and their ecological and evolutionary consequences, offer new frontiers for phage ecology research, which may be pursued within the contexts of naturally complex aquatic and terrestrial environments (Chapters 10 and 11). In addition, this research can and should be undertaken in the laboratory (e.g., see Chapter 3). As laboratory experiments often impose multiple adsorption and coinfection of a single phage strain without appreciating the effects of the resulting interactions, further understanding of intraspecific interactions will help in the interpretation of current phage research. It will also facilitate the study of interspecific interactions, such as within bacterial hosts, as researchers can build up the levels of community complexity within the controlled laboratory environment. Such research will also help elucidate the extent of coinfection occurring in nature, and its potential to influence phage evolution.

Many phages are well understood molecularly, offering suitable candidates for this line of research. The *Pseudomonas* phage ϕ6, which undergoes genetic exchange through reassortment, not recombination, is a particularly elegant model in which to tease apart the various proximate and ultimate consequences of intra- as well as interspecific coinfection. Phage ϕ6 and other facultatively sexual RNA viruses evolve very rapidly, so that laboratory experimental adaptation in these systems can quickly lead to genetic variation through mutation and mixis, and to changes in allele frequencies within populations. The evolutionary ecology of multi-phage infections thus offers one of the most dynamic and exciting forums for bacteriophage ecology research.

8.7 ACKNOWLEDGMENTS

This work was supported by the Woodrow Wilson Foundation and the US National Science Foundation grant DEB-04–52163 to P.E.T.; S.D. was supported by a predoctoral fellowship from the Howard Hughes Medical Institute.

REFERENCES

Abedon, S. T. 1994. Lysis and the interaction between free phages and infected cells. In J. D. Karam (ed.), *The Molecular Biology of Bacteriophage T4*. Washington, DC: ASM Press, pp. 397–405.

Brown, S. P. 2001. Collective action in an RNA virus. *J. Evol. Biol.* **14**: 821–8.

Bull, J. J., J. Millstein, J. Orcutt, and H. A. Wichman. 2006. Evolutionary feedback mediated through population density, illustrated with viruses in chemostats. *Am. Nat.* **167**: E39–E51.

Chao, L. 1990. Fitness of RNA virus decreased by Muller's ratchet. *Nature* **348**: 454–5.

Chao, L., T. Tran, and C. Matthews. 1992. Muller's ratchet and the advantage of sex in the RNA virus ϕ6. *Evolution* **46**: 289–99.

Chao, L., T. T. Tran, and T. T. Tran. 1997. The advantage of sex in the RNA virus ϕ6. *Genetics* **147**: 953–9.

Chao, L., K. A. Hanley, C. L. Burch, C. Dahlberg, and P. E. Turner. 2000. Kin selection and parasite evolution: higher and lower virulence with hard and soft selection. *Q. Rev. Biol.* **75**: 261–75.

Darwin, C. 1859. *On the Origin of Species by Means of Natural Selection, or the Preservation of Favoured Races in the Struggle for Life*. London: John Murray.

Dawkins, R. 1976. *The Selfish Gene*. Oxford: Oxford University Press.

Dehó, G., and D. Ghisotti. 2006. The satellite phage P4. In R. L. Calendar and S. T. Abedon (eds.), *The Bacteriophages*, 2nd edn. Oxford: Oxford University Press, pp. 391–408.

Delbrück, M., and W. T. Bailey, Jr. 1946. Induced mutations in bacterial viruses. Cold Spring Harbor Symp. *Quant. Biol.* **11**: 33–7.

Delbrück, M., and S. E. Luria. 1942. Interference between bacterial viruses. 1. Interference between two bacterial viruses acting upon the same host, and the mechanism of virus growth. *Arch. Biochem.* **1**: 111–41.

Dennehy, J. J., and P. E. Turner. 2004. Reduced fecundity is the cost of cheating in RNA virus φ6. *Proc. Biol. Sci.* **271**: 2275–82.

de Visser, J. A. G. M., J. Hermisson, G. P. Wagner, *et al.* 2003. Perspective: evolution and detection of genetic robustness. *Evolution* **57**: 1959–72.

Doermann, A. H. 1948. Lysis and lysis inhibition with *Escherichia coli* bacteriophage. *J. Bacteriol.* **55**: 257–276.

Duffy, S., P. E. Turner, and C. L. Burch. 2006. Pleiotropic costs of niche expansion in the dsRNA bacteriophage φ6. *Genetics* **172**: 751–7.

Duffy, S., C. L. Burch and P. E. Turner. 2007. Evolution of host specificity drives reproductive isolation among RNA viruses. *Evolution* **61**: 2614–22.

Eshel, I., and M. W. Feldman. 1970. On the evolutionary effect of recombination. *Theor. Pop. Biol.* **1**: 88–100.

Felsenstein, J. 1974. The evolutionary advantage of recombination. *Genetics* **78**: 737–56.

Fisher, R. A. 1930. *The Genetical Theory of Natural Selection*. Oxford: Oxford University Press.

Froissart, R., C. O. Wilke, R. Montville, S. K. Remold, L. Chao, and P. E. Turner. 2004. Co-infection weakens selection against epistatic mutations in RNA viruses. *Genetics* **168**: 9–19.

Futuyma, D. J., and G. Moreno. 1988. The evolution of ecological specialization. *Ann. Rev. Ecol. Syst.* **19**: 207–33.

García-Arriaza, J., S. C. Manrubia, M. Toja, E. Domingo, and C. Escarmís. 2004. Evolutionary transition toward defective RNAs that are infectious by complementation. *J. Virol.* **78**: 11678–85.

Hartl, D. L., and A. G. Clark. 1997. *Principles of Population Genetics*. Sutherland, MA: Sinauer Associates.

Heeney, J. L., A. G. Dalgleish, and R. A. Weiss. 2006. Origins of HIV and the evolution of resistance to AIDS. *Science* **313**: 462–6.

Hershey, A. D., and R. Rotman. 1949. Genetic recombination between host-range and plaque-type mutants of bacteriophage in single bacterial cells. *Genetics* **34**: 44–71.

Horiuchi, K. 1983. Co-evolution of a filamentous bacteriophage and its defective interfering particles. *J. Mol. Biol.* **169**: 389–407.

Kondrashov, A. S. 1993. Classification of hypotheses on the advantage of amphimixis. *J. Hered.* **84**: 372–87.

Lawrence, J. G., G. F. Hatfull, and R. W. Hendrix. 2002. Imbroglios of viral taxonomy: genetic exchange and failings of phenetic approaches. *J. Bacteriol.* **184**: 4891–905.

Lewontin, R. C. 1970. The units of selection. *Ann. Rev. Ecol. Syst.* **1**: 1–18.

Luria, S. E. 1947. Reactivation of irradiated bacteriophage by transfer of self-reproducing units. *Proc. Natl. Acad. Sci. U.S.A.* **33**: 253–64.

Mahy, B. W. J. 2001. *A Dictionary of Virology*. San Diego, CA: Academic Press.

Malmberg, R. L. 1977. The evolution of epistasis and the advantage of recombination in populations of bacteriophage T4. *Genetics* **86**: 607–21.

Maynard Smith, J. 1982. *Evolution and the Theory of Games*. Cambridge: Cambridge University Press.

Mayr, E. 1942. *Systematics and the Origin of Species*. New York, NY: Columbia University Press.

McGrath, S., G. F. Fitzgerald, and D. van Sinderen. 2002. Identification and characterization of phage-resistance gene in temperate lactococcal bacteriophages. *Mol. Microbiol.* **43**: 509–20.

Mindich, L. 2006. Phages with segmented double-stranded RNA genomes. In R. L. Calendar and S. T. Abedon (eds.), *The Bacteriophages*, 2nd edn. Oxford: Oxford University Press, pp. 197–207.

Montville, R., R. Froissart, S. K. Remold, O. Tenaillon, and P. E. Turner. 2005. Evolution of mutational robustness in RNA viruses. *PLoS Biol.* **3**: 1939–45.

Mosig, G., S. Yu, H. Myung, *et al.* 1997. A novel mechanism of virus–virus interactions: bacteriophage P2 Tin protein inhibits phage T4 DNA synthesis by poisoning the T4 single-stranded DNA binding protein. *Virology* **230**: 72–81.

Muller, H. J. 1932. Some genetic aspects of sex. *Am. Nat.* **66**: 118–38.

Nee, S. 2000. Mutualism, parasitism and competition in the evolution of coviruses. *Philos. Trans. R. Soc. Lond. Ser. B* **355**: 1607–13.

Nee, S., and J. Maynard Smith. 1990. The evolutionary biology of molecular parasites. *Parasitology* **100**: S5–S18.

Nilsson, A. S. and E. H. Ljungquist. 2006. The P2-like bacteriophages. In R. L. Calendar and S. T. Abedon (eds.), *The Bacteriophages*, 2nd edn. Oxford: Oxford University Press, pp. 365–90.

Olkkonen, V. M., and D. H. Bamford. 1989. Quantitation of the adsorption and penetration stages of the bacteriophage φ6 infection. *J. Virol.* **62**: 1180–5.

Pianka, E. 1999. *Evolutionary Ecology*, 6th edn. San Francisco, CA: Benjamin Cummings.

Roux, L., A. E. Simon, and J. J. Holland. 1991. Effects of defective interfering viruses on virus replication and pathogenesis. *Adv. Virus. Res.* **40**: 181–211.

Sevilla, N., C. M. Ruiz-Jarabo, G. Gomez-Mariano, E. Baranowski, and E. Domingo. 1998. An RNA virus can adapt to the multiplicity of infection. *J. Gen. Virol.* **79**: 2971–80.

Silander, O. K., D. M. Weinreich, K. M. Wright, *et al.* 2005. Widespread genetic exchange among terrestrial bacteriophages. *Proc. Natl. Acad. Sci. U.S.A.* **102**: 19009–14.

Steinhauer, D. A., and J. J. Skehel. 2002. Genetics of influenza viruses. *Annu. Rev. Genet.* **306**: 305–32.

Streisinger, G. 1956. Phenotypic mixing of host range and serological specificities in bacteriophages T2 and T4. *Virology* **2**: 388–98.

Susskind, M. M., and D. Botstein. 1980. Superinfection exclusion by λ prophage in lysogens of *Salmonella typhimurium. Virology* **100**: 212–16.

Susskind, M. M., D. Botstein, and A. Wright. 1974. Superinfection exclusion by P22 prophage in lysogens of *Salmonella typhimurium. Virology* **62**: 350–66.

Turner, P. E. 2003. Searching for the advantages of virus sex. *Origins Life Evol. Biosphere* **33**: 95–108.

Turner, P. E., and L. Chao. 1998. Sex and the evolution of intrahost competition in RNA virus ϕ6. *Genetics* **150**: 523–32.

Turner, P. E., and L. Chao. 1999. Prisoner's dilemma in an RNA virus. *Nature* **398**: 441–3.

Turner, P. E., and L. Chao. 2003. Escape from prisoner's dilemma in RNA phage ϕ6. *Am. Nat.* **161**: 497–505.

Turner, P. E., C. L. Burch, K. A. Hanley, and L. Chao. 1999. Hybrid frequencies confirm the limit to coinfection in the RNA bacteriophage ϕ6. *J. Virol.* **73**: 2420–4.

Varma, A., and V. G. Malathi. 2003. Emerging geminivirus problems: a serious threat to crop production. *Ann. Appl. Biol.* **142**: 145–64.

Wagner, G. P., G. Booth, and H. Bagheri-Chaichian. 1997.A population genetic theory of canalization. *Evolution* **51**: 329–47.

Wilke, C. O., J. L. Wang, C. Ofria, R. E. Lenski, and C. Adami. 2001. Evolution of digital organisms at high mutation rates leads to survival of the flattest. *Nature* **412**: 331–3.

CHAPTER 9

Patterns in phage experimental adaptation

J. J. Bull

9.1 INTRODUCTION

From the perspective of evolutionary studies, phages offer benefits of short generation times, small genomes, and ease of propagation. These advantages apply chiefly to the use of phages in laboratory experiments, however. Although some understanding of phage evolution in a natural environment can be inferred from genome sequences, we have precious little understanding of the natural environment of any phage and are even ignorant of what constitutes the population of a phage. So when phages are used as model systems of evolution, with tight controls on host strains, host density, media, temperature, and population sizes, it must be assumed that we are hoping to discover generalities that transcend the specific context of the lab environment. It certainly cannot be pretended that our systems create miniature replicas of what a phage experiences naturally, which would have too many uncontrolled variables to do science. In this chapter I explore phage evolutionary biology from the perspective of laboratory experimentation, considering in particular a statistical fitness perspective on phage evolution rather than the more common functional perspective as considered, for example, in Part I (Phage ecology) of this monograph. See Chapter 6 for a general introduction to many of the concepts presented here.

9.1.1 Why phages? Easy manipulation, rapid evolution, sequencing

Phages have been successfully used in a variety of experiments to elucidate mechanisms of evolution. It should be realized, of course, that

Bacteriophage Ecology: Population Growth, Evolution, and Impact of Bacterial Viruses, ed. Stephen T. Abedon. Published by Cambridge University Press.

even much of the early genetics work on phages was often inherently "evolutionary" (Chapter 6), exploiting host-range variants and other phenotypes important to lab fitness, even though that work was not cast as evolutionary. Originally, the big advantages of phages were the ease of manipulation and the short generation time. Soon after DNA sequencing methods became commonplace, the smaller-genome phages provided the additional advantage of easy determination of complete genome sequences. This latter advantage meant that several findings of a genomic nature came first or only from work on phages, even though similar processes likely accrued to phenotypic evolution observed in other microbial experiments. Work on bacteria and even yeast has reached the point that it is now feasible to obtain genome sequences of those organisms, but intensive sampling of complete genome sequences is currently economically feasible only with phages.

9.1.2 Functional versus fitness perspectives

There are different philosophies that can motivate studies of evolution. One commonly used one is a functional perspective, interpreting outcomes in light of the biological details of the system. Thus, a bacterium subjected to antibiotic treatment is likely to evolve mutations that increase drug resistance, and the details of the genes involved in that adaptation provide a basis for explaining evolution. Or a phage selected to grow on a new host may evolve changes in attachment proteins and other elements that interact with the host. An alternative approach is to neglect the functional or biological aspects of the system and describe it in terms of fitness patterns and statistics that may apply to other systems with different biologies: fitness effects of mutations that arise, numbers of beneficial mutations, how fast mutations accumulate, and how much fitness is improved at the end of experimental adaptation. The functional perspective is the more common. The fitness perspective offers the promise of generality that transcends the details, but whether generalities exist at this level is unresolved. The effort is certainly justified, however.

9.1.3 Patterns of adaptation and fitness change

The perspective adopted here is the latter one – patterns of adaptation and fitness change inferred from experiments that can be compared across diverse systems, and hence do not rely on the biological context. This area of evolutionary biology is still so young that the framework is fragmentary and the evidence scant. Indeed, this chapter reads more like a wish list than a set of accomplishments. Perhaps because so little has been done, this area is exciting and ripe for pioneers. At the same time, I feel some guilt that, for me, the most exciting work with phage experimental evolution to date has used a phenotypic and functional context, and that this chapter neglects that work. By contrast, the many studies of Chao have been particularly inspiring in their use of experimental evolution to address the classic concepts in evolutionary biology, and there is now a second generation of experimenters using phages, including (in no particular order) Abedon, Wichman, Turner, and Burch, as well as a few phage geneticists bending their ways toward evolution (Molineux and Fane). In this broad context, my emphasis here is a rather narrow and largely unproven slice at the interface between evolution and phage experiments. Furthermore, most of the emphasis requires knowledge of the molecular bases of evolution, and hence is heavily biased toward studies in which genomes have been sequenced.

9.2 ISSUES IN EXPERIMENTAL ADAPTATION

In studies of experimental evolution, the typical protocol is to choose some convenient organism and apply a selective regime that limits survival and/or breeding to a subset of the population. The protocol may be as simple as letting all individuals breed and choosing a fraction of all progeny for the next generation, thereby favoring those with the most offspring or highest reproductive rate under the experimental conditions. The evolved population may be assayed, at the end of the adaptation and also at intermediate time points, for changes of a phenotypic nature or a molecular nature. As a final step, these outcomes may then be compared to expectations based on the relevant theories.

9.2.1 Matching design with theory

Phages have obvious advantages in studies where the goal is to obtain rapid evolution: they can be maintained indefinitely as free viruses and they have enormous growth potential (e.g., 10^{12}-fold in one or a few hours). Large population sizes permit the application of stringent selection and ensure the presentation of a wide array of different mutations (especially so for RNA phages; Chapter 6). Finally, many phages are amenable to molecular genetic manipulations, such as site-directed mutagenesis and full-genome sequencing. In practice, the experimenter wishing to achieve a fast response to selection will use the largest population size feasible, but lesser population sizes may be chosen deliberately to achieve certain effects, such as genetic drift (Chapter 6).

As models of natural evolution, experimental evolution protocols can be designed to look at many types of questions – rates of evolution, effects of population size, the ecological bases of selection, epistasis, and so on. Yet they can offer fruitful tests only to the extent that the experimental conditions match the assumptions of the theory being tested, and this match is often difficult. An important concern is that, for tractability, nearly all theories consider extremes, and it is usually difficult to obtain exact theoretical results even at extremes. The impact of relaxing these extremes is rarely considered, even though the extremes do not usually apply in nature and may not apply in experiments. For example, one commonly assumed extreme in models of adaptation is a strict absence of recombination, which is much easier to model than a small amount of recombination. Alternatively, recombination becomes irrelevant if a different convenient extreme is assumed – a mutation rate so low that beneficial mutations arise one at a time and are either lost or fixed before the next one arises (Section 9.4.3.1). In practice, experiments with phages cannot strictly achieve these and other extremes, and in most cases the extent of an empirical deviation from an extreme is unknown. Thus, when attempting to test theory experimentally, one is often left with considerable uncertainty in knowing what can be legitimately concluded.

Even when extremes are approximated in experiments, unanticipated aspects of a system's biology can impact results. An interesting example was the effort by Turner and Chao (1998) to evaluate the advantage of recombination by comparing adaptation in a treatment that plated coinfected cells (thus recombination-positive) versus singly infected cells (thus recombination-negative), both at low density. Though designed to consider the effects of recombination, it turned out that the dominant feature of the study was an unforeseen selection for intracellular competitive ability (Section 9.2.2, immediately below). The study was re-interpreted and has become a classic in the experimental demonstration of the "prisoner's dilemma" game-theory model (Turner and Chao, 1999), although not a demonstration of the effects of recombination (Chapter 8).

A paper by Wahl and Krakauer (2000) stands in contrast to the usual disconnect between theory and empirical reality and provides an example of the potential that exists for fruitful collaborations between biologists and mathematical colleagues. We may hope that their paper inspires a more comprehensive synthesis and review of the issues in developing predictions that accommodate the realities of experimental adaptations. The next few paragraphs discuss some of the larger issues in matching experimental design to the theoretical frameworks.

9.2.2 Preventing direct interactions in the evolving population

Many theories that concern adaptation assume that selective conditions remain constant, or, equivalently, assume a fixed fitness landscape (fitness landscapes are described in Section 9.3.1 and Chapter 6). For example, a progressively changing environment is expected to favor more adaptive evolution in the long term than a constant environment. Selective constancy requires at least that the environment used for adaptation be maintained the same throughout the experiment. A constant experimental environment is necessary but not sufficient, however, since whenever organisms evolve, that evolution may alter the environment in some way that affects the landscape.

When the population reaches high density, interactions between individuals can affect fitness. If those interactions evolve, which they may by virtue of their influence on fitness, then the selective environment will be changing even if the abiotic environment and host physiology is static (Paquin and Adams, 1983; Rozen and Lenski, 2000). For example, it is common in serial passage designs to let phage density reach high levels before transfer. If all of the adaptation in such designs is to the low-density phase of growth, then the high-density phase may not matter. But if some of the adaptation occurs in response to the high density, then the selective conditions at high density will possibly change over time (see, for example, Chapters 3 and 8). Thus, deliberate effort is required to achieve a constant environment, and maintaining phages at low density is one important design feature in achieving this goal.

9.2.3 Biases against certain mechanisms

Comparisons of genome sequences among naturally occurring species reveal several types of evolutionary processes at the genomic level (Li, 1997), most commonly an accumulation of mutations (substitutions) at individual nucleotide positions throughout a genome. These substitutions may be selectively neutral, slightly deleterious (because greatly deleterious mutations would be expected to be quickly lost from populations), or beneficial. Presently, it is thought that only a small fraction of the substitutions in natural populations are beneficial, with near-neutral changes being the most common, although this dogma is being challenged (Andolfatto, 2005). In addition to nucleotide substitutions, genomes in nature have undergone wholesale rearrangements – losses and duplications of genes and larger regions – as well as the gain of new genes horizontally and the invention of genes with novel properties.

The short-term and "closed-system" features of experimental evolution means that it captures only a limited spectrum of natural evolutionary processes, chiefly fixation of highly beneficial mutations (Gerrish and Lenski, 1998; Wahl and Krakauer, 2000; Rozen *et al.*, 2002). In the typical evolution experiment, the strain of interest is propagated in some fashion in tightly controlled environmental conditions. Genetic variation may be arranged to be present initially, but even if not, if population sizes of a million or more are used, then a variety of mutations will be quickly generated for selection to act on. Yet unless the founding population is already near the equilibrium for neutral and deleterious mutation frequencies, those equilibria will probably not be reached by the end of an experiment.

Genetic drift leading to fixation of neutral changes often will be rare or absent in experiments, since the time from the mutational origin of a neutral mutation to its fixation requires many generations (i.e., several times the effective population size). Thus, neutral fixations will occur only when extreme population bottlenecks are used and then only at the neutral mutation rate (Crow and Kimura, 1970). In natural populations, even of large size, neutral and deleterious changes may hitchhike to high frequency with beneficial changes. However, when founding populations are recently derived from an isolate, even hitchhiking will not commonly occur due to a relative absence of segregating, non-beneficial mutations. Likewise, even beneficial changes of small effect are unlikely to accumulate because of insufficient time.

At the other extreme, many macro-genomic changes are also excluded. Novel genes and other DNAs cannot be acquired by horizontal transfer in single-species experiments unless they are engineered into the genome, although deletions and rearrangements can and often do evolve. So the potential for evolution of new functions is greatly restricted in the typical experiment. Thus, the types of natural processes represented by an experimental adaptation protocol are probably limited to relatively large fitness-effect beneficial point mutations, rearrangements, and deletions. This limited spectrum is not all bad, however, since it simplifies modeling the processes of experimental evolution.

9.2.4 Competing beneficial mutations and recombination

A variety of theories have been worked out for the fixation of beneficial mutations – especially for rates of substitution and fitness effects. Many aspects of those theories are sensitive to whether multiple beneficial mutations are arising and ascending at the same time and also to whether recombination exists (Gerrish and Lenski, 1998; Rozen and Lenski, 2000; Wahl and Krakauer, 2000; Hegreness *et al.*, 2006).

9.2.4.1 Clonal interference

In the absence of recombination, multiple beneficial mutations that have arisen in different lineages can compete with each other in a phenomenon known as clonal interference (Chapter 6). This occurs because, without recombination between lineages, the various mutations are unable to occupy the same genome except by mutational accumulation. Clonal interference, by forcing most independent beneficial mutations to compete instead of combining forces, impedes allele fixation. Its effects have been worked out theoretically for the extreme case of no recombination, both at low (Gerrish and Lenski, 1998) and at high (Wahl and Krakauer, 2000) mutation limits. Though limited amounts of recombination can still give rise to a weak form of clonal interference (Hill and Robertson, 1966), there is no corresponding theory of substitution rates across an entire genome with small to intermediate amounts of recombination.

9.2.4.2 Limitations in controlling phage recombination

The extremes of mutation or recombination assumed by most models are not easily achieved in experimental work with phages. Recombination is known for many phages (whether as an exchange mediated by recombination enzymes or by polymerase strand-hopping), but even when a phage is capable of recombination its frequency among different genomes depends on the incidence of coinfection (Chapter 8), which is something determined by experimental conditions. High levels of coinfection will enhance recombination, but, as noted in Section 9.2.2, coinfection also introduces a type of selection that may be undesirable. The evolutionary impact of intermediate levels of recombination, meanwhile, depends on the rates at which mutations are sweeping through the population, and maintaining population sizes small enough that beneficial mutations arise singly (as required by some theories) is tedious and risks neglecting mutations of small effect for lack of a noticeable change in fitness. Finally, low levels of coinfection reduce but do not completely exclude recombination.

9.2.4.3 Interference from epistasis

With clonal interference, the reference to clonality acknowledges asexuality as the impediment to adaptation, and restoring recombination should eliminate the clonal-interference "speed-limit" by eliminating this clonality

(Arjan *et al.*, 1999; Miralles *et al.*, 2000; Colegrave, 2002). There is, however, a second impediment to adaptation that can operate with or without recombination and produce the same effect: epistasis (Chapter 6). When epistasis operates antagonistically, so that the fitness of combined mutations is less than the sum of separate effects, increasing the input of beneficial mutations will not proportionally increase the rate of adaptation. In such cases, the methods used to measure the effect of clonal interference will overestimate the impact of asexuality, because some of the interference would remain even if recombination were restored.

In the extreme case of what is known as sign epistasis (Otto and Feldman, 1997), two mutations that are separately beneficial are worse in combination than the best of the singles. As is well known, recombination actually impedes adaptation in this case, and clonality is a benefit (Maynard Smith, 1978). The tests of clonal interference, however, do not discriminate between the absence of recombination and epistasis as separate causes of the impediment (for further discussion on this topic, see Holder and Bull, 2001). A welcome addition to this topic would thus be a theory that partitions clonal interference from epistatic interference. Empirically, when the fate of competing mutations is observed in genome sequences, then site-directed mutation can be used to determine the extent of antagonistic epistasis.

9.3 BASIC PRINCIPLES OF EXPERIMENTAL ADAPTATION

In this section I consider some basic principles of evolution (introduced in Chapter 6) that are commonly assumed to underlie adaptation.

9.3.1 Fitness landscapes

A common metaphor in evolutionary biology is that of a multidimensional fitness landscape, in which each possible genotype is represented by a single fitness (Kauffman and Levin, 1987). Because of their small genome sizes and relatively simple laboratory ecologies, microbes are well suited to answer interesting questions about fitness evolution in the context of fitness landscapes.

9.3.1.1 Adaptive walks

There are four possible states at each of the x sites making up a sequence (i.e., the nucleotides: A,C,G, and T or U) for a total of 4^x possible genotypes. This genotype "lattice" defining the plane of an adaptive landscape is rigid, due to the simple, discrete structure of nucleic acids (neglecting epigenetic modifications such as methylation). Rising from this multidimensional plane is a dimension on which the fitness assigned to each genotype is mapped. This fitness dimension gives the landscape its shape or ruggedness. Fitness necessarily depends on the environment, which for a phage includes the host genotype, host density, host quality, temperature, medium, and even the density of phages (e.g., Chapters 2, 3, 5, 8, and 10). As noted above, the fitness landscape will be constant throughout the adaptation only if the experimental environment is held constant, but not all designs allow for a reasonably constant environment.

An experimental adaptation using a constant environment likely results in a population's ascent of a local fitness peak in the fitness landscape. The population is a cloud of genotypes clustered around a region in this space (the population average), with variation in the population extending the cloud perhaps several nodes in many dimensions. Adaptation is the progressive shift of the population average to regions of ever-higher fitness. The trajectory the population follows when evolving through this space is termed an adaptive "walk."

9.3.1.2 Landscape topology

We have little direct evidence on the topographic details of any fitness landscape (although see Weinreich *et al.*, 2006), so it might seem impossible to develop any rules of adaptive evolution except by empirical observation. However, we do know that landscapes lie between the extremes of completely smooth and totally uncorrelated. In a perfectly smooth landscape, the effects of mutations do not depend on the background genotype. That is, the benefit of the mutation at a site might depend on the site, but its effect would not depend on the genotype at other sites. We know that this model is not realistic, of course. Some degree of ruggedness in real landscapes is implied by the vast wealth of genetic studies revealing mutational effects that depend on background genotype, such as compensatory mutations, second-site suppressors, and other forms of epistasis (Chapter 6). More trivially, mutations in coding sequences can have profoundly different effects according to the other two bases in the same codon. So landscapes are not perfectly smooth. The other extreme is also not realistic. Despite some expectation of ruggedness, molecular phenotype does not change profoundly for many point mutations from the wild type (genetic robustness; Chapter 6). Hence landscapes are somewhat smooth. But owing to the complexity of organisms and environments, there is a rather large range of ruggedness compatible with these observations of relative smoothness.

Fortunately, some theory has been derived that does not require knowing the details of landscape ruggedness, as reviewed below (Section 9.4.3.1). In other cases, results have been derived specific to particular landscape shapes (Fisher's geometric model of a smooth landscape, and uncorrelated landscapes). Yet there remains a desperate need for theory that can apply the types of data that are typically obtained in experimental adaptation to properties of landscape ruggedness. Site-directed mutagenesis of viral genomes has become so routine that it is feasible to measure fitness effects of large numbers of individual random mutations (Sanjuan *et al.*, 2004), and those data should provide important insights to landscape ruggedness.

9.3.2 Increasing rates of adaptation: $N \cdot \mu \cdot s$

The most basic principles of population genetics theory indicate that the rate of adaptive evolution should increase with population size (N), with the advantage of the allele (s), and with mutation rate (μ) up to a point known as the error threshold (Eigen, 1971; Eigen and McCaskill, 1988). This realization leads most experimentalists interested in speed of adaptation to use the largest population sizes feasible; likewise, geneticists have commonly used mutagenesis to recover rare mutations. The industry that uses "directed" evolution to create useful molecules goes to great lengths to enhance mutation rates and library sizes, based on the assumption that short-term evolution is enhanced by both N and μ (Voigt *et al.*, 2000).

Perhaps because the conventional wisdom seems so reasonable, there are few efforts to test the fundamental importance of population size, mutation rate, and strength of selection to rates of adaptation. The classic study of the importance of population size to adaptation is by Chao (1990), in which phage ϕ6 was subjected to plaque-to-plaque transfers (single plaques streaked onto a plate to serially isolate individual plaques), hence an effective population size close to one. Fitness declined, due to the accumulation of deleterious mutations, which (with the exception of lethals) are not efficiently selected against in such small populations. Although fixation of deleterious mutations is always a possibility in populations of any size (less so as population size increases), our main interest is the effect of population size on fixation of beneficial mutations. For large populations, the main effect of size is on the number of beneficial mutations introduced: the number should scale directly with N. For this reason alone, one might expect a direct scaling between rate of adaptation and population size.

As noted above, this conventional wisdom about $N\mu s$ is not correct for adaptation of asexual species and other genomes with low levels of recombination (Gerrish and Lenski, 1998; Rozen *et al.*, 2002). The few studies of viruses that assessed rates of fitness increase at different population sizes, in fact, actually looked for an absence of a direct relationship between population size and rate of adaptation as a test of clonal interference (Arjan *et al.*, 1999; Miralles *et al.*, 1999, 2000). However, two studies have used increases in population size as a way of obtaining progressively better mutants (Burch and Chao, 1999; Sanjuan *et al.*, 2005).

Testing the effect of N, μ, and s should be easy with phages. Some assays using strong selection are able to easily recover isolates carrying mutations of large effect (as opposed to waiting for beneficial mutations to sweep through a population). For these types of single-step adaptations, the potential complication of clonal interference is not an issue. These systems, therefore, should lead to especially fast answers about the impacts of changes in mutation rate and population size on the numbers of alternative mutations available, which is an important piece in adaptation. A really interesting study to be done is to compare rates of adaptation with and without recombination at different population sizes. Results from that comparison would elucidate the importance of population size and recombination in rates of adaptation. The direct comparison of adaptation with and without recombination has been done with directed evolution of plasmids (Rowe *et al.*, 2003), but population size was not a variable.

9.4 EMPIRICAL PROPERTIES OF ADAPTIVE WALKS

A "walk" is a trajectory of population fitness and genotype over time, and a walk is "adaptive" if the population evolves toward higher fitness within a landscape. An adaptive walk has several properties that can be quantified easily and that are of special interest in evolution: the final fitness, the shape of the fitness trajectory over time, the number of steps, and their fitness effects. These and other properties have been addressed empirically, theoretically, or both. In virtually all cases, more work is justified before we can be confident of generality.

9.4.1 *Fitness limits*

In most applications, the final fitness is more important than the amount of molecular evolution, for example as in the evolution of drug resistance, vaccine escape, and pesticide/herbicide resistance. However, there is as yet no specific theory (that does not require empirical input) regarding absolute fitness limits at the end of an adaptive walk (some theory requires one empirically determined parameter to give a limit; Wahl and Krakauer, 2000). Likewise, there are almost no empirical studies that might motivate such theories. Nevertheless, we know that in natural populations there are many processes that may *prevent* an accumulation of the best sequence of beneficial mutations available. These impediments include migration, small population size, and a changing environment. In lab experiments, designs can easily mitigate all problems except mutation limitation (all populations are finite, after all). Even here, mutation rates may be temporarily enhanced via exposure to mutagens, though in the absence of recombination, clonal interference can still confound most of the straightforward efforts to accelerate the speed of adaptation in the laboratory (Gerrish and Lenski, 1998; but see Wahl and Krakauer, 2000).

9.4.1.1 The simplest limit: evolution back to the starting point

One special case of expected fitness limits is trivial (and also not very useful for developing a general theory): if a well-adapted organism is subsequently impaired by one or more deleterious mutations, then the expected fitness limit after recovery (by compensatory evolution) is the original fitness. This limit could obviously be achieved by a reversal of the deleterious mutation(s), but it might also be attainable by compensatory mutations at "second" sites that overcame the defect (Chapter 6). This expected limit underlies the study of the fitness costs of drug resistance in bacteria and viruses, which could help select against resistance mutations in the absence of the drug. However, compensatory mutations often arise and reduce the fitness cost of resistance without compromising resistance, and the standard against which compensated fitness is compared is the original fitness (Schrag *et al.*, 1997). A similar framework has been used to study the recovery of viruses from deleterious mutations (Burch and Chao, 1999).

9.4.1.2 Comparison between different genome types

A different perspective on the limits of adaptation was posed in a comparative study of the maximum fitness achieved by different phages separately adapted to the same environment, including the same host (Bull *et al.*, 2004). Serial passage favored high phage growth rates as virtually the sole fitness component, and an absolute measure of fitness was obtained that could be compared across all phages. The possible models that could be entertained were that the highest fitnesses would be achieved by those phages displaying (1) genomes with high mutation rates (RNA phages), (2) the smallest genomes (RNA and ssDNA phages), (3) the largest genomes, with the most genes (T-even phages), or (4) an ability to evolve the most extreme value of some life-history trait (e.g., adsorption). Alternatively, no type of phage may have achieved limits to adaptation significantly higher than the others.

Not supporting any of the above models, the phages that achieved the highest growth rates were DNA phages with medium genomes. The *post hoc* explanation offered for this pattern was that those phages struck the best balance between small genome size and encoding functions needed for speedy replication. Thus, small genome phages had to rely on host processes for transcription and DNA replication, and did not suppress transcription of host genes, and so could not evolve to accelerate these steps. T-even phages, on the other hand, carried large numbers of "contingency" genes that were transcribed and translated but served no or at best a very limited purpose in the employed host. It is also possible that historical factors determined which phages achieved the highest growth rates, because the ranking of final fitnesses differed only slightly from the ranking of initial fitnesses. Of course, even if historical factors are important, it still does not answer the question why the T7 types have such higher fitnesses.

The suggestion from this latter study was that genome characteristics impose an approximate fitness limit for the phage. Where the phage sits in the landscape relative to peaks of various heights and valleys of various depths imposes additional, local limits. This initial study used related pairs of four types of phages, so only limited information could be obtained about the range of limits within a single type.

9.4.1.3 Possible additional approaches

A big unresolved technical question is how to decide when a fitness limit has been reached, or has at least been neared. This matter overlaps with the question of the shape of adaptive walks (Section 9.4.3), because if fitness increases progressively diminish, then a limit is easy to identify. If walks commonly show major jumps after periods of small increases, then even an approximate limit may not be knowingly attainable in the short term. Thus some careful long-term studies are needed (like those of Lenski: Lenski and Travisano, 1994; Papadopoulos *et al.*, 1999; Rozen and Lenski, 2000) in which fitness changes are monitored throughout.

It is difficult to imagine a general theory of fitness limits that does not depend on biological details, but a theory based on broad knowledge of gene interactions and genome function could possibly be attained as we acquire more knowledge from genomics (as in Endy *et al.*, 2000). Systems biology approaches may be especially useful in this endeavor. The general question has obvious relevance to the creation of engineered genomes, in which it is unlikely that the assembly of genes from different sources will co-function optimally at the start: Evolution will be needed to correct engineering imperfections, and it would be useful to know in advance how much fitness gain could be had.

9.4.2 Routes toward higher fitness: redundancy, divergence

A characteristic of adaptive walks that is deeply embedded in landscape topology is the number of alternative pathways to higher fitness from a single starting genotype. At one extreme, there might be a single possible pathway – one possible first mutation, followed by a single possible second mutation, and so on, to the point that a local optimum is reached. Toward another extreme, there could be thousands of alternative beneficial mutations at each step.

Replicate adaptations to the same conditions can shed light on the number of pathways, but results may be difficult to interpret unless the study has been designed to avoid clonal interference and other potential problems. It is clear that the number of pathways observed in replicate adaptations is a minimum of the possible number, and little overlap in the mutations observed across multiple adaptations is a strong indication of many pathways. The interpretation is complicated when the same mutations are observed repeatedly. If experimental adaptations were conducted such that each beneficial mutation's origin and fate was determined before the next beneficial mutation arose, then the observed incidence of parallel evolution would be an only slightly biased estimate of the number of alternative beneficial mutations (Orr, 2005). However, adaptation at large population sizes will introduce biases from clonal interference in favor of the best mutations, so it is possible to observe few beneficial substitutions despite a large number of alternatives (Wahl and Krakauer, 2000). On the other hand, clonal interference itself is a demonstration of multiple beneficial mutations (see Section 9.5).

There is considerable evidence from phages (and other viruses) of multiple pathways. Most commonly, these demonstrations apply to the first step of adaptation and merely show that there are multiple beneficial mutations. The classic work of Benzer that used fine-scale mapping to reveal the genic nature of recombination in phage T4 (e.g., Benzer, 1955) relied on multiple mutations with the same phenotypic effect (effecting a host range change). In isometric phages (ϕX174 and a G4-like phage), Bull *et al.* (2000) and Rokyta *et al.* (2005) identified multiple beneficial mutations in single-step adaptations (see Section 9.4.3.1), observed different fitness effects of those mutations, and observed parallel evolution of single steps in the absence of clonal interference. When long-term adaptations are carried out, it is of course possible that mutations beneficial as first steps will evolve in later steps if they did not fix earlier. In the extreme, the same k mutations could fix in arbitrary order, so that there are $k!$ pathways but one final genotype.

In contrast to alternative pathways, several studies using phages have demonstrated extensive parallel evolution at the nucleotide level in adaptations conducted with large population sizes in which the extent of clonal interference was usually unknown (Bull *et al.*, 1997, 2000; Cunningham *et al.*, 1997; Wichman *et al.*, 1999, 2000; Rokyta *et al.*, 2002; Wood *et al.*, 2005; Pepin and Wichman, 2007). What is remarkable about some of those observations is that the parallel evolution (1) spans hundreds to thousands of generations and (2) occurs across genomes that have diverged by several percent. Since large population sizes were used in most of those studies, thereby ensuring multiple introductions of the best mutations, the high level of parallel evolution probably does not imply few pathways (Wahl and Krakauer, 2000).

9.4.3 The shape of adaptive walks: fitnesses and substitutions

The data that are easiest to obtain from adaptive walks are (1) fitness at different times, and, if genomes are sequenced, (2) the number of substitutions, perhaps also at different times. The number of "substitutions" may be difficult to quantify exactly because of polymorphism: many generations are required for mutations to increase from detectable levels to fixation, so some mutations will invariably be polymorphic at the time of sampling (Chapter 6). (Clonal interference can also lead to some paradoxical results, in which mutations once present at high frequency disappear.)

A general expectation is that fitness will ultimately plateau due to attainment of an optimum, mutation limitation, or from a quasi-equilibrium in which both beneficial and (mildly) deleterious mutations are fixed at similar rates (Poon and Otto, 2000). Thus, if a study is carried out long enough, the tendency must be for greater fitness increases earlier than later in the adaptation (Chapter 6). The chief interest lies in specific properties of shape, such as the distribution of step sizes, the number of steps, and their order. With phages, fitness can increase profoundly, so some of these properties are rewarding to study.

9.4.3.1 First steps

One of the most exciting and surprising results for adaptive evolution is a largely biology-free prediction from the work of Gillespie (1991) and Orr (2002) that gives the distribution of fitness effects for the first step in an adaptive walk; this theory assumes that independent mutations do not compete (a low mutation extreme, $N \cdot \mu \ll 1$). This result requires knowledge of the total number of different beneficial mutations that could serve as the first step in an adaptive walk (call this number M, which might be 50 or less but must be very small relative to the possible number of deleterious and neutral mutations). When M is known and is sufficiently small relative to deleterious mutations, then results from a branch of statistics known as "extreme value theory" may apply. On average, the distribution of fitness effects among the M good mutations should be exponential: most effects should be relatively small, and mutations of increasing benefit should be progressively rarer. Furthermore, given empirical input for one additional parameter, the fitness differences between successively fitter mutations are also specified in this theory: If the average difference between the fittest and next most-fit mutation is C, then $C/2$ is the average difference between the second and third best mutations, is $C/3$ for the next pair, and so on.

This relationship allows several properties to be derived about the first step in an adaptive walk (Orr, 2002). In particular, the average first step of an adaptive walk will be to allele $(M + 3)/2$, for an average fitness effect of $2M \cdot C/(M + 1)$. Therefore, although this result requires knowledge of two important biological parameters (M and C) before the phenotypic effect can be predicted, the result is significant in showing that some properties can be derived without explicit knowledge of system-specific molecular details.

The Gillespie–Orr result need not be true. Extreme value theory will not apply, for example, if the available mutations for a genotype do not represent a random draw from the tail of a distribution. There are good biological reasons to question this assumption of randomness – the good mutations available to a particular genotype may all be constrained in some way by the genetic background and not represent the fitness effects that might be available to a slightly different genotype. The theory applies to averages, not necessarily to individual genomes. Thus, empirical tests are essential to decide whether this theory is useful. Last, the exponential distribution is not the only one that may apply to extreme values, so extreme value theory could apply to the first-step mutations but the distribution of effects could be wildly different than exponential: the distribution could be truncated and even exhibit an increasing probability of large effects (Beisel *et al.*, 2007).

The sole unambiguous test of this theory (of exponential effects) has been done with one phage isolate, an ssDNA phage isolated from the wild but related to ϕX174 (Rokyta *et al.*, 2005). Among 20 independent single-mutation adaptations whose genomes were fully sequenced, nine different nucleotide changes were identified (one per genome, thus with considerable parallel evolution), and fitnesses were measured. Assuming that these nine mutations comprised the entire set of possible beneficial mutations (unlikely, since mutations of very small effect would be lost or not lead to noticeable fitness increases), the fitnesses of the mutations showed an acceptable fit to Orr's model, but only when adjusted for the higher mutation rate of transitions than transversions.

Other tests are in order. An important and unresolved matter is how widely the same fitness parameter, *C*, applies across related genotypes. If this parameter is conserved, then it will allow assays on one genome to be extended to other genomes that have not been studied, which is potentially useful. But a test of even a single genome is not trivial. A precise test requires: (1) the same strain must be adapted multiple times to the same environmental conditions, with enough replications that most or all of the beneficial, first-step mutations are sampled, (2) the fitness effect of each first substitution must be measured, and (3) the nucleotide basis of each substitution must be determined to identify repeated cases of the same mutation and to verify that only one substitution swept through. Step 3 is also necessary to ascertain whether most mutations have been sampled, while step 1 requires that the adaptations be performed at small enough population sizes so that only single beneficial mutations are sampled; large populations tend to pull out the best (see Gerrish and Lenski, 1998; Wahl and Krakauer, 2000, for a consideration of step sizes with clonal interference; Rozen *et al.*, 2002; Hegreness *et al.*, 2006).

In addition, some theoretical issues must be resolved. There will invariably be beneficial mutations that were not sampled, so *M* is underestimated. It also needs to be decided whether mutations in coding regions that lead to the same amino acid substitution by different nucleotide changes should be treated as the same mutation. Third, since any one study is likely to identify only a handful of beneficial mutations, theory needs to be developed that allows combining results from different studies. The extended theoretical results of Beisel *et al.* (2007) have taken important steps in addressing these difficulties.

9.4.3.2 Steps across the entire walk: sizes and order

The increase in fitness across a walk depends both on how many beneficial mutations are ascending and how much each mutation improves fitness. The suggested generalities from work to date are that early substitutions confer the largest fitness increases (Section 9.4.3.1) and that rates of substitution tend to remain constant (Wahl and Krakauer, 2000). However, there are no empirical studies that fully identify the mutations and their individual fitness effects throughout the adaptation, beyond first steps. To do so is somewhat tedious, but such studies are otherwise practical with phages and would be informative. The closest empirical work we have toward this objective consists of observations of substitutions and fitnesses at different times during adaptation, mostly without regard for the effects of individual substitutions (Holder and Bull, 2001; Rokyta *et al.*, 2002); there is also a study estimating fitness effects of different mutations fixed in bacterial adaptations (Rozen *et al.*, 2002).

A variety of theories have been offered for steps across adaptive walks (Gerrish and Lenski, 1998; Orr, 1998, 1999; Wahl and Krakauer, 2000; Rozen *et al.*, 2002; Hegreness *et al.*, 2006). As noted above, these theories differ in the nature of their assumptions, and a welcome addition to this field would be a review that compares these theories as well as describes how to decide which ones apply to an experiment. More basically, we need to know whether the distribution of mutational effects follows the distribution assumed in some of those theories (e.g., exponential) to know whether there is hope for a truly general theory in this arena. One important step on the theoretical side is an extension of the single-step theory to a full walk, assuming that the distribution of fitness effects does not change in subsequent steps (Rokyta *et al.*, 2006).

9.4.3.3 The largest step is often big

A long history of opinion and argument in evolutionary biology proposes that most of evolution by natural selection in so-called quantitative traits is due to alleles with small effects (Lande, 1983). It is acknowledged that this small-effect model may not apply with human-imposed selection and with some special traits (e.g., mimicry), but there is no clear-cut resolution of the extent to which large-effect mutations have been important in evolution. As new techniques have been developed that enable investigation of the effect size or fitness effect of individual mutations or loci, there has been renewed interest in whether the small-effect dogma is true (Bradshaw *et al.*, 1998). Orr (1998, 1999) showed that adaptation on a fixed landscape is in fact biased toward the mutations of largest effect, but "large" is always measured relative to the distribution of available mutations, and is not on an absolute scale. As noted (Section 9.4.2), clonal interference also creates a strong bias toward fixation of mutations of large effect (Gerrish and Lenski, 1998; Wahl and Krakauer, 2000).

Phages are obviously well suited for the study of large mutational effects (Section 9.6.2). Host-range mutants constitute the most prevalent example, but temperature extremes and other forms of inhibition are also studied (Adams, 1959; Hashemolhosseini *et al.*, 1994a; Whang *et al.*, 1996; Bull *et al.*, 2000). In some cases the nature of inhibition may not be obvious *a priori*, but the phage merely adapts to the conditions used for propagation (Rokyta *et al.*, 2005). The large population sizes attainable with phages facilitate the search for extremely rare genotypes, and the consensus opinion seems to be that, among those of high fitness, those with the very highest fitness are also the rarest (an assertion formalized by application of extreme value theory).

That some mutants offer profound fitness gains in novel environments is not an argument that evolution typically occurs by big fitness jumps; it merely demonstrates the possibility of large fitness jumps. Even when these "large-effect" mutations evolve, they may not confer a large *net* increase on fitness, if the environment in which the mutations are useful is only rarely encountered. The experimenter can, of course, design the environment to be so extreme that only mutants survive. Not surprisingly, studies employing environmental extremes observe extremely large fitness effects (in many cases the fitness effects have not been quantified, but a large value can be inferred). Of course, it is easy to imagine that phages commonly do get trapped in extreme environments, where survival requires the right mutation.

9.4.3.4 Rates of substitution

An interesting and straightforward question is the rate at which beneficial substitutions accumulate over periods of adaptation. The most direct way of ascertaining this property is from sequences (though if all beneficial mutations boosted fitness to the same extent, then relative rates of substitution could be inferred merely from the fitness at different times of the adaptation). Furthermore, sequences can be obtained from the population at the end of the adaptation to identify all substitutions, and the timing of those substitutions can then be determined from limited sequencing at earlier time.

The rate of substitution depends on various issues such as population size, recombination (hence clonal interference), the number of possible beneficial mutations, their fitness effects, and the mutation rate. All else equal, large-benefit mutations should evolve faster than small-benefit mutations because they ascend faster; the large-benefit ones will displace the small-benefit ones either absent recombination or if epistasis acts such that multiple-mutant combinations are less fit than single-mutant genotypes (Wahl and Krakauer, 2000). Furthermore, beneficial mutations with weak effects may not ascend fast enough to evolve during the few hundred generations of experimental adaptation. Beyond this, expected rates of substitution are sensitive to many details, and generalities from the theory are not evident.

The general finding from most studies of experimental adaptation is that relatively few changes accumulate – typically no more than a couple dozen – if the environment is held constant and high-density interactions are avoided. This number is not a constraint of laboratory evolution. For example, one study of a phage that used extreme bottlenecks and mutagenesis to fix arbitrary mutations reported 300 and 400 mutations in two independently evolved genomes, but subsequent adaptation of those genomes to recover fitness resulted in 14 and 23 beneficial substitutions (Bull *et al.*, 2003). Many other studies have reported numbers of presumed beneficial mutations this low or lower. However, relatively few studies describe any temporal pattern in substitution rate. The most extensive work of this sort has been done with the isometric phage ϕX174 (Bull *et al.*, 1997; Wichman *et al.*, 1999, 2005). In these studies, the phage was grown in a two-stage chemostat, which allowed high rates of coinfection (Section 9.6.1.2). In an initial study, phages adapted for up to 11–33 days showed a higher rate of substitution during the initial stage of adaptation than during a later stage if the environment was held constant, but a change in host type resulted in an accelerated rate of nucleotide substitution during the second phase (Bull *et al.*, 1997).

A subsequent study monitored the frequencies of mutations daily over an 11-day period and observed mostly monotonic increases in mutation frequency (of those that reached high frequency by the end). Those studies then motivated a much longer study of the phage, spanning 120 days (13 000 generations): the rate was slow during the first 2000 generations but quickly increased to a virtually linear accumulation of substitutions over the next 10 000 generations. The final population had a genome-wide substitution rate of 1.8%, and over 100 nucleotide positions were observed to have changed during the adaptation. The characteristics of those substitutions also changed over the period of adaptation, however (e.g., silent versus missense, genic location). The clocklike behavior of molecular evolution in this latter study was inferred to have resulted from an ongoing arms race among the phages, due to the high coinfection rates, rather than from adaptation to a fixed optimum (Wichman *et al.*, 2005; Bull *et al.*, 2006).

Other studies, for which a single optimum can be inferred, have not been as extensive, but sometimes appear to slow down in the rate of substitution (Holder and Bull, 2001; Rokyta *et al.*, 2002). Part of this slowdown may stem from the fact that these experimental adaptations often continue until the rate of fitness gain slows, and a lack of fitness progress may be coupled with a slowdown in molecular substitution. However, we lack a long-term study conducted at low density in which rates of substitution are monitored long after fitness has reached a plateau, such that we can confidently compare substitution rates with fitness change; such work is very feasible but perhaps tedious.

9.5 RECOMBINATION

One of the classic questions in evolutionary biology is the evolutionary significance of recombination. This topic was heralded by two books in the 1970s (Williams, 1975; Maynard Smith, 1978), and ever since has remained an active area of research in evolutionary biology (Chapter 8). Experimental work on the evolutionary advantage of recombination, however, has lagged theoretical work, though phage-based experiments could be employed to fill at least some of this gap. For example, as noted in Section 9.2.1, a study with phage φ6 attempted to address the significance of recombination by comparing fitnesses evolution with and without recombination (Turner and Chao, 1998), though competition among coinfecting phage turned out to be the more dominant theme (Turner and Chao, 1999). The study could be redesigned to more specifically test the effect of recombination, but this has not been done. For instance, if the coinfected cells were allowed to lyse before plating, then plaque outgrowth would not experience competition from multiple genomes, and recombination would still be introduced by the coinfections.

Although it is difficult to significantly vary levels of recombination in phages throughout the full course of an experiment without introducing unwanted selection, it should be possible to conduct short-term studies in which recombination is the primary variable: a set of defined genotypes could be evolved at low density (as the zero recombination treatment) and compared to an experimental adaptation in which the same set of genotypes was initially shuffled before adaptation. Likewise, as suggested above for the Turner and Chao design, high levels of recombination (even with accompanying unwanted selection) could be introduced infrequently during an adaptation by using high levels of coinfection; the effects of recombination should last longer than the momentary change in selection.

Special controls are required for many experimental tests of the evolutionary impact of recombination, however, because some types of epistasis can lead to patterns similar to those resulting from clonal interference (Section 9.2.4.3). Recombination eliminates clonal interference but not epistasis. In particular, special care must be taken to shuffle mutations that occur close together (e.g., in the same gene): they are inherently difficult to shuffle by natural recombination (and thus subject to strong clonal interference), yet they may also be especially prone to epistasis by virtue of their proximity. Although coinfection may be an effective way of shuffling distant sites in phage genomes, site-directed mutagenesis may be the most effective means of shuffling close mutations.

The impact of zero recombination has been studied from the perspective of clonal interference. The earliest studies used eukaryotic viruses and evaluated clonal interference indirectly, as the impact of population size on the rate of fitness increase (Arjan *et al.*, 1999; Miralles *et al.*, 1999, 2000). A recent study using phages has looked directly at clonal interference based on genome sequences at different times during adaptation, finding that many mutations present at intermediate times had disappeared by the end (Pepin and Wichman, 2007).

Phages are also suited to demonstrate Muller's ratchet, another mechanism considered to be a disadvantage of clonality (Chapter 6). The essence of this process is that the population comes to consist of individuals, each carrying one or more deleterious mutations, but no one of the deleterious mutations is found in all individuals. Thus, recombination could restore the wild type. The experimental challenge in this case is to sample the entire population and show that every genotype carries one or more deleterious mutations, but none of them occurs in all genotypes. A design similar to that of Chao's (1990) bottleneck study with an RNA virus would be appropriate, although bottlenecks must be larger than one individual.

9.6 METHODS OF PROPAGATION, SELECTION OF PHENOTYPES, AND FITNESS MEASURES

Any experimental evolution study implicitly or explicitly involves a phenotype that is selected, a method of propagation, and an assay of fitness. A central but often underappreciated dimension in experimental adaptation studies is the interplay between these design properties. In this section, I offer a brief list of different methods of phage propagation, and the phenotypes that can be selected (and their assays), and highlight some of the drawbacks and advantages of each.

9.6.1 Methods of propagation

Phages can be propagated in any of four basic environments, differing in how bacteria are provided. Assume phages are obligately lytic (*sensu* Chapter 1).

9.6.1.1 Liquid culture, serial transfer

In this method, phages are added to a bacterial culture and allowed to replicate (ideally, cells are in exponential phase and at a defined density). Phage growth may be continued for a fixed time, or continued until culture lysis or phage population growth otherwise stops (e.g., as discussed in Chapter 3). An aliquot is transferred to a fresh culture at the same cell density, and the process repeated. Aliquots may be treated with chloroform between transfers, but direct transfer of an aliquot of the culture (pre-lysis and with cells at various stages of infection) helps avoid selection for synchronization of the life cycle to the transfer interval. The advantages of this method include: phage density and phage number can be maintained within desired limits, large phage population sizes can be used, the host does not evolve (it is replaced at every transfer), the infection process is appropriately modeled by mass-action dynamics (Chapter 15), and fitness can be measured in the same conditions used for passage (as either an absolute growth rate or relative to a competitor; Chapter 6). The main drawback is that the method is tedious and, if the phage grows rapidly, the transfers must use strong bottlenecks that result in a loss of rare mutations (Wahl and Gerrish, 2001; Wahl *et al.*, 2002). This method has been used in some recent studies (Bull *et al.*, 2004; Rokyta *et al.*, 2005; Pepin and Wichman, 2007).

9.6.1.2 Liquid culture, two-tube chemostat

This style of "chemostat" (Husimi *et al.*, 1982) is to a phage what the normal, one-tube chemostat is to a bacterium. Medium is fed into a tube with cells only, and outflow from it goes unidirectionally into a tube with phages. Thus, the first tube serves as a reservoir of "food" for phages and if the cell concentration maintained in the first tube is high, then a high phage density will be maintained in the second tube. The advantages of this system include very high densities of phages and ease of maintenance. The drawbacks of two-tube chemostats are several for understanding adaptive walks. First, and most importantly, the dynamic equilibrium of the system is often at a phage density so high that hosts are multiply infected (Bull *et al.*, 2006; Chapter 8, but see also Chapter 3). Thus, phages are evolving in a direct-competition environment, and the selective environment of the phage is itself evolving. Second, fitness is not easily measured, because it is possibly density-dependent and frequency-dependent, requiring use of a competitor genotype (and such measures of fitness commonly reveal non-transitivity; Paquin and Adams, 1983). Third, attachment of the cells to the wall of the chemostat complicates understanding of the dynamics of phage washout rate (Dykhuizen, 1993). Nonetheless, the system can be used to achieve high levels of molecular evolution (Bull *et al.*, 1997; Wichman *et al.*, 2005).

9.6.1.3 Liquid culture, one-tube chemostat

Phages and cells are maintained in a single tube, with medium input (and outflow) at a constant rate (see Chapters 2 and 15 for discussion of the modeling of such a chemostat). As discussed in Chapter 2, several dynamic and evolutionary outcomes are possible in this system. As a phage evolution environment, one-tube systems have the advantage that, at equilibrium, multiple phages rarely infect the same host, so there is little direct competition among phages (oscillations, however, will lead to periods with potentially high levels of coinfection as described for batch culture in Chapter 3). However, phage population sizes may be small, hence slow to experience beneficial mutations. Also, the phage environment may change if the sensitive host evolves mechanisms to partially resist the phage (which is an experimental advantage if the interest is in coevolution). This method has been used to study the phenotypic evolution of bacterial and phage resistance evolution, but not at a molecular level (Bohannan and Lenski, 2000; Buckling and Rainey, 2002; Mizoguchi *et al.*, 2003; Chapter 2).

9.6.1.4 Plates

Plates are universally used to assay phage densities (Adams, 1959; see Chapters 4 and 16 for discussion of the ecology and modeling of phage growth in such environments). They can obviously be used as the environment for propagation. The most common method involves plating 100–1000 plaque-forming units (PFU), allowing individual plaques to form, recovering (and diluting) the phages before inoculating the subsequent plate. Bacteria are mixed with phages in a thin layer of "soft agar" medium on top of a larger volume of solidified, dense agar, so the phages are chiefly confined to a small volume that can easily be scraped into liquid medium to allow the phages to diffuse. Evolutionary changes arise within plaques and, if they increase the phage numbers in the plaque, can spread throughout the population as they produce progressively more progeny per plaque in subsequent platings (Burch and Chao, 1999, 2004; Poon and Chao, 2005). A second plate-propagation method is only known to work with phage T7: 10 or more milliliters of buffered medium, agar, cells, and 1–5 phages are mixed and poured into an empty plate. The solution solidifies, and plaques form and grow for several days, reaching several centimeters in diameter. Here measurable evolution occurs within giant plaques (Yin, 1993).

The advantage of the first plate method for propagation is both minimal effort and the ability to measure fitness in the same environment as the selection. Furthermore, fitness can be measured against a competitor or in absolute terms, and it has been possible to use plaque diameter as an accurate surrogate of phage number per plaque (Burch and Chao, 1999, 2004; but see also the discussion by Abedon and Yin, in press). Although the phage environment changes during outgrowth in plaques, the environment can be maintained constant across platings. The drawbacks include (1) the physical dynamics of phage–host interactions are not easily understood on plates, (2) the population sizes at transfer must be low enough that plaques do not routinely grow into each other (if plaques do grow together, then phage–phage interactions can favor competitive ability and thus lead to effects that violate environmental constancy). This requirement for low population size weakens selection by enhancing stochastic loss of new mutations.

9.6.2 Phage phenotypes that may be selected

There are several properties of phages that may be selected. In general, selection can be applied to specific individual phenotypes that apply to only a small part of the life history, or the phenotype can be the overall ability to grow or rate of growth.

9.6.2.1 Selection for propagation

This type of design merely uses some method of propagation, and the propagation is continued long enough that variants able to grow better eventually take over the population. Thus, serial passage in liquid culture (Section 9.6.1.1) automatically favors those able to grow faster, at least so long as transfer occurs prior to phage-mediated lysis of the bacterial population (see Chapter 3 for discussion of the importance of this distinction). Serial passage on plates (Section 9.6.1.4) automatically favors those that create more progeny per plaque (though again there could exist transfer-timing issues, plus this effect might be modified by harvesting solely toward a plaque's periphery). When the right conditions are used (e.g., densities are maintained low enough that multiple, unrelated phages are not allowed to coinfect cells often), then fitness is easily interpreted as an absolute measure of growth rate. However, if propagation is done at high phage densities, then the more-fit phage types may simply be good competitors, and absolute measures of fitness no longer apply. This method's drawback is that selection is only so strong as there are variants capable of growing faster – that is, "*s*" is not controlled by the experimenter, except perhaps by reducing population size to weaken selection.

9.6.2.2 Selection on part of the life history

Phages have simple life histories, but they can nonetheless be selected for several specific phenotypes: host range (e.g., ability to infect and reproduce in a new host; Adams, 1959; Hashemolhosseini *et al.*, 1994a, 1994b), attachment and binding (sticking to a surface, which has been the basis of the biotechnology method of phage display; Smith, 1991), virion survival in extreme environments (high temperature, caustic agents; Whang *et al.*, 1996; Bull *et al.*, 2000), and timing of lysis of the host (Roof and Young, 1995). Geneticists commonly extend this approach by creating a phage genotype that is defective in some critical function and then selecting mutants that overcome the defect (compensatory mutations; Chapter 6).

The utility of this general method is that it is informative about specific phage functions (hence its utility to geneticists), and, in many cases, the strength of selection can be varied arbitrarily. The main drawback is that phage success in the phenotypic selection may not translate into total fitness. Since all of these selections involve, by definition, just part of the life history, the survivors must be propagated to continue the selection or to assay the survivors. That propagation can introduce additional, unwanted selection. For example, if phage virus particles are selected to survive heat, and there are multiple genotypes that survive, the best high-temperature survivors may be disadvantaged in subsequent growth so that other (poorer survivor) genotypes are recovered most often.

9.7 CONCLUSION

The ease of conducting experimental adaptations of microbes and analyzing the molecular details of those processes is a relatively recent advance. Accompanying these technical advances is a growing interest in the details of adaptation. Although much of the recent theory pertains to fitness effects, molecular analyses are essential to interpreting how many mutations have occurred and which ones are unique. So far, we have enough pieces of a broad theory to inspire further work, but it should not be assumed that the exciting findings have already been revealed. We know that lab adaptation typically results in fitness increases, and that nucleotide substitutions accompany those fitness gains. It is also known that most of these adaptations involve mutations of large effect. But the connection between theories of adaptation and the experiments is still in need of rigorous testing. Even the simple task of quantifying the fitness effect of individual substitutions during adaptation and identifying those substitutions remains to be done in a thorough manner. Beyond that, many challenges and opportunities lie in confining the experiments to the realms of the models.

Most adaptations (intentionally) involve large population sizes and likely involve some possibly low level of recombination. The theories so far do not address those intermediate conditions. Adaptations in small population sizes will allow us to get around the main problems for one set of theories, but a strict absence of recombination is difficult to enforce with phages. Nonetheless, some of the most important work to be done involves merely identifying different mutations and their fitness effects from a given starting genotype. This work is feasible with most phages, and it can be done unambiguously with phages, in contrast to bacteria and other microbes. Thus, we have the systems in place to answer some important questions, and we may expect a flurry of activity to answer them in the near future.

ACKNOWLEDGMENTS

I thank Holly Wichman for comments that improved the disposition of this chapter. My phage work has been supported for some years now by NIH GM57756 and the Miescher Regents Professorship at the University of Texas.

REFERENCES

Abedon, S. T., and J. Yin, in press. Bacteriophage plaques: theory and analysis. In M. Clokie and A. Kropinski (eds.), *Bacteriophages: Methods and Protocols.* Totowa, NJ: Humana Press.

Adams, M. H. 1959. *Bacteriophages.* New York, NY: Interscience.

Andolfatto, P. 2005. Adaptive evolution of non-coding DNA in Drosophila. *Nature* **437**: 1149–52.

Arjan, J. A., M. Visser, C. W. Zeyl, P. J. Gerrish, J. L. Blanchard, and R. E. Lenski. 1999. Diminishing returns from mutation supply rate in asexual populations. *Science* **283**: 404–6.

Beisel, C. J., D. R. Rokyta, H. A. Wichman, and P. Joyce. 2007. Testing the extreme value domain of attraction for distributions of beneficial fitness effects. *Genetics* epub ahead of print.

Benzer, S. 1955. Fine structure of a genetic region in a bacteriophage. *Proc. Natl. Acad. Sci. U. S. A.* **41**: 344–54.

Bohannan, B.J. M., and R. E. Lenski. 2000. Linking genetic change to community evolution: Insights from studies of bacteria and bacteriophage. *Ecol. Lett.* **3**: 362–77.

Bradshaw, H. D. Jr., K. G. Otto, B. E. Frewen, J. K. McKay, and D. W. Schemske. 1998. Quantitative trait loci affecting differences in floral morphology between two species of monkeyflower (*Mimulus*). *Genetics* **149**: 367–82.

Buckling, A., and P. B. Rainey. 2002. Antagonistic coevolution between a bacterium and a bacteriophage. *Proc. Biol. Sci.* **269**: 931–6.

Bull, J. J., M. R. Badgett, H. A. Wichman, *et al.* 1997. Exceptional convergent evolution in a virus. *Genetics* **147**: 1497–507.

Bull, J. J., M. R. Badgett, and H. A. Wichman. 2000. Big-benefit mutations in a bacteriophage inhibited with heat. *Mol. Biol. Evol.* **17**: 942–50.

Bull, J. J., M. R. Badgett, D. Rokyta, and I. J. Molineux. 2003. Experimental evolution yields hundreds of mutations in a functional viral genome. *J. Mol. Evol.* **57**: 241–248.

Bull, J. J., M. R. Badgett, R. Springman, and I. J. Molineux. 2004. Genome properties and the limits of adaptation in bacteriophages. *Evolution* **58**: 692–701.

Bull, J. J., J. Millstein, J. Orcutt, and H. A. Wichman. 2006. Evolutionary feedback mediated through population density, illustrated with viruses in chemostats. *Am. Nat.* **167**: E39–E51. www.journals.uchicago.edu/AN/journal/issues/v167n2/40945/40945.html

Burch, C. L., and L. Chao. 1999. Evolution by small steps and rugged landscapes in the RNA virus φ6. *Genetics* **151**: 921–7.

Burch, C. L., and L. Chao. 2004. Epistasis and its relationship to canalization in the RNA virus φ6. *Genetics* **167**: 559–67.

Chao, L. 1990. Fitness of RNA virus decreased by Muller's ratchet. *Nature* **348**: 454–5.

Colegrave, N. 2002. Sex releases the speed limit on evolution. *Nature* **420**: 664–6.

Crow, J. F., and M. Kimura. 1970. *An Introduction to Population Genetics Theory*. New York, NY: Harper & Row.

Cunningham, C. W., K. Jeng, J. Husti, *et al.* 1997. Parallel molecular evolution of deletions and nonsense mutations in bacteriophage T7. *Mol. Biol. Evol.* **14**: 113–16.

Dykhuizen, D. E. 1993. Chemostats used for studying natural selection and adaptive evolution. *Meth. Enzymol.* **224**: 613–31.

Eigen, M. 1971. Self organization of matter and the evolution of biological macromolecules. *Naturwissenschaften* **58**: 465–523.

Eigen, M., and J. McCaskill. 1988. Molecular quasispecies. *J. Phys. Chem.* **92**: 6881–91.

Endy, D., L. You, J. Yin, and I. J. Molineux. 2000. Computation, prediction, and experimental tests of fitness for bacteriophage T7 mutants with permuted genomes. *Proc. Natl. Acad. Sci. U. S. A.* **97**: 5375–80.

Gerrish, P. J., and R. E. Lenski. 1998. The fate of competing beneficial mutations in an asexual population. *Genetica* **102–103**: 127–44.

Gillespie, J. H. 1991. *The Causes of Molecular Evolution*. New York, NY: Oxford University Press.

Hashemolhosseini, S., Z. Holmes, B. Mutschler, and U. Henning. 1994a. Alterations of receptor specificities of coliphages of the T2 family. *J. Mol. Biol.* **240**: 105–10.

Hashemolhosseini, S., D. Montag, L. Krämer, and U. Henning. 1994b. Determinants of receptor specificity of coliphages of the T4 family: a chaperone alters the host range. *J. Mol. Biol.* **241**: 524–33.

Hegreness, M., N. Shoresh, D. Hartl, and R. Kishony. 2006. An equivalence principle for the incorporation of favorable mutations in asexual populations. *Science* **311**: 1615–17.

Hill, W. G., and A. Robertson. 1966. The effect of linkage on limits to artificial selection. *Genet. Res.* **8**: 269–94.

Holder, K. K., and J. J. Bull. 2001. Profiles of adaptation in two similar viruses. *Genetics* **159**: 1393–404.

Husimi, Y., K. Nishigaki, Y. Kinoshita, and T. Tanaka. 1982. Cellstat: a continuous culture system of a bacteriophage for the study of the mutation rate and the selection process at the DNA level. *Rev. Sci. Instrum.* **53**: 517–22.

Kauffman, S., and S. Levin. 1987. Towards a general theory of adaptive walks on rugged landscapes. *J. Theor. Biol.* **128**: 11–45.

Lande, R. 1983. The response to selection on major and minor mutations affecting a metrical trait. *Heredity* **50**: 47–65.

Lenski, R. E., and M. Travisano. 1994. Dynamics of adaptation and diversification: a 10,000-generation experiment with bacterial populations. *Proc. Natl. Acad. Sci. U. S. A.* **91**: 6808–14.

Li, W.-H. 1997. *Molecular Evolution*. Sunderland, MA: Sinauer Associates.

Maynard Smith, J. 1978. *The Evolution of Sex*. Cambridge: Cambridge University Press.

Miralles, R., P. J. Gerrish, A. Moya, and S. F. Elena. 1999. Clonal interference and the evolution of RNA viruses. *Science* **285**: 1745–7.

Miralles, R., A. Moya, and S. F. Elena. 2000. Diminishing returns of population size in the rate of RNA virus adaptation. *J. Virol.* **74**: 3566–71.

Mizoguchi, K., M. Morita, C. R. Fischer, M. Yoichi, Y. Tanji, and H. Unno. 2003. Coevolution of bacteriophage PP01 and *Escherichia coli* O157: H7 in continuous culture. *Appl. Environ. Microbiol.* **69**: 170–6.

Orr, H. A. 1998. The population genetics of adaptation: the distribution of factors fixed during adaptive evolution. *Evolution* **52**: 935–49.

Orr, H. A. 1999. The evolutionary genetics of adaptation: a simulation study. *Genet. Res.* **74**: 207–214.

Orr, H. A. 2002. The population genetics of adaptation: the adaptation of DNA sequences. *Evolution* **56**: 1317–30.

Orr, H. A. 2005. The probability of parallel evolution. *Evolution* **59**: 216–20.

Otto, S. P., and M. W. Feldman. 1997. Deleterious mutations, variable epistatic interactions, and the evolution of recombination. *Theor. Popul. Biol.* **51**: 134–47.

Papadopoulos, D., D. Schneider, J. Meier-Eiss, W. Arber, R. E. Lenski, and M. Blot. 1999. Genomic evolution during a 10,000-generation experiment with bacteria. *Proc. Natl. Acad. Sci. U. S. A.* **96**: 3807–12.

Paquin, C. E., and J. Adams. 1983. Relative fitness can decrease in evolving asexual populations of *S. cerevisiae*. *Nature* **306**: 368–71.

Pepin, K. M., and H. A. Wichman. 2007. Variable epistatic effects between mutations at host recognition sites in ϕX174 bacteriophage. *Evolution* **61**: 1710–24.

Poon, A., and L. Chao. 2005. The rate of compensatory mutation in the DNA bacteriophage ϕX174. *Genetics* **170**: 989–99.

Poon, A., and S. P. Otto. 2000. Compensating for our load of mutations: freezing the meltdown of small populations. *Evolution* **54**: 1467–79.

Rokyta, D., M. R. Badgett, I. J. Molineux, and J. J. Bull. 2002. Experimental genomic evolution: extensive compensation for loss of DNA ligase activity in a virus. *Mol. Biol. Evol.* **19**: 230–8.

Rokyta, D. R., P. Joyce, S. B. Caudle, and H. A. Wichman. 2005. An empirical test of the mutational landscape model of adaptation using a single-stranded DNA virus. *Nat. Genet.* **37**: 441–4.

Rokyta, D. R., C. J. Beisel, and P. Joyce. 2006. Properties of adaptive walks on uncorrelated landscapes under strong selection weak mutation. *J. Theor. Biol.* **243**: 114–20.

Roof, W. D., and R. Young. 1995. ϕX174 lysis requires slyD, a host gene which is related to the FKBP family of peptidyl-prolyl cis-trans isomerases. *FEMS Microbiol. Rev.* **17**: 213–18.

Rowe, L. A., M. L. Geddie, O. B. Alexander, and I. Matsumura. 2003. A comparison of directed evolution approaches using the β-glucuronidase model system. *J. Mol. Biol.* **332**: 851–60.

Rozen, D. E., and R. E. Lenski. 2000. Long-term experimental evolution in *Escherichia coli*. VIII. Dynamics of a balanced polymorphism. *Am. Nat.* **155**: 24–35.

Rozen, D. E., J. A. de Visser, and P. J. Gerrish. 2002. Fitness effects of fixed beneficial mutations in microbial populations. *Curr. Biol.* **12**: 1040–5.

Sanjuan, R., A. Moya, and S. F. Elena. 2004. The distribution of fitness effects caused by single-nucleotide substitutions in an RNA virus. *Proc. Natl. Acad. Sci. U. S. A.* **101**: 8396–401.

Sanjuan, R., J. M. Cuevas, A. Moya, and S. F. Elena. 2005. Epistasis and the adaptability of an RNA virus. *Genetics* **170**: 1001–8.

Schrag, S. J., V. Perrot, and B. R. Levin. 1997. Adaptation to the fitness costs of antibiotic resistance in *Escherichia coli*. *Proc. Biol. Sci.* **264**: 1287–91.

Smith, G. P. 1991. Surface presentation of protein epitopes using bacteriophage expression systems. *Curr. Opin. Biotechnol.* **2**: 668–73.

Turner, P. E., and L. Chao. 1998. Sex and the evolution of intrahost competition in RNA virus φ6. *Genetics* **150**: 523–32.

Turner, P. E., and L. Chao. 1999. Prisoner's dilemma in an RNA virus. *Nature* **398**: 441–3.

Voigt, C. A., S. Kauffman, and Z. G. Wang. 2000. Rational evolutionary design: the theory of in vitro protein evolution. *Adv. Protein. Chem.* **55**: 79–160.

Wahl, L. M., and P. J. Gerrish. 2001. The probability that beneficial mutations are lost in populations with periodic bottlenecks. *Evolution* **55**: 2606–10.

Wahl, L. M., and D. C. Krakauer. 2000. Models of experimental evolution: the role of genetic chance and selective necessity. *Genetics* **156**: 1437–48.

Wahl, L. M., P. J. Gerrish, and I. Saika-Voivod. 2002. Evaluating the impact of population bottlenecks in experimental evolution. *Genetics* **162**: 961–71.

Weinreich, D. M., N. F. Delaney, M. A. Depristo, and D. L. Hartl. 2006. Darwinian evolution can follow only very few mutational paths to fitter proteins. *Science* **312**: 111–114.

Whang, T., B. Daly, and J. Yin. 1996. Metal-ion discrimination by phage T7. *J. Inorg. Biochem.* **63**: 1–7.

Wichman, H. A., M. R. Badgett, L. A. Scott, C. M. Boulianne, and J. J. Bull. 1999. Different trajectories of parallel evolution during viral adaptation. *Science* **285**: 422–4.

Wichman, H. A., L. A. Scott, C. D. Yarber, and J. J. Bull. 2000. Experimental evolution recapitulates natural evolution. *Philos. Trans. R. Soc. Lond. Ser. B* **355**: 1677–84.

Wichman, H. A., J. Millstein, and J. J. Bull. 2005. Adaptive molecular evolution for 13,000 phage generations: a possible arms race. *Genetics* **170**: 19–31.

Williams, G. C. 1975. *Sex and Evolution*. Princeton, NJ: Princeton University Press.

Wood, T. E., J. M. Burke, and L. H. Rieseberg. 2005. Parallel genotypic adaptation: when evolution repeats itself. *Genetica* **123**: 157–70.

Yin, J. 1993. Evolution of bacteriophage T7 in a growing plaque. *J. Bacteriol.* **175**: 1272–7.

Part III Phage ecology in environments

CHAPTER 10

Aquatic phage ecology

T. Frede Thingstad,* Gunnar Bratbak, and Mikal Heldal

10.1 INTRODUCTION

Phage ecology, considering the ecological significance of native viruses in natural aquatic ecosystems, has had a momentous development since the early 1990s, when it was acknowledged that the abundance of viruses in natural aquatic ecosystems was much higher than had been anticipated. During that time new methods and approaches have been applied to the study of viral communities as well as different aspects of viral activity in natural ecosystems. It has been necessary to develop new concepts and models to interpret these new data and knowledge within the context of ecosystem structure and function. Several excellent reviews provide a comprehensive summary and analysis of the data and information that has accumulated (Fuhrman, 1999; Wilhelm and Suttle, 1999; Wommack and Colwell, 2000; Paul *et al.*, 2002; Weinbauer, 2004; Suttle, 2005) and the reader should consult these for a more complete review of the literature. The purpose of the present chapter is to provide a brief overview of contemporary aquatic phage ecology, including a closer look at the extent to which present theory can be claimed to explain the observations now available.

Based on the numerical dominance of prokaryote over eukaryote unicellular organisms in the pelagic environment (a factor of 2–3 orders of magnitude), the usual assumption is that the population of free viruses is dominated by bacteriophages. Many of the numbers available, however, are based on techniques that do not allow identification of the type of host. We

* Corresponding author

Bacteriophage Ecology: Population Growth, Evolution, and Impact of Bacterial Viruses, ed. Stephen T. Abedon. Published by Cambridge University Press.

will therefore refer to viruses, rather than phages, in instances where it is not positively known that we are dealing with bacteriophages.

10.2 VIRAL ABUNDANCE

Although for a long time we have been aware that viruses were present in marine ecosystems (Spencer, 1955, 1960; Valentine and Chapman, 1966; Valentine *et al.*, 1966; Wiebe and Liston, 1968; Torrella and Morita, 1979), the possibility that they could play a significant quantitative role there was not considered before the late 1980s or early 1990s, when the high phage abundance and high frequency of infected bacterial cells were first reported (Bergh *et al.*, 1989; Proctor and Fuhrman, 1990; Suttle *et al.*, 1990). These first studies were based on electron microscopy, a facility which is not amply available to microbial ecology research groups. Within a few years, however, development of new techniques employing specific and highly fluorescent dyes and 0.02 μm pore size aluminum filters allowed viruses to be counted by standard epifluorescence microscopy (Hara *et al.*, 1991; Hennes and Suttle, 1995; Noble and Fuhrman, 1998). This made it possible for many more research groups to include viruses in their studies. Later, flow cytometry was also employed for counting of viruses (Marie *et al.*, 1999; Brussaard *et al.*, 2000).

When a new field of research is explored with new methods, the quality of the first data obtained is often questioned by later methodological improvements and control experiments. We know from many studies that the abundance of bacteria decreases in preserved samples (Turley and Hughes, 1992; Gundersen *et al.*, 1996) and this has also been shown to be true for viruses (Chen *et al.*, 2001; Brussaard, 2004; Wen *et al.*, 2004). The total abundance of viruses in natural seawater is typically reported to be in the range of 10^4 – 2×10^7 mL^{-1} (Wilhelm and Suttle, 1999; Paul and Kellogg, 2000; Wommack and Colwell, 2000; Weinbauer, 2004). Many of these data, due to improper sample preservation and long storage times, may be underestimates (Wen *et al.*, 2004) and the correct values are probably in the range of 3×10^6 – 10^8 mL^{-1} (Suttle, 2005). Even given these losses, observations of variation in *relative* virus abundance in time and space – from high values during productive seasons, in surface waters, in near-shore waters, and in eutrophic waters, to low values during unproductive seasons and in deep, offshore, and oligotrophic waters – are still correct.

Problems with preservation and storage can be avoided with proper protocols, but our estimates of total virus abundance may nevertheless still be in error. The methods used for total counting of viruses are dependent on our ability to recognize and detect viral particles in the electron microscope, the epifluorescence microscope, or the flow cytometer by their nucleic acid content, electron density, structure, morphology, or size. The fraction of particles that is below the detection limit for nucleic acid content, electron density, and size by these methods is unknown, but it is probably variable and we may presume that it is dependent on both method and operator. The fraction of viruses that have structures and morphologies that are not easily detected or recognized by routine methods also is not known. Structures very similar to Ff phages, which are about 900 nm long and 9 nm in diameter (Stubbs, 2001), for example, are often seen in aquatic samples prepared for counting of viruses by transmission electron microscopy, but we do not know if these structures are viruses and they are not included in any total counts of viruses that we are aware of.

10.3 VIRAL ACTIVITY

The activity of viruses manifests itself in various ways in the pelagic ecosystem. In addition to the production and removal processes of viruses themselves, viruses are believed to influence the cycling of matter, the diversity of the host community, and the transfer of genetic material between hosts.

10.3.1 Detection of viral production

A number of different approaches have been used to assess viral activity in natural waters, but they all suffer from drawbacks such as extensive sample manipulation or poorly constrained assumptions. Most methods aim at measuring virus particle production, and from these data researchers infer the viral impact on host populations. This approach is inherently difficult, however, because the actual number of viruses produced per lytic event (burst size) is variable and often unknown. A robust and reliable method that provides accurate estimates of viral activity is thus still lacking (Suttle, 2005). The methods developed and applied so far are as follows:

(1) Observation of net change in viral abundance over time in enclosed water masses (Heldal and Bratbak, 1991; Bratbak *et al.*, 1993) or embottled water samples (Bratbak *et al.*, 1996). Since what is observed is only the difference between production and removal, only minimum estimates for the two rates can be obtained.

(2) Viral production rates estimated from observed viral decay or loss rates in samples where the virus-producing organisms (i.e., bacteria along with eukaryotic organisms) have been removed or inhibited (Heldal and Bratbak, 1991; Suttle and Chen, 1992; Wommack *et al.*, 1996).

(3) Incorporation of radiotracer (^{32}P-phosphate or ^{3}H-thymidine) has been used to estimate production of viral DNA (Steward *et al.*, 1992a, 1992b)

but has proven difficult to calibrate (Noble and Steward, 2001) and consequently has not been widely adopted.

(4) Rates of virus production and removal may be determined simultaneously using fluorescently labeled viral tracers in an approach that is mathematically similar to the isotope dilution technique employed to measure release and uptake of radiolabeled nutrients (Noble and Fuhrman, 2000). The critical assumption in this approach is that the tracer viruses are representative of the *in situ* viral communities, though this assumption is difficult to test and verify.

(5) Virus production may be estimated from host–virus contact rates when the abundances of both host and virus are known (Waterbury and Valois, 1993; Suttle and Chan, 1994). The calculation is based on the assumption that by multiplying the number of infections with burst size one gets the number of viruses produced (after a delay given by the latency time). This method thus relies heavily on assumptions about host specificity, effective adsorption constants, and burst size.

(6) Frequency of visibly infected cells (FVIC) as observed by transmission electron microscopy (TEM) (Bratbak *et al.*, 1992; Proctor *et al.*, 1993; Weinbauer *et al.*, 1993) has the advantage of being a non-invasive technique, one not requiring incubation of the sample. Conversion factors are required, based on assumptions of how large a part of lytic cycles intracellular viruses are visible, but are not well defined (Binder, 1999; Weinbauer *et al.*, 2002).

(7) A dilution technique where virus-free water is added to reduce the contact rate between host and virus has been developed by Evans *et al.* (2003) as an equivalent to the method introduced by Landry and Hassett (1982) for quantifying microzooplankton grazing. The rationale for this equivalency may be questioned because of the intrinsic delay between virus infection, subsequent host-cell death, and bacterium removal by lysis, as compared to the instantaneous death and removal of prey organisms during a grazing event. The critical factor relates to the length of the latent period for the host-virus systems involved as a fraction of the experimental incubation time. The underlying assumptions for this approach thus need further testing.

(8) A technique based on differential filtration to reduce the viral concentration while maintaining host concentration allows virus production rates to be monitored as changes in viral concentration over time (Weinbauer *et al.*, 2002; Wilhelm *et al.*, 2002). The production of viruses in this approach is assumed to result only from cells infected prior to filtration, so the simultaneous loss of (newly produced) viral

particles is assumed to be negligible. The method requires extensive sample manipulation, and the effect of this manipulation is unpredictable and requires further investigation and testing.

10.3.2 Impact of virus-induced cell lysis

The prevailing view is that viruses, on a daily basis, kill some 10–40% of the standing stock of bacteria in marine ecosystems (Suttle, 1994, 2005; Fuhrman, 1999). As such, these values are comparable to the amount of bacterial biomass grazed by protozoa per day. The individual values reported for viral lysis range from undetectable to >100% per day, and although methodological problems are obvious, these studies suggest that viral activity is highly variable in time and space. Evidence suggesting that viral activity indeed may change rapidly over time may be inferred from observations of native bacterial assemblages where a sizable fraction of bacteria (10–40%) contain mature viral particles, i.e. that are close to lysis (Heldal and Bratbak, 1991; Bratbak *et al.*, 1992), and from short-time incubation studies showing transient increases in viral abundance (Bratbak *et al.*, 1996). One interpretation of this variability is that viral activity, at least in some cases, might be synchronized (that is, set by approximately simultaneous initiation of standing infections; Heldal and Bratbak, 1991; Bratbak *et al.*, 1996) and thus that the rates of viral lysis measured depend on where in the lytic cycle the synchronized population is when an experiment is conducted.

10.3.3 Lytic or lysogenic?

Viral production may be due to lytic infection or to induction of viral reproduction in cells carrying a virus (lysogenic, pseudolysogenic, or otherwise; see Chapter 5). Some studies strongly suggest that lytic infection is the predominating mode of viral infection in bacterial communities (Wilcox and Fuhrman, 1994), while other studies show that the fraction of cells in which phage production may be induced is very high (Cochran *et al.*, 1998), suggesting that induction of lysogenic phages also may be important (Weinbauer, 2004). As has been shown for cyanobacteria (Sullivan *et al.*, 2003; Comeau *et al.*, 2004), bacteria are probably susceptible to more than one type of phage. Most bacteria may also be lysogens and at the same time subject to lytic infection by other phages. Observed variations in viral abundance in natural waters and results obtained from different methods used for measuring virus production should be interpreted in light of properties of viruses and their modes of replication (lytic infection or induction).

Methods of viral-activity determination based on host–virus contact rates (e.g., methods 5 and 7; Section 10.3.1) will obviously exclude virus production caused by induction. Methods based on net increase in viral abundance over time (e.g., methods 1, 3, 4, and 8) assume that the particles produced are stable and that simultaneous removal rates are insignificant. This may be a valid assumption for obligately lytic virus particles, which tend to be durable, but not for temperate virus particles, which at least in some cases seem to lose integrity quite rapidly (Bratbak *et al.*, 1994). These methods may thus be biased to measure the production of obligately lytic viruses. The length of incubation required by a protocol for assessing viral activity and the timing of the observation may be critical if virus production and release are synchronized, transient phenomena, as has been indicated by some studies (Heldal and Bratbak, 1991; Bratbak *et al.*, 1996). Signs of viral activity may be averaged out if the incubation time is too long. Alternatively, the active period may be missed if the incubation time is too short.

The nature of viral activity is diverse, and a common feature that allows detection and quantification of all viral production, both instantaneous and average over time, remains to be identified. Although individual methods may be developed and tested to give reliable and reproducible results relating to aspects of viral activity, the search for a single all-inclusive method has so far been in vain.

10.3.4 Factors controlling viral activity

The factors controlling viral activity and abundance in natural waters are elusive, but it seems reasonable to presume that the physiological condition and activity of host populations plays an important role (Weinbauer, 2004). Inferior growth conditions such as nutrient or carbon limitation will in general impede lytic processes so that infected cells do not produce new viruses or lyse, or so that cell lysis will be delayed and burst size reduced (Chapter 5).

A few case studies may be interpreted to support the view that availability of energy as well as organic and inorganic nutrients, and thus the physiological condition and activity of the host cells, are key factors for understanding viral production in natural communities. Bratbak *et al.* (1990), for example, described a spring phytoplankton bloom that was succeeded by an increase in bacterial abundance. The subsequent increase in viral abundance and collapse of the bacterial (i.e., non-phytoplankton) abundance may have been triggered by the bloom's input of carbon and energy to the heterotrophic system. Tuomi *et al.* (1995) made a comparable observation when they added amino acids and phosphate to a seawater microcosm. The increase in carbon and energy availability stimulated virus production more than bacterial biomass production, while the increase in phosphate availability stimulated biomass production rather than virus production. Reduced burst size and delayed cell lysis has been shown for virus-infected cyanobacteria growing under phosphorous limitation (Wilson *et al.*, 1996), but whether this is true for native heterotrophic bacteria is unknown.

In the open sea, intermittent aperiodic changes in dissolved oxygen saturation show that net autotrophy (carbon fixation) is variable over time (Karl *et al.*, 2003), suggesting that the input of carbon and energy to the heterotrophic community also may be highly variable as a function of time. Such short-term changes occurring in many systems may influence viral production and hence provide a possible explanation for several phenomena including reported rapid changes in viral abundance, large variations in virus-to-bacteria ratios, and highly variable frequencies of infected cells over time (e.g., Bratbak *et al.*, 1990, 1992; Heldal and Bratbak, 1991; Jiang and Paul, 1994; Hennes and Simon, 1995; Weinbauer *et al.*, 1995; Rodriguez *et al.*, 2000).

10.3.5 Aquatic viral diversity

One obvious aspect of viral diversity is the difference *between* different environments. Extreme environments with specialized host populations, such as hot springs, also have specialized viral communities, some of which (e.g., those with crenarchaean hosts) are even evolutionary distinct (Ortmann *et al.*, 2006). More obscure are the differences that may exist between different samples from relatively similar environments and, on a local spatial scale, what diversity to expect within a single, homogeneous environment (if such a thing even exists). We here primarily focus on local aspects within natural marine ecosystems, and are thus consciously leaving aside a debate on the differences between the many different aquatic environments. In the pelagic marine environment, viruses are not only active and abundant, they are also very diverse (Suttle, 2005). Molecular studies focusing on specific viral genes have qualitatively shown that the genetic variation in viral communities in natural waters is very large (Chen *et al.*, 1996; Short and Suttle, 2002; Zhong *et al.*, 2002), but also that the same genes might be found in very different environments (Breitbart *et al.*, 2004; Short and Suttle, 2005). Studies employing pulsed field gel electrophoresis (PFGE) to characterize and analyze viral communities by the size of the viral genomes often conclude that 10–30 discrete populations may be distinguished by genome size (Wommack *et al.*, 1999; Steward *et al.*, 2000; Castberg *et al.*, 2001; Larsen *et al.*, 2001; Riemann and

Middelboe, 2002). Each of these discrete bands or virus populations typically makes up 0.5–5% of the total DNA or 0.1–5% of the total viral abundance, and together they comprise 10–20% of the viral community in terms of both DNA and particle abundance (unpublished analysis). The remaining 80% of viral DNA appears as an unresolved background presumably made up of many minor populations with a range of genome sizes producing a smear on the gel. These observations are in accordance with Steward's (2001) estimate that the PFGE method is capable of detecting viruses that make up 0.1–1% of the community, assuming that virus from 100 mL water is loaded on the gel and that their genome sizes are in the range of 30–300 kbp. These PFGE data are corroborated by metagenome analysis of samples (200 L) from coastal waters suggesting that the viral communities comprise several thousand genotypes, none of which on its own accounts for more than 5% of the total virus abundance (Breitbart and Rohwer, 2005; Edwards and Rohwer, 2005).

10.4 ECOLOGICAL ROLE

The possible ecological functions of viruses may be inferred from what is known about viruses in general. The attainment of these functions in natural aquatic ecosystems is qualitatively confirmed by examples from a great number of studies focusing on different aspects of viral ecology (see reviews by Fuhrman, 1999; Weinbauer, 2004; Weinbauer and Rassoulzadegan, 2004). The quantitative significance of these functions, however, is still vague.

10.4.1 Viruses and organic-carbon release

Viruses lyse their host cells and convert particulate organic material and nutrients that otherwise might have been consumed in the grazing food chain into a dissolved form that is available only to osmotrophs, which are organisms that take up nutrients directly across their cell membranes (contrast with phagotroph). From a biogeochemical perspective, predators and parasites thus have opposite effects. Predation has a net effect of moving energy and material from smaller to larger while parasitism moves energy from larger to smaller particles as well as to dissolved forms (Fig. 10.1). Since lytic viruses are believed to be highly active also in unicellular eukaryotic communities, this effect is not restricted solely to the bacterium–bacteriophage interactions in the pelagic food web.

The recycling of organic matter from particulate form back to the dissolved pool from where it can be re-incorporated into particulate form by

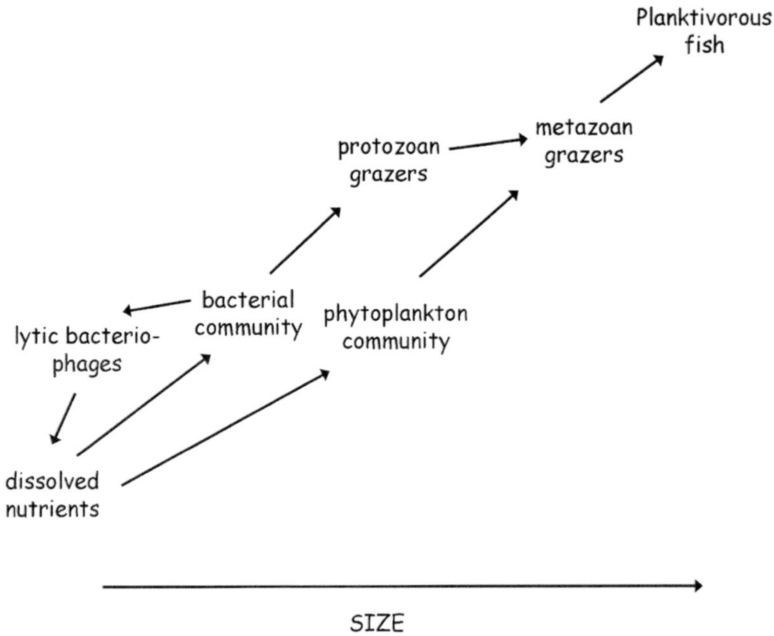

Figure 10.1 Idealized flow scheme of matter for the planktonic food web, illustrating how predatory processes move material and energy "up" the food web towards larger organisms. Lysis by bacteriophages is one of the processes that divert material from this flux, back to the dissolved-nutrient pool.

heterotrophic bacteria has been termed the "viral loop" (Bratbak *et al.*, 1992). The mechanisms involved may be seen as a subset of the general set of mechanisms belonging to the "microbial loop" (Azam *et al.*, 1983) whereby (parts of) the organic material released from the food web is re-incorporated into the food web via the osmotrophic bacteria. Since a lot of the organic material may in principle be respired away as it circles numerous times through the viral loop, the consequence could be that little energy and carbon from the consumed dissolved organic material is channeled up the predatory food chain to higher trophic levels. Several studies show that virus-induced lysis indeed takes place in natural systems and that the organic material and nutrients released are readily available to heterotrophic bacteria and phytoplankton (Gobler *et al.*, 1997; Bratbak *et al.*, 1998). Although viral activity obviously affects carbon and nutrient flow and thus the whole structure of microbial food webs, quantitative data are scarce and our present conceptual models (see Section 10.5) are at a stage where the mechanisms controlling this flux are only poorly understood.

10.4.2 Viruses versus grazers

Viruses and grazers are the main factors of mortality in microbial systems, but their effect on population dynamics, community composition, and nutrient flow are very different. Viruses, in particular, are host-specific at some level, and a lytic viral attack will only affect a strain, a species, or a few related species, but not an entire cellular community. Induction of viral activity in lysogenic cells may also be specific as far as host–virus systems differ in susceptibility to the inducing factor. The populations removed by viruses may be replaced by other populations, and when the system is at equilibrium the net outcome is that although viruses affect the size of individual populations, they do not affect the size and biomass of the whole community. In contrast, microbial grazers typically feed on particles within a certain size range, and though they may be selective with respect to food quality (Gonzalez *et al.*, 1993; Simek *et al.*, 1994; Gonzalez, 1996) and prey morphology, they are not particularly species-specific. Grazers will thus typically affect the size of the community rather than the size of individual populations within a community.

In addition to host specificity and prey selectivity, there are at least two additional major functional differences between viruses and grazers. Grazers represent a trophic level that consumes, accumulates, and transfers organic material and nutrients in the food web. Viruses are small, contain little biomass, and are hardly grazed (Gonzalez and Suttle, 1993; Bettarel *et al.*, 2005). Their functional role, consequently, can be considered more as a process than as a trophic level. While grazers are dynamically controlled by larger predators, viral abundance and activity is presumably controlled by a more intrinsic rate of decay and inactivation (Heldal and Bratbak, 1991; Suttle and Chen, 1992). In addition, unlike grazers, the activity of some viruses is independent of virus concentration as the virus is carried by the host and induced by some internal cellular or external environmental factors. Besides host abundance, we know very little about the factors that control the activity of lytic viruses in natural aquatic ecosystems (Section 10.3.4; Chapters 2 and 3). The same is true for the factors that control the activity of temperate viruses in lysogenic cells (Chapter 5), where we still have to fall back on general textbook knowledge.

10.5 ECOLOGICAL THEORY OF AQUATIC FOOD WEBS

The experimental work reviewed above gives insight into quantitative aspects of viral ecology: how many there are, how large a fraction of bacterial loss is accounted for by viral lysis, etc. The underlying control mechanisms are, however, not immediately obvious from such numbers. One is still left with such questions as, for example, what is it that determines viral abundance and what is it that determines the partitioning of bacterial loss between grazing and viral lysis? Before about 1990, the standard scientific folklore for aquatic systems was that, since bacterial density was low and bacterial diversity probably was very high ("everything is everywhere"), the probability of a virus locating and infecting a new host before being inactivated would be small, and the process could therefore not be of any ecological importance. When it was realized that there were abundant viral populations in pelagic environments (Bergh *et al.*, 1989; Proctor and Fuhrman, 1990), that these viruses had no general ability to withstand destruction in the natural environment (Heldal and Bratbak, 1991), and that, at least in some environments, the lytic process seemed to dominate over the lysogenic (Wilcox and Fuhrman, 1994), the whole argument for a low ecological significance of lytic viruses fell apart. This led not only to a need for experimental and descriptive work to understand the ecological role of viruses, but also to a need for a better theoretical understanding of how they interact with all the other processes of the microbial food web.

10.5.1 Viral impact on biodiversity

In addition to natural transduction (see Chapter 11), phages can impact the diversity of bacterial communities by differentially killing bacteria that otherwise are especially successful ("killing the winner").

10.5.1.1 Hutchinson's paradox

At the root of the biodiversity issue lies the fundamental question of whether species or organism groups coexist because they have each specialized for one resource. Alternatively, they share a common limiting resource (Chapter 2), in which case the "principle of competitive exclusion" may seem to be contradicted. In the case of phytoplankton, this problem is best known in the form of Hutchinson's paradox (Hutchinson, 1961): How can so many species coexist in an apparently stable environment, all apparently limited by the same resources (see also Chapter 6)? There is a vast literature on suggested solutions to this paradox, many based on spatial or temporal variability in the environment, but also in the more subtle form of chaotic behavior in the internal dynamics of the competitors (Huisman and Weissing, 1999). A potential problem with many of the proposed models of this kind is that they have a tendency to impose relatively strict conditions on the range in parameter values that allow coexistence, leading one to wonder whether coexistence is a series of special cases, rather than rooted in a general phenomenon based on a common, generic explanation.

10.5.1.2 A viral solution to Hutchinson's paradox

It is possible to propose a very simple and very general principle that solves Hutchinson's paradox. In connection with lytic viruses, this has been termed the "killing the winner" principle (Thingstad and Lignell, 1997). The phrase is meant to illustrate the intuitively logical idea that *any* mechanism you add to a competition model that will prevent the best competitor from sequestering all of the limiting resource will tend to allow other, less efficient competitors to become established, thus promoting higher diversity. The generic principle is illustrated in Fig. 10.2. As may be illustrated using a simple Lotka–Volterra type of mathematical formulation for the food-web structure of Fig. 10.2 (unpublished calculations), the balance between the two specialists is a function of the "total limiting resource," with oligotrophic conditions favoring the competition specialist (e.g., as for nutrients), and eutrophic conditions favoring the defense specialist (e.g., as against viruses). Those who in general feel that ecosystem models tend to be very speculative and founded on meager experimental or observational support should be aware that the structure in Fig. 10.2 actually has a range of supporting investigations, although these investigations come in many forms of disguise (Table 10.1). The generic system in Fig. 10.2 is thus perhaps one of the few examples of a general ecological theory that can claim extensive experimental back-up.

Fundamental to the establishment of structures such as that in Fig. 10.2 is the existence of a trade-off between the two competitor strategies. Should an organism invent a strategy that allows it to simultaneously optimize defense and competition (termed "Winnie-the-Pooh" strategists by Thingstad *et al.*, 2005), it would be able to replace the other three organisms presented in Fig. 10.2, reducing strongly the biodiversity of this idealized system. In the case of phage-bacterium interactions, however, there typically are trade-offs in the choice between optimizing for competition and for defense (Bohannan *et al.*, 2002; Chapter 2), likely to prevent such simultaneous optimization. The perspective is interesting because it identifies the physiological and molecular mechanisms behind such trade-offs as central in shaping the structure of the pelagic microbial food web. In the case of viruses, the observation that virus attachment sites may be associated with the porins required for transport of potentially limiting substrates into the cells (Nikaido and Vaara, 1987; Schwartz, 1987) suggests one such mechanism: an uptake system

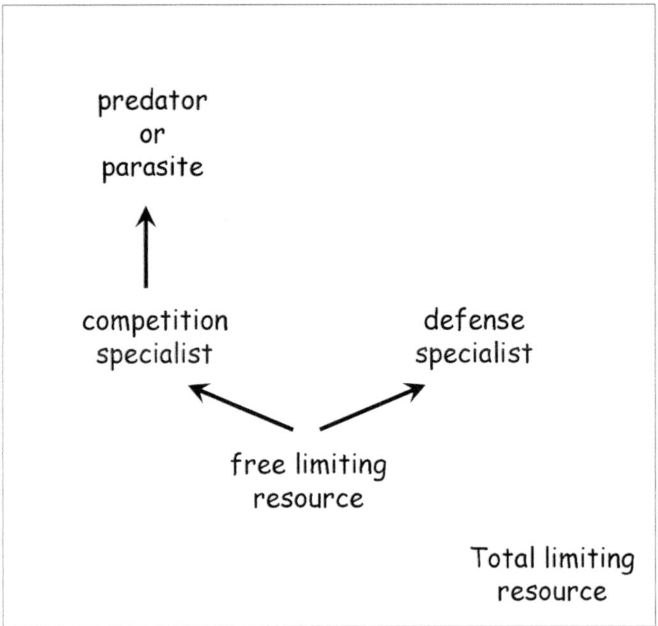

Figure 10.2 "Killing the winner" principle in generalized and idealized form. Two generalized organism types, an idealized *competition specialist* and an idealized *defense specialist*, compete for the *free limiting resource*. *Total limiting resource* is the amount of resource (e.g., phosphorous) available for sharing between the four components of the food web. Implicit in this structure is a trade-off between the two strategies so that, in this idealized structure, only the competition specialist is susceptible to loss caused by the *predator or parasite* group representing the generalization of grazers and viruses discussed in the text. For bacteria, grazing from mainly (but not only) size-selective heterotrophic flagellates and lysis by host-specific phages are believed to be the main loss factors. Coexistence of host-groups of bacteria (Fig. 10.3), the occurrence of grazing-resistant morphological forms of bacteria, and the coexistence of mineral-nutrient-limited bacteria and phytoplankton, may all be explained as special cases of this generic principle.

optimized for nutrient capture is likely to become less effective if modified in order to reduce the probability of viral attachment or nucleic acid injection. The potential implications for our interpretation of data on bacterial diversity may be profound. Among fast-growing competition specialists it may well be that a high nutrient affinity is coupled to a highly effective phage adsorption coefficient, with low host abundance as the result. Slow-growing defense specialists with low effective phage adsorption coefficients coupled to low nutrient affinity would be able to maintain populations with higher host abundance (Chapter 2). Dominant bacterial populations as (presumably)

Table 10.1 *Examples of studies focusing on various aspects of systems representing variations over the general "killing the winner" principle illustrated in Fig. 10.2.*

Competition/ nutrient-uptake specialist	Predator- or parasite-defense specialist	Predator or parasite	Shared limiting resource	System tested	Reference
Edible phytoplankton species	Inedible phytoplankton species	Meso-zooplankton	Mineral nutrients (?)	Mesocosms manipulated with reduction in zooplankton	(McCauley and Briand, 1979)
Heterotrophic bacterial species	Diatom (*Skeletonema costatum*)	Heterotrophic flagellate	Phosphate	Laboratory chemostats	(Pengerud *et al.*, 1987)
Bacterial host (*E. coli*)	Resistant strain of host	Virus (T4; T2)	Glucose	Laboratory	(Bohannan and Lenski, 1999, 2000)
Edible bacteria ($<1\mu m$)	Inedible bacteria (filaments or aggregates)	Protozoa	Glucose or phosphate	Laboratory, mixed bacterial communities	(Madz and Jürgens, 2003)
Edible algae ($<35\mu m$)	Inedible algae ($>35\mu m$)	Cladoceran	Nitrate and phosphate	Laboratory, mixed algal community	(Steiner, 2003)
Rod shaped bacterium *Pedobacter*	Coccus-formed bacterium *Brevundimonas*	Ciliate (*Tetrahymena pyriformis*)	?	Laboratory. Demonstrated the possibility of chaotic behavior	(Becks *et al.*, 2005)
Small phytoplankton	Large phytoplankton	Natural zooplankton communities	?	Meta-analysis of oceanic data	(Irigoien *et al.*, 2004)

determined, for example, by DGGE (denaturing gradient gel electrophoresis) techniques, may thus not necessarily provide a picture of which species are most active.

More general phage protection mechanisms are probably also at work. A large fraction of marine bacteria are known to produce polysaccharide capsules (Stoderegger and Herndl, 2001), a process potentially costly in terms of diffusive nutrient transport, and certainly costly in terms of energy and carbon requirements. This investment may be speculated to have a function in host defense, in which case the finding of phages with enzymes to penetrate polysaccharide capsules (e.g. Hänfling *et al.*, 1996) may be seen as evidence of the ongoing phage–host arms race. Capsules have, however, also been speculated to actually "invite" viral infection (Weinbauer, 2004).

Although rather innocent looking, the structure in Fig. 10.2 thus combines many aspects often studied by different research traditions. In addition to the links between cell physiology and food-web structure, and between viral lysis and diversity, it contains a biogeochemical aspect linked to the diversifying of routes for material transport through the food web, and it contains an evolutionary aspect linked to the arms race between the three strategies involved.

10.5.2 Factors affecting viral impact on natural food webs

One conceptual approach to a natural food web is to look at it as a set of basic units of the type in Fig. 10.2, linked together in series and in parallel, or, as in the case of combining protozoan grazing with lytic viruses, in a hierarchical structure (Fig. 10.3), with boxes within boxes. The structure in Fig. 10.3 can also be seen as a special case of the generalized case combining a low-specificity loss mechanism (in this instance size- and shape-specific grazing) with a high-specificity loss mechanism (host-specific viral lysis). The steady state for this construction, assuming Lotka–Volterra type equations, has been analyzed (Thingstad, 2000). As presented in this section, some of the main results can be derived in a heuristic manner.

10.5.2.1 Number of simultaneously dominant host populations

By dividing the size of the total bacterial community by the average size of a host population, one gets a number, N, for bacterial diversity measured as the number of simultaneously dominant "species" (actually "host groups" defined as the group of hosts susceptible to a specific type of phage). This type of theory may allow for an order of magnitude of ~100 simultaneously coexisting host-groups ("species"), leading to the hypothesis (hypothesis H1)

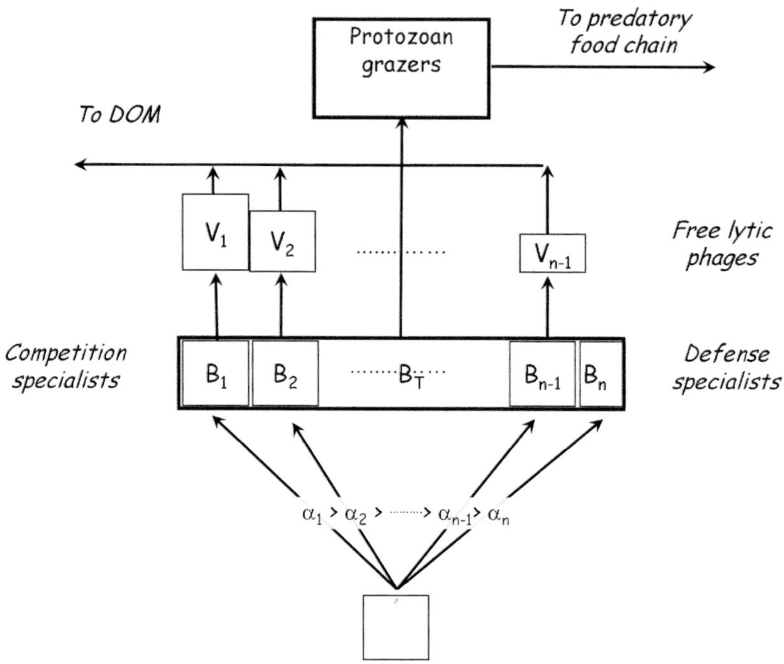

Figure 10.3 An idealized model combining loss of bacteria to lytic phages and to protozoan grazing. Bacterial hosts ($B_1 \ldots B_n$), together comprising the total bacterial community, B_{T}, are arranged from left (competition specialists) to right (defense specialists) according to their affinities ($\alpha_1 \ldots \alpha_n$) for the common limiting substrate, *S*. The community (B_{T}) is grazed upon by protozoan grazers that are assumed (in this idealized version) not to be selective among bacteria. Protozoan grazing shunts part of the bacterial production "up" the food chain towards larger predators, while lytic viruses ($V_1 \ldots V_{n-1}$) shunt material "down" towards the pool of dissolved organic material (DOM). Similar mechanisms are believed to be at work in the phytoplankton community, and most likely also in protozoan communities. See text for further discussion. (Modified from Thingstad, 2000.)

in Table 10.2 that there should be a moderate number of simultaneously dominating bacterial species in a natural pelagic sample. Such a theory thus separates the control of the number of available niches (species richness, N) from the question of *who* in a given system occupies these niches. The former is here a primarily top-down (i.e., grazer and viral lysis) controlled phenomenon. The latter presumably is more a question of bottom-up (i.e., food-source) control (Chapter 2). For this bottom-up control, an organism can only occupy a niche if, at the environmental conditions given, it can grow at a rate that compensates for the sum of losses by grazing and viral lysis. If the

Table 10.2 *Four (in principle) testable hypotheses derived from the "killing the winner" hypothesis as outlined in Fig. 10.3.*

	Hypothesis	Relevant observations (see text for discussion)
H1	Relatively low number of dominant host groups	Moderate number (5–15) bands on DGGE gels
H2	Comparable and co-varying richness in host and virus groups	Comparable number of bands on DGGE and PFGE gels
H3	Changes in host community composition are accompanied by changes in viral community composition	Both in natural environments and in mesocosm experiments, changes in DGGE and PFGE patterns have been found to co-occur.
H4	High virus : host ratios in systems with large differences in growth rate between fast-growing "competition type" and slow-growing "defense type" coexisting hosts	No experimental corroboration

host's nutrient uptake and viral defense are coupled, then the consequence of a moderate defense strategy (low nutrient affinity combined with low viral adsorption coefficient) would be low growth rate, but high biomass, and this possibly would have only minor effects on the production (growth rate × biomass) of the host. In the simple world of Fig. 10.3, the ultimate danger associated with a defense strategy against viruses would be if growth rate is reduced below that of protozoan grazing rate, in which case the result would be extinction of the host.

10.5.2.2 Number of simultaneously dominant virus groups

In the formalized world of this simple model, the ratio between the grazer-controlled size of the community and the virus-controlled size of host populations will not be an exact integer. As illustrated in Fig. 10.3, the consequence would be a steady state with one host-type not reaching a level sufficient to carry a population of free lytic viruses. The number of simultaneously dominant free virus types should therefore be $N - 1$. With the precision and the methodological complications involved in measuring this for a natural system, this boils down to the prediction that the number of dominant host-groups and virus-groups should be comparable and co-vary (hypothesis H2). It also implies that if, for some reason, the community composition should change, then one would expect a rapid subsequent shift in the composition of the viral community with new types replacing the former (hypothesis H3), but not necessarily any dramatic changes in the number of viral types.

10.5.2.3 Phage abundance

Even in the simple model of Fig. 10.3, it is not immediately obvious what regulates phage abundance. That is, there presently is no known "killing the winner" mechanism for aquatic viruses. To our knowledge, the only suggestion for a theory aspiring to contain a mechanism for phage abundance (Thingstad, 2000) builds on the following lines of argument. Assume non-selective grazing from protozoa and host-selective lysis to be the only two loss factors for bacteria. Assume that different hosts have different competitive abilities for the shared limiting resource, and therefore grow at different growth rates. Arrange the hosts according to growth rate. The slowest-growing host, host number N, would have to be the one which had no viral lysis, and would have to grow at a rate that balances the grazing loss. Host $N-1$, which grows slightly faster, will need lysis to compensate for this faster growth. Host number $N-2$, growing faster still, is subject to additional viral lysis, and so on. Consequently, the more *different* the hosts are in terms of growth rate, the more viral lysis is needed to get a balanced system. More lysis would require a higher infection rate and so in general would require a higher phage abundance.

Reformulated, the reciprocal interaction between phages, diversity, and biogeochemistry suggested by this theory appears quite intriguing. It is lytic phages that allow different bacterial hosts to coexist. Phages thus control bacterial diversity (in terms of number of simultaneously coexisting host-groups). However, it is host diversity (in terms of differences in growth rates) that determines phage abundance and thus also the partitioning of bacterial production between fluxes "up" the food chain to larger particles via grazing, and fluxes "down" to dissolved material via lysis. The theory thus leads to the expectation (hypothesis H4) of high ratios of phage-to-host abundance in systems where there are large differences in growth rate between fast-growing "competition-type" and slow-growing "defense-type" coexisting hosts. Interpreted in this framework, a relatively fixed virus : bacterium ratio of 10–20, as sometimes observed in the water column (e.g. Drake *et al.*, 1998), would suggest that host diversity in terms of differences in growth rate may be a relatively constant feature.

10.5.3 More biogeochemical impacts

A frequently noted biogeochemical consequence of viruses is their potential role in retaining nutrients and carbon in the top layer of the ocean (see, e.g., Suttle, 2005), implying that viruses potentially interfere with the biological transport of carbon from the atmosphere to the ocean's interior. In such discussions, it is often forgotten that biogeochemical processes occur on time scales different from those usually considered in experimental microbiology. To illustrate the importance of taking into account the correct time scale, consider the balance between import into the photic zone of limiting nutrients like nitrate-N and phosphate-P and their export, predominantly in the form of sinking biogenic particles containing N and P. To obtain such a balance, a food web must form in the photic zone that, over the appropriate time scale (for example, seasons), exports as much nutrient as is imported. With this as a starting point, we can do the hypothetical experiment of assuming that we have a photic-zone food web without viruses and then ask the question: what is the consequence of adding viruses?

Our answer is that viruses will shunt material away from the predatory pathway that leads to larger organisms such as copepods that would otherwise lead to production of particles that can sink out of the photic zone. On a short time scale (days – weeks) one would then expect less food for copepods, given viral activity against microorganisms, and thus less export. This will, however, shift the import–export balance towards positive net import. The nutrient content of the upper ocean will increase and therefore support more microorganisms (phytoplankton, bacteria, and protozoa). The import–export balance will be restored when a new food-web structure – one with a higher nutrient content in the photic zone, presumably containing more microorganisms, and exporting the same amount of nutrients as before – has been established. The expectation would thus be that viruses promote an enrichment of the lower part of the photic-zone food web, more regenerated production (*sensu* Dugdale and Goering, 1967), but neither more nor less export of the limiting nutrient (e.g., N or P). Phages, however, may in principle influence the associated carbon export if they somehow influence the stoichiometric composition of the material exported such that more or less C is associated with the mineral nutrient export.

Theoretically the efficiency of the biological C-pump in moving C from the atmosphere to the ocean's interior depends on the relative depths over which C and the limiting element are released to non-sinking forms (Thingstad, 1993). If the limiting nutrient is released to non-sinking forms over depths more shallow than for C, this will generally allow a more rapid return of the limiting nutrient than of C to the photic zone. Phytoplankton growth on the returned mineral nutrients will then give a photosynthesis that sequesters C from the atmosphere. With sufficient separation in the mean depths for release, mineral nutrients can thus cycle between the photic zone and shallow intermediate layers, pumping C to deep waters for each cycle.

Attached bacteria are believed to dissolve sinking particles via extracellular enzymatic splitting of the macromolecules in the particles (Azam *et al.*, 1992). Since bacteria on sinking particles also become phage-infected (Proctor and Fuhrman, 1991), lytic phages potentially influence the release of elements. The complexity of the process may be illustrated by the example of a sinking carbohydrate particle which would initially absorb N and P while being colonized by bacteria, release C by extracellular glucosidase activity of the attached bacteria, and release C, N, and P according to the stoichiometry of bacterial biomass by lysis. Sinking velocity is modified by changing particle size, and other processes, such as repackaging by copepods that feed on and produce sinking particles, further complicate any detailed description of this central biogeochemical process in the ocean.

10.5.4 Experimental and observational evidence

While the "killing the winner" hypothesis in its simple form (Fig. 10.2) has been confirmed by experimental work in laboratory model systems containing bacteria and lytic viruses (Bohannan and Lenski, 1997, 1999, 2000; Bohannan *et al.*, 2002; Chapter 2), this does not prove that its more elaborate form (Fig. 10.3) provides us with a helpful tool for understanding natural systems. Even if the underlying principles may represent a correct understanding, it may very well be that the transition to such complex biological systems adds so many additional factors in the form of different environmental conditions, differences in the surrounding communities of phytoplankton and protozoa, adaptation mechanisms working on different time scales, etc., that any strict comparison with nature becomes impossible.

10.5.4.1 Support for hypothesis H1

Support for hypothesis H1 (low or moderate number of simultaneously coexisting hosts; Table 10.2) comes from the recent use of molecular methods such as DGGE (Muyzer and Ramsing, 1995). DGGE gels typically show 10–25 bands in samples from aquatic environments (Riemann *et al.*, 1999; Fandino *et al.*, 2001). This is in accordance with hypothesis H1, although with a value about an order of magnitude below the estimate in Section 10.5.2.1. If number of bands on DGGE gels is a reasonably accurate estimate for the number of host species simultaneously dominating a natural sample, then we are actually at present in a situation with Hutchinson's paradox reversed: why is natural diversity *smaller* than we predict from simple steady-state models? One tempting answer is that bacteria in nature (or, more precisely, those bacteria that become dominant in nature) evolve resistance mechanisms. This would mean that a lower value should be used for the effective adsorption coefficient, α_V, than the value for adsorption of phage T4 to *Escherichia coli* (Stent, 1963). Experimental illustration of the probable importance of such defense mechanisms can be found in the work of Middelboe *et al.* (2001), where phage-resistant clones rapidly became dominating when continuous cultures containing four strains of bacteria were inoculated with viruses.

10.5.4.2 Support for hypothesis H2

Support for hypothesis H2 (comparable species richness in the bacterial and viral communities; Table 10.2) has primarily been obtained by comparing the number of bands on DGGE gels with those from PFGE gels (see Section 10.3.5). Although there are many possible sources of error – such as lack of precision due to weak bands, the possibility of one host producing several bands on the DGGE gels, and the possible biases in the PCR step of the DGGE procedure – the similarity in the order of magnitude of bands ($\sim$10) is at least indicative.

10.5.4.3 Support for hypothesis H3

Support for hypothesis H3 (covariation of shifts in composition of the bacterial and viral communities; Table 10.2) can be found in investigations combining the use of DGGE and PFGE on the same samples. Riemann and Middelboe (2002) found the band patterns to be stable over large areas, while both changed when crossing hydrographical fronts. In mesocosm experiments, manipulations were found either to have minor effects on both types of band patterns or to induce changes in both (Øvreås *et al.*, 2003).

10.5.4.4 Support (or lack thereof) for hypothesis H4

We are not aware of experiments or observations made to explicitly test hypothesis H4 (high virus : host ratios in systems with large interspecific differences in bacterial growth rate; Table 10.2). A direct test would seem to require species-specific growth-rate measurements in combination with determinations of viral abundance. From an investigation in the North Sea, Winter *et al.* (2005) concluded, in accordance with hypotheses H1–H3, that high prokaryotic productivity of the total community was sustained by a relatively small number of highly active bacterial host populations that also maintained high populations of phages. Some of their observations, such as low bacterial species richness being associated with high viral abundance, and also with high bacterial production, are not immediate inferences from the simple model. Not too surprisingly, a more elaborate theory, including perhaps aspects such as a dynamic variation in adsorption coefficients, may turn out to be required in order to satisfactorily explain the relationships between activity levels in natural bacterial communities and their lytic viruses.

10.6 CONCLUSION

Some 15 years have elapsed since it was recognized that viruses are partners in the web of trophic interactions constituting microbial aquatic ecosystems. Research in this period has confirmed the early expectations that this should have consequences over a large spectrum of interrelated issues: on biodiversity, gene transfer, and evolution; on food-web structure and population dynamics; and on the biogeochemistry of energy and material flows in the microbial food web. Together, these relationships constitute the microbial ecology of the pelagic ecosystem. It is thus evident that detailed understanding of the microbial ecology of this system requires the integration of viruses into our conceptual and numerical models. Yet theories in the field are still highly idealized and for some basic issues still largely untested by experiment or observation. A key element lacking in building these models from first principles is an incorporation of the trade-off faced by the hosts in optimizing nutrient-uptake competitive abilities and phage resistance. Although this has been studied by some research groups (Bohannan *et al.*, 2002; Chapter 2), experimental data allowing quantitative description of such trade-offs are largely missing.

ACKNOWLEDGMENTS

This work was financed by the Research Council of Norway through project 158936, Patterns in Biodiversity.

REFERENCES

Azam, F., T. Fenchel, J. G. Field, J. S. Gray, L. A. Meyer-Reil, and T. F. Thingstad. 1983. The ecological role of water-column microbes in the sea. *Mar. Ecol. Prog. Ser.* **10**: 257–63.

Azam, F., D. C. Smith, and A. F. Carlucci. 1992. Bacterial transformation and transport of organic-matter in the southern California bight. *Prog. Oceanogr.* **30**: 151–66.

Becks, L., F. M. Hilker, H. Malchow, K. Jürgens, and H. Arndt. 2005. Experimental demonstration of chaos in a microbial food web. *Nature* **435**: 1226–9.

Bergh, Ø., K. Y. Børsheim, G. Bratbak, and M. Heldal. 1989. High abundance of viruses found in aquatic environments. *Nature* **340**: 467–8.

Bettarel, Y., T. Sime-Ngando, M. Bouvy, R. Arfi, and C. Amblard. 2005. Low consumption of virus-sized particles by heterotrophic nanoflagellates in two lakes of the French Massif Central. *Aquat. Microb. Ecol.* **39**: 205–9.

Binder, B. 1999. Reconsidering the relationship between virally induced bacterial mortality and frequency of infected cells. *Aquat. Microb. Ecol.* **18**: 207–15.

Bohannan, B. J. M., and R. E. Lenski. 1997. Effect of resource enrichment on a chemostat community of bacteria and bacteriophage. *Ecology* **78**: 2303–15.

Bohannan, B. J. M., and R. E. Lenski. 1999. Effect of prey heterogeneity on the response of a model food chain to resource enrichment. *Am. Nat.* **153**: 73–82.

Bohannan, B. J. M., and R. E. Lenski. 2000. The relative importance of competition and predation varies with productivity in a model community. *Am. Nat.* **156**: 329–40.

Bohannan, B. J. M., B. Kerr, C. M. Jessup, J. B. Hughes, and G. Sandvik. 2002. Trade-offs and coexistence in microbial microcosms. *Antonie Van Leeuwenhoek* **81**: 107–15.

Bratbak, G., M. Heldal, S. Norland, and T. F. Thingstad. 1990. Viruses as partners in spring bloom microbial trophodynamics. *Appl. Environ. Microbiol.* **56**: 1400–5.

Bratbak, G., M. Heldal, T. F. Thingstad, B. Riemann, and O. H. Haslund. 1992. Incorporation of viruses into the budget of microbial C-transfer: a first approach. *Mar. Ecol. Prog. Ser.* **83**: 273–80.

Bratbak, G., J. K. Egge, and M. Heldal. 1993. Viral mortality of the marine alga *Emiliania huxleyi* (Haptophyceae) and termination of algal blooms. *Mar. Ecol.Prog. Ser.* **93**: 39–48.

Bratbak, G., F. Thingstad, and M. Heldal. 1994. Viruses and the microbial loop. *Microbial Ecol.* **28**: 209–21.

Bratbak, G., M. Heldal, T. F. Thingstad, and P. Tuomi. 1996. Dynamics of virus abundance in coastal seawater. *FEMS Microbiol. Ecol.* **19**: 263–9.

Bratbak, G., A. Jacobsen, and M. Heldal. 1998. Viral lysis of *Phaeocystis pouchetii* and bacterial secondary production. *Aquat. Microb. Ecol.* **16**: 11–16.

Breitbart, M., and F. Rohwer. 2005. Here a virus, there a virus, everywhere the same virus? *Trends Microbiol.* **13**: 278–84.

Breitbart, M., J. H. Miyake, and F. Rohwer. 2004. Global distribution of nearly identical phage-encoded DNA sequences. *FEMS Microbiol. Lett.* **236**: 249–56.

Brussaard, C. P. D. 2004. Optimization of procedures for counting viruses by flow cytometry. *Appl. Environ. Microbiol.* **70**: 1506–13.

Brussaard, C. P. D., D. Marie, and G. Bratbak. 2000. Flow cytometric detection of viruses. *J. Virol. Methods* **85**: 175–82.

Castberg, T., A. Larsen, R. A. Sandaa, *et al.* 2001. Microbial population dynamics and diversity during a bloom of the marine coccolithophorid *Emiliania huxleyi* (Haptophyta). *Mar. Ecol. Prog. Ser.* **221**: 39–46.

Chen, F., C. A. Suttle, and S. M. Short. 1996. Genetic diversity in marine algal virus communities as revealed by sequence analysis of DNA polymerase genes. *Appl. Environ. Microbiol.* **62**: 2869–74.

Chen, F., J. R. Lu, B. J. Binder, Y. C. Liu, and R. E. Hodson. 2001. Application of digital image analysis and flow cytometry to enumerate marine viruses stained with SYBR gold. *Appl. Environ. Microbiol.* **67**: 539–45.

Cochran, P. K., C. A. Kellogg, and J. H. Paul. 1998. Prophage induction of indigenous marine lysogenic bacteria by environmental pollutants. *Mar. Ecol. Prog. Ser.* **164**: 125–33.

Comeau, A. M., S. Short, and C. A. Suttle. 2004. The use of degenerate-primed random amplification of polymorphic DNA (DP-RAPD) for strain-typing and inferring the genetic similarity among closely related viruses. *J. Virol. Methods* **118**: 95–100.

Drake, L. A., K. H. Choi, A. G. E. Haskell, and F. C. Dobbs. 1998. Vertical profiles of virus-like particles and bacteria in the water column and sediments of Chesapeake Bay, USA. *Aquat. Microb. Ecol.* **16**: 17–25.

Dugdale, R. C., and J. J. Goering. 1967. Uptake of new and regenerated forms of nitrogen in primary productivity. *Limnol. Oceanogr.* **12**: 196–206.

Edwards, R. A., and F. Rohwer. 2005. Viral metagenomics. *Nat. Rev. Microbiol.* **3**: 504–10.

Evans, C., S. D. Archer, S. Jacquet, and W. H. Wilson. 2003. Direct estimates of the contribution of viral lysis and microzooplankton grazing to the decline of a *Micromonas* spp. population. *Aquat. Microb. Ecol.* **30**: 207–19.

Fandino, L. B., L. Riemann, G. F. Steward, R. A. Long, and F. Azam. 2001. Variations in bacterial community structure during a dinoflagellate bloom analyzed by DGGE and 16S rDNA sequencing. *Aquat. Microb. Ecol.* **23**: 119–30.

Fuhrman, J. A. 1999. Marine viruses and their biogeochemical and ecological effects. *Nature* **399**: 541–8.

Gobler, C. J., D. A. Hutchins, N. S. Fisher, E. M. Cosper, and S. A. Sanudo-Wilhelmy. 1997. Release and bioavailability of C, N, P, Se, and Fe following viral lysis of a marine chrysophyte. *Limnol. Oceanogr.* **42**: 1492–504.

Gonzalez, J. M. 1996. Efficient size-selective bacterivory by phagotrophic nanoflagellates in aquatic ecosystems. *Mar. Biol.* **126**: 785–9.

Gonzalez, J. M., and C. A. Suttle. 1993. Grazing by marine nanoflagellates on viruses and virus-sized particles: ingestion and digestion. *Mar. Ecol. Prog. Ser.* **94**: 1–10.

Gonzalez, J. M., E. B. Sherr, and B. F. Sherr. 1993. Differential feeding by marine flagellates on growing versus starving, and on motile versus nonmotile, bacterial prey. *Mar. Ecol. Prog. Ser.* **102**: 257–67.

Gundersen, K., G. Bratbak, and M. Heldal. 1996. Factors influencing the loss of bacteria in preserved seawater samples. *Mar. Ecol. Prog. Ser.* **137**: 305–10.

Hänfling, P., A. S. Shashkov, B. Jann and K. Jann. 1996. Analysis of the enzymatic cleavage (ß elimination) of the capsular K5 polysaccharide of *Escherichia coli* by the K5-specific coliphage: a reexamination. *J. Bacteriol.* **178**: 4747–50.

Hara, S., K. Terauchi, and I. Koike. 1991. Abundance of viruses in marine waters – assessment by epifluorescence and transmission electron-microscopy. *Appl. Environ. Microbiol.* **57**: 2731–4.

Heldal, M., and G. Bratbak. 1991. Production and decay of viruses in aquatic environments. *Mar. Ecol. Prog. Ser.* **72**: 205–12.

Hennes, K. P., and M. Simon. 1995. Significance of bacteriophages for controlling bacterioplankton growth in a mesotrophic lake. *Appl. Environ. Microbiol.* **61**: 333–40.

Hennes, K. P., and C. A. Suttle. 1995. Direct counts of viruses in natural-waters and laboratory cultures by epifluorescence microscopy. *Limnol. Oceanogr.* **40**: 1050–5.

Huisman, J., and F. J. Weissing. 1999. Biodiversity of plankton by species oscillations and chaos. *Nature* **402**: 407–10.

Hutchinson, G. E. 1961. The paradox of the plankton. *Am. Nat.* **95**: 137–45.

Irigoien, X., J. Huisman, and R. P. Harris. 2004. Global biodiversity patterns of marine phytoplankton and zooplankton. *Nature* **429**: 863–7.

Jiang, S. C., and J. H. Paul. 1994. Seasonal and diel abundance of viruses and occurrence of lysogeny/bacteriocinogeny in the marine-environment. *Mar. Ecol. Prog. Ser.* **104**: 163–72.

Karl, D. M., E. A. Laws, P. Morris, P. J. L. Williams, and S. Emerson. 2003. Global carbon cycle: metabolic balance of the open sea. *Nature* **426**: 32.

Landry, M. R., and R. P. Hassett. 1982. Estimating the grazing impact of marine micro-zooplankton. *Mar. Biol.* **67**: 283–8.

Larsen, A., T. Castberg, R. A. Sandaa, *et al.* 2001. Population dynamics and diversity of phytoplankton, bacteria and viruses in a seawater enclosure. *Mar. Ecol. Prog. Ser.* **221**: 47–57.

Madz, C., and K. Jürgens. 2003. Interaction of nutrient limitation and protozoan grazing determines the phenotypic structure of a bacterial community. *Microb. Ecol.* **101**: 384–98.

Marie, D., C. P. D. Brussaard, R. Thyrhaug, G. Bratbak, and D. Vaulot. 1999. Enumeration of marine viruses in culture and natural samples by flow cytometry. *Appl. Environ. Microbiol.* **65**: 45–52.

McCauley, E., and F. Briand. 1979. Zooplankton grazing and phytoplankton species richness. *Limnol. Oceanogr.* **24**: 243–52.

Middelboe, M., A. Hagström, N. Blackburn, *et al.* 2001. Effects of bacteriophages on the population dynamics of four strains of pelagic marine bacteria. *Microb. Ecol.* **42**: 395–406.

Muyzer, G., and N. Ramsing. 1995. Molecular methods to study the organization of microbial communities. *Water Sci. Technol.* **32**: 1–9.

Nikaido, H., and M. Vaara. 1987. Outer membrane. In J. Ingraham, K. Low, B. Magasanik, M. Schaechter, and H. Umbarger (eds.), Escherichia coli *and* Salmonella typhimurium*: Cellular and Molecular Biology*. Washington, DC: American Society for Microbiology, pp. 7–23.

Noble, R. T., and J. A. Fuhrman. 1998. Use of SYBR Green I for rapid epifluorescence counts of marine viruses and bacteria. *Aquat. Microb. Ecol.* **14**: 113–18.

Noble, R. T., and J. A. Fuhrman. 2000. Rapid virus production and removal as measured with fluorescently labeled viruses as tracers. *Appl. Environ. Microbiol.* **66**: 3790–7.

Noble, R. T., and G. F. Steward. 2001. Estimating viral proliferation in aquatic samples. In J. H. Paul (ed.), *Marine Microbiology*. Methods in Microbiology **30**. New York, NY: Academic Press, pp. 67–84.

Ortmann, A. C., B. Wiedenhef, T. Douglas, and M. Young. 2006. Hot crenarchaeal viruses reveal deep evolutionary connections. *Nat. Rev. Microbiol.* **4**: 520–8.

Øvreås, L., D. Bourne, R. A. Sandaa, *et al.* 2003. Response of bacterial and viral communities to nutrient manipulations in seawater mesocosms. *Aquat. Microb. Ecol.* **31**: 109–21.

Paul, J. H., and C. A. Kellogg. 2000. Ecology of bacteriophages in nature. In C. Hurst (ed.), *Viral Ecology*. Academic Press, pp. 211–46.

Paul, J. H., M. B. Sullivan, A. M. Segall, and F. Rohwer. 2002. Marine phage genomics. *Comp. Biochem. Phys. B* **133**: 463–76.

Pengerud, B., E. F. Skjoldal, and T. F. Thingstad. 1987. The reciprocal interaction between degradation of glucose and ecosystem structure: studies in mixed chemostat cultures of marine bacteria, algae, and bacterivorous nanoflagellates. *Mar. Ecol. Prog. Ser.* **35**: 111–17.

Proctor, L. M., and J. A. Fuhrman. 1990. Viral mortality of marine bacteria and cyanobacteria. *Nature* **343**: 60–2.

Proctor, L. M., and J. A. Fuhrman. 1991. Roles of viral infection in organic particle flux. *Mar. Ecol. Prog. Ser.* **69**: 133–42.

Proctor, L. M., A. Okubo, and J. A. Fuhrman. 1993. Calibrating estimates of phage-induced mortality in marine bacteria: ultrastructural studies of marine bacteriophage development from one-step growth experiments. *Microbial Ecol.* **25**: 161–82.

Riemann, L., and M. Middelboe. 2002. Stability of bacterial and viral community compositions in Danish coastal waters as depicted by DNA fingerprinting techniques. *Aquat. Microb. Ecol.* **27**: 219–32.

Riemann, L., G. F. Steward, L. B. Fandino, L. Campbell, M. R. Landry, and F. Azam. 1999. Bacterial community composition during two consecutive NE monsoon periods in the Arabian Sea studied by denaturing gradient gel electrophoresis (DGGE) of rRNA genes. *Deep-Sea Res. II Top. Stud. Oceanogr.* **46**: 1791–811.

Rodriguez, F., E. Fernandez, R. N. Head, *et al.* 2000. Temporal variability of viruses, bacteria, phytoplankton and zooplankton in the western English Channel off Plymouth. *J. Mar. Biol. Ass. UK* **80**: 575–86.

Schwartz, M. 1987. The maltose regulon. In J. Ingraham, K. Low, B. Magasanik, M. Schaechter, and H. Umbarger (eds.), Escherichia coli *and* Salmonella typhimurium: *Cellular and Molecular Biology*. Washington, DC: American Society for Microbiology, pp. 1482–502.

Short, S. M., and C. A. Suttle. 2002. Sequence analysis of marine virus communities reveals that groups of related algal viruses are widely distributed in nature. *Appl. Environ. Microbiol.* **68**: 1290–6.

Short, C. M., and C. A. Suttle. 2005. Nearly identical bacteriophage structural gene sequences are widely distributed in both marine and freshwater environments. *Appl. Environ. Microbiol.* **71**: 480–6.

Simek, K., J. Vrba, and P. Hartman. 1994. Size-selective feeding by *Cyclidium* sp on bacterioplankton and various sizes of cultured bacteria. *FEMS Microbiol. Ecol.* **14**: 157–67.

Spencer, R. 1955. Marine bacteriophage. *Nature* **175**: 690–1.

Spencer, R. 1960. Indigenous marine bacteriophages. *J. Bacteriol.* **79**: 614.

Steiner, C. F. 2003. Keystone predator effects and grazer control of planktonic primary production. *Oikos* **101**: 569–77.

Stent, G. S. 1963. *Molecular Biology of Bacterial Viruses.* San Francisco, CA: W. H. Freeman.

Steward, G. F. 2001. Fingerprinting viral assemblages by pulsed field gel electrophoresis (PFGE). In J. H. Paul (ed.), *Marine Microbiology*. Methods in Microbiology **30**. New York, NY: Academic Press, pp. 85–103.

Steward, G. F., J. Wikner, S. D. C., W. P. Cochlan, and F. Azam. 1992a. Estimation of virus production in the sea: I. Method development. *Mar. Microbial Food Webs* **6**: 57–78.

Steward, G. F., J. Wikner, W. P. Cochlan, D. C. Smith, and F. Azam. 1992b. Estimation of virus production in the sea: II. Field results. *Marine Microbial Food Webs* **6**: 79–90.

Steward, G. F., J. L. Montiel, and F. Azam. 2000. Genome size distributions indicate variability and similarities among marine viral assemblages from diverse environments. *Limnol. Oceanogr.* **45**: 1697–706.

Stoderegger, K. E., and G. J. Herndl. 2001. Visualization of the exopolysaccharide bacterial capsule and its distribution in oceanic environments. *Aquat. Microb. Ecol.* **26**: 195–9.

Stubbs, G. 2001. Fibre diffraction studies of filamentous viruses. *Rep. Prog. Phys.* **64**: 1389–425.

Sullivan, M. B., J. B. Waterbury and S. W. Chisholm. 2003. Cyanophages infecting the oceanic cyanobacterium *Prochlorococcus. Nature* **424**: 1047–51.

Suttle, C. A. 1994. The significance of viruses to mortality in aquatic microbial communities. *Microb. Ecol.* **28**: 237–43.

Suttle, C. A. 2005. Viruses in the sea. *Nature* **437**: 356–61.

Suttle, C. A., and A. M. Chan. 1994. Dynamics and distribution of cyanophages and their effect on marine *Synechococcus* spp. *Appl. Environ. Microbiol.* **60**: 3167–74.

Suttle, C. A., and F. Chen. 1992. Mechanisms and rates of decay of marine viruses in seawater. *Appl. Environ. Microbiol.* **58**: 3721–9.

Suttle, C. A., A. M. Chan, and M. T. Cottrell. 1990. Infection of phytoplankton by viruses and reduction of primary productivity. *Nature* **347**: 467–9.

Thingstad, T. F. 1993. Microbial processes and the biological carbon pump. In G. T. Evans and M. J. R. Fasham (eds.), *Towards a Model of Ocean Biogeochemical Processes.* NATO ASI Series Vol I, 10. Heidelberg: Springer, pp.193–208.

Thingstad, T. F. 2000. Elements of a theory for the mechanisms controlling abundance, diversity, and biogeochemical role of lytic bacterial viruses in aquatic systems. *Limnol. Oceanogr.* **45**: 1320–8.

Thingstad, T. F., and R. Lignell. 1997. Theoretical models for the control of bacterial growth rate, abundance, diversity and carbon demand. *Aquat. Microb. Ecol.* **13**: 19–27.

Thingstad, T. F., L. Øvreås, J. K. Egge, T. Løvdal, and M. Heldal. 2005. Use of non-limiting substrates to increase size; a generic strategy to simultaneously

optimize uptake and minimize predation in pelagic osmotrophs? *Ecol. Lett.* **8**: 675–82.
Torrella, F., and R. Y. Morita. 1979. Evidence by electron micrographs for a high-incidence of bacteriophage particles in the waters of Yaquina Bay, Oregon: ecological and taxonomical implications. *Appl. Environ. Microbiol.* **37**: 774–8.
Tuomi, P., K. M. Fagerbakke, G. Bratbak, and M. Heldal. 1995. Nutritional enrichment of a microbial community – the effects on activity, elemental composition, community structure and virus production. *FEMS Microbiol. Ecol.* **16**: 123–34.
Turley, C. M., and D. J. Hughes. 1992. Effects of storage on direct estimates of bacterial numbers of preserved seawater samples. *Deep-Sea Res. A Oceanogr. Res. Pap.* **39**: 375–94.
Valentine, A. F., and G. V. Chapman. 1966. Fine structure and host–virus relationship of a marine bacterium and its bacteriophage. *J. Bacteriol.* **92**: 1535–54.
Valentine, A. F., P. K. Chen, R. R. Colwell, and G. B. Chapman. 1966. Structure of a marine bacteriophage as revealed by negative-staining technique. *J. Bacteriol.* **91**: 819–22.
Waterbury, J. B., and F. W. Valois. 1993. Resistance to co-occurring phages enables marine *Synechococcus* communities to coexist with cyanophages abundant in seawater. *Appl. Environ. Microbiol.* **59**: 3393–9.
Weinbauer, M. G. 2004. Ecology of prokaryotic viruses. *FEMS Microbiol. Rev.* **28**: 127–81.
Weinbauer, M. G., and F. Rassoulzadegan. 2004. Are viruses driving microbial diversification and diversity? *Environ. Microbiol.* **6**: 1–11.
Weinbauer, M. G., D. Fuks, and P. Peduzzi. 1993. Distribution of viruses and dissolved DNA along a coastal trophic gradient in the northern Adriatic Sea. *Appl. Environ. Microbiol.* **59**: 4074–82.
Weinbauer, M. G., D. Fuks, S. Puskaric, and P. Peduzzi. 1995. Diel, seasonal, and depth-related variability of viruses and dissolved DNA in the northern Adriatic Sea. *Microb. Ecol.* **30**: 25–41.
Weinbauer, M. G., C. Winter, and M. G. Höfle. 2002. Reconsidering transmission electron microscopy based estimates of viral infection of bacterioplankton using conversion factors derived from natural communities. *Aquat. Microb. Ecol.* **27**: 103–10.
Wen, K., A. C. Ortmann, and C. A. Suttle. 2004. Accurate estimation of viral abundance by epifluorescence microscopy. *Appl. Environ. Microbiol.* **70**: 3862–7.
Wiebe, W. J., and J. Liston. 1968. Isolation and characterization of a marine bacteriophage. *Mar. Biol.* **1**: 244–9.
Wilcox, R. M., and J. A. Fuhrman. 1994. Bacterial-viruses in coastal seawater: lytic rather than lysogenic production. *Mar. Ecol. Prog. Ser.* **114**: 35–45.

Wilhelm, S. W., and C. A. Suttle. 1999. Viruses and nutrient cycles in the sea – Viruses play critical roles in the structure and function of aquatic food webs. *Bioscience* **49**: 781–8.

Wilhelm, S. W., S. M. Brigden, and C. A. Suttle. 2002. A dilution technique for the direct measurement of viral production: A comparison in stratified and tidally mixed coastal waters. *Microbial Ecol.* **43**: 168–73.

Wilson, W. H., N. G. Carr, and N. H. Mann. 1996. The effect of phosphate status on the kinetics of cyanophage infection in the oceanic cyanobacterium *Synechococcus* sp WH7803. *J. Phycol.* **32**: 506–16.

Winter, C., A. Smit, G. J. Herndl, and M. G. Weinbauer. 2005. Linking bacterial richness with viral abundance and prokaryotic activity. *Limnol. Oceanogr.* **50**: 968–77.

Wommack, K. E., and R. R. Colwell. 2000. Virioplankton: viruses in aquatic ecosystems. *Microbiol. Mol. Biol. Rev.* **64**: 69–114.

Wommack, K. E., R. T. Hill, T. A. Muller, and R. R. Colwell. 1996. Effects of sunlight on bacteriophage viability and structure. *Appl. Environ. Microbiol.* **62**: 1336–41.

Wommack, K. E., J. Ravel, R. T. Hill, J. S. Chun, and R. R. Colwell. 1999. Population dynamics of Chesapeake Bay virioplankton: Total-community analysis by pulsed-field gel electrophoresis. *Appl. Environ. Microbiol.* **65**: 231–40.

Zhong, Y., F. Chen, S. W. Wilhelm, L. Poorvin, and R. E. Hodson. 2002. Phylogenetic diversity of marine cyanophage isolates and natural virus communities as revealed by sequences of viral capsid assembly protein gene g20. *Appl. Environ. Microbiol.* **68**: 1576–84.

CHAPTER 11

Phage ecology of terrestrial environments

Martin J. Day* and Robert V. Miller

11.1 INTRODUCTION

Terrestrial environments differ from aquatic environments in terms of the extent and persistence of inundation, with the terrestrial world, of course, drier than the aquatic one. Despite this basic difference between aquatic and terrestrial environments, the survival and reproductive problems faced by bacteriophages in the terrestrial world, such as within soils, can be similar to those faced in aquatic habitats or, for that matter, can be similar to those faced in or on plants or animals. Indeed, to the extent that bacterial hosts grow in biofilms, it matters little – periodic desiccation aside – whether we think of these bacteria as inhabiting rivers, puddles, sediment, soil, or, alternatively, other organisms. Therefore, although this chapter concentrates on soil as the environment, we will review studies performed in various non-terrestrial environments, to the extent that they illuminate our understanding of phage ecology in the terrestrial world. We will briefly consider phage and bacterial host life histories as they are integrally associated, drawing themes together using relevant information derived from the study of associations and interactions between phages and prokaryotes in other environments to see what is consistent, possible, or improbable. In addition, we review common themes associated with phage-mediated horizontal (also described as lateral) gene transfer between bacteria (transduction) as it occurs in the wild.

We note that the term "terrestrial" may be defined broadly to include anything except seas and oceans, or even the Earth as opposed to other "extraterrestrial" planets. Here, however, we employ a much narrower definition of

* Corresponding author

Bacteriophage Ecology: Population Growth, Evolution, and Impact of Bacterial Viruses, ed. Stephen T. Abedon. Published by Cambridge University Press.

the term, meaning relatively dry land as opposed to aquatic environments (the latter are covered in Chapter 10). For consideration of phage interactions with land plants, see Gill and Abedon (2003), and for phage interactions with animals (mostly terrestrial), see Chapters 12, 13, 14, and 17. Also, for general concepts of the ecology of phage population growth within spatially structured environments, which includes terrestrial environments such as soils, see Chapters 4 and 16. Here we concentrate on phage–bacterial interactions at or below soil surfaces. Due to space limitations, not all phage–host combinations are considered; see Williams *et al.* (1987) for an older review on soil-phage ecology, or Burroughs *et al.* (2000), Kieser *et al.* (2000), and Burke *et al.* (2001) for more recent overviews of the literature concerning the phages and phage ecology of the common soil bacteria found in the genus *Streptomyces.*

11.1.1 The terrestrial environment

The terrestrial environment comprises interactions between inorganic and organic particles and organisms. It may be differentiated into soils and the environment found above soils. Ignoring organisms set aloft in the atmosphere, the majority of habitats above soils are found on non-root plant surfaces, otherwise known as the phyllosphere. These phyllosphere communities are generally smaller and exposed to a very different physical and chemical environment. Soils we can differentiate into those regions found at the surface of living plants (typically their roots) in a region called the rhizoplane, the surrounding soil (the rhizosphere), and soil which is beyond the rhizosphere, that is, not under the direct influence of living plants. It is the phage ecology of the rhizosphere that is the focus of this chapter.

11.1.1.1 Soil

Microorganisms are an important component of soils. They in particular impact the cycling of nutrients and elements (Janzen *et al.*, 1995), and phages have the potential to produce a major ecological impact if they grow on bacteria involved directly with this nutrient cycling. The actions of phages and bacteria give rise to dynamic interactions of a local nature (e.g., rhizosphere) but also on a global scale (element cycling). Thus, soil is partially an inert physical support, but it is also chemically and structurally modified by organisms living within and utilizing it. As animals travel through and over soil, for example, they actively transport and deposit matter, such as excreta, containing microbes and phages.

Plants, both living and deceased, are a major component of most soils. Plant roots invade the substratum, and around their roots exists a region of intense and localized microbiological activity, due to root exudates. This region is termed a rhizosphere (Whipps, 2001). Plant roots exude water and a range of cellular compounds (including amino acids, carbohydrates, sugars, vitamins, and proteins). These exudates promote bacterial growth and consequently influence interactions of phage with susceptible bacteria. Mineralizing plant material also contributes, often substantially, to the nutrient content and physical structure of soils. We can thus view plants (in their many guises), bacteria (of many types), and phages (along with animals) as forming, at a minimum, a food chain within soils.

Soils vary temporally, in depth, in their gross structure, and in chemical characteristics. This means that some are nutritionally rich and others deficient, some are fine-grained and some are acidic; others lack water. Soil structure is also changed by the growth of organisms (Karlen *et al.*, 1997). The result is that soil character imposes constraints on microbial populations and is modified by their interactions. Phages are produced by actively growing microbial populations (Chapter 5) and thus their presence, numbers, and diversity will reflect an environment providing suitable growth conditions for the reproductive success of their hosts.

11.1.1.2 Biome-to-biome diversity

There are clearly major regional groups of distinctive plant and animal communities, interacting and well adapted to a region's physical environment. Thus deserts, grasslands, rainforests, and oceans are discernible at a global scale and are termed biomes (Sano *et al.*, 2004). These are larger than a habitat, which can be a very small site occupied by a microbial community and associated plants and animals. Biomes are characterized by macro-environmental factors such as seasonal temperature, moisture levels, and light. These factors combine to influence the local habitat environment.

The challenges imposed on phages for survival and reproduction in one biome, such as in a rainforest rhizosphere, are different from those of another biome, such as for cacti in a desert. These differences can be illustrated by just considering water availability. In an environment where water is freely available, the survival characteristics required by free phages may well be different from those needed in an environment where hosts and water are more transiently present. But at whatever level we consider phages and their ecology, phages are dependent upon host cells for growth and upon their own structural characteristics for survival through periods while seeking another host. These stresses will clearly be different in different environments, and we can predict that the physical characteristics of individual phages have evolved to permit survival in specific environments. Thus, it is probable that there are niche-, habitat-, and biome-adapted survival characteristics, even though there is evidence for natural dispersal of phages over large distances, as the same phage can be isolated from different biomes (Breitbart and Rohwer, 2005).

11.2 COMPLEXITIES OF REAL-WORLD ECOSYSTEMS

Consideration of phage ecology is incomplete unless both host and environmental demands are considered. Given that we have a reasonable understanding of phages and their bacterial hosts from laboratory studies, what then is preventing us from extending this knowledge into a comprehensive understanding of phage ecology? In the following we discuss three areas where phage or ecosystem properties likely diverge greatly from those observed in the laboratory: (1) phage survival, inactivation, and the perhaps related issue of phage temperance (Section 11.2.1; see also Chapter 5); (2) spatial and temporal environmental heterogeneity (vs. the standard understanding of phage growth as it occurs within well-mixed broth (Section 11.2.2; see also Chapter 3); and (3) phage host-range breadth (vs. habitually expected host-range narrowness; Section 11.2.3).

11.2.1 Phage survival

Phages, whether as free virions or during their intracellular stage in a bacterium, are susceptible to inactivation. This inactivation can be a consequence of environmental insult (Section 11.2.1.1), a consequence of active anti-phage action on the part of adsorbed bacteria (Chapter 1), or a consequence of predation, such as of bacteria by protozoa (Chapter 10). Depending on circumstances, phage survival may be enhanced via ongoing infection (e.g., lysogeny; Section 11.2.1.2; Chapter 5) or, alternatively, by either acquiring or retaining the free-phage state. These various forms of phage population reduction complicate understanding of phage ecology, if only because laboratory conditions are typically at least somewhat optimized so as to maximize phage and infection survival.

11.2.1.1 Virion durability

To obtain an understanding of phage ecology in general, and phage terrestrial ecology specifically, we have to consider what adaptations and strategies phage have evolved to raise their likelihood of surviving passage between one host and the next. Since phages make this passage as virions (by definition, since they are viruses), we explore issues of relevance to virion survival (durability). One can explore how this survival may be curtailed, resulting in virion decay, in terms of how the virion is affected, namely by two differentiable mechanisms. First, the genome may be damaged, affecting survival post virion adsorption. Second, as a consequence of damage to the virion capsid, the ability of the phage to attach to the host bacterium, or deliver the phage genome once attached, may also be impaired. Both mechanisms will impact on apparent virion stability *in situ* by affecting the ability of the phage to replicate. We additionally can explore phage survival in terms the characteristics of the damaging agent, with phages inhabiting terrestrial environments expected to be exposed to four common insults: UV radiation, desiccation, temperature extremes (including freezing), and exposure to degradative chemistries including (but not limited to) pH extremes and hydrolytic enzymes released from various cellular organisms.

Study of phage-virion survival typically involves an exploration of the durability of one or only a limited number of phages under specific laboratory or environmental circumstances. Phage PRD1, for example, is used as a model for studying how microorganisms are transported through soil and groundwater environments (Harvey and Joseph, 2004). The phage is probably atypical as it has a high degree of structural stability over a range of temperatures and ionic conditions, has a low degree of attachment in aquifer sediments/soil, and consequently may be widely dispersed, intact. The presence of an internal membrane in PRD1 (Rydman and Bamford, 2002) may contribute to its excellent stability in the environment. Phage MS2 normally infects in relatively low-salt environments and is more sensitive to structural inactivation in high-salt environments (Trouwborst and De Jong, 1973). Phage predating halophiles, by contrast, are relatively durable in high-salt environments but are structurally sensitive and inactivated in a low-salt environment (Seaman and Day, unpublished observations). Acidic soils also can rapidly inactivate some phages (Pantastica-Caldas *et al.*, 1992). In general, in any environment – whether it is a high-saline or a high-radiation one, the rhizosphere or the gut – there will be challenges for phages to maintain virion integrity. See Chapter 12 for additional consideration of phage durability in environments.

The key distinction between terrestrial and aquatic ecosystems is that terrestrial ecosystems, by definition, are drier than aquatic ones. The resulting desiccation, in addition to potentially impacting phage capsid stability, can produce DNA lesions, leading to double-stranded breaks. These breaks typically are lethal and similar to the damage caused by high levels of radiation (Billi *et al.*, 2000). Thus, finding that some phage can carry DNA repair systems is not surprising (Goodman, 2002). But clarifying what mechanisms of virion inactivation are most prevalent in soils, over a range of phages and conditions, awaits further study.

11.2.1.2 Lysogeny

Free-virion durability is not the only means by which phages can effect survival within hostile environments. An alternative approach is to maintain the infected state for long periods, even indefinitely. Given semi-permanent periods of infection, such as during lysogeny, it would seem logical for the phage and the host to have coevolved systems to enable their co-reproductive success (Chapters 5, 6, and 14). Alternatively, the evolution of regulatory processes to enable the phage to decide the optimum time to exit or remain as a lysogen, based on the host's physiological condition, is not surprising (Chapter 5, but see also Chapters 2 and 3), given that ultimately phages have presumably evolved to maximize their own survival and reproductive success rather than bacterial survival and reproductive success (Chapter 6). Consistent with a role as effectors of phage survival, hybridization experiments have shown that about 5% of soil bacteria were lysogens, while in fresh water some 70% of *Pseudomonas aeruginosa* isolates were lysogenic (Miller *et al.*, 1992; see also Chapter 5).

The results from a study on dynamics of *Bacillus subtilis* and bacteriophages in soil microcosms are consistent with this idea that lysogeny can effect phage survival (Pantastica-Caldas *et al.*, 1992). Pantastica-Caldas asked two questions: What are the population dynamics of a phage–bacterium interactions in soil? Are the dynamics controlled more by the population biology of the phages or of the bacteria? They showed that an initial bloom of phage occurred and that the phage population then remained stable for periods lasting weeks to months. At equilibrium, both temperate and obligately lytic phages were much less abundant than the bacteria. However, the temperate phage did not depress the equilibrium host density while the obligately lytic phage lowered it by a factor of ten. A key to understanding the dynamics of this system was the acidic soil, which caused rapid and permanent inactivation of free phages. The conclusion was that this inactivation essentially drives selection for temperate phages, ones capable of forming a lysogenic association, as obligately lytic phages cannot hide out inside of bacteria indefinitely (Chapter 1).

The results of Pantastica-Caldas *et al.* (1992) lead to an expectation that a temperate life cycle would be more common in soil environments. Besides virion decay, what other aspects of this system might have lead to this result? One clue is that only low levels of bacterial resistance to phage were typically seen. Also, at equilibrium, host growth and phage reproduction in the soil microcosm were slow, relative to broth culture. In addition, the sporulating stage in the life history of *B. subtilis* adds a complexity (Pantastica-Caldas *et al.*, 1992). These observations suggest that at least some of the interactions between the *B. subtilis* host and phage in soil were different from those seen in chemostat studies of *Escherichia coli* and obligately lytic phages (Chapters 2 and 15) where, given sufficient resource densities, phage populations can be dominating and phage-resistant bacteria often evolve (Chapter 2). So, since we know there are many obligately lytic phages in soil, we need to determine why, and also whether this relationship holds true for a diversity of environments.

11.2.2 Spatial heterogeneity

A phage will only reproduce and generate more phages by infecting an appropriate host cell. In a typical ecosystem, however, host cells do not grow, as they do in pure culture, in a consistent or otherwise standardized manner. The host instead is usually just one of many species of bacteria present. For example, the terrestrial density of phage host bacteria is typically lower by several orders of magnitude than the total bacterial numbers (Ashelford *et al.*, 2000), and total bacterial densities can vary considerably with nutrient levels (Janzen *et al.*, 1995), for example between bulk soil and rhizosphere regions. The absence of high densities of specific host bacteria suggests that phage densities will also not be high, and so we might predict that the overall phage impact on bacterial populations will not be great (see Chapter 10 for additional consideration of this idea that bacterial densities drive degrees of phage impact within ecosystems).

These statements refer to bulk properties of terrestrial ecosystems, and to understand their phage ecology it is also important to take into account the spatial heterogeneity (non-random dispersion) of more solid-phase environments such as soils, as well as their temporal heterogeneity. That is, the environment that free phages occupy is clearly physically discontinuous and subject to constant change, in terms of abiotic parameters (e.g., temperature and hydration) as well as biotic parameters, most importantly host cell density. In particular, environments seem to have a patchy structure (as modeled in Chapter 2) where most bacteria probably live in physically isolated, slimy communities attached to surfaces as biofilms. See Chapters 2, 4, and 16 for basic considerations of the impact of this spatial structure on phage ecology.

11.2.3 Host-range breadth

Natural microbial communities are interacting assemblages of many bacterial species growing, for example, in a biofilm. It has been suggested, however, that phage replication would be most effective in ecosystems characterized by low host diversity (Pedrós-Alió, 1993) and that numerically dominant populations in this community would be constantly targeted by phages and consequently reduced by lysis (Thingstad *et al.*, 1993; Chapter 10). This idea is reinforced by the often-repeated fact that most phages have a limited host range, meaning they only infect closely related bacteria – as is consistent with their widespread use in the identification of clinical bacterial species and particular strains (Welkos *et al.*, 1974; see also Chapter 12). What is not appreciated, however, is that many phages fail to yield a dominant lytic interaction or immediate lysis. It is also true that many bacteria are naturally lysogens, and these are protected from lysis after infection by a related phage. Thus, the interpretation of host range based on lysis alone leads to an underestimation of host range in nature.

The work of Jensen and colleagues (1998) somewhat dispels the conclusion that phages found in nature are inherently highly limited in their host range. That is, some bacteriophages do have broad host ranges, productively infecting a range of bacterial species, though such phages are generally considered to be unusual. Prior to this work the major examples cited were the generalized transducing phage P1, which grows on several enteric species in addition to *E. coli* (Welkos *et al.*, 1974), and the phage Mu, as determined by the orientation of the invertible viral G segment region (van de Putte *et al.*, 1980), which produces progeny virions capable of adsorption to and producing plaque on different bacterial species (Harshey, 1988).

A bacteriophage with a broad host range (i.e., one with a common or cross-species receptor specificity) would be expected to be advantaged relative to phages with more narrow host ranges, since a broad host range allows a phage to utilize one prey species when another is not available, or even utilize several simultaneously. The ability to infect a variety of bacterial host species would be an ecologically and evolutionary sound strategy especially in diverse natural bacterial communities including (but not limited to) biofilms. In particular, on a per-virion basis, a broad-host-range phage would be more likely to encounter a susceptible host, *in situ*, than a narrow-host-range phage.

11.3 PHAGE EXISTENCE IN SOIL

The free phage reservoir is derived from the multiplication in and subsequent phage release from bacterial hosts. Thus, the occurrence of free phages will reflect virion viability, phage growth characteristics within a given host strain, the historical and current presence of hosts, and environmental factors affecting host physiology. Notwithstanding any limitations these requirements might imply, phages have been found literally everywhere: in marine, ground and fresh water, in sewage, on and in healthy and diseased plants and animals, in food and food products (Ackerman, 1997). Phages are ubiquitous. At what densities do they occur? What factors are important to their survival in the environment? We address these questions especially in terms of phage association with soil.

11.3.1 High soil phage densities

Given ecosystem heterogeneity, where might we find numerically high bacterial populations and thus expect to see high phage densities and maybe an impact caused by phage growth? High concentrations of phages, such as $\geq 10^9$ phages per gram, are present in marine sediment where bacterial host numbers are high (Danovaro *et al.*, 2002). So we can use this example to predict that in nutrient rich regions, even given spatial constraints, there will be an increase in the abundance of host cells and that this will promote the development of phage populations. Though not identical to sediments, soils nevertheless too can display high bacterial densities, also spatially constrained. However, until recently it was an open question whether soils could similarly sustain high phage densities.

The first analyses of phage prevalence in soils was done using vital count methods (e.g., plaque formation), which notoriously undercount total phage densities. Direct counts of bacterial viruses in soil, using transmission electron microscopy (TEM) – a far better means of determining total virion prevalence – were first performed by Ashelford *et al.* (2003). There was consistency with aquatic sediments in that they found high phage densities in both bulk soil and soil associated with plant roots (rhizosphere; both in sugar beet fields). Total counts of 10^7 viruses per gram were seen using TEM, showing that substantial populations of phage exist in soil (mean = 1.5×10^7 g^{-1}), at least 350-fold more than the highest numbers estimated from traditional viable plaque counts. Surprisingly, there was little difference in phage numbers between rhizosphere and bulk soil (2 m away with no plants), which indicates that phages are widely and apparently well dispersed.

The addition of a lysate of a *Serratia* phage to soil suggested that the direct counting methods with electron microscopy underestimated the added phage populations by at least eight-fold. So, assuming natural phages were similarly underestimated, virus numbers in soil averaged 1.5×10^8 g^{-1}, which is equivalent to 4% of the total population of bacteria. This is 10% of the estimate made for phages in sediment (Danovaro *et al.*, 2002) and is a number at least qualitatively consistent with the results of Pantastica-Caldas *et al.* (1992), discussed in Section 11.2.1.2. So the obvious conclusions are that phages are extremely common entities and their presence reflects their formation by host populations, together with survival and dispersal.

11.3.2 Impact of terrestrial environments on phage

Since most free phages are effectively inert particles until they encounter a host, they passively respond to the vagaries of the environment. Phage distribution therefore follows those paths common to particulate non-motile organisms. They can be carried by animals, blown by wind on dust and seeds, and carried on water currents and rain (Goto, 1992). In many terrestrial environments the soil may only be partially hydrated. Rainfall solubilizes nutrients and promotes bacterial growth and motility, and simultaneously activates phage interactions with hosts. In drier sites, the lack of a continuous aqueous phase complicates predictions, especially of localized phage movement otherwise associated with rainfall and hydration.

Another factor which complicates predictions is virion sorption to solid substrates, which reduces phage mobility but can also physically protect phages from damage (Kleczkowska, 1957; Williams *et al.*, 1987). Free phages, for example, can be trapped in biofilms or bound to clay surfaces by non-specific and often reversible adsorption (Chattopadhyay and Puls, 2000). Overall, association with clay particles appears to stimulate phage–host interactions (Ripp and Miller, 1995), at least in one aquatic environment, but virion interactions with soil may well be another story. Virions can then be eluted from clay sites under appropriate ionic conditions (Primrose and Day, 1977).

Seasonality may also impact the suitability of environments to phage growth. It has been recognized for years that phages show dynamic and seasonal fluctuations, responding to the density of their hosts in marine environments (Boyd and Brüssow, 2002). More recently these interactions have been confirmed to occur in soil by Ashelford *et al.* (1999). Ashelford found two types of *Serratia* phages that vary in prevalence seasonally within the rhizosphere of sugar beets. One, a *Siphoviridae,* was temperate with a long latent period and large burst size while the other, a *Podoviridae,* was obligately lytic and had a short latent period and small burst size (see Chapter 1 for discussion of phage morphotypes as well as infection strategies). The siphovirus population predominated in spring and early summer but was replaced by a phage population dominated by the podovirus in the later part of the growing season. It is tempting to interpret these differences in phage types as reflecting their host densities early and late in the growth season (see Chapters 2 and 3 for discussion of the impact of host densities on phage ecology). This suggests that the two phages are adapted to different niches that occur only temporally in the rhizosphere, as the plant's nutritional contribution modifies the types and physiological status of host bacteria over the growth season.

11.4 IMPACT OF PHAGES ON TERRESTRIAL BACTERIA

Bacteriophages are ubiquitous, and so it is logical to presume that they impact the success and failure of their host populations, both as predators/parasites of bacteria (this section; see also Chapters 2 and 10) and as effectors of horizontal gene transfer (Section 11.5). These interactions can directly influence host population densities, can affect interactions between microbial populations, can influence community diversity, and can culminate with changes in ecosystem processes (Marsh and Wellington, 1994; Ashelford *et al.*, 1999; Miller and Ripp, 2001).

11.4.1 Quantitative impact of phages on terrestrial bacteria

Can we model the ecological impact of bacteriophages? The answer depends on the type of ecology being discussed. If we consider health-related issues then we can see an impact through disease in plants, animals, and humans (see Section 11.5 on transduction). However, if we consider phage effects on non-health-related phenomena, then information is severely limited, particularly within terrestrial environments, though we do at least know that high phage concentrations can be present, whether as free phages (Section 11.3.1) or as lysogens (Section 11.2.1.2). Outside of these assessments of phage prevalence, much of what we understand of terrestrial phage ecology, especially in soils, comes from a limited amount of *in situ* or microcosm experimentation. In microcosm, but not *in situ* experiments, phages and bacteria are typically added to sterile soils.

What has this *in situ* and microcosm research taught us? Small-scale microcosm experiments have shown that rhizobiophages can reduce nodulation and nitrogen fixation and consequently can reduce plant growth (Ahmad and Morgan, 1994). Presumably the phages were antagonistic to a member of the bacterial consortium that fixed nitrogen. More significantly, Keel *et al.* (2002) examined the effect of the obligately lytic phage GP100 on the biocontrol efficacy of its host, *P. fluorescens*, which protects cucumber roots in soil

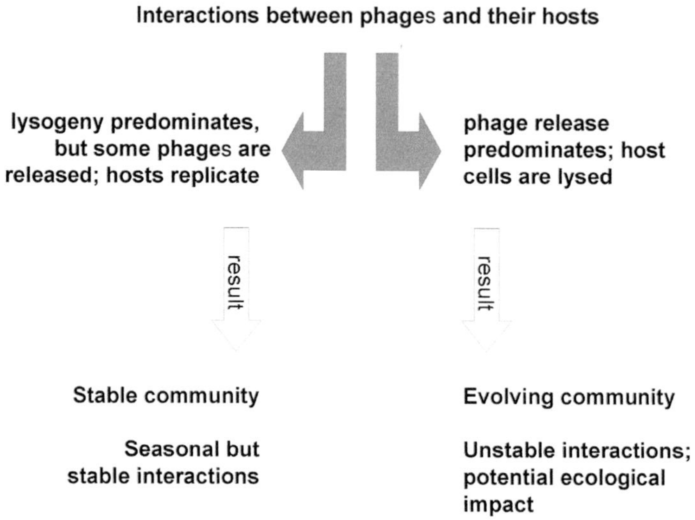

Figure 11.1 A summary of the potential interactions between phages and their hosts.

against the plant-pathogenic bacterium *Pythium ultimum*. The phage reduced the host population densities of *P. fluorescens* and thus increased the sensitivity of the plants to *P. ultimum* infection. This provides the first substantive evidence for an ecological impact resulting from phage intervention. In addition, in their soil microcosms Pantastica-Caldas *et al.* (1992) demonstrated that temperate phages did not depress the equilibrium host density, but that obligately lytic phage did (ten-fold).

Phages will always impact their host populations, but assessment of ecological impact requires demonstration of an effect resulting from phage–host interaction. Figure 11.1 briefly summarizes the principle. In observations of real-world ecosystems we see fluctuations in host and phage populations, but in some cases these are simply seasonal variations and community stability is maintained over the long run (e.g., Section 11.4.2). Equally important are the non-seasonal impacts that lead to evolutionary change in the community by effecting the density of the host and consequently its role in the ecosystem. Little is presently known about these effects, but the last example gives tantalizing clues to how phages may mediate ecosystem evolution, as we consider more fully in the following section.

11.4.2 Qualitative impact of phages on terrestrial bacteria

Presumably there is competition among phages for hosts, as these bacteria are a resource for phage growth. Ashelford *et al.* (1999, 2000, 2003) showed a seasonal effect on phage populations exploiting an indigenous host growing on a plant surface. Just what drives this variation in phage populations remains an open question, but the bacterial host responds to changes in the exudates from the host plant as it matures. So is it the plant exudates that make associated bacteria display different physiologies as the season proceeds? The hypothesis is that these different physiologies impact the competitiveness of different phage types differently.

The *in situ* experiments reported by Ashelford *et al.* showed that several unique phage types can coexist in a dynamic fashion in successive years in the rhizosphere of sugar beet (Section 11.3.2). Others have previously seen similar changes in phage population dynamics occurring in bulk soil microcosms over much shorter periods (e.g., Smit *et al.*, 1996). This means that if conditions are suitable for bacterial growth then a microbial bloom can occur, resulting in a massive numerical increase in a subset of the microbial species present. These will act as hosts for suitable phages and the phages will be amplified at the expense of the host population (Weinbauer and Rassoulzadegan, 2004; Chapter 10). Thus, the implication is that phages can affect bacterial community composition.

Buckling and Rainey (2002) examined the effect of an obligately lytic phage on the phenotypic diversification of the bacterium *Pseudomonas fluorescens*. Because of a retention of spatial structure in their broth microcosms (through an absence of stirring), this work is pertinent to the phage ecology of soils. In the presence of spatial structure but absence of phages, *P. fluorescens* rapidly diversifies through mutation into numerous mutant populations of spatial-niche specialists. In the presence of phages, however, Buckling and Rainey found that bacterial diversification into niche specialists was greatly curtailed, as also was overall bacterial density, with lower bacterial densities resulting in further reductions in bacterial diversity because decreased competition for resources reduced selection for niche specialization. Phages in these experiments also reduced physiological diversification in host populations through selection for resistant host clones, at least in the short term. That is, only a few of the many possible types of niche-specialized bacteria swept to high frequencies in these microcosms because their sweeps (Chapters 6 and 9) were a consequence of their phage resistance rather than their potential to display ecological competitiveness in the absence of phages.

In a more complicated experiment, Harcombe and Bull (2005) asked what occurs when two bacterial host species are interacting in a community and only one acts as a phage host. In the absence of phages specific for *E. coli* and *Salmonella enterica*, both species were stably maintained. When either of two *E. coli*-specific phages (T7 or T5) was added, the impact was evident: *E.*

coli became extinct or nearly so, whereas in the absence of *S. enterica*, *E. coli* phage-resistant mutants accumulated normally. In contrast, the addition of *Salmonella* phage SP6 to an *S. enterica* culture led to transient decreases in bacterial density whether *E. coli* was absent or present. These data demonstrate the difficulties in predicting interactions and their outcome between phages and hosts, but support the hypothesis that such interactions are capable of influencing population dynamics and hence community structure. Applying this knowledge to microbial populations in a soil environment suggests that we may be unable to predict effects of adding additional bacterial species to phage-containing environments: the consequences could range from no effect to phage-resistant mutants of the added bacteria dominating the environment. The impact of phage–bacterial interactions in microbial communities in terrestrial environments thus remains to be elucidated.

11.5 HORIZONTAL GENE TRANSFER

Horizontal gene transfer (HGT) is the movement of DNA from one bacterium to another. This movement can be between closely related bacteria or, alternatively, less closely related (even distantly related) bacteria. There are three mechanisms by which HGT is typically mediated between bacteria. These are transformation, conjugation, and transduction. Transformation is the taking up of naked DNA from the environment. Conjugation is the plasmid-mediated movement of DNA from an intact donor bacterium to a targeted recipient bacterium. Transduction, in turn, is the movement of bacterial DNA packaged within bacteriophage virions.

11.5.1 Transduction

There exist arguably three types of transduction, two generally recognized, the third well appreciated but not necessarily always associated with the word transduction (see, for example, Breitbart *et al.*, 2005). The two well-recognized forms of transduction are so-called generalized transduction and specialized transduction. Generalized transduction is so named because of the types of bacterial genes that may be transferred. These genes are not a specific subset but instead comprise all types of genes found within a given bacterial genome. Generalized transduction occurs as a consequence of accidentally packaging into a phage virion the DNA of the host rather than that of the phage. The resulting virions are non-infectious (unable to give rise to phage infections, because they are missing phage DNA), but are still infectious in the sense of being able to transfer their packaged DNA to a recipient bacterium. This DNA enters the recipient bacterium's cytoplasm and is able to enter the bacterial genome via homology-driven recombination (Miller, 2004).

Specialized transduction is much more biased in terms of the bacterial genes transported by a phage virion (Miller, 2004). These genes are typically limited to those adjacent to the phage when it is integrated into the host genome. These genes consequently become accidentally associated with phage DNA owing to imprecise prophage excision upon lysogen induction. The resulting phage, as a result of acquiring new genetic material, is not normally viable, as it loses an equivalent amount of phage DNA so that the wild-type phage genome size may be retained. The packaged bacterial DNA nevertheless may be carried to a recipient bacterium. Because specialized transduction by any given temperate phage is a process in which only specific genes can be transferred it is not considered further.

On rare occasions bacterial genes may become incorporated into phage genomes via illegitimate recombination, where there is little or no homology between genetic sequences. These genes that may become incorporated into a phage are not limited in the way that they are with specialized transduction. They are referred to as morons, a description of their status as representing more (than expected) DNA (more precisely, genes) within the phage genome (Chapter 7; see also Chapter 14).

Though transduction can vary appreciably in terms of how bacterial genes are carried – particularly within phage genomes, as against independent of phage DNA – all transduction mechanisms have the characteristic that the donor bacterium dies in the process of releasing the virion-carried bacterial DNA into the extracellular environment. As a consequence, it is not the donor bacterium that benefits from gene exchange, but instead it is the subset of their genes that are dispersed which gain. Additionally, it can be the bacterium receiving the novel phenotype which gains, since these genes may confer advantageous phenotypes. Thus transduction (Canchaya *et al.*, 2003) provides opportunity to transfer genes for phenotypes that have potential evolutionary value to the recipient bacterium (Miller, 1998). Selection provides the test for the value of transferred genes, which may be measured, ultimately, in terms of their persistence in an environment.

11.5.2 Transduction in terrestrial ecosystems, etc.

Generalized transduction has been recognized, for a long time, to occur in numerous environments (Miller, 2004). For example, phages of the soil bacteria *Rhizobium* (syn. *Sinorhizobium* and *Ensifer*) *meliloti* (Finan *et al.*, 1984) and in *Bradyrhizobium japonicum* (Shah *et al.*, 1981) have been shown to mediate transduction in laboratory experiments. Obligately lytic rhizobiophage can also transduce auxotrophic and antibiotic resistance markers in *Rhizobium leguminosarum* bv. *viciae* (Buchanan-Wollaston, 1979). Finally, specialized transduction occurs in *R. meliloti* (Svab *et al.*, 1978). Thus, the processes of generalized and specialized transduction seen in laboratory studies have their equivalents in soil bacteria.

One of us (Day, unpublished results) has tested over 25 phages isolated directly from the rhizosphere for their ability to transduce host (*Serratia* sp. CP6) antibiotic resistance phenotypes in laboratory experiments over a range of temperatures (15–35 °C). None did so (level of detection about 10^{-9}). Day also tested lysogens that routinely grew to 2×10^9 CFU mL^{-1} and released phages (about 10^4 PFU mL^{-1}), but none of these transduced. Thus, in at least one study, no evidence for generalized transduction was observed among a large set of terrestrial phages.

On the other hand, using an aquatic bacterium host, *Sphaerotilus natans,* Jensen *et al.* (1998) found two phages, SN-1 and SN-T, from just 10, that were capable of generalized transduction and showed their potential to exchange genes between diverse host species. Such broad-host-range phages could promote genetic diversity through genetic exchange in microbial communities in sediment and water. There clearly is a proportion of phages in the environment capable of transducing chromosomal genes, but while some people have found these phages, others have not. The reason for the observed variation in results remains obscure, but it is probably related to a combination of the strains used, the phage, the genes/phenotypes tested, and serendipity.

11.5.3 Transduction prevalence in the wild

As a general rule the frequency of transfer of any single gene, by a generalized transducing phage, is about one in a million recipient cells. Thus it is easy to see why transduction was considered to be environmentally insignificant until recently, when very high numbers of phages and host bacteria were established to occur in soil and water (see Section 11.3.1).

How common is the ability to transduce genes, and how significant to bacterial evolution might it be? A novel approach to achieving an understanding of the potential of transduction in a natural habitat was reported by Sander and Schmieger (2001). This approach determines the species participating in generalized transduction and allows an estimate of the host populations involved. Thus, an estimate of the transductional potential in a habitat is achieved. Briefly, phage-encapsulated bacterial DNA is isolated from an environmental sample and used as a template for PCR amplification of 16 S ribosomal DNA using primers specific for the 16 S rRNA genes of eubacteria. Sequencing the cloned amplification product identifies the host bacterial species. Since the DNA was derived from the contents of phage particles, by definition it has the potential to be transduced. Thus, this method identifies the members of the habitat that can potentially participate in horizontal gene transfer via transduction.

Regardless of the rate of occurrence of transduction, the potential for observation of a given transduction event goes up dramatically if the transductant may be selected, either naturally, or artificially in the laboratory. The latter is particularly easy to observe since it can involve selection for novel, selectable phenotypes in recipient bacteria. But what evidence exists for the transfer of wild genes by phages increasing the fitness of wild bacterial recipients? Such evidence is beginning to accumulate from medical and environmental studies, although this evidence has been demonstrated directly in sediment rather than in soil (Waldor and Mekalanos, 1996). An excellent example is the phage conversion of *Vibrio cholerae* that leads to this bacterium's ability to cause cholera (Chapter 14). This conversion is initiated by the infection of the host bacterium with a temperate phage, CTXΦ, that encodes genes for the cholera toxin (Waldor and Mekalanos, 1996). The new phenotype increases the fitness of *Vibrio* cells and allows colonization of a new ecological site. Although this is considered lysogenic conversion, as the host cell now carries a phage with a pathogenicity gene in it, it can also be considered a special case of gene exchange mediated by a phage (Section 11.5.1).

11.6 CONCLUSIONS AND UNANSWERED QUESTIONS

We are only beginning to learn about the ecology of terrestrial phages. Our knowledge has lagged behind knowledge of phages in aquatic environments because of the perceived greater difficulty of working in less-liquid environments. What we have learned provides us with intriguing hints at an exciting field of research that is sure to provide interesting answers and pose even more interesting questions as its study is expanded. To a large extent, however, what we observe happening to phages in the terrestrial environment, including the interactions between phages and hosts, mirrors what we see in the aquatic environment. In particular, bacteria often live in biofilms (Pedrós-Alió, 1993; Chattopadhyay and Puls, 2000), which are microenvironments that provide opportunity for interactions between host bacteria and their phages.

There are suggestions that phages can play a pivotal role in regulating bacterial numbers and their genetic and physiological diversity. They cycle on a seasonal succession that may or may not mimic their host's growth pattern (Ashelford *et al.*, 1999, 2000; Section 11.4). As has been clearly documented in the aquatic environment (Wommack and Colwell, 2000), bacteriophages are likely to be important players in the terrestrial ecosystem, regulating the bottom of the food chain and thereby the potential of the ecosystem to support various functions and species diversity, possibly impacting on the functioning of a biome. Even with these observations, we currently have more questions than we have answers.

ACKNOWLEDGMENTS

This work was supported in part by the Microbial Observatories program of the National Science Foundation (MCB-0132097).

REFERENCES

Ackermann, H.-W. 1997. Bacteriophage ecology. In M. T. Martins, M. I. Z. Sato, J. M. Tiedje, L. C. N. Hagler, J. Döbereiner, and P. S. Sanchez (eds.), *Progress in Microbial Ecology* (Proceedings of the Seventh International Symposium on Microbial Ecology). São Paulo: Brazilian Society for Microbiology/International Committee on Microbial Ecology, pp. 335–9

Ahmad, M. H., and V. Morgan. 1994. Characterization of a cowpea (*Vigna unguiculata*) rhizobiophage and its effects on cowpea nodulation and growth. *Biol. Fertil. Soils* **18**: 297–301.

Ashelford, K. E., J. C. Fry, M. J. Bailey, A. R. Jeffries, and M. J. Day. 1999. Characterization of six bacteriophages of *Serratia liquefaciens* CP6 isolated from the sugar beet phytosphere. *Appl. Environ. Microbiol.* **65**: 1959–65.

Ashelford, K. E., S. Norris, J. C. Fry, M. J. Bailey, and M. J. Day. 2000. Seasonal population dynamics and interactions of competing bacteriophages and their host in the rhizosphere. *Appl. Environ. Microbiol.* **66**: 4193–9.

Ashelford, K. E., M. J. Day, and J. C. Fry. 2003. Elevated abundance of bacteriophage infecting bacteria in soil. *Appl. Environ. Microbiol.* **69**: 285–9.

Billi, D., E. I. Friedmann, H. G. Hofer, G. G. Caiola, and R. Ocampo-Friedman. 2000. Ionizing-radiation resistance in the desiccation-tolerant *Cyanobacterium chroococcidiopsis*. *Appl. Environ. Microbiol.* **66**: 1489–92.

Boyd, E. F., and H. Brüssow. 2002. Common themes among bacteriophage-encoded virulence factors and diversity among the bacteriophages involved. *Trends Microbiol.* **10**: 521–9.

Breitbart, M., and F. Rohwer. 2005. Here a virus, there a virus, everywhere the same virus? *Trends Microbiol.* **13**: 278–84.

Breitbart, M., F. Rohwer, and S. T. Abedon. 2005. Phage ecology and bacterial pathogenesis. In M. K. Waldor, D. I. Friedman, and S. L. Adhya (eds.), *Phages: Their Role in Bacterial Pathogenesis and Biotechnology*. Washington DC: ASM Press, pp. 66–91.

Buchanan-Wollaston, V. 1979. Generalized transduction in *Rhizobium leguminosarum*. *J. Gen. Microbiol.* **112**: 135–42.

Buckling, A., and P. B. Rainey. 2002. The role of parasites in sympatric and allopatric host diversification. *Nature* **420**: 496–9.

Burke, J., D. Schneider, and J. Westpheling. 2001. Generalized transduction in *Streptomyces coelicolor*. *Proc. Natl. Acad. Sci. U.S.A.* **98**: 6289–94.

Burroughs, N. J., P. Marsh, and E. M. H. Wellington. 2000. Mathematical analysis of growth and interaction dynamics of streptomycetes and a bacteriophage in soil. *Appl. Environ. Microbiol.* **66**: 3868–77.

Canchaya, C., G. Fournou, S. Chibani-Chennoufi, M. L. Dillmann, and H. Brüssow. 2003. Phage as agents of lateral gene transfer. *Curr. Opin. Microbiol.* **6**: 417–24.

Chattopadhyay, D., and R. W. Puls. 2000. Forces dictating colloidal interactions between viruses and soil. *Chemosphere* **41**: 1279–86.

Danovaro, R., E. Manini, and A. Dell'Anno. 2002. Higher abundance of bacteria than of viruses in deep Mediterranean sediments. *Appl. Environ. Microbiol.* **68**: 1468–72.

Finan, T. M., E. Hartwieg, K. LeMieux, K. Bergman, G. C. Walker, and E. R. Signer. 1984. Generalized transduction in *Rhizobium meliloti*. *J. Bacteriol.* **159**: 120–4.

Gill, J. J., and S. T. Abedon. 2003. Bacteriophage ecology and plants. APSnet Feature, November 2003. www.apsnet.org/online/feature/phages.

Goodman, M. F. 2002. Error-prone repair DNA polymerase in prokaryotes and eukaryotes. *Annu. Rev. Biochem.* **71**: 17–50.

Goto, M. 1992. *Fundamentals of Bacterial Plant Physiology*. New York, NY: Academic Press.

Harcombe, W. R., and J. J. Bull. 2005. Impact of phages on two-species bacterial communities. *Appl. Environ. Microbiol.* **71**: 5254–9.

Harshey, R. M. 1988. Phage Mu. In R. Calendar (ed.), *The Bacteriophages*, vol. 1. New York, NY: Plenum Press, pp. 193–234.

Harvey, R. W. and N. R. Joseph. 2004. Use of PRD1 bacteriophage in groundwater viral transport, inactivation, and attachment studies. *FEMS Microb. Ecol.* **49**: 3–16.

Janzen, R. A., F. D. Cook, and W. B. McGill. 1995. Compost extract added to microcosms may simulate community-level controls on soil microorganisms involved in element cycling. *Soil Biol. Biochem.* **27**: 181–8.

Jensen, E. C., H. S. Schrader, B. Rieland, *et al.* 1998. Prevalence of broad-host-range lytic bacteriophages of *Sphaerotilus natans, Escherichia coli*, and *Pseudomonas aeruginosa*. *Appl. Environ. Microbiol.* **64**: 575–80.

Karlen, D. L., M. J. Mausbach, J. W. Doran, R. G. Cline, R. F. Harris, and G. E. Schuman. 1997. Soil quality: a concept, definition, and framework for evaluation. *Soil Sci. Soc. America J.* **61**: 4–10.

Keel C., Z. Ucrurum, P. Michaux, M. Adrian, and D. Haas. 2002. Deleterious impact of a virulent bacteriophage on survival and biocontrol activity of *Pseudomonas flourescens* strain CHAO in natural soil. *Mol. Plant-Microbe Interact.* **15**: 567–76.

Kieser, T., M. J. Bibb, M. Buttner, K. Chater, and D. Hopwood. 2000. *Practical* Streptomyces *Genetics*. Norwich: John Innes Foundation.

Kleczkowska, J. 1957. A study of the distribution and the effects of bacteriophage of root nodule bacteria in the soil. *Can. J. Microbiol.* **3**: 171–80.

Marsh, P., and E. M. Wellington. 1994. Phage–host interactions in soil. *FEMS Microbiol. Ecol.* **15**: 99–108.

Miller, R. V. 1998. Bacterial gene swapping in nature. *Sci. Am.* **278**: 66–71.

Miller, R. V. 2004. Bacteriophage-mediated transduction: an engine for change and evolution. In R. V. Miller and M. J. Day (eds.), *Microbial Evolution: Gene Establishment, Survival, and Exchange*. Washington, DC: ASM Press, pp. 144–57.

Miller, R. V., and S. A. Ripp. 2001. Pseudolysogeny: a bacteriophage strategy for increasing longevity in situ. In M. Syvanen and C. I. Kado (eds.), *Horizontal Gene Transfer*. London: Academic Press, pp. 79–89.

Miller, R. V., S. Ripp, J. Replicon, O. A. Ogunseitan, and T. A. Kokjohn. 1992. Virus-mediated gene transfer in freshwater environments. In M. J. Gauthier (ed.), *Gene Transfers and Environment*. Springer-Verlag, Berlin: Springer, pp. 50–62.

Pantastica-Caldas, M., K. E. Duncan, C. A. Istock, and J. A. Bell. 1992. Population dynamics of bacteriophage and *Bacillus subtilis* in soil. *Ecology* **73**: 1888–902.

Pedrós-Alió, C. 1993. Diversity of bacterioplankton. *Trends Ecol. Evol.* **8**: 86–90.

Primrose, S. B. and M. J. Day. 1977. Rapid concentration of bacteriophage from aquatic habitats. *J. Appl. Bacteriol.* **42**: 417–21.

Ripp, S., and R. V. Miller. 1995. Effects of suspended particulates on the frequency of transduction among *Pseudomonas aeruginosa* in a freshwater environment. *Appl. Environ. Microbiol.* **61**: 1214–19.

Rydman, P. S., and D. H. Bamford. 2002. Phage enzymes digest eptidoglycan to deliver DNA. *ASM News* **68**: 330–5.

Sander, M., and H. Schmieger. 2001. Method for host-independent detection of generalized transducing bacteriophages in natural habitats. *Appl. Environ. Microbiol.* 64: 1490–3.

Sano, E., S. Carlson, L. Wegley, and F. Rohwer. 2004. Movement of viruses between biomes. *Appl. Environ. Microbiol.* **70**: 5842–6.

Shah, K., S. de Sousa, and V. V. Modi. 1981. Studies in transducing phage M-1 for *Rhizobium japonicum* D211. *Arch. Microbiol.* **130**: 262–6.

Smit, E., A. C. Wolters, H. Lee, J. T. Trevors, and J. D. van Elsas. 1996. Interactions between a genetically marked Pseudomonas fluorescens strain and bacteriophage ϕR2f in soil: effects of nutrients, alginate encapsulation, and the wheat rhizosphere. *Microbial Ecol.* **31**: 125–40.

Svab, Z., A. Kondorosi, and L. Orosz. 1978. Specialized transduction of a cysteine marker by *Rhizobium meliloti* phage 16–3. *J. Gen. Microbiol.* **106**: 321–7.

Thingstad, T. F., M. Heldal, G. Bratbak, and I. Dundas. 1993. Are viruses important partners in pelagic food webs? *Trends Ecol. Evol.* **8**: 209–13.

Trouwborst, T. and J. C. De Jong. 1973. Interaction of some factors in the mechanism of inactivation of bacteriophage MS2 in aerosols. *Appl. Environ. Microbiol.* **26**: 252–7.

van de Putte, P., S. Cramer, and M. Giphart-Gassler. 1980. Invertible DNA determines host specificity of bacteriophage Mu. *Nature* **286**: 218–22.

Waldor, M. K. and J. J. Mekalanos. 1996. Lysogenic conversion by a filamentous phage encoding cholera toxin. *Science* **272**: 1910–14.

Weinbauer, M. G., and F. Rassoulzadegan. 2004. Are viruses driving microbial diversification and diversity? *Environ. Microbiol.* **6**: 1–11.

Welkos, S., M. Schreiber, and H. Baer. 1974. Identification of *Salmonella* with the O-1 bacteriophage. *Appl. Microbiol.* **28**: 618–22.

Whipps, J. M. 2001. Microbial interactions and biocontrol in the rhizosphere. *J. Exper. Botany* **52**: 487–511.

Williams, S. T., A. M. Mortimer, and L. Manchester. 1987. Ecology of soil bacteriophages. In S. M. Goyal, C. P. Gerba, and G. Bitton (eds.), *Phage Ecology*. New York, NY: Wiley, pp. 157–79.

Wommack, K. E., and R. R. Colwell. 2000. Virioplankton: viruses in aquatic ecosystems. *Microbiol. Mol. Biol. Rev.* **64**: 69–114.

CHAPTER 12

Phages, bacteria, and food

Lawrence D. Goodridge

12.1 INTRODUCTION

Bacteriophages have been observed to occupy virtually every ecological niche, and it should be no surprise that they are naturally present in food. While most phage–host interactions in food are inferred, but not routinely or knowingly observed, the phage-mediated destruction of starter cultures during fermentation of dairy products is a potent reminder of the ability of bacterial viruses to alter the microbial ecology of food. Alternatively, in recent years researchers have developed applied uses of bacteriophages, especially to solve agricultural problems. For example, bacteriophages have been proposed as indicators of possible viral and fecal contamination of food, while the ability of lytic bacteriophages to infect and kill their hosts has been exploited in the biocontrol of foodborne pathogenic and spoilage bacteria. Other workers have developed phage-based methods to rapidly detect bacterial pathogens. Most of these methods take advantage of the ability of phages to amplify themselves to large numbers during bacterial infection. Nevertheless, it is clear that several challenges must be overcome before phages will be routinely used as components of food production. These include the problem of bacterial resistance to phage infection and the need for separation of target bacteria from the food matrix during phage-mediated bacterial detection. Finally, the recent Food and Drug Administration (FDA) approval of the use of phages as antimicrobials to control bacterial growth in ready-to-eat meats is an exciting development that should lead to increased commercial interest in phage-based biocontrol of bacteria in food (i.e., phage "therapy" of food). This chapter reviews issues and technologies associated with phage–bacterial interactions

Bacteriophage Ecology: Population Growth, Evolution, and Impact of Bacterial Viruses, ed. Stephen T. Abedon. Published by Cambridge University Press.

in food, taking as a premise that many of these interactions may be fruitfully explored from a phage-ecological perspective.

Kennedy and Bitton (1987) provide an earlier, general review of the phage ecology of food. Additional, more recent and narrowly focused reviews on the diverse nature of phage–bacterial interactions that occur in food are indicated in the text. For additional consideration of phage therapy, see Chapters 13 and 17.

12.2 PHAGES AND FOOD

Bacteriophages have been isolated from a variety of agricultural environments, including soil, sewage, water, farm and food-processing effluents, manure, retail foods (Greer, 2005), and foods of dairy origin (Brüssow, 2001). Studies outlining the presence of phages in foods have been carried out, in part as a preface toward understanding the ecological roles that phages play in influencing the complex food microflora. Somatic and F^+ coliphages also have been proposed as indicators of the presence of fecal contamination in foods of animal and vegetable origin. In addition, as indicated above, phages also may be employed to selectively reduce bacterial numbers in food.

12.2.1 Dairy phages

Dairy phages may be readily isolated from fermented foods including yogurt (e.g., Brüssow *et al.*, 1994). They are a major source of faulty fermentations and may be of enormous economic importance. Whitehead and Cox (1935) discovered that bacteriophages, by killing bacteria, could cause complete cessation of acid production, especially of fermentations employing a single bacterial starter culture. Since then there have been numerous reports of phage-associated starter-culture failure, ranging from production delays when relatively few phages are present to total loss of product when phage densities are high (Brüssow, 2001). Exacerbating the problem, dairy phages can be difficult to inactivate, often surviving pasteurization and spray-drying (Labrie and Moineau, 2000). Alternative approaches toward limiting phage growth consequently are important, especially the employment of phage-resistant bacterial starter cultures (Sanders, 1988; Kim *et al.*, 1992). The identification of highly conserved genes in distinct groups of phages is also of interest because these regions represent potential targets in the development of phage-resistant starter strains. The ecology of phages in dairy foods and their role in fermentation failure have been the subject of several reviews (Sanders, 1987; Wünsche, 1989).

Due to their impact, dairy phages of the lactic acid bacteria (LAB) *Lactococcus lactis* and *Streptococcus thermophilus* are the most thoroughly studied and sequenced of food-isolated phages (Brüssow, 2001). *L. lactis* phages comprise three main species, all belonging to the family *Siphoviridae*: 936, c2 (which are lytic but not temperate; Chapter 1), and P335 (which are also lytic, but with both temperate and non-temperate members; Labrie and Moineau, 2002). *L. lactis* phages share various degrees of DNA homology. Members of the same species are closely related, and their genomic DNA sequences differ mainly by point mutations and small insertions or deletions (Desiere *et al.*, 2002). Phages of the 936 species are most often the cause of disrupted fermentation, followed by the c2 and P335 species (Labrie and Moineau, 2002). *S. thermophilus* phages are closely related to each other, illustrating their evolution from a common ancestor (Desiere *et al.*, 2002), with the two most common species designated DT1 and Q1. Sequence data suggest that spontaneous deletions, insertions, and point mutations in the lysogeny module probably led to a loss of temperance by phages such as DT1. There is no correlation between phage host range and DNA homology, implying that host range is determined by parameters that do not reflect genetic relationships (Le Marrec *et al.*, 1997).

12.2.2 Indigenous phages in other fermented foods

Several reports have described phages from a variety of non-dairy but fermented foods including sauerkraut (Lu *et al.*, 2003a), pickles (Lu *et al.*, 2003b), and salami (Chibani-Chennoufi *et al.*, 2004). These foods support complex microbial ecosystems in which large populations of bacteria may be found. Understanding phage ecology can be important when fermenting vegetables, since these fermentations often rely on LAB that are naturally present on fresh vegetables. Taken collectively, these studies highlight the diversity of phages that can be isolated from fermented foods and are suggestive of a role for phages in modulating the species of bacterial microflora found in these foods.

Several studies have highlighted the diversity of phages in fermented foods. Lu *et al.* (2003a), for example, isolated from a commercial sauerkraut fermentation a total of 171 phages, including at least 26 distinct phages, plus 28 distinct LAB strains. Their data suggest that phages may play an important role in the microbial ecology and succession of LAB species in vegetable fermentations. Lu *et al.* (2003b) describe the isolation of a phage, ϕJL-1, from a commercial cucumber fermentation. This phage infects strains of *Lactobacillus plantarum* that had been evaluated both as starter cultures for controlled cucumber fermentation and as biocontrol microorganisms for minimally processed vegetable products. They suggest that knowledge of the properties of phage ϕJL-1 may be important for the development of controlled vegetable fermentations. Chibani-Chennoufi *et al.* (2004) isolated an obligately lytic *L. plantarum* myophage, LP65, from an industrial meat fermentation. Sequence analysis indicated that LP65 was chimeric, sharing homology with T4-like *Myoviridae* from Gram-negative bacteria, as well as phages with Gram-positive hosts including the *Listeria* phage A511, *Staphylococcus* phage K, and *Bacillus* phage SPO1 (A511 and K are SPO1-related; see Chapter 7 for further discussion on the mosaicism of tailed phages). The genome organization of the structural module in LP65 resembled that of *Siphoviridae* from the lambda supergroup.

Pauling (1992) identified two phages, Hh-1 and Hh-3, that infect extremely halophilic bacteria of the genus *Halobacterium*. These phages were isolated from fermented anchovy sauce. Characterization revealed that both Hh-1 and Hh-3 were more tolerant of suspension in solutions of low ionic strength than their host bacteria, and that bacterial cells infected with both phages remained viable, produced phages, were immune to superinfection with homologous phages, and remained fully capable of supporting heterologous phages. The authors concluded from the observed properties that halophages are well adapted to survival in environments in which the salinity is subject to rapid changes of considerable magnitude.

12.2.3 Phages as indicators of fecal contamination

Studies have been conducted on the use of certain phages that can be isolated from foods as indicators of fecal contamination. These isolatable, otherwise feces-associated phages, in the parlance of food microbiology, serve as "microbial indicators of food safety" – which in this chapter I will refer to simply as "indicators" – as opposed to their serving as "microbial indicators of food quality" (such as could be provided, in principle, by the phages of spoilage bacteria, or by the spoilage bacteria themselves). While coliphages and phages that infect the anaerobic bacterium *Bacteroides fragilis* have been proposed as fecal indicators, the coliphages have emerged as the most likely candidates due to the fact that they are relatively resistant to desiccation and are viable over a wide range of temperatures (Seeley and Primrose, 1980). In this scenario, the presence of phages would indicate the presence of their bacterial hosts, which may otherwise be present in too low concentrations to be easily detected. Coliphages are also present in human wastewater at high concentrations (Gerba, 1987). Their presence in fecally contaminated food and water (Hsu *et al.*, 2002) is suggestive of their usefulness as indicators of food safety and quality.

Kennedy *et al.* (1986) investigated the distribution of coliphages in several foods. Coliphages were detected in 56% of the samples, and in all foods with the exception of roasted turkey breast. The authors reported significant nonparametric correlations between coliphages and bacterial indicators of fecal contamination, including *Escherichia coli*, fecal coliforms, and total coliforms. Characterizing for phage prevalence alone, Hsu *et al.* (2002) investigated the presence of F^+ coliphage (those adsorbing to host F pili) and somatic coliphages in five market samples of ground beef and three of chicken meat. F^+ coliphages were found in three of the five beef samples, and in two of the three chicken samples. Somatic coliphages (i.e., those that adsorb to bacterial structures other than F pili) were found in four of the five beef samples and in all of the chicken samples.

Hirotani *et al.* (2001) obtained 24 vegetables, including tomatoes, peppers, root crops, and cole (mustard) herbs, from local markets in Sonora, Mexico, and Arizona, USA. Each vegetable was tested for the presence of total coliforms, fecal coliforms, fecal streptococci, and coliphages. All samples were detached from the vegetables with a solution of 0.1% Tween 80 in distilled water. Coliphages were detected in all vegetables, though at only low levels in tomatoes. However, little correlation between prevalence of coliphages and coliform bacterial indicators was found. The researchers concluded that the bacterial strain used as the host for the coliphages (*E. coli* 15597), as well as the detachment procedures used to isolate the phages from the vegetables, may have affected coliphage isolation. More recently, Endley *et al.* (2003) evaluated the efficacy of F^+ coliphages as fecal contamination indicators for fresh carrots. Coliphages were detected in a higher percentage of the carrot samples (24%) than either *E. coli* (8%) or *Salmonella* (4%), leading Endley and coworkers (2003) to suggest that the use of F^+ coliphages as indicators of fecal contamination on vegetables should be employed in addition to screening for bacterial indicators.

Researchers have also evaluated the use of enteric bacteriophages as indicators for monitoring microbiological levels at critical control points as part of a Hazard Analysis Critical Control Point (HACCP) plan in a poultry slaughter plant. It was observed that the levels of F^+ phages were reduced by 1.6 $\log_{10}$ PFU (plaque-forming units) per gram through the poultry processing steps of evisceration, washing, and chilling. It was concluded that F^+ RNA coliphages may be effective indicators for monitoring the microbiological quality of meat and poultry during processing (Hsu *et al.*, 2002).

Contradictions between studies are likely explained by the use of different bacterial hosts to isolate the coliphages, as well as differences in the ecological specificity of the phages recovered. For example, three distinct physiological types of coliphages have been identified based on their infectivity at different temperatures (Seeley and Primrose, 1980): high-temperature (HT), mid-temperature (MT), and low-temperature (LT) coliphage types, possessing the ability to proliferate between 30 and 42 °C (HT), at 15–45 °C (MT), and at 15–30 °C (LT). The distribution of coliphages in water closely reflected the temperature of the source of fecal contamination in the environments from which they were isolated. Based on these observations, it was theorized that HT phages originated from the gastrointestinal tract of warm-blooded animals, while MT phages likely originated in either the enteric or aquatic environments. LT phages are thought to be present only in the aquatic environment.

In an elegant study, Kennedy *et al.* (1986) followed up on the work of Seeley and Primrose (1980) by isolating coliphages from 10 different food samples, and characterizing the phages with respect to temperature of infectivity. The authors reported temperatures for the three classes of phages that were different from those employed by Seeley and Primrose (1980). For example, HT phages were able to proliferate at temperatures $\geq$30 °C, while MT phages were infective over a range of 20–42 °C, and LT phages only proliferated at temperatures $\leq$20 °C. In addition, the researchers observed that HT phages were isolated in much higher percentages when *E. coli* C-3000 was used as the host strain, as compared to *E. coli* strain C. In contrast, the MT and LT phages were isolated in higher percentages when *E. coli* C was used as the host (Kennedy *et al.*, 1986). Collectively, these results demonstrate that the choice of bacterial host, the method in which the coliphages are detached from the food sample, and the temperature of incubation will all influence the concentrations and characteristics of coliphages that are isolated from food.

12.2.4 Phages as animal-virus surrogates

Several researchers have proposed the use of phages as indicators of the presence of enteric viruses in filter-feeding shellfish. The utility of these phages is due to the fact that they are similar in size and morphology to enteric viruses including foodborne viruses such as hepatitis A virus and the noroviruses (Havelaar *et al.*, 1993). To serve as an effective animal-virus surrogate, the resistance of both the indicator organism and the pathogen to environmental factors (pH, temperature, dessication; Section 12.4.3) should be similar. Based on this statement, Havelaar and colleagues (1993) have described a number of general criteria for the ideal enteric virus model in water: They should (1) occur exclusively and consistently in human feces, (2) not occur in animal feces, (3) not multiply in natural waters, (4) outnumber human viruses in fecally polluted water by several orders of magnitude, (5) behave like human viruses in fecally polluted water treatment processes, and (6) be detectable by simple, inexpensive, and rapid methods. See studies, for example, by Dore *et al.* (2000, 2003) for the use of phages as indicators of fecal contamination in shellfish. Phages have also been proposed as surrogates to study within-food inactivation of enteric viruses using non-thermal processing methods such as hydrostatic pressure (San Martin *et al.*, 2002; Smiddy *et al.*, 2006).

12.3 DETECTION OF FOODBORNE PATHOGENS

The use of phages to detect their bacterial hosts presents what is likely the most commercially viable application for bacteriophages. Historically, phages have been used to subtype bacterial species, and phage typing schemes now exist for numerous bacterial pathogens, including foodborne pathogens (Krylova, 1963; Ibrahim, 1969; van der Mee-Marquet and Audurier, 1995; Sechter *et al.*, 2000). More recently, the trend has been to develop rapid assays based on the utilization of a specific component or characteristic of a phage. For example, assays have been based on genetic modification of the phage with a reporter gene. Given infection of viable bacteria, expression of the reporter gene leads to a detectable signal (Section 12.3.1). Other tests have been developed in which the phage particle is labeled with a fluorescent dye, followed by specific attachment to the host for identification (Section 12.3.2). Finally, the virion-amplifying nature of phage infections has been harnessed to produce assays in which a low concentration of phage is added to an unknown sample. Phage amplification (as detected by plaque assay, or some other endpoint) is indicative of the presence of the host bacteria (Section 12.3.3). A number of more general reviews exist covering phage-mediated bacterial detection (Goodridge and Griffiths, 2002; Kuhn *et al.*, 2002; Petrenko and Vodyanoy, 2003; Kalantri *et al.*, 2005).

Note that in terms of the impact of phage ecology, phage-based bacterial detection is highly dependent on three things: bacterial physiology, bacterial numbers, and phage density. That is, labeled phages can detect if they can adsorb, while reporter phages can only detect if they infect a metabolizing bacterium. Both of these issues may be addressed by enriching target bacteria in order to both amplify their number and to make sure that they are physiologically healthy, and then by employing large numbers of phage virions to assure sufficient bacterial adsorption (Section 12.6). Enrichment in broth also decreases problems with detection due to phage interactions with the food matrix (e.g., the presence of biofilms or non-specific binding of phages to food particles). For phage amplification, there must also be release of sufficient phage progeny to significantly raise phage titers above the titer of added phages, a hurdle that can be mitigated somewhat by a virion-inactivation step post phage adsorption and prior to phage-induced bacterial lysis.

12.3.1 Reporter bacteriophage assays

The majority of phage-based detection methods have been founded upon the concept of the reporter phage. Ulitzer and Kuhn (1987) were the first to demonstrate that phages carrying a reporter gene could be used to detect bacteria in a rapid manner. In this work, the researchers genetically engineered phage λ Charon 30 to carry the *Vibrio fischeri* bioluminescence (*lux*) genes, and showed that it was possible to detect between 10 and 100 *E. coli* cells per mL of milk within one hour (Ulitzur and Kuhn, 1987). Subsequent work has concentrated on the development of qualitative phage-based reporter gene assays for foodborne bacteria (Section 12.4.1.1), though there has additionally been some effort to develop phage-based enumeration (quantitative) assays for foodborne bacteria (Section 12.3.1.2).

12.3.1.1 Qualitative detection of foodborne bacteria

Since the pioneering work of Ulitzur and Kuhn, several reporter bacteriophage assays have been developed for detection of foodborne pathogens. For example, Waddell and Poppe (2000) constructed a *lux* reporter phage and used it to detect *E. coli* O157:H7. The *lux*$^+$ phage was capable of detecting a wide range of *E. coli* O157:H7 phage types, and when the assay was tested on broth cultures, bioluminescence was easily measured within one hour.

Chen and Griffiths (1996) used a cocktail of three *lux*$^+$ phages to rapidly detect *Salmonella* spp. in food. Phage P22 and two uncharacterized temperate phages were modified to carry the *V. fischeri lux* genes, and the researchers showed that the assay was capable of detecting *Salmonella* isolates from groups B, C, and D. This study was unique in that it demonstrated that *Salmonella* could be detected in whole eggs by the direct introduction of the reporter phages into the eggs. After 24 hours of incubation, as few as 10 colony-forming units of bacteria (CFU) per egg could be detected, allowing identification of *Salmonella*-contaminated eggs, and also indicating the location where the bacteria were growing. Bioluminescence could be detected using a variety of instruments, including a photon-counting charge-coupled device (CCD) camera, a luminometer, and X-ray film (Chen and Griffiths, 1996).

A commercially available test for *Salmonella* spp. was developed based on the use of an ice nucleation (*ina*) reporter gene. Certain bacteria possess proteins that enable them to initiate ice formation in supercooled water. These ice-nucleation proteins are thought to produce templates for the assembly of very small seed crystals of ice. Introduction of ice nucleation genes into a bacterial cell, and subsequent expression, leads to the formation of ice nuclei, which results in a catalyzed reaction, culminating in ice formation and rapid freezing (Warren 1997). Wolber and Green (1990) utilized this principle to develop the Bacterial Ice Nucleation Diagnostic (BIND). The assay included infecting *Salmonella* cells with an *ina*$^{+}$ reporter phage. The *Salmonella* cells produced ice nuclei, which caused the cells to freeze when the temperature was lowered to less than –9.3 °C. Ice formation was easily detected with the use of an indicator dye, which changed color when the sample began to freeze (Wolber and Green, 1990).

The isolation of a listeriophage (A511) with a wide host range enabled the construction of a *lux*$^{+}$ reporter phage assay for *L. monocytogenes*. Loessner and coworkers (1996) constructed a recombinant phage with the *lux* genes, and the reporter listeriophage was tested for its ability to detect *L. monocytogenes* in pure culture and in a variety of foods. The results indicated that as low as 500 cells were detected in pure culture, and with an enrichment step, the presence of less than 1 CFU per gram of salad was detectable within 24 hours. The reporter listeriophage was also used to demonstrate the rapid detection of *L. monocytogenes* in a variety of other foods including meat, poultry, and various cheeses (Loessner *et al.*, 1997). In addition to the work described here, reporter phage assays have been developed for other foodborne pathogens including *Staphylococcus aureus* (Pagotto *et al.*, 1996) and *Mycobacterium bovis* (Jacobs *et al.*, 1993), and other than foodborne bacteria including *Mycobacterium tuberculosis* and *M. bovis* (Sarkis *et al.* 1995; Pearson *et al.*, 1996).

12.3.1.2 Quantification of foodborne bacteria

All of the reporter phage assays described so far have been developed for qualitative detection of bacteria. Several researchers have developed quantitative reporter assays based on the use of reporter phages. For example, Turpin *et al.* (1993) developed a reporter phage most probable number (MPN) method which allowed for detection and enumeration of *Salmonella* Typhimurium in environmental samples within 24 hours, based upon light-producing ability transduced to the *Salmonella* cells by a *lux*$^{+}$ phage. In the assay, positive samples were defined as those that produced bioluminescence approximately 100–1000 times higher than background.

Based on Turpin's work, Loessner and colleagues (1997) developed a reporter phage MPN assay for *L. monocytogenes*. In this study, *Listeria*-free samples of four different foods were thawed and spiked with low numbers of *L. monocytogenes* strain Scott A. The food samples were enriched for either 20 or 40 hours, serially diluted, mixed with a lux^+ reporter listeriophage, and incubated at 20 °C for 2 hours. Bioluminescence was measured in a photon-counting single-tube luminometer. The results showed that the sensitivity of the assay was one viable *Listeria* cell per tube. The authors concluded that the *lux* MPN method was simple and enabled rapid enumeration of *Listeria* cells in a variety of foods. Reporter phage MPN assays also appear to be useful in situations in which direct plating for determination of cell count is not feasible, such as in cases where a strong competing microflora can overwhelm the selective agents used for isolation of the target organism.

12.3.1.3 Reporter-labeled phage

In a novel study, Oda *et al.* (2004) labeled bacteriophages with a reporter gene and used this combined strategy to detect *E. coli* O157:H7. In this work, the researchers fused the green fluorescent protein (GFP) to the phage small outer capsid (SOC) protein, enabling the capsid to become fluorescent. Adsorption of the GFP-labeled PP01 phage to the *E. coli* O157:H7 cell surface enabled visualization of cells under a fluorescence microscope. The GFP-labeled PP01 phage adsorbed not only culturable *E. coli* cells but also viable but nonculturable or pasteurized cells. The presence of insensitive *E. coli* strain K-12 cells did not influence the specificity and affinity of GFP-labeled PP01 adsorption on *E. coli* O157:H7. After a 10-minute incubation with GFP-labeled PP01 phages at an MOI (multiplicity of infection) of 10^3 at 4 °C, *E. coli* O157:H7 cells could be visualized by fluorescence microscopy. The authors concluded that the GFP-labeled PP01 phage could be a rapid and sensitive tool for *E. coli* O157:H7 detection (Oda *et al.*, 2004). More generally, phages may be labeled chemically, as reviewed in Section 12.3.2. See Section 12.6, however, for general discussion of critical caveats associated with employing MOI as a gauge of phage density.

12.3.2 Labeled bacteriophage assays

While the literature is full of many studies outlining the development of reporter phages, the technique has not been widely accepted. One possible drawback to the use of reporter phages is the difficulty in modifying phages that have not been genetically well characterized. Also, there is some concern in various parts of the world regarding the use of genetically modified organisms in food production. Several reports have detailed the use of bacteriophages as fluorescent probes, in a manner analogous to antibodies. This approach attempts to harness the specificity of phages for their target bacteria, without the need to genetically modify the phages.

Goodridge *et al.* (1999a), for example, describe a fluorescent bacteriophage assay (FBA) in which a fluorescently labeled coliphage (LG1) was combined with immunomagnetic separation (IMS) to effect rapid detection of *E. coli* O157:H7. A sample to be tested was incubated with *E. coli* O157:H7-specific IMS beads, which isolated the target bacteria from the food matrix and background microflora, followed by addition of the fluorescently labeled phages to the sample. The fluorescent phages attached to the target cells, conferring a fluorescent halo on the *E. coli* bacteria that was easily visualized by fluorescent microscopy. When it was combined with flow cytometry, the FBA was capable of detecting 10^4 cells mL^{-1}. A modified direct epifluorescent-filter technique (DEFT) was employed in an attempt to estimate bacterial concentrations. Using regression analysis, the lower detection limit was calculated to be between 10^2 and 10^3 cells mL^{-1}; however, the modified DEFT was found to be an unreliable method for determining bacterial concentrations.

When used in conjunction with flow cytometry, the FBA was able to detect 2.2 CFU g^{-1} of *E. coli* O157:H7 in artificially contaminated ground beef following a 6-hour enrichment (Goodridge *et al.* 1999b). Between 10^1 and 10^2 CFU mL^{-1} in artificially contaminated raw milk were detectable after a 10-hour enrichment step. The results showed that fluorescent bacteriophage assays are a rapid technique for the preliminary detection of *E. coli* O157:H7 in food; however, the high cost associated with flow cytometry would seem to limit the practicality of this assay.

12.3.3 Bacteriophage amplification

One of the simpler methods for bacterial detection takes advantage of the fact that, upon infection of a target cell, one phage will produce multiple progeny (e.g., Fig. 2.1 of Chapter 2). The amplified phage can then be detected by a variety of end points, including plaque assay or growth on a helper bacterial strain. Favrin *et al.* (2001) developed a simple assay that utilized the infection cycle of bacteriophage SJ2 for detection of *Salmonella* Enteritidis in broth. The assay consisted of four steps including (1) capture and concentration of target cells by IMS, (2) infection of the target bacterium with phages,

(3) amplification and recovery of progeny phages, and (4) assay of progeny phages on the basis of their effect on an indicator strain of bacteria. The in-broth end point of the assay was determined by using either fluorescence or optical density measurements. The detection limit of the assay was determined to be less than 10^4 CFU mL^{-1}, and the assay could be performed in 4–5 hours. When the assay was tested using artificially inoculated skimmed milk powder, chicken rinse, and ground beef, the bacteriophage amplification assay was capable of detecting an average of 3 CFU of *S.* Enteritidis in 25 g or 25 mL of food sample. An enrichment step was necessary for detection of the bacteria, and the total assay time including pre-enrichment was approximately 20 hours (Favrin *et al.*, 2001).

The assay was adapted to enable detection of *E. coli* O157:H7 and permitted the detection of 2 CFU of *E. coli* O157:H7 per gram of ground beef. The assay was further adapted to allow for simultaneous detection of *S.* Enteritidis and *E. coli* O157:H7. The researchers concluded that the bacteriophage amplification assay showed promise for the simultaneous detection of foodborne pathogens, but further development work would be necessary to improve sensitivity and produce reliable results at low inoculation levels (Favrin *et al.*, 2003). A possible explanation for the erratic results obtained at low bacterial inoculation levels is the fact that a combination of too-low bacterial densities and too-low phage densities will result essentially in no infection and therefore no amplification (Section 12.6).

12.4 PHAGE-MEDIATED BIOCONTROL OF BACTERIA

The continued threat of foodborne illness has necessitated the development of novel strategies to reduce the presence of foodborne bacterial pathogens on and in foods. Bacteriophage biocontrol has been suggested as one method to control pathogens in foods, and a great deal of research in this area is currently under way. In addition to pathogen control, others have investigated extending the shelf life of foods through phage-based reductions in concentrations of spoilage microflora. The obvious benefits of using phages to control bacteria in foods, as opposed to other antimicrobials, lies in the facts that phages are naturally present in a variety of foods (Ackermann, 1997) and that phages with the proper host-range specificity should be able to reduce the target pathogen and spoilage bacteria, without compromising the normal flora of the food (Goodridge and Abedon, 2003) or food quality. While the results of many studies have been promising, it is apparent that several challenges to phage-based, food-grade antimicrobials must be met before their use can be accepted. These challenges range from the phage-ecological – including stability of the phages to various physical stresses such as pH, temperature, and osmotic stress (Section 12.4.3), and the issue of bacterial resistance – to the regulatory, i.e., FDA approval of phage-based food additives.

12.4.1 Reduction of foodborne pathogens

The use of bacteriophage therapy as a biocontrol method to control pathogenic bacteria in food continues to gain attention, and two excellent reviews have been recently published in this area (Greer, 2005; Hudson *et al.*, 2005). In general, phage-based approaches have focused on four of the most common foodborne pathogens: *E. coli* O157:H7, *Salmonella* spp., *Campylobacter* spp., and *Listeria monocytogenes*.

12.4.1.1 Control of *Escherichia coli* O157:H7

Kudva *et al.* (1999) were the first to show that *E. coli* O157:H7 could be controlled via the addition of lytic bacteriophages. *E. coli* O157:H7 is commonly associated with foods of livestock origin, as well as fresh produce (Rangel *et al.* 2005). The reduction of this pathogen in foods is extremely important, because serotype O157:H7 has been implicated in numerous outbreaks of foodborne illness (Chapter 14). The lack of any defined treatment, combined with poor prognosis for the systemic sequelae, has led to pronounced efforts to eliminate this pathogen from its sources (Tarr, 1995).

Kudva and coworkers (1999) isolated three *E. coli* O157 O-antigen-specific phages, combined them into a mixture, and evaluated them for their ability to lyse broth cultures of *E. coli* O157:H7 at 37 °C and 4 °C. The three-strain phage mixture lysed all 15 of the *E. coli* O157 isolates that were tested, and did not lyse any of the non-O157 *E. coli* or non-*E. coli* bacteria tested. The researchers observed that the *E. coli* broth cultures had to be aerated, and a high multiplicity of infection (MOI) had to be present for lysis to occur. In the absence of aeration, complete lysis of the bacterial cells took five days, and was only observed at 4 °C. The researchers concluded that *E. coli* O157-specific bacteriophages could be useful as biocontrol agents to reduce or eliminate *E. coli* O157:H7 in foods that support the survival of this pathogen (Kudva *et al.*, 1999).

Other workers have evaluated phages for their ability to control *E. coli* O157:H7 on food. O'Flynn *et al.* (2004) investigated the use of three phages (e11/2, e4/1c, and PP01), individually, and combined as a cocktail, for their ability to lyse *E. coli* O157:H7 in broth and on meat. In broth culture, at both 30 and 37 °C, the cocktail or individual phages e11/2 and PP01 reduced log-phase bacterial cells to undetectable levels. To assess efficacy of the phage treatment in meat, the three-phage cocktail (MOI of 10^6) was pipetted evenly onto pieces of beef that were inoculated with 100 μL of a rifampin-resistant *E. coli* O157:H7 strain at 10^3 CFU mL^{-1}. Following incubation for 1 hour, the treated as well as untreated meat samples were enriched in BHI broth at 37 °C for 2 hours. Seven of the nine phage-treated samples were devoid of *E. coli* O157:H7, while two of the samples had *E. coli* O157:H7 counts of less than 10 CFU mL^{-1}. In contrast, the control samples had *E. coli* O157:H7 concentrations of 10^5 CFU mL^{-1} (O'Flynn *et al.*, 2004).

Taken collectively, the use of phages to reduce the presence of *E. coli* O157:H7 in food appears feasible, if a large phage concentration can be achieved. Presumably, the need for high concentrations of phages reflects the diffusion limitation that phages encounter on the surface of meat products that are contaminated with low levels of pathogens (Brüssow, 2005). The issue of bacterial resistance is more complicated, but given the low numbers of *E. coli* O157:H7 in foods, and the frequency of resistance developing, which O'Flynn and coworkers (2004) calculated in their study to be approximately 10^{-6} CFU, it seems unlikely that bacterial resistance to phage cocktails would occur. The development of a cocktail comprised of phages that use different receptors to attach to the bacterial surface would also decrease the frequency at which phage-resistant mutants develop (Tanji *et al.*, 2004).

12.4.1.2 Control of *Salmonella* spp.

Bacteriophages have been evaluated as a method to control the presence of *Salmonella* spp. and *Campylobacter jejuni* on poultry products and other foods. For example, Goode and coworkers (2003) artificially contaminated portions of chicken skin with *Salmonella* Enteritidis to achieve a final density of 10^3 CFU cm^{-2}, and half of the contaminated samples were inoculated with *Salmonella* typing phage 12 at MOI of 1. (Note: the bacterial and therefore phage densities employed in this study are greater than they "appear" since they are per a volume that is somewhat smaller than 1 mL, i.e., 1 cm^2 on the surface of chicken skin). The samples were incubated at 4 °C and swabs were taken from three separate 10 cm^2 areas of each chicken skin sample before phage treatment and 24 and 48 hours following addition of the phage. The result was statistically more decline in *Salmonella* density with versus without phage treatment, plus increases in phage density, with the latter consistent with bacterial death being due to phage infection.

The authors' observation of an increase in phage density when host bacteria were present is a result at least potentially consistent with what Brüssow (2005) describes as "active" biocontrol, where the effect is due to the replication of the phages in the presence of the target bacteria. Alternatively, killing can be a consequence of "passive" action, where the initial phage concentration alone removes the pathogen. Goode *et al.* (2003), for example, showed that when the density of *Salmonella* was low ($<\log_{10} 2.0\ \text{cm}^{-2}$) and the MOI was 10^5, no bacteria were recovered. Also, when the MOI was increased to 10^7, it was possible to eliminate *Salmonella* strains that showed high levels of resistance because of restriction, but to which the phages were able to attach (Goode *et al.*, 2003).

Several studies have investigated the use of phage therapy to control *Salmonella* in non-muscle foods. Modi *et al.* (2001) evaluated the effects of phages on the survival of *S.* Enteritidis during the manufacture and storage of Cheddar cheese. The authors concluded that the addition of phages to raw and pasteurized milk significantly reduced the *S.* Enteritidis concentration in Cheddar cheeses made from these milks. In a study employing fresh fruits, Leverentz *et al.* (2001) artificially inoculated melon slices with *S.* Enteritidis and demonstrated that a cocktail of four phages led to a significant reduction in the *S.* Enteritidis concentration when the melon samples were stored at refrigeration as well as at elevated temperatures. However, the phage cocktail was unable to reduce the *S.* Enteritidis concentration on artificially inoculated apple slices. This result was attributed to the inability of the phage to survive in the acid environment on the surface of the apple slices (pH 4.4), which was supported by the fact that initial phage numbers declined to below detectable levels within 24 hours (Section 12.4.3). Pao and coworkers (2004) conducted trials aimed at establishing whether phages could be used to control *Salmonella* in sprouting seeds. In this work, the researchers isolated and characterized two bacteriophages with different host ranges, and showed that a mixture of both phages resulted in a 1.5 log reduction in the numbers of *Salmonella* in the soaking water of broccoli seeds.

12.4.1.3 Control of *Campylobacter jejuni*

Campylobacter jejuni, a Gram-negative microaerophilic, curved, and motile rod, is the leading cause of bacterial diarrheal illness in the United States, causing more foodborne illness than *Shigella* spp. and *Salmonella* spp. combined (Mead *et al.*, 1999). Goode *et al.* (2003) inoculated chicken skin samples with *C. jejuni* followed by addition of a *C. jejuni* typing phage in a manner similar to the *Salmonella* experiment that these researchers conducted (described in Section 12.4.1.2). The results indicated that the *C. jejuni* counts on the chicken skin samples that were treated with the phage were significantly lower than on the controls. Atterbury *et al.* (2003) also conducted experiments aimed at using phages to reduce *C. jejuni* on chicken skin, and they reported results that agreed with the Goode *et al.* (2003) study. At 4 °C, *Campylobacter* recovery from controls inoculated with 10^6 and 10^4 CFU remained constant through the entire course of the experiment. In chicken samples inoculated with the lowest phage titer (10^3 PFU), no significant reduction in *C. jejuni* numbers was observed. In contrast, when the highest phage titer (10^7 PFU) was inoculated onto the chicken skin, there was a significant reduction of *C. jejuni* at all sampling points. The efficacy of the phage treatment was more pronounced in frozen chicken samples, leading the authors to conclude that the use of bacteriophage therapy, when coupled with a freeze step, could be an effective treatment to reduce *C. jejuni* on poultry (Atterbury *et al.*, 2003).

12.4.1.4 Control of *Listeria monocytogenes*

Listeria monocytogenes is a Gram-positive foodborne pathogen that has received a plethora of public health concern due to the high mortality rate in at-risk individuals (Farber and Peterkin, 1991). Pregnant women are particularly susceptible to *L. monocyogenes* infection, which can result in damage to the fetus and fetal death (Farber and Peterkin, 1991). *L. monocytogenes* tends to contaminate ready-to-eat foods that are unlikely to undergo further processing. Bacteriophages have been investigated as a non-thermal method to control the spread of *L. monocytogenes* in such foods.

Leverentz and colleagues (2003) investigated the ability of two phage cocktails to reduce concentrations of *L. monocytogenes* on fresh-cut apples and honeydew melons. The researchers determined that the phage cocktails reduced *L. monocytogenes* concentrations on honeydew melons by 2.0 to 4.6 logs as compared to the control. On the fresh-cut apples, the phage cocktail reduced the *L. monocytogenes* concentration by less than 0.4 log units. The efficacy of the cocktail was also shown to increase when phages were employed in combination with the bacteriocin nisin. In another study with honeydew melons, the authors showed that a mixture of six phages resulted in a larger reduction in the *L. monocytogenes* concentration when the phages were applied at higher concentrations (Leverentz *et al.*, 2004).

Concerning the control of *L. monocytogenes* in meat, it was shown that the combination of a *L. monocytogenes* phage and nisin provided an antimicrobial effect against *L. monocytogenes* in broth, but not in buffer or raw beef (Dykes and Moorhead, 2002). In broth, nisin alone reduced or prevented the growth of the two *L. monocytogenes* strains, but regrowth to levels equivalent to the controls (untreated cells) occurred. When listeriophage LH7 was used alone, there was no effect on the *L. monocytogenes* concentration in broth. However, when both antimicrobials were used together, a combined effect in broth was observed, with the antimicrobials reducing the concentration of *L. monocytogenes* without regrowth (Dykes and Moorhead, 2002). The researchers concluded that the use of nisin and bacteriophages has potential to control *L. monocytogenes* in foods, but much research detailing the ecological aspects of complex systems like foods must be achieved before any practical use of these treatments can be realized (Dykes and Moorhead, 2002). For example, bacterial pathogens probably grow within biofilms in food, and the ability of phages to destroy bacteria in the biofilm will be dependent on the susceptibility of these bacteria to the phages, as well as the availability of bacterial receptor sites for the phages to attach to (Sutherland *et al.*, 2004). See Chapter 17 for consideration of constraints on phage bacterial killing ability within complex environments.

Carlton *et al.* (2005) described the characterization of the virulent listeriophage P100, which infects and kills a majority of *L. monocytogenes* strains. The authors also demonstrated its use as an antimicrobial by producing surface-ripened red-smear soft cheese and contaminating the cheese with low concentrations of *L. monocytogenes* at the beginning of the ripening period. Phage P100 was applied to the surface during the rind washings. Depending on the time points, frequency, and dose of phage applications, the researchers observed a significant reduction ranging from 3.5 logs to complete elimination of the *L. monocytogenes* concentration. The authors did not recover any resistant *L. monocytogenes* cells from the samples. This study indicated the possibility of using bacteriophages to control *L. monocytogenes* surface contamination of soft cheeses.

12.4.2 Bacteriophage biocontrol of food spoilage

While the control of foodborne pathogens via the use of bacteriophages has received much attention, it is important to note that the majority of work in the use of phages as antimicrobials has centered around control of food spoilage and plant pathogenic bacteria. For example, phages have been isolated and used to control several plant diseases including fire blight on apple trees (Schnaebel *et al.*, 1999), bacterial spot on tomatoes (Balogh *et al.*, 2003), and bacterial spot disease of peaches (Randhawa and Civerolo, 1986). Munsch and colleagues (1991) showed that phages could control *Pseudomonas tolaasii*, the causative agent of bacterial blotch, in cultivated mushrooms. This study was noteworthy because it demonstrated the ability of phage treatment to increase the quality of the treated food. For example, the authors observed a 70% reduction in crop loss when the mushrooms were treated with phages (Munsch *et al.*, 1991).

Most research on phage biocontrol of bacteria in meat has centered around increasing the shelf life of foods by controlling spoilage bacteria. The majority of work in this area has been accomplished by Greer, who focused on *Pseudomonas* spp., which play a major role in the spoilage of meat (Greer, 1982, 1983, 1986, 1988; Greer and Dilts,1990). An important conclusion of these studies is that while beef shelf life may be extended by phage application, the host range of the phages employed often are not wide enough to inhibit the majority of spoilage flora, which can continue to grow and subsequently spoil the meat.

The phage-based inhibition of *Brochothrix thermosphacta* growth in meat has been investigated. Greer and Dilts (2002) demonstrated that *B. thermosphacta* phages inhibited the growth of *B. thermosphacta* on pork adipose tissue, and also prevented the development of objectionable off-odors. These results indicate that the control of foodborne bacteria by phages is possible, at least under carefully controlled conditions. Nevertheless, the inconsistent results in the literature demonstrate the ongoing need for detailed experiments in which the ecological interactions between bacteria and phages in food are fully investigated.

12.4.3 Phage stability within the food matrix

It is clear that a thorough understanding of the stability of phages within complex environments is necessary in order to develop effective phage-based biocontrol strategies. There are many characteristics of a given food, for example, that can affect the ability of the phage to remain viable. These intrinsic characteristics include the structure and composition of the food, the water activity, redox potential, antimicrobial agents present within the food, and pH. The structure and composition of food can affect phage survival by allowing the food to resist penetration of the phages into the food matrix, causing the phages to remain on the surface of the food, where they may become desiccated. High osmolarity may also decrease viability of any phages present. Antimicrobial agents can affect phage survival indirectly, by decreasing or eliminating concentrations of the host bacteria in the food, which would decrease the amount of phage amplification that could occur.

Most of the work accomplished on phage stability in food has centered around pH stability studies. Early phage studies indicated that phages were fairly stable at pH values between 5 and 8. At low temperatures, phage stability could be expanded to pH values between 4 and 10 (Adams, 1959). In one study, Smith *et al.* (1987) determined the stability of phages in milk whey adjusted to various pH values for 1 hour at 37 °C. Most of the phages were stable over a range of pH values from 3.5 to 6.8. At a pH of 3.0, a decrease in the phage concentration was observed; below pH 3.0, the phage concentration decreased dramatically. Other researchers have reported the instability of phages below pH 4.4 (Alisky *et al.*, 1998; Leverentz *et al.*, 2001, 2003). It is obvious that better approaches to dealing with the low pH of certain foods will need to be developed in order for phages to be utilized as antimicrobials. These solutions could include adding a buffering agent such as $CaCO_3$ to the phage cocktail (Hudson *et al.*, 2005). Alternatively, the isolation of low-pH-resistant phages would also solve the problem of rapid phage inactivation.

12.5 THE FUTURE IS NOW

The use of phages as biocontrol agents of pathogenic and spoilage bacteria in food has gained governmental acceptance in the USA with the recent regulatory approval of two diverse phage products. In 2002, the Environmental Protection Agency (EPA) approved the use of a pesticide, for control of bacterial spot (rot) of tomatoes and peppers, that consists of two bacteriophages for *Xanthomonas campestris* pv. *vesicatoria* and *Pseudomonas syringae* as the active ingredients. The *Xanthomonas* phage controls bacterial spot on tomatoes and peppers and the *Pseudomonas* phage controls bacterial speck on tomatoes. The pesticide is approved for various uses including direct application to plants and the surrounding soil throughout the growing season (USEPA, 2002).

More recently, the Food and Drug Administration approved the use of a bacteriophage preparation made from six individually purified phages to be used on RTE (ready-to-eat) meat and poultry products as an antimicrobial agent against *L. monocytogenes* (FDA, 2006). These developments are exciting, bode well for the continued future development of phage biocontrol strategies, and indicate the increasing acceptance of phages as natural antimicrobials.

Originally, the ecological impact of phages in food was studied purely from an economical viewpoint, with the goal of decreasing or eliminating fermentation failure in dairy products. As knowledge of phage biology has increased, so too have applied uses of phages to aid in the production of safe food, either as antimicrobials, as indicators of safety, or as sentinels that rapidly indicate the presence of bacterial pathogens. Phages have also found use in extending the shelf life of foods through destruction of spoilage microflora. All of this work has led to increased knowledge regarding the ecological roles that phages play in affecting the diverse microbial flora within food. The importance, nearly endless variety, and the spatial and chemical complexity of foods will ensure that these environments, as unique as any on this planet, will continue to be important subjects of phage ecological study.

12.6 APPENDIX: RATE OF ADSORPTION IS A FUNCTION OF PHAGE DENSITY*

Phages may be employed to detect the presence of bacteria (Section 12.3) and/or to reduce the presence of bacteria (Section 12.4). In either process, phage adsorption to bacteria is required and increased efficacy occurs if more bacteria are adsorbed. During bacterial detection, for example, a 1000-fold increase in signal should occur if 1000-fold more bacteria are phage-adsorbed. For the sake of real-world efficacy, it also is best if those bacteria become adsorbed sooner rather than later. Similarly, for phage-based biocontrol of bacteria, the more bacteria that are adsorbed (and killed), the better, and it is preferable that adsorption occurs over periods that are reasonably short relative to the replication rates of the target bacteria, or the handling time of the sample.

What approaches should one take to assure that a majority of bacteria become phage-adsorbed relatively quickly? Obvious answers are to employ phages that have a reasonably high affinity for the problem bacteria, especially under the conditions encountered, and then to provide large numbers of phages. These points should be fairly apparent, but supplying sufficient phage numbers can be problematic because many individuals are seemingly of the mistaken opinion that what determines likelihood of phage adsorption to a population of bacteria is the ratio of the densities of phages to bacteria (multiplicity of infection, MOI; Chapter 3). However, it is phage density alone that determines rates of phage adsorption to individual bacteria. The notion of MOI consequently can obscure one's understanding of adsorption efficacy, as this appendix will elaborate.

12.6.1 Bacterial density and free-phage loss rates

Though phage densities are key to determining rates of phage adsorption to individual bacteria, in fact bacterial densities are not irrelevant to understanding rates of phage adsorption. This impact of bacterial density, however, is *the* reason not to employ the concept of MOI in the context of phage-based detection or biocontrol protocols. To illustrate why, note that the average length of time it takes for a population of free phage to adsorb a given density

* This section was written by S. T. Abedon.

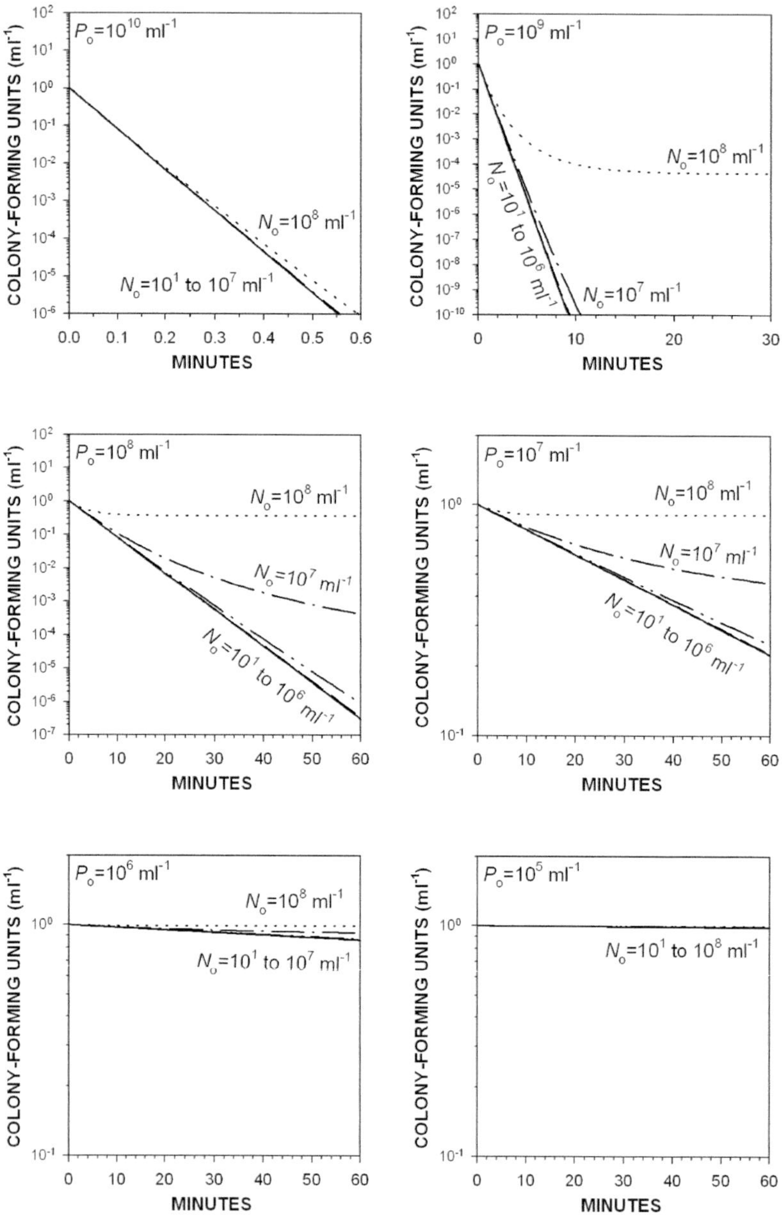

Figure 12.1 Relative reduction in colony-forming units as a function of starting phage (P_o) and bacterial (N_o) densities. Starting phage densities are as indicated in the upper left of each panel. Starting bacterial densities, N_o, are indicated, with 10^8 mL^{-1} (dotted line), 10^7 mL^{-1} (dash-dot-dash), and 10^6 mL^{-1} (dash-dot-dot-dash). 10^5 down to 10^1 bacteria mL^{-1} are all seen as an overlapping solid line. Note that x and y axes vary in range

of bacteria is equal to $1/(k \cdot N)$, where k is the phage adsorption constant and N is bacterial density. This is the time until 1/e free phage have *failed* to adsorb, meaning that 63% ($= 100 \times (1 - 1/e)$) of a free-phage population *has* adsorbed (Abedon *et al.*, 2001). If $N = 10^5$ bacteria mL^{-1} and $k = 2.5 \times 10^{-9}$ mL min^{-1} (Stent, 1963), then the average phage will take 67 hours ($= 1/(2.5 \times 10^{-9} \times 10^5 \times 60)$) to adsorb!

If it takes 67 hours for half of the phages in a sample to adsorb, then it will take 67 hours regardless of MOI, so long as bacterial density is held constant. This is because the rate of free-phage loss to adsorption is a function of bacterial density rather than free-phage density. Determining phage adsorption rates in terms of free-phage loss is problematic, however. This is because (1) bacterial densities in samples are not always known (especially during application of phage-based bacterial detection or biocontrol); (2) when bacterial densities are low, then the fraction of free phage lost as a function of time can be trivially low (below); and (3) calculating rates of phage adsorption in terms of free-phage loss does not directly address a more fundamental issue – how fast are bacteria being adsorbed?

12.6.2 Phage density and bacterial loss rates

The only practical way to assure rapid bacterial adsorption is to employ so many phages that significant adsorption occurs in absolute terms even if fractionally few free phages are adsorbing per unit time. This solution holds even if the problem is environmental shielding of bacteria from phage attack, e.g., as in biofilms. If bacterial susceptibility is not reduced to zero, then employing higher phage densities should still give rise to more bacterial

Figure 12.1 (*continued*)
between the different panels. The presented curves were generated based on the model presented in Fig. 15.1, Chapter 15: $N_t = N \cdot e^{-(P \cdot (1 - e^{-t \cdot k \cdot N})/N)}$. where N_t is the bacterial density after t minutes of adsorption (presented as N_t / N_o = "colony-forming units" in the figure), $N = N_o$ is the starting bacterial density, $P = P_o$ is the starting phage density, k is the phage adsorption constant which here has been set to 2.5×10^{-9} mL min^{-1} after Stent (1963), and t is time (indicated as "minutes" in the figure). Multiplicity of infection (MOI) as defined by P_o / N_o is readily calculated from the presented parameter values. For example, when $P_o = 10^7$ mL^{-1} then MOI = 0.1, 1.0, 10, 10^2, 10^3, 10^4, 10^5, and 10^6 for $N_o = 10^8$, 10^7, 10^6, 10^5, 10^4, 10^3, 10^2, and 10^1 mL^{-1}, respectively. See Chapter 15 for modeling details. Note that the number of viable, uninfected bacteria lost to phage adsorption is equal to the gain in adsorbed or infected bacteria (i.e., $N_a = N_o - N_t$ where N_a is the number of bacteria adsorbed at time, t).

adsorption and therefore more bacterial detection or killing (Abedon, 2006). However, it is when applying this excess of phages that the concept of MOI is misleading. If there are only 10^3 bacteria per mL – a reasonably large target density for either detection or biocontrol – then the average phage will take 6667 hours ($= 1/(2.5 \times 10^{-9} \times 10^{3} \times 60)$) to adsorb! Employing an MOI of 10 means having 10 phages for every bacterium, but how much of a difference does that make if each of those phages takes on average 40 weeks ($\approx 6667/24/7$) to adsorb?

When detecting bacteria it should be the bacteria that are limiting, not the phages. Similarly, when killing bacteria, there is no reason to depend on bacterial density to assure rapid phage adsorption. In both cases, rapid adsorption of individual bacteria may be achieved by employing large densities of free phages. In Fig. 12.1, the degree of adsorption to bacteria is indicated as functions of length of exposure of bacteria to phages, total density of bacteria present, and beginning density of phages present. Note that not until there are 10^{7} phages per mL is appreciable killing of bacteria observed, with as much as 80% of bacteria adsorbed in one hour. Furthermore, at 10^{7} phages per mL, MOI *does* matter, but only because you can't adsorb any more bacteria than you have phages. But, strikingly, even at 10^{6} phages per mL, the MOI – which ranges from 0.01 to 10^{5} in Fig. 12.1 – makes little difference in terms of total numbers of bacteria adsorbed, which at best is approximately 20% after one hour. Thus, in situations where the bacterial population densities are low, attempts should be made to increase phage density, independent of the actual bacterial density, such that the bacterial target is more rapidly found (Abedon, 1990; Kasman *et al.*, 2002; Hyman and Abedon, in press). As a caution, note that too much adsorption of certain kinds of phage (e.g., T4-like) can result in bacterial death without phage infection, a process known as lysis from without (Abedon, 1994; 1999).

ACKNOWLEDGMENTS

We (L.D.G. and S.T.A.) would like to thank Michael Dubow, whose 2000 Millennial Phage Meeting (in Montreal) marked the start of our collaboration.

REFERENCES

Abedon, S. T. 1990. Selection for lysis inhibition in bacteriophage. *J. Theor. Biol.* **146**: 501–511.

Abedon, S. T. 1994. Lysis and the interaction between free phages and infected cells. In J. D. Karam (ed.), *The Molecular Biology of Bacteriophage T4*. Washington, DC: ASM Press, pp. 397–405.

Abedon, S. T. 1999. Bacteriophage T4 resistance to lysis-inhibition collapse. *Genet. Res.* **74**: 1–11.

Abedon, S. T. 2006. Phage ecology. In R. L. Calendar and S. T. Abedon (eds.), *The Bacteriophages*, 2nd edn. Oxford: Oxford University Press, pp. 37–46.

Abedon, S. T., T. D. Herschler, and D. Stopar. 2001. Bacteriophage latent-period evolution as a response to resource availability. *Appl. Environ. Microbiol.* **67**: 4233–41.

Ackermann, H.-W. 1997. Bacteriophage ecology. In M. T. Martins, M. I. Z. Sato, J. M. Tiedje, L. C. N. Hagler, J. Döbereiner, and P. S. Sanchez (eds.), *Progress in Microbial Ecology* (Proceedings of the Seventh International Symposium on Microbial Ecology). São Paulo: Brazilian Society for Microbiology/International Committee on Microbial Ecology, pp. 335–9

Adams, M. H. 1959. *Bacteriophages*. New York, NY: Interscience.

Alisky, J., K. Iczkowski, A. Rappaport, and N. Troitsky. 1998. Bacteriophages show promise as antimicrobial agents. *J. Infect.* **36**: 5–15.

Atterbury, R. J., P. L. Connerton, C. E. R. Dodd, C. E. D. Rees, and I. F. Connerton. 2003. Application of host-specific bacteriophages to the surface of chicken skin leads to a reduction in recovery of *Campylobacter jejuni*. *Appl. Environ. Microbiol.* **69**: 6302–6.

Balogh, B., J. B. Jones, M. T. Momol, *et al.* 2003. Improved efficacy of newly formulated bacteriophages for management of bacterial spot on tomato. *Plant Dis.* **87**: 949–54.

Brüssow, H. 2001. Phages of dairy bacteria. *Annu. Rev. Microbiol.* **55**: 283–303.

Brüssow, H. 2005. Phage therapy: the *Escherichia coli* experience. *Microbiology* **151**: 2133–40.

Brüssow, H., M., Fremont, A. Bruttin, J. Sidoti, A. Constable, and V. Fryder. 1994. Detection and classification of *Streptococcus thermophilus* bacteriophages isolated from industrial milk fermentation. *Appl. Environ. Microbiol.* **60**: 4537–43.

Carlton, R. M., W. H. Noordman, B. Biswas, E. D. de Meester, and M. J. Loessner. 2005. Bacteriophage P100 for control of *Listeria monocytogenes* in foods: genome sequence, bioinformatic analyses, oral toxicity study, and application. *Regul. Toxicol. Pharmacol.* **43**: 301–12.

Chen, J., and M. W. Griffiths. 1996. Detection in eggs using Lux+ bacteriophages. *J. Food Prot.* **59**: 908–14.

Chibani-Chennoufi, S., M. L. Dillmann, L. Marvin-Guy, S. Rami-Shojaei, and H. Brüssow. 2004. *Lactobacillus plantarum* bacteriophage LP65: a new member of the SPO1-like genus of the family *Myoviridae*. *J. Bacteriol.* **186**: 7069–83.

Desiere, F., S. Lucchini, C. Canchaya, M. Ventura, and H. Brüssow. 2002. Comparative genomics of phages and prophages in lactic acid bacteria. *Antonie Van Leeuwenhoek* **82**: 73–91.

Dore, W. J., K. Henshilwood, and D. N. Lees. 2000. Evaluation of F-specific bacteriophage as a candidate human enteric virus indicator for bivalve molluscan shellfish. *Appl. Environ. Microbiol.* **66**: 1280–5.

Dore, W. J., M. Mackie, and D. N. Lees. 2003. Levels of male specific RNA bacteriophage and *Escherichia coli* in molluscan bivalve shellfish from commercial harvesting areas. *Lett. Appl. Microbiol.* **36**: 92–6.

Dykes, G. A., and S. M. Moorhead. 2002. Combined antimicrobial effect of nisin and listeriophage against *Listeria monocytogenes* in broth but not in buffer or on raw beef. *Int. J. Food Microbiol.* **73**: 71–81.

Endley, S., L. Lu, E. Vega, M. E. Hume, and S. D. Pillai. 2003. Male-specific coliphages as an additional fecal contamination indicator for screening fresh carrots. *J. Food Prot.* **66**: 88–93.

Farber, J. M., and P. I. Peterkin. 1991. *Listeria monocytogenes,* a food-borne pathogen. *Microbiol. Rev.* **55**: 476–511.

Favrin, S. J., S. A. Jassim, and M. W. Griffiths. 2001. Development and optimization of a novel immunomagnetic separation- bacteriophage assay for detection of *Salmonella enterica* serovar *enteritidis* in broth. *Appl. Environ. Microbiol.* **67**: 217–24.

Favrin, S. J., S. A. Jassim, and M. W. Griffiths. 2003. Application of a novel immunomagnetic separation-bacteriophage assay for the detection of *Salmonella enteritidis* and *Escherichia coli* O157:H7 in food. *Int. J. Food Microbiol.* **85**: 63–71.

Food and Drug Administration. 2006. Food additives: bacteriophage preparation. *Fed. Regist.* **71**: 47729–32.

Gerba, C. P. 1987. Phage as indicators of fecal pollution. In S. M. Goyal, C. P. Gerba, and G. Bitton (eds.), *Phage Ecology*. New York, NY: Wiley, pp. 197–209.

Goode, D., V. M. Allen, and P. A. Barrow. 2003. Reduction of experimental *Salmonella* and *Campylobacter* contamination of chicken skin by application of lytic bacteriophages. *Appl. Environ. Microbiol.* **69**: 5032–6.

Goodridge, L., and S. T. Abedon. 2003. Bacteriophage biocontrol and bioprocessing: application of phage therapy to industry. *SIM News* **53**: 254–262.

Goodridge, L., and M. Griffiths. 2002. Reporter bacteriophage assays as a means to detect foodborne pathogenic bacteria. *Food Res. Int.* **35**: 863–70.

Goodridge, L., J. Chen, and M. Griffiths. 1999a. Development and characterization of a fluorescent-bacteriophage assay for detection of *Escherichia coli* O157:H7. *Appl. Environ. Microbiol.* **65**: 1397–404.

Goodridge, L., J. Chen, and M. Griffiths. 1999b. The use of a fluorescent bacteriophage assay for detection of *Escherichia coli* O157:H7 in inoculated ground beef and raw milk. *Int. J. Food Microbiol.* **47**: 43–50.

Greer, G. G. 1982. Psychrotrophic bacteriophages for beef spoilage pseudomonads. *J. Food Prot.* **45**: 1318–25.

Greer, G. G. 1983. Psychrotrophic *Brochothrix thermosphacta* bacteriophages isolated from beef. *Appl. Environ. Microbiol.* **46**: 245–251.

Greer, G. G. 1986. Homologous bacteriophage control of *Pseudomonas* growth and beef spoilage. *J. Food Prot.* **49**: 104–9.

Greer, G. G. 1988. Effect of phage concentration, bacterial density, and temperature on phage control of beef spoilage. *J. Food Sci.* **53**: 1226–7.

Greer, G. G. 2005. Bacteriophage control of foodborne bacteria. *J. Food Prot.* **68**: 1102–11.

Greer, G. G., and B. D. Dilts. 1990. Inability of a bacteriophage pool to control beef spoilage. *Int. J. Food Microbiol.* **10**: 331–42.

Greer, G. G., and B. D. Dilts. 2002. Control of *Brochothrix thermosphacta* spoilage of pork adipose tissue using bacteriophages. *J. Food Prot.* **65**: 861–3.

Havelaar, A. H., M. van Olphen, and Y. C. Drost. 1993. F-specific RNA bacteriophage are adequate model organisms for enteric viruses in fresh water. *Appl. Environ. Microbiol.* **59**: 2956–62.

Hirotani, H., J. Naranjo, P. G. Moroyoqui, and C. P. Gerba. 2001 Demonstration of indicator microorganisms on surface of vegetables on the market in the United States and Mexico. *J. Food Sci.* **67**: 1847–50.

Hsu, F. C., C. Y. S. Shieh, and M. D. Sobsey. 2002. Enteric bacteriophages as potential fecal indicators in ground beef and poultry meat. *J. Food Prot.* **65**: 93–9.

Hudson, J. A., C. Billington, G. Carey-Smith, and G. Greening. 2005. Bacteriophages as biocontrol agents in food. *J. Food Prot.* **68**: 426–37.

Hyman, P. and S. T. Abedon, in press. Practical methods for determining phage growth parameters, M. Clokie and A. Kropinski (eds.), *Bacteriophages: Methods and Protocols.* Totowa, NJ: Humana Press.

Ibrahim, A. A. 1969. Bacteriophage typing of *Salmonella.* I. Isolation and host range study of bacteriophages. *Appl. Microbiol.* **18**: 444–7.

Jacobs, W. R., R. G. Barletta, R. Udani, *et al.* 1993. Rapid assessment of drug susceptibilities of *Mycobacterium tuberculosis* by means of luciferase reporter phages. *Science* **260**: 819–22.

Kalantri, S., M. Pai, L. Pascopella, L. Riley, and A. Reingold. 2005. Bacteriophage-based tests for the detection of *Mycobacterium tuberculosis* in clinical specimens: a systematic review and meta-analysis. *BMC Infect. Dis.* **5**: 59–72.

Kasman, L. M., A. Kasman, C. Westwater, J. Dolan, M. G. Schmidt, and J. S. Norris. 2002. Overcoming the phage replication threshold: a mathematical model with implications for phage therapy. *J. Virol.* **76**: 5557–64.

Kennedy, J. E., and G. Bitton. 1987. Bacteriophages in food. In S. M. Goyal, C. P. Gerba, and G. Bitton (eds.), *Phage Ecology*. New York, NY: Wiley, pp. 289–316.

Kennedy, J. E., C. I. Wei, and J. L. Oblinger. 1986. Distribution of coliphages in various foods. *J. Food Prot.* **49**: 944–51.

Kim, J. H., S. G. Kim, D. K., Chung, Y. C., Bor, and C. A. Batt. 1992. Use of antisense RNA to confer bacteriophage resistance in dairy starter cultures. *J. Ind. Microbiol.* **10**: 71–8.

Krylova, M. D. 1963. Phage typing of staphylococci in food poisoning. *Vopr. Pitan.* **22**: 89–93.

Kudva, I. T., S. Jelacic, P. I. Tarr, P. Youderian, and C. J. Hovde. 1999. Biocontrol of *Escherichia coli* O157 with O157-specific bacteriophages. *Appl. Environ. Microbiol.* **65**: 3767–73.

Kuhn, J., M. Suissa, J. Wyse, *et al.* 2002. Detection of bacteria using foreign DNA: the development of a bacteriophage reagent for *Salmonella*. *Int. J. Food Microbiol.* **74**: 229–38.

Labrie, S., and S. Moineau. 2000. Multiplex PCR for detection and identification of lactococcal bacteriophages. *Appl. Environ. Microbiol.* **66**: 987–94.

Labrie, S., and S. Moineau. 2002. Complete genomic sequence of bacteriophage ul36: demonstration of phage heterogeneity within the P335 quasi-species of lactococcal phages. *Virology* **296**: 308–20.

Le Marrec, C., D. van Sinderen, L. Walsh, *et al.* 1997. Two groups of bacteriophages infecting *Streptococcus thermophilus* can be distinguished on the basis of mode of packaging and genetic determinants for major structural proteins. *Appl. Environ. Microbiol.* **63**: 3246–53.

Leverentz, B., W. S. Conway, Z. Alavidze, *et al.* 2001. Examination of bacteriophage as a biocontrol method for *Salmonella* on fresh-cut fruit: a model study. *J. Food Prot.* **64**: 1116–21.

Leverentz, B., W. S. Conway, M. J. Camp, *et al.* 2003. Biocontrol of *Listeria monocytogenes* on fresh-cut produce by treatment with lytic bacteriophages and a bacteriocin. *Appl. Environ. Microbiol.* **69**: 4519–26.

Leverentz, B., W. S. Conway, W. Janisiewicz, and M. J. Camp. 2004. Optimizing concentration and timing of a phage spray application to reduce *Listeria monocytogenes* on honeydew melon tissue. *J. Food Prot.* **67**: 1682–6.

Loessner, M. J., C. E. D. Rees, G. S. A. B. Stewart, and S. Scherer. 1996. Construction of luciferase reporter bacteriophage A511: : luxAB for rapid and sensitive detection of viable *Listeria* cells. *Appl. Environ. Microbiol.* **62**: 1133–40.

Loessner, M. J., M. Rudolf, and S. Scherer. 1997. Evaluation of luciferase reporter bacteriophage A511: : luxAB for detection of *Listeria monocytogenes* in contaminated foods. *Appl. Environ. Microbiol.* **63**: 2961–5.

Lu, Z., F. Breidt, V. Plengvidhya, and H. P. Fleming. 2003a. Bacteriophage ecology in commercial sauerkraut fermentations. *Appl. Environ. Microbiol.* **69**: 3192–202.

Lu, Z., F. Breidt, Jr., H. P. Fleming, E. Altermann, and T. R. Klaenhammer. 2003b. Isolation and characterization of a *Lactobacillus plantarum* bacteriophage, phiJL-1, from a cucumber fermentation. *Int. J. Food Microbiol.* **84**: 225–35.

Mead, P. S., L. Slutsker, V. Dietz, *et al.* 1999. Food-related illness and death in the United States. *Emerg. Infect. Dis.* **5**: 607–25.

Modi, R., Y. Hirvi, A. Hill, and M. W. Griffiths. 2001. Effect of phage on survival of *Salmonella* Enteritidis during manufacture and storage of Cheddar cheese made from raw and pasteurized milk. *J. Food Prot.* **64**: 927–33.

Munsch, P., J. M. Olivier, and G. Houdeau. 1991. Experimental control of bacterial blotch by bacteriophages. In M. J. Maher (ed.), *Science and Cultivation of Edible Fungi*. Rotterdam: Balkema, pp. 389–96.

Oda, M., M. Morita, H. Unno, and Y. Tanji. 2004. Rapid detection of *Escherichia coli* O157:H7 by using green fluorescent protein-labeled PP01 bacteriophage. *Appl. Environ. Microbiol.* **70**: 527–34.

O'Flynn, G., R. P. Ross, G. F. Fitzgerald, and A. Coffey. 2004. Evaluation of a cocktail of three bacteriophages for biocontrol of *Escherichia coli* O157:H7. *Appl. Environ. Microbiol.* **70**: 3417–21.

Pagotto, F., L. Brovko, and M. W. Griffiths. 1996. Phage-mediated detection of *Staphylococcus aureus* and *Escherichia coli* O157:H7 using bioluminescence. In *Bacteriological Quality of Raw Milk*. IDF Special Issue 9601. Brussels: International Dairy Federation, pp. 152–6.

Pao, S., S. P. Randolph, E. W. Westbrook, and H. Shen. 2004. Use of bacteriophages to control *Salmonella* in experimentally contaminated sprout seeds. *J. Food Sci.* **69**: 127–30.

Pauling C. 1982. Bacteriophages of *Halobacterium halobium*: isolated from fermented fish sauce and primary characterization. *Can. J. Microbiol.* **28**: 916–21.

Pearson, R. E., S. Jurgensen, G. G. J. Sarkis, G. F. Hatfull, and W. R. Jacobs, Jr. 1996. Construction of D29 shuttle phasmids and luciferase reporter phages for detection of mycobacteria. *Gene* **183**: 129–36.

Petrenko, V. A., and V. J. Vodyanoy. 2003. Phage display for detection of biological threat agents. *J. Microbiol. Methods* **53**: 253–62.

Randhawa, P. S., and E. L. Civerolo. 1986. Interaction of *Xanthomonas campestris* pv. *pruni* with pruniphage and epiphytic bacteria on detached peach leaves. *Phytopathology* **76**: 549–53.

Rangel, J. M., P. H. Sparling, C. Crowe, P. M. Griffin, and D. L. Swerdlow. 2005. Epidemiology of *Escherichia coli* O157:H7 outbreaks, United States, 1982–2002. *Emerg. Infect. Dis.* **11**: 603–9.

Sanders, M. 1987. Bacteriophages of industrial importance. In S. M. Goyal, C. P. Gerba, and G. Bitton (eds.), *Phage Ecology*. New York, NY: Wiley, pp. 221–44.

Sanders, M. E. 1988. Phage resistance in lactic acid bacteria. *Biochimie* **70**: 411–22.

San Martin, M. F., G. V. Barbosa-Canovas, and B. G. Swanson. 2002. Food processing by high hydrostatic pressure. *Crit. Rev. Food Sci. Nutr.* **42**: 627–45.

Sarkis, G. J., W. R. Jacobs, Jr., and G. F. Hatfull. 1995. L5 luciferase reporter mycobacteriophages: a sensitive tool for the detection and assay of live mycobacteria. *Molecular Microbiol.* **15**: 1055–67.

Schnaebel, F. L., W. G. D. Fernando, M. P. Meyer, and A. I. Jones. 1999. Bacteriophages of *Erwinia amylovara* and their potential for biocontrol. In M. T. Mornol and H. Saygili (eds.), *Proceedings of the 8th International Workshop on Fireblight*. Leuven: International Society of Horticultural Science, pp. 649–53.

Sechter, I., F. Mestre, and D. S. Hansen. 2000. Twenty-three years of *Klebsiella* phage typing: a review of phage typing of 12 clusters of nosocomial infections, and a comparison of phage typing with K serotyping. *Clin. Microbiol. Infect.* **6**: 233–8.

Seeley, N. D., and S. B. Primrose. 1980. The effect of temperature on the ecology of aquatic bacteriophages. *J. Gen. Virol.* **46**: 87–95.

Smiddy, M., A. L. Kelly, M. F. Patterson, and C. Hill. 2006. High pressure-induced inactivation of Qbeta coliphage and c2 phage in oysters and in culture media. *Int. J. Food Microbiol.* **106**: 105–10.

Smith, H. W., M. B. Huggins, and K. M. Shaw. 1987. Factors influencing the survival and multiplication of bacteriophages in calves and in their environment. *J. Gen. Microbiol.* **133**: 1127–35.

Stent, G. S. 1963. *Molecular Biology of Bacterial Viruses*. San Francisco, CA: W. H. Freeman.

Sutherland, I. W., K. A. Hughes, L. C. Skillman, and K. Tait. 2004. The interaction of phage and biofilms. *FEMS Microbiol. Lett.* **232**: 1–6.

Tanji, Y., T. Shimada, M. Yoichi, K. Miyanaga, K. Hori, and H. Unno. 2004. Toward rational control of *Escherichia coli* O157:H7 by a phage cocktail. *Appl. Microbiol. Biotechnol.* **64**: 270–4.

Tarr, P. I. 1995. *Escherichia coli* O157:H7: clinical, diagnostic, and epidemiological aspects of human infection. *Clin. Infect. Dis.* **20**: 1–8.

Turpin, P. E., K. A. Maycroft, J. Bedford, C. L. Rowlands, and E. M. H. Wellington. 1993. A rapid luminescent-phage based MPN method for the enumeration of *Salmonella typhimurium* in environmental samples, *Lett. Appl. Microbiol.* **16**: 24–7.

Ulitzur, S., and J. Kuhn. 1987. Introduction of lux genes into bacteria, a new approach for specific determination of bacteria and their antibiotic susceptibility. In J. Schölmerich, R. Anreesen, A. Kapp, M. Ernst, and W. G. Woods (eds.), *Bioluminescence: New Perspectives.* New York, NY: Wiley, pp. 463–72.

United States Environmental Protection Agency. 2002. Biopesticide Fact Sheet. Bacteriophages of *Xanthomonas campestris* pv. *vesicatoria* (006449) & Bacteriophages of *Pseudomonas syringae* pv. *tomato* (006521). www.epa.gov/pesticides/biopesticides/ingredients/factsheets/factsheet˙006449-006521.htm

van der Mee-Marquet, N., and A. Audurier. 1995. Proposals for optimization of the international phage typing system for *Listeria monocytogenes*: combined analysis of phage lytic spectrum and variability of typing results. *Appl. Environ. Microbiol.* **61**: 303–9.

Waddell, T. E., and C. Poppe. 2000. Construction of mini-Tn10luxABcam/Ptac-ATS and its use for developing a bacteriophage that transduces bioluminescence to *Escherichia coli* O157:H7. *FEMS Microbiol. Lett.* **182**: 285–9.

Warren, G. J. 1987. Bacterial ice nucleation: molecular biology and applications. In G. E. Russell (ed.), *Biotechnology and Genetic Engineering Reviews.* Wimbrome, Dorset, UK: Intercept, pp. 107–35.

Whitehead, H. R. and G. A. Cox. 1935. The occurrence of bacteriophage in cultures of lactic streptococci. *N.Z. J. Sci. Technol.* **16**: 319–20.

Wolber, P. K., and R. L. Green. 1990. Detection of bacteria by transduction of ice nucleation genes. *Trends Biotech.* **8**: 276–9.

Wünsche, L. 1989. Importance of bacteriophages in fermentation processes. *Acta Biotechnol.* **9**: 395–419.

CHAPTER 13

Interaction of bacteriophages with animals

Carl R. Merril

13.1 INTRODUCTION

A number of bacterial virus strains interact with eukaryotes, including with humans. The study of these interactions is still in an early discovery phase. To put this situation in perspective, it is estimated that there are about 10^{12} cells in a human, while the number of bacteria, and their viruses (most of which inhabit the large intestine) outnumber these human cells by a factor of at least 10 (Bäckhed *et al.*, 2005). Study of the interactions of bacterial viruses, their genes, and their gene products on their animal "co-hosts" has only really begun. Early studies concentrated on their anatomical distribution in animals. These studies were followed by investigations of interactions with the immune system, particularly the adaptive immune system. More recently, studies have been initiated on the interactions of these viruses with the innate immune system and mammalian cells. Unfortunately, we currently do not have the technology to culture many of the bacterial strains associated with humans and other animals, which limits our ability to study the viruses associated with these organisms. However, the development of new molecular methods, particularly high-throughput genomic sequencing, are now providing tools to characterize the human "microbiome" in both normal and disease states (Relman, 2002). Despite the current dearth of knowledge concerning phage interactions with animals, there are already significant efforts to use the knowledge we have, to engineer the few phage strains that have been studied for antibacterial, vaccine, and gene therapy applications. These efforts will also provide data on the capacity of phages to interact with animals and humans in addition to their contributions to medical therapies.

Bacteriophage Ecology: Population Growth, Evolution, and Impact of Bacterial Viruses, ed. Stephen T. Abedon. Published by Cambridge University Press.

13.2 POTENTIAL FOR PHAGE–ANIMAL INTERACTION

Bacterial viruses, bacteriophages or phages, are by definition viruses that replicate in and interact with bacteria. As these viruses presumably arose in the same time fame as bacteria, they have been evolving and interacting with their bacterial hosts for about 4 billion years, or for the 22 trips the solar system has made around our Milky Way galaxy. As each of these galactic trips takes 225 million years, animals have been present for only three trips, and placental animals for only one-half of a trip. Humans have made less than 1/100 of a trip (Cockell, 2003). However, the 600 million years that animals have had on the Earth provides sufficient time for the generation of considerable diversity in bacteriophage–bacterium–animal interactions. This is especially so given that the sum of phage hosts in the 6.5 billion humans that currently inhabit the Earth, alone, is between 10^{23} to 10^{24} bacteria (Ley *et al.*, 2006).

Most viral strains that have been studied, including bacteriophages, have narrow host ranges. However, there are some viruses that are capable of infecting widely divergent hosts (Rice, 1996). The single-stranded RNA flaviviruses provide a classic example of a virus genus whose members posses wide host ranges. There are over 68 members of this genus, which include the viral etiologies of dengue, Japanese encephalitis, and yellow fever. Most of these flaviviruses are transmitted through insect vectors and many of them individually are able to replicate in avian, mammalian, and insect hosts (Rice, 1996). Given this divergent host-range potential for viruses, and the large number of bacteriophages on the Earth (10^{30} or more; Suttle, 2005), there is the possibility that there are some phage strains with a host range that spans both pro- and eukaryotes. Although the capacity of a "bacteriophage" to replicate and interact with bacteria is, by definition, a necessary trait, the actual host-range limits of individual bacteriophage strains can only be determined experimentally. Thus, we cannot definitively rule out, except empirically, an ability of phages to replicate during "infection" of eukaryotic cells, unlikely as that possibility might seem.

Despite this restatement of the typical impossibility of proving negatives, we nevertheless should note Hendrix's suggestion that "it seems very unlikely that a bacterial virus that found itself able to infect eukaryotic cells would function well enough in the new environment to establish itself in competition with viruses already present. (The caveat here, of course, is that even very low-probability events become likely if they are given enough time and opportunity to occur)" (Hendrix, 2002). However, as phages were present at the origin of the eukaryotes, they have had ample opportunity to expand their host ranges to include eukaryotes. A second caveat is that individual viruses are not necessarily competing against each other so much as against host defenses, in this case those of the eukaryote host – which is not to imply that overcoming host defenses is necessarily any easier than out-competing fellow pathogens.

With regard to the genes of bacterial viruses crossing domain boundaries, at least on some occasions during their presumably multiple-billion-year history, it is of interest to note that a nucleus-encoded mitochondrial RNA polymerase of yeast, and a number of other eukaryotes, has been found to be homologous to bacteriophage T3 and T7 RNA polymerase (Cermakian *et al.*, 1996). That is, exchange of genes between phages and eukaryotes remains a distinct possibility, just as there appear to have been mechanisms for movement of bacterial genes into phagotrophic eukaryotes perhaps for much of eukaryote existence (Andersson, 2005). Furthermore, it has been clearly established that prokaryotic genes in a segment of the Ti plasmid of *Agrobacterium tumerfaciens* can function in a eukaryote. They have been shown to have the capacity to transform plant cells of numerous plant species, resulting in the induction of crown gall disease, the cause of the commonly observed crown gall tumors in plants. In addition, in laboratory experiments, *Agrobacterium*-mediated transformation can be extended to non-plant eukaryotes including yeast, filamentous fungi, cultivated mushrooms, and human cells in culture (Lacroix *et al.*, 2006).

In addition, it has been known for decades that bacteriophages can interact with animal hosts phenotypically without replication in or direct genetic interaction with mammalian cells. For example, certain bacterial viruses, such as the β phage of *Corynebacterium diphtheria*, code for toxins that are responsible for pathological effects in human diseases (see Chapter 14 for additional examples). In this case the phage toxin gene codes for an enzyme, an ADP-ribosylase, which shuts down mammalian protein synthesis by ADP-ribosylating an amino acid in mammalian elongation factor 2. It is of interest that this enzyme has no known substrate in bacterial metabolic pathways and therefore serves as a fairly strong example of an evolved interaction between phage product and eukaryote.

There are also innate and adaptive immune responses that occur when animals and phages interact. The innate immune response relies on pattern recognition receptors that include membrane-encased foreign organisms, components of bacterial and fungal cell walls and numerous other patterns (Meylan *et al.*, 2006). The innate immune response can also detect specific genomic sequence patterns, such as bacterial DNA sequences with multiple CpG nucleotides (Goodnow, 2006). Mammalian DNA contains fewer such sequences and those present are often modified, most commonly by methylation (Kariko *et al.*, 2005; Akira *et al.*, 2006). These recognition factors could provide for the initial response of animals to phages by triggering production of adhesion molecules, antiviral cytokines, inflammatory mediators, and other host defense factors (Meylan *et al.*, 2006).

13.3 THE FATE OF PHAGES ADMINISTERED TO ANIMALS

In the 1920s, the early practitioners of phage therapy (Section 13.4.1) recognized the need to learn the fate of phages injected into animals. However, as these early researchers generally only employed qualitative methods, they could only report whether lysis had occurred following incubation in liquid culture of ground-up tissue or drops of blood with the host bacteria. Despite such limitations, these efforts led to the observation that phages injected into the circulatory system of rabbits could still be found in the spleen long after all phage traces had vanished from other organs and the blood (Appelmans, 1921). This finding was corroborated twelve years later by another experiment using rabbits. Three days after intravenous phage injection, animals were sacrificed and the liver, spleen, and blood were examined for phages. On the third day phages could not be detected in the blood or the liver, but were found in the spleen (Evans, 1933).

One of the first quantitative studies of the fate of phages in animals was performed by Nungester and Watrous (1934). They reported that following the intravenous inoculation of 10^9 PFU (plaque-forming units) of a *Staphylococcus* phage into albino rats, the titer in the blood dropped to 10^5 PFU in five minutes and to 4×10^1 PFU in two hours. It should be noted that such a large drop in titer could not be due to dilution effects, as the average blood volume of a rat is 16 mL (Diehl *et al.*, 2001). The authors suggested this rapid elimination of phages from the circulatory system was due to functions associated with the organs of the reticulo-endothelial system, primarily the liver and spleen. Inchley (1969) confirmed their interpretation three decades later. He found that the liver phagocytosed more than 99% of the phages within 30 minutes after inoculation and that it removed 12 times as much phage as the spleen, as measured by the uptake of ^{51}Cr-labelled T4 phage. Inchley also demonstrated that the liver inactivated phages at a higher rate than the spleen, as measured by phage PFUs that could be detected in these organs and the rate of loss of ^{131}I-labeled T4 phage in these organs. (The ^{51}Cr-labelled T4 phage was initially used to measure the tissue distribution of the phage while ^{131}I-labeled T4 phage was used to study the fate of the phage because, as Inchely noted, "the elimination of ^{131}I-label from the tissues is held to be an accurate indication of the breakdown of the protein to which it is attached.")

13.3.1 Phage interactions with the adaptive immune system

When a specific phage strain is first inoculated into an animal or human, if there are no detectable pre-existing antibodies for that specific phage strain, then it is defined as a neoantigen. Whether a phage strain serves as a neoantigen within a given individual depends on the phage-exposure and physiological history of that individual. As an example, in one clinical study, Kucharewica-Krukowska and Slopek (1987) reported that 11% of healthy controls and 23% of their patients had antibodies against a *Staphylococcus* phage strain before administration of phage therapy. Such phage immunogenicity is important for two major reasons: first, because of potential adverse reactions, such as anaphylactic shock, and, second, because the presence of adaptive antibodies can have significant effects on the pharmacokinetics of phage experiments and/or therapy.

The ability of phage strains to stimulate the immune system has been used by Hans Ochs and his colleagues for the past three decades to study normal individuals and patients with immune deficiencies (Ochs *et al.*, 1971, 1993). They demonstrated that normal individuals injected with the highly immunogenic phage ϕX174 display a primary IgM response that peaks 2 weeks after the initial injection (or immunization) and that IgM and IgG antibody titers can be further increased with subsequent phage injections. They also found that patients with severe combined immunodeficiency disease (SCID), characterized by absence of both B and T cell functions, display a prolonged period for the clearance of phage ϕX174, with detectable phage present in the circulation for up to 4–6 weeks after the initial injection. In addition, SCID patients do not develop detectable antibody responses to phages, even after repeated injections (Ochs *et al.*, 1971). Ochs and his colleagues found no recognized toxic effects in their experiments with ϕX174 phage, despite its potent antigenic properties in normal individuals (Ching *et al.*, 1966; Ochs *et al.*, 1971).

13.3.2 Phage interactions with the innate immune system

Despite the lack of detectable antibodies to phages that present as neoantigens, such phages are still rapidly eliminated from the circulatory system. This was demonstrated in experiments using germ-free mice (Geier *et al.*, 1973). In these experiments, performed within a gnotobiotic environment isolator, 2×10^{12} PFU of phages were administered by intravenous, intramuscular, or intraperitoneal injection, or by oral administration. The gross anatomic distribution of phages was tissue-specific with a rapid drop in circulating titer, by 3–4 orders of magnitude, within one hour of administration for phages given by injection. However, in that first-hour time interval the titer of phages given orally decreased by 9–10 orders of magnitude. In animals administered phage by the oral route, detectable phages one hour after administration were found only in the spleen, liver, and kidney, suggesting that oral phage administration is not an effective method for delivery of phages to systemic sites, particularly if significant titers are needed, as in certain antibacterial phage-therapy applications (Section 13.4.1). In addition to the tissue-specific decrease in phage titers, the germ-free animal study also demonstrated that the ability of animals to trap and destroy phages is not dependent on antibodies, as prior to phage administration the germ-free animals displayed no detectable antibodies to the phage λ used in the experiments (Geier *et al.*, 1973).

Using serial passage techniques, λ phage mutants were selected that possessed a 13 000- to 16 000-fold greater capacity to remain in the mouse circulatory system 24 hours after intraperitoneal injection than that achieved by the parental wild-type λ phage. The phages resulting from selection for prolonged survival in the mouse circulatory system displayed a substitution of a single amino acid in a major head protein. This observation suggested that rate of elimination of λ phage from the mouse circulation is highly dependent on the composition of its surface (head or capsid) proteins (Merril *et al.*, 1996). It was later proven, through the use of isogenic strains of λ phage, that a single mutational change that results in substitution of a lysine for the normal glutamatic acid residue at amino acid 158 of the lambda capsid E protein (which is present in about 450 copies per virion; Daniels *et al.* 1983) is sufficient to increase the circulatory survival of the phage in mice and that no other mutations are needed for this phenotype (Vitiello *et al.*, 2005). These results raise the question of whether this single mutation, which results in a three-orders-of-magnitude increase in the capacity of the phage to remain in the circulatory system, represents the phage as it exists in a "native" environment rather than the "wild type" phage, which has been cultured for decades in the laboratory. Whether this speculation is correct or not, the findings demonstrated that phage survival in the vascular circulation is under phage-genetic control.

Recently, it has been reported that the rapid elimination of T7 phage from the circulatory system of the mouse is diminished in immunocompromized B-cell-deficient mice (Srivastava *et al.*, 2004). As B lymphocytes can secrete innate immune-response "natural" antibodies, in addition to adaptive antibodies, it may be useful to study the "wild type" and long circulating isogenic λ phage strains developed by Vitiello *et al.* (2005) in the parental and immunocompromized B-cell-deficient mice developed by Srivastava *et al.* (2004). Apects of phage inactivation and elimination from the circulatory system may also be mediated by additional soluble factors, such as complement in the blood. In a study examining T7 phage survival in the rat circulatory system it was reported (Sokoloff *et al.*, 2000) that phages with peptides containing carboxy-terminal lysine or arginine residues displayed on the T7 phage coat protein, 10B, had longer circulating half-lives. In addition, this enhanced phage circulatory-system survival was correlated with the ability of T7 phages displaying these carboxy-terminal lysine or orginine residues to be protected from complement-mediated inactivation by binding to C-reactive protein, which is normally elevated in rats and mice (Sokoloff *et al.*, 2000).

13.3.3 Fate of orally administered phages

It should be noted that some investigators who have administered phages orally report the subsequent detection of phages in the circulatory system (Dabrowska *et al.*, 2005a; Górski *et al.*, 2006). While these findings are compatible with the finding of a trace amount of phage in the spleen, liver, and kidney in the germ-free animal experiments, it may be difficult, as previously noted (Section 13.3.2), to achieve the titers needed for some acute infection therapies by the oral route. This concern with limitations of oral phage administration was reinforced by the observations that when healthy adult volunteers were administered T4 phage in their drinking water, neither T4 phages nor T4-specific antibodies were found in the serum of the volunteers by the end of the study (Bruttin and Brüssow, 2005).

13.3.4 Phages in the nasopharynx and central nervous system

Following the observation by Draghia *et al.* (1995) that a replication-defective animal-virus (adenoviral) vector could be used to deliver the *Escherichia coli lacZ* gene into the central nervous system of rats by nasal administration, this method was employed in 2004 by Frenkel *et al.* with a filamentous phage displaying an anti-β amyloid antibody fragment, which was then visualized using thioflavin-S and fluorescent-labeled antiphage antibodies. This phage system was shown to be effective for monitoring amyloid plaque formation in Alzheimer's transgenic mice. Viruses are apparently transported into the central nervous system via olfactory neurons or by adjacent neuronal tissue. The olfactory receptor neurons, in particular, are bipolar cells with axons that normally traverse the cribriform plate of the skull, the bone which separates the brain from the nasal cavity. This nasopharynx pathway thus provides phages with access to the central nervous system (specifically the brain). As these researchers reported, genetically engineered filamentous bacteriophages delivered nasally offer the capacity to serve as efficient, nontoxic, viral delivery vectors for central nervous system studies.

In addition to using phages to visualize amyloid plaques in the brain, these researchers also employed a filamentous phage which displayed an epitope of the Aβ amyloid protein associated with Alzheimer's disease (AD) as a vaccine. Immunization of AD transgenic mice with the phage, again delivered via the nasopharynx, resulted in a reduction of Aβ amyloid plaques in the brain of these animals (Frenkel *et al.*, 2003). In independent studies, nasal administration of a filamentous phage displaying a cocaine-binding protein proved to be effective, in a locomotor-activity rodent model, in blocking the psychoactive effects of cocaine (Carrera *et al.*, 2004; Dickerson *et al.*, 2005). These experiments clearly demonstrate the ability of phages to gain entrance to the brain from the nasal cavity.

13.4 PHAGE APPLICATIONS IN ANIMALS

Despite the discovery of phages in the early part of the twentieth century, applications of this discovery are still in their early phase. The development of antibacterial therapies was inhibited primarily because of inadequate knowledge of phage-strain host range, effective phage purification methods, and phage physiology and pharmacokinetics. Carefully controlled animal experiments have clearly demonstrated the potential for phage antibacterial therapies. These findings, coupled with increases in antibiotic resistance of infectious organisms, may result in a resurgence of efforts to use phage therapy (Section 13.4.1; see also Chapters 12 and 17).

While direct applications of phage antibacterial therapies may have been delayed, the use of phages as a foundation for molecular biology has provided both the knowledge and the methods for using these viruses as gene-delivery and vaccine vectors. Their use in this manner has been facilitated by their general lack of tissue tropism, apart from their recognition by host defense systems such as the innate immune system. By manipulating the surface proteins of phage strains it thus becomes possible to target phages for specific tissues. In addition, by using the tools of molecular biology that were developed in a large part through the study of phages it is now possible to incorporate eukaryotic promoters into the phage genome so that selected genes can be efficiently expressed in eukaryotic cells that take up either the whole phage or the capsid-free phage genomes. The use of such engineered phages will provide not only useful vectors for gene therapy but tools to explore potential phage interactions with eukaryotic systems.

13.4.1 Phage antibacterial therapies

The ability of phages to replicate exponentially and lyse bacteria suggests that they should play a vital role in our treatment of infectious diseases. In spite of initial enthusiasm in the 1920s and 1930s, problems reported in early clinical applications resulted in a negative shift of opinion concerning the therapeutic potential of phages. In retrospect, the major factors were an initial lack of understanding of the narrow host range of many or most phages, lack of knowledge of the pharmacokinetics of phages, and an inability to purify phage preparations from bacterial products and debris (see Chapter 17 for additional discussion of the complications affecting the phage therapy of animals). These contaminating materials often include bacterial exo- and endotoxins, along with bacterial cellular components that tend to inactivate phage preparations when they are stored without further purification.

By the late 1940s and early 1950s the introduction of antibiotics effective against a broad range of bacterial strains was the final element responsible for this rejection of the use of phages as antibacterial therapeutic agents, particularly in countries that require certification based on the results of clinical trials designed to demonstrate efficacy. The broad host range of clinically employed antibiotics has encouraged physicians to treat infections even before they determine the causative bacterial strain. The narrow host range of phages would make such a practice questionable at best for phage antibacterial therapy (Merril *et al.*, 2003, 2006). It would, however, be possible to address this latter problem by using a library of clinically adapted antibacterial phages carrying marker genes to facilitate the rapid identification of infectious organisms and to simultaneously select the proper therapeutic phage.

Clinical applications of phage therapy have been practiced since the initial discoveries of these viruses at the beginning of the twentieth century (Chapter 1). However, these clinical applications have never faced the scrutiny now mandatory in countries that require certification of pharmacological agents. In addition, only a limited number of carefully controlled phage-therapy experiments utilizing animal models of acute infections were initiated in the 1940s. Despite initial limitations of supporting data concerning the efficacy of phage therapy, experiments have continued, and with a recent increase in frequency. The results obtained from these recent well-controlled animal studies suggest that phage therapy deserves careful review, since phage therapy may provide an ideal therapeutic agent for the treatment of emerging antibiotic-resistant bacterial strains, and for the treatment of epidemics such as cholera in refugee camps (Lederberg, 1996).

As noted, there are concerns over the views of regulatory agencies. However, these agencies are aware of the problems associated with antibiotic-resistant organisms and the need for new approaches to dealing with this problem. In addition, they realize that we are normally in contact with phages throughout our lives, with the complex interactions of bacteria and these viruses in our colon, upper respiratory system, and on our skin. Indeed, most of the phages that have been extensively studied were isolated from human waste, and therefore presumably originated in bacteria associated with our bodies. While some phages carry toxin (Chapter 14) and possibly antibiotic-resistance genes, such phages or their genes can be eliminated prior to their use in therapeutic applications. As regards the safety of invasive phage exposure, Hans Ochs and his colleagues have been using phages as a means to determine the extent of immune deficiencies and as a probe of the immune system in human studies for the past three decades (Section 13.3.1). In addition, the fetal calf serum used to produce vaccines, and the resulting vaccines, were found to be contaminated with phages (Merril *et al.*, 1972). Consistent with the relative safety of phage administration to humans, the continued use of such phage-contaminated vaccines was allowed by an executive order issued in the 1970s (Merril, 1975). Finally, see Chapter 12 for discussion of a recent US Food and Drug Administration (FDA) ruling allowing the addition of certain bacteriophages (of *Listeria*) to food.

Even given extensive cooperation by regulatory bodies, the further development of phage therapy as a treatment for bacterial infections still depends on a commitment to fulfill the scientific requirements required for the certification of pharmaceutical agents. The years of experience gained from the use of phages to discover the basic tenets of molecular biology, however, should prove to be an asset toward meeting these requirements. The encouraging results reported in recent carefully controlled animal experiments, demonstrating the capacity of phage therapy to rescue animals with acute life-threatening infections, suggest that such an effort could result in the development of phages into clinically useful antibacterial therapeutic agents.

13.4.2 Phage-based vaccines

While the study of the interaction of phages and animals is still in its infancy, there are a number of groups investigating the capacity of a few of the well-studied phage strains to deliver antigenic gene products. These applications were initiated by Smith's demonstration, in 1985, that it was possible to produce phages that display foreign proteins fused to their normal coat proteins: phage display. Following the introduction of this method, its power was greatly expanded when Smith and others incorporated the use of affinity selection to isolate phages that displayed specific peptides from random peptide libraries (Scott and Smith, 1990; Devlin *et al.*, 1990; Cwirla *et al.*, 1990). The use of this technology for the development of vaccines and diagnostics has become an industry (Irving *et al.*, 2001; Curiel *et al.* 2004; Wang and Yu, 2004). In addition to using random selection and affinity screening, it is also possible to use specific epitopes that have been chosen on the basis of biological experiments, as illustrated by the use of a recombinant phage, fd, displaying a pIII capsid protein fused to an epitope of the glycoprotein G of the human respiratory syncytial virus. A phage display vaccine using this system provided complete immunological protection to mice nasally challenged with live respiratory syncytial virus (Bastien *et al.*, 1997).

A similar approach was used to engineer a phage carrying a gene product protective against *Yersinia pestis* infections (the V antigen). This recombinant phage was compared with a plasmid carrying this gene. The phage IgG2 a responses were shown to be significantly higher following intramuscular phage vaccinations with the phage vaccine than with the plasmid DNA vaccine (March *et al.*, 2004). In addition, as Clark and March (2006) note, phage vaccine vectors might be made more efficient by decorating their coat with phage fusion proteins displaying immunogenic antigens for professional immune cells, such as dendritic cells. As an example, this could be accomplished using a peptide that has a binding epitope for the CD40 receptor, which has numerous functions in the activation of antigen-presenting cells (Richards *et al.*, 2003). This requirement for peptide display by phage vaccines can be enhanced by the use of a recently developed dual expression system that permits the generation of phages with mosaic heads that can accommodate recalcitrant recombinant fusion proteins. This system uses bacteria carrying a prophage that is deficient in a major head protein, along with plasmids carrying foreign head protein genes for both wild-type and recombinant phage (Zanghi *et al.*, 2005).

13.4.3 Phage-based gene delivery systems

Phages can also be used as gene delivery vectors to deliver DNA vaccines, as has been demonstrated in studies using λ phage to immunize against hepatitis B (March *et al.*, 2004; Clark and March, 2004). In this application the researchers used a λ phage with a "vaccine expression cassette" incorporating a eukaryotic promoter (the CMV promoter) and either genes for enhanced green fluorescent protein or the hepatitis B surface antigen. Mice injected with a phage construct containing the gene for enhanced green fluorescent protein developed anti-EGFP antibodies 28 days after intramuscular vaccination, while both mice and rabbits vaccinated with lambda carrying the gene for the hepatitis B surface antigen developed anti-hepatitis B surface antigen responses. Of interest, while the animals exhibited an anti-λ phage antibody response after the first inoculation, this anti-phage antibody did not interfere with subsequent responses to genes expressed by the "vaccine expression cassette."

In one study employing phages with marker genes associated with mammalian promoters, the researchers reported marker gene expression in 1–2% of transfected cells. However, after optimizing the display to create multivalent phagemid-based vectors combined with genotoxic treatments, they achieved transduction efficiencies of up to 45% in certain cell lines (Larocca *et al.*, 2002, 2005). The coupling of phage display technologies, for targeting specific cell types or tissues, with gene delivery cassettes containing eukaryotic promoters offers possibilities for the development of effective new gene-therapy methods.

13.4.4 Phage interactions with the vascular system

In addition to phage capabilities derived from natural evolutionary processes, additional phage capabilities have resulted from phage-display molecular manipulations. Technology, using phage display libraries, has been developed to determine whether there are regional vascular endothelium-specific binding epitopes in particular anatomical regions (Trepel *et al.*, 2002). This approach has been extended to humans in an experiment with a terminally ill patient in which a phage library was injected just prior to the termination of life support. Tissue from the patient was then assayed for phage strains from the library that were bound at anatomically specific regions following death. This experiment provided data for an initial molecular map of specific binding epitopes for regions of the human vasculature (Arap *et al.*, 2002a). This approach has also been used to identify specific targets on angiogenic tumor endothelium associated with cancer (Hajitou *et al.*, 2006a).

One example of the application of this technology concerns a study in which a phage was selected from a phage-display library that recognizes the vasculature in the prostate (the selected phage displayed an order-of-magnitude greater affinity for prostate than for other tissue). When a proapoptotic peptide that disrupts mitochondrial membranes was added to the display on this phage it was found to result in the tissue destruction of the prostate (Arap *et al.*, 2002b). A similar phage display system has been developed for a phage that displays a peptide that targets white-fat vasculature and that also displays a proapoptotic peptide. This phage system results in damage to the vasculature of white-fat tissue and the rapid reversal of obesity in genetically obese mice (Kolonin *et al.*, 2004). The capacity of phage display to bind to select targets has also been combined with the functional genomes of eukaryotic viruses, such as the adeno-associated viruses, to produce chimeric entities for tumor targeting and imaging (Hajitou *et al.*, 2006b).

13.5 PHAGE INTERACTIONS WITH MAMMALIAN CELLS

Given the large number of phages in intimate contact with humans, it is surprising that efforts to study their possible interactions with mammalian cells are so limited. These limited efforts may have been due, at least in part, to the narrow host range found in most prokaryotic and eukaryotic viral strains. However, we now know that there are exceptions, that in fact some prokaryotic genes preferentially function in eukaryotic systems – as in the case of *Agrobacterium tumerfaciens*, which contains a plasmid that can tranform plant and animal cells (Lacroix *et al.*, 2006). Phages carrying genes for specific enzymes and structural proteins have now become standard tools, with the advent of genetically engineered phages carrying eukaryotic promoters for reporter genes. There also have been a number of studies indicating the uptake of phage DNA in mammalian cells. Clearly phages can both gain access to eukaryote cells and express appropriately controlled genes there.

13.5.1 Phage metabolic impact on mammalian cells

Traditionally virologists studied viral host range and effects in cell cultures by observing for one or two viral-associated effects: cell killing or a gross change in cellular behavior (so-called cytopathic effects such as loss of cellular contact inhibition). While no such changes have been reported for naturally occurring phage strains, increasingly sensitive assays for alterations in metabolic pathways have permitted researchers to look for more subtle effects. One of the first adaptations of such a system was to screen for phage effects in mammalian cells employing transducing phages carrying genes for specific enzymes. Such phages were administered to human cells known be deficient in those particular enzyme activities. These initial efforts resulted in the induction of gal-transferase and β-galactosidase in human cells that were deficient respectively in these enzymes (Merril *et al.*, 1971; Horst *et al.*, 1975).

While these early studies reported trace amounts of enzymes, by employing eukaryotic promoters the amount of enzyme activity achieved by phage delivery vectors has now made routine the use of phages as a vector in mammalian cell and animal studies. Phage gene delivery and expression in non-engineered/wild-type phages may be a relatively rare event in eukaryotic organisms. However, current genetic-engineering efforts are making phage vectors increasingly efficient. For example, phages have been engineered to contain eukaryotic vital promoters in their genomes and to display epitopes that bind to specific receptors on mammalian cells. Together these enhance the capacity of phages to serve as vectors for targeted gene delivery or vaccines in mammalian systems (Section 13.4.2; Larocca *et al.*, 2002, 2005).

Phages have also been reported to interfere with mammalian viral infections and tumors (Bloch, 1940; Miedzbrodzki *et al.*, 2005; Dabrowska *et al.*, 2005b). However, such experiments involve a number of variables including non-specific induction of an innate immune response. For example, the immune-response effects of the unmodified CpG sequence motif carried by bacterial viruses, or any of the other myriad components carried by the phages or as residuals from the bacterial hosts used to produce the phages (Section 13.2), could interfere with mammalian viral infections or tumor growth. To determine whether the phage is directly responsible for viral interference or tumor growth it will be necessary to determine the molecular mechanisms of such effects. The determination of such molecular mechanisms could be facilitated through the use of mammalian genomic microarrays to determine which if any mammalian cellular metabolic pathways are perturbed by phages and the use of phage genomic microarrays to determine which if any phage genes are being transcribed in the mammalian cells. The use of isogenic phage strains, in which specific phage genes are mutated, could be used to determine which phage genes are responsible for direct mammalian cellular effects.

It should be noted that some phage strains have been found to carry toxin genes associated with carcinogenesis. So far, the toxins produced are made in the classic manner, by phage-infected bacteria. For example, a phage-encoded toxin produced by some strains of *Pasteurella mutocida* (a bacterium identified in chronic respiratory infections in humans) can enter mammalian cells and stimulate signaling pathways associated with the regulation of cell growth and carcinogenesis (Lax, 2005). Phage interactions with more complex mammalian systems such as viral interference and tumor progression have also been reported, but some of the results may be due to nonspecific responses. With recently developed genomic microarrays for both mammalian cells and bacterial viruses, it should now be possible to monitor for specific genetic effects in such cases (DeFilippis *et al.*, 2003; Kash *et al.*, 2006).

13.5.2 Phage genomes in mammalian cells

Early evidence that phage genomes gain direct entrance to mammalian cells was reported when phage genomic fragments were found in mammalian cells following oral exposure to phage DNA (Doerfler *et al.*, 1995; Schubbert *et al.*, 1997, 1998). This DNA, from M13 and λ phages, was detected in the cells of Peyer's patches of the gastrointestinal tract, peripheral white bloods, and the cells of the liver and spleen, using PCR. Phage DNA could be detected for up to 24 hours in the spleen and liver following a single oral feeding of phage M13 DNA.

When phage M13 DNA was fed daily for one week, Doerfler and his colleagues were able to isolate, by re-cloning DNA from the mouse spleen, clones containing M13 DNA. One of these clones contained a 1299 bp fragment of M13 DNA covalently linked to an 80 bp DNA segment with 70% homology to the mouse IgE receptor gene (Schubbert *et al.*, 1997). In addition, when pregnant mice were regularly fed phage M13 DNA, evidence of M13 DNA could be detected in the fetuses with *in situ* hybridization methods, and in some rare fetal cells this M13 DNA appeared to be associated with the chromosomes (Schubbert *et al.*, 1998). With the advent of microarray technology for gene transcription and detection it should be possible to investigate the capacities of phages or phage genes to "function" within mammalian cells in a manner that was not previously thought possible.

13.6 CONCLUSION

Our knowledge of phage interactions with animals is still in its infancy. The extent of the interactions between phage strains and animal cells is limited to those few phage strains that have been carefully studied for their molecular biology and physiology in bacterial hosts. However, it is clear that within just a little over two decades, technologies have been developed that have permitted the engineering of phages that can bind to specific tissues and cell receptors and phages that can express specific genes in mammalian cells. This success brings into question whether similar types of viruses could have evolved naturally over the 600 million years since animals first came into existence in the biosphere. Even now, with a biosphere bacterial population of over 10^{30} cells (Whitman *et al.*, 1998) and with between 10^{23} and 10^{24} of these microbial cells in close proximity with human cells (Savage, 1977), there are ample opportunities for interactions between the bacterial viruses and human cells, particularly given the relatively rapid replication times of the phages and their bacterial hosts in the large intestine.

Our knowledge of the capabilities of naturally occurring phages that can interact with animals is expanding, as exemplified by the discovery of phages that encode genes for toxins that only affect animals and not their bacterial hosts (for instance, the phage-encoded diphtheria toxin which inactivates protein synthesis elongation factor 2 in mammalian cells; Chapter 14). More recently it has been recognized that phages and other viruses can gain entrance to the central nervous system from the nasopharnyx. Given that the nasopharnyx is the normal site of *C. diphtheria* colonization, it may be of interest that there is a decrease in activity of elongation factor 2 in regions of the brain affected in Alzheimer's disease (Johnson *et al.*, 1992). Could chronic leakage of a phage-encoded toxin affect protein synthesis in the brain, resulting in cumulative damage that only becomes apparent in the elderly? Whether this scenario is correct or just fiction, it is now clear – with newly gained technologies and genetic engineering – that even if bacterial viruses have not evolved naturally with extended host range and effects, they exist now in the laboratory.

The ability to engineer phages and monitor individual gene responses in mammalian systems will play an important part in furthering our knowledge of the immune system and the way specific immune cells react to specific extrinsic genomic stimuli. Such knowledge will permit the development of more effective vaccines and methods for gene therapy. These efforts will extend our understanding, but they will also lead to a greatly expanded role for phages in human health and well-being, with impact potentially far beyond their past successes as tools in the study of molecular genetics.

REFERENCES

Akira, S., S. Uematsu, and O. Takeuchi. 2006. Pathogen recognition and innate immunity. *Cell* **124**: 783–801.

Andersson, J. O. 2005. Lateral gene transfer in eukaryotes. *Cell. Mol. Life Sci.* **62**: 1182–97.

Appelmans, R. 1921. Le bacteriophage dans l'organisme. *Comp. Rend. Soc. de biol. (Paris)* **85**: 722–4.

Arap, W., M. G. Kolonin, M. Trepel, *et al.* 2002a. Steps toward mapping the human vasculature by phage display. *Nat. Med.* **8**: 121–7.

Arap, W., W. Haedicke, M. Bernasconi, *et al.* 2002b. Targeting the prostate for destruction through a vascular address. *Proc. Natl. Acad. Sci. U.S.A.* **99**: 1527–31.

Bäckhed, F., R. F. Ley, J. L. Sonnenburg, D. A. Peterson, and J. I. Gordon. 2005. Host–bacterial mutualism in the human intestine. *Science* **307**: 1915–20.

Bastien, N., M. Trudel, and C. Simard. 1997. Protective immune responses induced by the immunization of mice with a recombinant bacteriophage displaying an epitope of the human respiratory syncytial virus. *Virology* **234**: 118–22.

Bloch, H. 1940. Experimental investigation on the relationships between bacteriophages and malignant tumors. *Arch. Virol.* **1**: 481–96.

Bruttin, A., and H. Brüssow. 2005. Human volunteers receiving Escherichia coli phage T4 orally: A safety test of phage therapy. *Antimicrob. Agents Chemother.* **49**: 2874–8.

Carrera, M. R. A., G. F. Kaufmann, J. M. Mee, M. M. Meijler, G. F. Koob, and K. D. Janda. 2004. Treating cocaine addiction with viruses. *Proc. Natl. Acad. Sci. U.S.A.* **101**: 10416–21.

Cermakian, N., T. M. Ikeda, R. Cedergren, and M. W. Gray. 1996. Sequences homologous to yeast mitochondrial and bacteriophage T3and T7RNA polymerases are widespread throughout the eukaryotic lineage *Nucl. Acids Res.* **24**: 648–54.

Ching, Y.-C., S. D. Davis, and R. J. Wedgwood. 1966. Antibody studies in hypogammaglobulinemia. *J. Clin. Invest.* **45**: 1593–600.

Clark, J. R., and J. B. March. 2004. Bacteriophage-mediated nucleic acid immunization. *FEMS Immunol. Med. Microbiol.* **40**: 21–6.

Clark, J. R., and J. B. March. 2006. Bacteriophages and biotechnology: vaccines, gene therapy and antibacterials. *Trends Biotechnol.* **24**: 212–18.

Cockell, C. S. 2003. *Impossible Extinction: Natural Catastrophes and the Supremacy of the Microbial World.* Cambridge: Cambridge University Press.

Curiel, T. J., C. Morris, M. Brumlik, *et al.* 2004. Peptides identified through phage display direct immunogenic antigen to dendritic cells.*J. Immunol.* **172**: 7425–31.

Cwirla, S. E., E. A Peters, R. W. Barrett, and W. J. Dower. 1990. Peptides on phage: A vast library of peptides for identifying ligands. *Proc. Natl. Acad. Sci. U.S.A.* **87**: 6378–82.

Dabrowska, K., K. Switaa-Jelen, A. Opolski, B. Weber-Dabrowska, and A. Gorski. 2005a. A review: bacteriophage penetration in vertebrates. *J. Appl. Microbiol.* **98**: 7–13.

Dabrowska, K., A. Opolski, J. Wietrzyk, *et al.* 2005b. Activity of bacteriophages in murine tumor models depends on the route of phage administration. *Oncol. Res.* **15**: 183–7.

Daniels, D. L., J. L. Schroeder, W. Szybalski, F. Sanger, and F. R. Blattner. 1983. Appendix: a molecular map of coliphage lambda. In R. W. Hendrix, J. W. Roberts, F. W. Stahl, and R. A. Weisberg (eds.), *Lambda II.* Cold Spring Harbor, NY: Cold Spring Harbor Laboratory Press, pp. 519–676.

DeFilippis, V., C. Raggo, A. Moses, and K. Früh. 2003. Functional genomics in virology and antiviral drug discovery. *Trends Biotechnol.* **21**: 425–57.

Devlin, J. J., L. C. Panganiban, and P. E. Devlin. 1990. Random peptide libraries: A source of specific protein binding molecules. *Science* **249**: 404–6.

Dickerson, T. J., G. F. Kaufmann, and K. D. Janda. 2005. Bacteriophage-mediated protein delivery into the central nervous system and its application in immunopharmacotherapy. *Expert Opin. Biol. Ther.* **5**: 773–81.

Diehl, K.-H., R. Hull, D. Morton, *et al.* 2001.A good practice guide to the administration of substances and removal of blood, including routes and volumes. *J. Appl. Toxicol.* **21**: 15–23.

Doerfler W., G. Orend, R. Schubbert, *et al.* 1995. On the insertion of foreign DNA into mammalian genomes: mechanism and consequences. *Gene* **157**: 241–5.

Draghia, R., C. Caillaud, R. Manicom, A. Pavirani, A. Kahn, and L. Poenaru. 1995. Gene delivery into the central nervous system by nasal instillation in rats. *Gene Ther.* **2**: 418–23.

Evans, A. C. 1933. Inactivation of antistreptococcus bacteriophage by animal fluids. *Public Health Rep.* **48**: 411–26.

Frenkel, D., I. Dewachter, F. Van Leuven, and B. Solomon. 2003. Reduction of β-amyloid plaques in brain of transgenic mouse model of Alzheimer's disease by EFRH-phage immunization. *Vaccine* **21**: 1060–5.

Frenkel, D., M. Dori, and B. Solomon. 2004. Generation of anti-β-amyloid antibodies via phage display technology. *Vaccine* **22**: 2505–8.

Geier, M. R., M. E. Trigg, and C. R. Merril. 1973. The fate of bacteriophage lambda in non-immune germfree mice. *Nature* **246**: 221–3.

Goodnow, C. C. 2006. Discriminating microbe from self suffers a double toll. *Science* **312**: 1606–8.

Górski, A., E. Wazna, B.-W. Dabrowska, K. Dabrowska, K. Świtała-Jeleń, and R. Miedzybrodzki. 2006. Bacteriophage translocation. *FEMS Immunol. Med. Microbiol.* **46**: 313–19.

Hajitou, A., R. Pasqualini, and W. Arap. 2006a. Vascular targeting: recent advances and therapeutic perspectives. *Trends Cardiovasc. Med.* **16**: 80–8.

Hajitou, A., M. Trepel, C. E. Lilley, *et al.* 2006b. A hybrid vector for ligand-directed tumor targeting and molecular imaging. *Cell* **125**: 385–98.

Hendrix, R. W. 2002. Bacteriophages: evolution of the majority. *Theor. Popul. Biol.* **61**: 471–80.

Horst, J., F. Kluge, K. Beyreuther, and W. Gerok. 1975. Gene transfer to human cells: transducing phage Lambda plac gene expression in GM1-gangliosidosis fibroblasts. *Proc. Natl. Acad. Sci. U.S.A.* **72**: 3531–5.

Inchley, C. J. 1969. The activity of mouse kupffer cells following intravenous injection of t4 bacteriophage. *Clin. Exp. Immunol.* **5**: 173–87.

Irving, M. B., O. Pan, and J. K. Scott. 2001. Random-peptide libraries and antigen-fragment libraries for epitope mapping and the development of vaccines and diagnostics. *Curr. Opin. Chem. Biol.* **5**: 314–24.

Johnson, G., J. Gotlib, V. Haroutunian, *et al.* 1992. Increased phosphorylation of elongation factor 2 in Alzheimers's disease.*Mol. Brain Res.* **15**: 319–26.

Kariko, K., M. Buckstein, H. Ni, and D. Weissman. 2005. Suppression of RNA recognition by Toll-like receptors: The impact of nucleoside modification and the evolutionary origin of RNA. *Immunity* **23**: 165–75.

Kash, J. C., Goodman, A. G., Korth, M. J., and M. G.Katze. 2006. Hijacking of the host-cell response and translational control during influenza virus infection. *Virus Res.* **119**: 111–20.

Kolonin, M. G., P. K. Saha, L. Chan, R. Pasqualini, and W. Arap. 2004. Reversal of obesity by targeted ablation of adipose tissue. *Nat. Med.* **10**: 625–32.

Kucharewica-Krukowska, A., and S. Slopek. 1987. Immunogenic effect of bacteriophage in patients subjected to phage therapy. *Arch. Immunol. Ther. Exp. (Warsz)* **35**: 553–61.

Lacroix, B., T. Tzfira, A. Vainstein, and V. Citovsky. 2006. A case of promiscuity: agrobacterium's endless hunt for new partners. *Trends Genet.* **22**: 29–37.

Larocca, D., M. A. Burg, K. Jensen-Pergakes, E. P. Ravey, A. M. Gonzalez, and A. Baird. 2002. Evolving phage vectors for cell targeted gene delivery. *Curr. Pharm. Biotechnol.* **3**: 45–57.

Larocca, D., M. A. Burg, and A. Baird. 2005. Evolving phage vectors for cell targeted gene delivery: an update. *Med. Chem. Rev. Online* **2**: 111–14.

Lax, A. J. 2005. Bacterial toxins and cancer: a case to answer? *Nat. Rev. Microbiol.* **3**: 343–9.

Lederberg, J. 1996. Smaller fleasad infinitum: therapeutic bacteriophage redux. *Proc. Natl. Acad. Sci. U.S.A.* **93**: 3167–8.

Ley, R. E., D. A. Peterson, and J. I. Gordon. 2006. Ecological and evolutionary forces shaping microbial diversity in the human intestine. *Cell* **124**: 837–48.

March, J. B., J. R. Clark, and C. D. Jepson. 2004. Genetic immunization against hepatitis B using whole bacteriophage lambda particles. *Vaccine* **22**: 1666–71.

Merril, C. R. 1975. Phage in human vaccines. *Science* **188**: 8.

Merril, C. R., M. R. Geier, and J. C. Petricciani. 1971. Bacterial virus gene expression in human cells. *Nature* **233**: 398–400.

Merril, C. R., T. B. Friedman, A. Attallah, M. R. Geier, K. Krell, and R. Yarkin. 1972. Isolation of bacteriophages from commercial sera. *In Vitro* **8**: 91–3.

Merril, C. R., B. Biswas, R. Carlton, *et al.* 1996. Long-circulating bacteriophage as antibacterial agents. *Proc. Natl. Acad. Sci. U.S.A.* **93**: 3188–92.

Merril, C. R., D. Scholl, and S. L. Adhya. 2003. The prospect for bacteriophage therapy in Western medicine. *Nat. Rev. Drug Discov.* **2**: 489–97.

Merril, C. R., D. Scholl and S. Adhya. 2006. Phage therapy. In R. L. Calendar and S. T. Abedon (eds.), *The Bacteriophages*, 2nd edn. Oxford: Oxford University Press, pp. 725–41.

Meylan, E., J. Tschopp, and M. Karin. 2006. Intracellular pattern recognition receptors in the host response. *Nature* **442**: 39–44.

Miedzybrodzki R., W. Fortuna, B. Weber-Dabrowska, and A. Gorski. 2005. Bacterial viruses against viruses pathogenic for man? *Virus Res.* **110**: 1–8.

Nungester, W. J., and R. M. Watrous. 1934. Accumulation of bacteriophage in spleen and liver following its intravenous inoculation. *Proc. Soc.Exper. Biol. Med.* **31**: 901–5.

Ochs H. D., S. D. Davis, and R. J. Wedgwood. 1971. Immunologic responses to bacteriophage ϕX174 in immunodeficiency diseases.*J. Clin. Invest.* **50**: 2559–68.

Ochs H. D., S. Nonoyama, Q. Zhu, M. Farrington, and R. J. Wedgwood. 1993. Regulation of antibody responses: the role of complement and adhesion molecules. *Clin. Immunol. Immunopathol.* **3** (2): S33–40.

Relman, D. A. 2002. New technologies, human–microbe interactions, and the search for previously unrecognized pathogens. *J. Infect. Dis.* **186** (Suppl. 2): S254–8.

Rice, C. M. 1996. Flaviviridae: the viruses and their replication. In B. N. Fields, D. M. Knipe, P. M. Howley, *et al.* (eds.), *Fields Virology*, 3rd edn. Philadelphia, PA: Lippincott-Raven, pp. 931–59.

Richards, J. L., J. R. Abend, M. L. Miller, S. Chakraborty-Sett, S. Dewhurst, and L. E. Whetter. 2003.A peptide containing a novel FPGN CD40-binding sequence enhances adenoviral infection of murine and human dendritic cells. *Eur.J. Biochem.* **270**: 2287–94.

Savage, D. C. 1977. Microbial ecology of the gastrointestinal tract. *Annu. Rev. Microbiol.* **31**: 107–33.

Scott, J. K., and G. P. Smith, 1990. Searching for peptide ligands with an epitope library. *Science* **249**: 386–90.

Schubbert R., D. Renz, B. Schmitz and W. Doerfler. 1997. Foreign (M13) DNA ingested by mice reaches peripheral leukocytes, spleen, and liver via the intestinal wall mucosa and can be covalently linked to mouse DNA. *Proc. Natl. Acad. Sci. U.S.A.* **94**: 961–6.

Schubbert R., U. Hohlweg, D. Renz, and W. Doerfler. 1998. On the fate of orally ingested foreign DNA in mice: chromosomal association and placental transmission to the fetus.*Mol. Gen. Genet.* **259**: 569–76.

Smith, G. P. 1985. Filamentous fusion phage: novel expression vectors that display cloned antigens on the virion surface. *Science* **228**: 1315–17.

Sokoloff, A. V., I. Bock, G. Zhang, M. G. Sebestyen, and J. A. Wolff. 2000. The interactions of peptides with the innate immune system studied with use of T7 phage peptide display. *Mol. Ther.* **2**: 131–9.

Srivastava, A. S., T. Kaido, and E. Carrier. 2004. Immunological factors that affect the in vivo fate of T7 phage in the mouse. *J. Virol. Methods* **115**: 99–104.

Suttle, C. A. 2005. Viruses in the sea. *Nature* **437**: 356–61.

Trepel, M., W. Arap, and R. Pasqualini. 2002. In vivo phage display and vascular heterogeneity: implications for targeted medicine. *Curr. Opin. Chem. Biol.* **6**: 399–404.

Vitiello, C. L., C. R. Merril, and S. Adhya. 2005. An amino acid substitution in a capsid protein enhances phage survival in mouse circulatory system more than a thousand fold. *Virus Res.* **114**: 101–3.

Wang, F., and M. Yu. 2004. Epitope identification and discovery using phage display libraries: applications in vaccine development and diagnostics. *Curr. Drug Targets* **5**: 1–15.

Whitman, W. B., D. C. Coleman, and W. J. Wiebe. 1998. Prokaryotes: the unseen majority. *Proc. Natl. Acad. Sci. U.S.A.* **95**: 6578–83.

Zanghi, C. N., H. A. Lankes, B. Bradel-Tretheway, J. Wegman, and S. Dewhurst. 2005. A simple method for displaying recalcitrant proteins on the surface of bacteriophage lambda. *Nucleic Acids Res.* **33**: 1–7.

CHAPTER 14

Phage ecology of bacterial pathogenesis

Paul Hyman* and Stephen T. Abedon

14.1 INTRODUCTION

Found in *Clostridium botulinum*, *Corynebacterium diphtheriae*, *Escherichia coli*, *Pseudomonas aeruginosa*, *Staphylococcus aureus*, *Streptococcus pyogenes*, *Vibrio cholerae*, etc., prophage-encoded bacterial virulence factors (ϕVFs) provide an additional level of interaction between phages, bacteria, and environments (Langley *et al.*, 2003; Brüssow *et al.*, 2004; ϕVF, which stands for phage-encoded virulence factor, we pronounce "phee-vee-eff."). In some cases multiple prophages are present in the same cell and each may encode a separate virulence factor (VF). For example, some strains of enteropathogenic *E. coli* encode several variants of the VF, Shiga toxin, and in a number of *Salmonella* strains about 5% of the genome consists of prophages, most of which contain VF genes. Despite their unambiguous genomic association with prophages, a question still under debate is whether these VF genes should be considered phage genes or bacterial genes. Supporting the concept that they are simply unusually located bacterial genes, in some bacterial strains the prophage is clearly defective and no progeny phage will ever be produced. On the other hand, there are some VFs that cannot be released from the bacterial cell except following prophage induction, implying a great deal of phage involvement in VF expression. In this chapter we consider the evolutionary ecology and ecological impact – on phages, bacteria, and animals – of phage encoding of bacterial VF genes, especially prophage-encoded exotoxins.

See Chapter 13 for additional consideration of phage interaction with animals, and Chapter 5 for a broader discussion of the changes phages can

* Corresponding author

Bacteriophage Ecology: Population Growth, Evolution, and Impact of Bacterial Viruses, ed. Stephen T. Abedon. Published by Cambridge University Press.

impart to bacteria upon entrance into lysogeny (lysogenic conversion). For additional reviews of the ecology and evolution of VF-encoding bacteriophages, see Krylov (2003), Brüssow *et al.* (2004), Chibani-Chennoufi *et al.* (2004a), Abedon and LeJeune (2005), and Breitbart *et al.* (2005). For recent general reviews of phage and bacterial virulence, see Boyd (2005), Waldor *et al.* (2005), and Wagner and Waldor (2002).

14.2 LYSOGENY, LINKAGE, AND CO-REPLICATION

Pathogenic bacterial lysogens comprise multiple genetic entities including the bacterium and the prophage, both of which can encode VFs as well as other genes. Post-induction, the former prophage will continue to harbor phage genes, but ones which are no longer linked to bacterial genes. In this section we consider ideas on cooperation and genetic recombination (Chapters 3, 6, 7, and 9), to develop a general understanding of how, and when, the action of phage genes – particularly VF genes – might benefit the encoding phage versus the harboring lysogen.

14.2.1 Linkage and linkage disequilibrium

Two pieces of genetic information – such as a ϕVF and another prophage-encoded gene – may become separated from one another over the course of transmission to the next generation. The degree to which this separation will occur depends upon the pieces of information in question, where and how these two pieces of information are encoded, and to what degree attempts at separation have been made. At one extreme are two pieces of information that exist on a single chromosome within a system in which molecular recombination within that chromosome is impossible. These loci are absolutely linked and therefore do not replicate independently. Alternatively, if these two loci are unlinked and freely mix within a population, then they are said to be in linkage equilibrium. Intermediate between these extremes, there may be biases such that two alleles found at two distinct loci are more likely found together than apart. This is linkage disequilibrium. A VF that is usually or always phage-encoded can be said to be encoded by a gene that is experiencing a linkage disequilibrium with the rest of the phage genome. That is, there is a bias toward association of a phage genome with the VF gene.

In addition to reduced opportunity for outcrossing or poor recombinational machinery, another means toward achieving linkage disequilibrium is via selection for or against certain allele combinations (Chapter 7). For example, coevolved genes – those producing products that physically interact – tend to congregate on the phage chromosome, e.g., the tail fiber genes of T4-like phages (genes *34, 35, 36,* and *37*). The explanation for this congregation is that it reduces likelihoods of separation of coevolved genes during outcrossing between diverse phage types (Stahl and Murray, 1966). That is, coevolved alleles that are able to remain clustered are less likely to go extinct (Lawrence, 2003). Similarly, genes encoding related functions, such as pathogenicity, often congregate on bacterial chromosomes (e.g., into pathogenicity islands; Section 14.5.3 and Chapter 11). Those units that are more closely linked tend to be transferred functionally intact with greater likelihood (e.g., via generalized transduction; see also Chapter 11) and as a consequence are more likely to be represented among the bacteria we study today (Lawrence, 1997, 1999). In bacteria, as in phages, linkage additionally allows facile gene coregulation within operons (Stahl and Murray, 1966; Lawrence, 2003). A combination of these forces presumably explains the clustering of at least some ϕVF genes with prophage genes, such as the clustering of the two genes required to produce the multisubunit Shiga toxin and clustering of both with the phage-encoded lysis genes (Section 14.5.2.1).

14.2.2 Prophage–bacterial co-replication

In bacteria, routes toward outcrossing include transformation, transduction, and conjugation (e.g., Redfield, 2001). Because these mechanisms are not nearly as robust as outcrossing in eukaryotes, the genes present within bacterial populations can display significant linkage disequilibrium (Feil and Spratt, 2001). This linkage disequilibrium ties together the fates of bacterial genes within individual clones. As a consequence, the evolution of cheater genes, i.e., those that replicate at the expense of the rest of the bacterial genome, is somewhat constrained (see levels of selection, Chapter 6). This point is a restatement of the ideas of Levin and Bergstrom (2000), who declared that accessory genetic elements (such as plasmids) have to display a "niceness" toward their host bacteria if they are to be retained within populations. That is, a somewhat obligatory co-replication limits the utility of selfish reproduction by the various genetic elements making up a bacterium.

Unlike mobile genetic elements such as transposons or plasmids (Eberhard, 1990; Paulsson, 2002), which in principle can exist as populations within individual bacteria, there actually is little reason for most uninduced prophages to increase their numbers, via duplication, within the bacterial chromosome. Indeed, Levin and Lenski (1985) go so far as to argue that temperate phages are at a competitive *advantage* versus otherwise identical virulent phages only so long as the lysogens they give rise to are not at a competitive disadvantage relative to isogenic non-lysogens. Consequently, once established, a prophage should *not* have a tendency to display an increased cost of retention within the host bacterium (that is, when prophage is good, it is very, very good). Alternatively, prophage induction is so catastrophic to the bacterium, due to the ensuing lytic cycle, as to completely eliminate any common ground in phage versus bacterial interests (but when prophage is bad, it is horrid). We can use considerations of these conflicting prophage tendencies to excogitate on which genetic entities ϕVF expression may benefit.

14.2.3 VF gene–phage co-replication

Different VFs have different properties (Table 14.1). Based upon these properties, as well as degrees of co-replication, we can postulate how different genetic entities – phage, prophage, or bacterium – may benefit from the expression of different ϕVFs:

(i) *VFs that are functional only so long as a bacterium remains intact* – e.g., pili or other attached VFs – should be more beneficial during lysogenic cycles than to the free virions of phages which encode these factors. This is especially so if time from transcription to benefit is long relative to the duration of a lytic infection. It is certainly possible, though, that these attached VFs could modify lysogen behavior in such a way – e.g., in terms of biasing lysogen location within animals or enhancing lysogen physiology – so as to increase the number or fitness of virions produced upon prophage induction.

(ii) *VFs, including many exotoxins, that are secreted from bacteria and can remain functional independent of the existence of the expressing bacterium.* These exotoxins may serve phage interests during lysogenic infection or following prophage induction, plus may facilitate lytic-cycle-mediated phage population growth by benefiting other bacteria that could potentially serve as future phage hosts. Also, by disrupting animal tissues, exotoxins may promote phage dispersion during the search for a new host.

(iii) *Exotoxins that are secreted from lysogens, along with phage virions, without killing the bacterial cell.* These phages, such as CTXΦ (Section 14.5.3), may display interests that are simultaneously similar to those of both phage and bacterium because their lysogens release virions chronically (Chapter 1), i.e., without terminating either bacterial replication or ongoing phage infection. In this case, bacterial interests are similar to

Table 14.1 *Phage-encoded virulence factors (ϕVFs) source, function, expression with regard to solubility, and host-lysis requirement.*

Category[a]	ϕVF name (function)[b]	Source	Exotoxin[c]	Host lysis?[d]	Enviromod?[e]	Reference
i	Bor (outer membrane protein effecting serum resistance)	*Escherichia coli*/lambdoid phage	no	no	no	Barondess and Beckwith (1995)
i	Lom (outer membrane protein involved in adhesion)	*E. coli*/lambdoid phage	no	no	no	Vaca Pacheco *et al.* (1997)
i	macrolide resistance, mefA adhesin, PblA/B platelet adhesin	*Streptococcus pyogenes* phage	no	no	no	Banks *et al.* (2005)
i or iv	Glo (?)	*V. cholerae* phage K139 lysogen	yes	no?	?	Nesper *et al.* (1999)
ii	$EspF_u$, NleA, Cif (functions unknown)	*E. coli* phage	yes	no?	?	Tyler *et al.* (2005)
ii	many exotoxins varying between serovars	*Salmonella enterica* sv. Typhimurium	varies	no	varies	Bossi and Figueroa-Bossi (2005)
ii	streptokinase, enterotoxin A, exfoliative toxin A, Panton-Valentine leucocidin (all cytotoxic)	*Staphylococcus aureus* phage	yes	no	yes	Matthews and Novick (2005)
ii	diphtheria toxin	*Corynebacterium diphtheriae* phage	yes	no	yes	Johnson (2005)

(continued)

Table 14.1 *(continued)*

Category[a]	ϕVF name (function)[b]	Source	Exotoxin[c]	Host lysis?[d]	Enviromod?[e]	Reference
ii	botulinum toxins C_1 and D (neurotoxins)	*Clostridium botulinum* phage	yes	no	no	Johnson (2005)
ii	C_2 (hemagglutinin) and C_3 (ADP-ribosylating toxin)	*C. botulinum* phage	yes	no	yes	Johnson (2005)
ii	superantigens, phospholipase, DNase	*S. pyogenes* phage	yes	no	varies	Banks *et al.* (2005)
ii?	hemolysin[f]	*Vibrio harveyi* phage?	yes	?	yes	Austin *et al.* (2003)
iii	cholera toxin (cytotoxin)	*Vibrio cholerae* CTXΦ	yes	no	yes	Davis and Waldor (2005)
iii	Zot, Ace (cytopathic)	*V. cholerae* CTXΦ	yes	no	yes	Davis and Waldor (2005)
iv	Shiga toxins (cytotoxic)	*E. coli* lambdoid phage	yes	yes	yes	Tyler *et al.* (2005)
iv	hyaluronidase	*S. pyogenes* phage	yes	yes	yes	Hynes *et al.* (1995)
iv or ii	*Pasteurella multocida* toxin (mitogen but also causes necrotic lesions)	*Pasteurella multocida* phage	yes	yes?	yes?	Pullinger *et al.* (2004)

[a] Roman numerals correspond to categories of VF release discussed in section 14.2.3.
[b] In all cases virulence factors are active against the animal host, resulting in or contributing to bacterial-mediated animal disease.
[c] Exotoxins are soluble (that is, free of the host bacterium) and can function independent of host-bacterium survival.
[d] If the exotoxin requires host-bacterium lysis to be effective then this is indicated with "yes."
[e] Enviro-mod meaning "environmental modification". Does the ϕVF function in a manner that directly causes modification of the animal host, i.e., in a manner that does not require concomitant presence of the expressing bacterium?
[f] Presence of hemolysin gene in phage genome has not been definitively demonstrated.

those of uninduced prophages (Section 14.2.2), but phage induction does not necessarily decouple these interests.

(iv) *Exotoxins that are not functional without phage-induced lysis of the expressing bacterium.* These exotoxins more likely serve the interests of phages displaying their lytic cycle – following prophage induction or independent of lysogenic cycles – rather than serve the interests of the prophages still residing within the expressing bacterium. Indeed, such VFs could have a positive impact on phage replication or survival while detrimentally impacting not just the VF-expressing bacterium (which is lysed) but also related bacteria located in the same environment (which, if not already lysogenic, may also be subject to phage attack and lysis). These ϕVFs thus may facilitate virion acquisition and infection of commensal bacteria while enhancing the growth of as-yet uninduced lysogens (whether clonal to the parental lysogen or newly created by released phages). The produced virions, meanwhile, display allelopathic effects against otherwise bystander bacteria (Section 14.3.2). We can envisage, therefore, that lysis-released VFs could positively impact VF-encoding prophages, perhaps even more so than the bacterial host.

14.2.4 Lack of ϕVFs in obligately lytic phages

A common assertion is that obligately lytic phages generally do not encode bacterial VFs (obligately lytic, *sensu* Barksdale and Arden, 1974; see Chapter 1). However, any conceptual framework capable of addressing benefits that temperate phages may obtain from VF expression (Section 14.2.3) ought to be able to explain why this expression would not similarly result in a lack of encoding of ϕVFs by obligately lytic phages. We therefore propose two possible mechanisms: (1) obligately lytic phages may not encode non-exotoxin VFs because of the relatively short durations of their infections, assuming that VFs such as adhesins take time to export and then function; (2) bacterial lysogens, as much more genetically complex entities than phages, may be far better equipped than free phages at targeting their growth to those environments in which specific VFs might be useful. In both instances, temperate phages would be more readily capable of taking advantage of VF expression because they infect for much longer periods. They are able to infect for long enough to utilize especially non-exotoxin VFs and they are better able to hitchhike on bacteria to locations where the ϕVFs may be appropriately expressed.

Despite these arguments, we nonetheless see no reason that tissue-modifying exotoxins (or, more generally, environment-modifying extracellular factors) could not in principle aid lytic-phage population growth (Section 14.3.3). We therefore are unwilling to rule out the possibility of evolutionary retention of VF genes even by obligately lytic phages. Indeed, a possibility exists that the essentially universal association of ϕVFs with temperate rather than obligately lytic phages is a consequence of sampling bias, with VF-encoding phages typically isolated as lysogens and therefore by definition temperate. Alternatively, biases would exist if VF-encoding phages were isolated as free virions but then studied in a manner such that potential to form lysogens is assumed rather than rigorously addressed.

14.3 ϕVF EVOLUTIONARY ECOLOGY

Evolutionary ecology is the branch of evolutionary biology (and ecology) that is concerned with ecological functionality of adaptations (i.e., functional contribution to Darwinian fitness; Chapters 1, 2, 3, 8, and 9). In considering VF-encoding bacteriophages from an evolutionary ecological perspective, Abedon and LeJeune (2005) describe ten mechanisms for selecting for VF association with prophage genomes. The first five of these are mechanisms that benefit the ϕVF gene as an independently selected entity, and we will not consider these here. The remaining five, discussed in this section, address interactions between the VF-encoding phage, the host bacterium, and the environment that surrounds both (i.e., the animal).

14.3.1 Linkage and epistasis

"Epistasis linking VF and phage genes" (mechanism number 6 of Abedon and LeJeune, 2005) refers to the observation that other phage genes may supply functions that improve VF activity or function (for more on phages and epistasis, see Chapter 6). For example, Shiga toxin-producing enteropathogenic *E. coli* lack a mechanism to secrete or otherwise release the toxin from the intact bacterial cell (Section 14.5.2.2). Instead, the toxin is released when the bacterium is lysed upon completion of the phage lytic cycle (Tyler *et al.*, 2005). Likewise, although some VFs are controlled by bacterial regulatory factors (e.g., cholera toxin; Quinones *et al.*, 2006), others are regulated by bacteriophage elements (e.g., *Streptococcus pyogenes* phage hyaluronidase; Hynes *et al.*, 1995). For prophages that only express a VF during the lytic cycle, phage induction determines the timing and environmental location of VF expression (Wagner and Waldor, 2002). A final set of examples are VF-expressing defective prophages that depend on helper phages for induction and successful growth (see Boyd *et al.*, 2001 for several examples; see also phage P4, Chapter 6).

14.3.2 Roles during infection *in situ*

"Dissemination of an effective toxin dose" (mechanism 7 of Abedon and LeJeune, 2005) refers to the possibility that induced phages may "recruit" (by infection) nonlysogenized bacteria, especially established normal flora bacterial commensals, as VF producers. For Shiga-toxigenic *E. coli* (STEC), for example, there is some evidence that the co-inoculation of STEC along with bacteria susceptible to infection by the Shiga-toxin-encoding phage can lead to higher levels of toxin production within animals (Gamage *et al.*, 2003; Section 14.6.2). This mechanism has also been suggested for pseudolysogenic VF-encoding phages of *C. botulinum* (Section 14.5.6), where the levels of toxin production might be a function of the ratio of lysogenized to nonlysogenized bacteria within a particular population (Hariharan and Mitchell, 1976; the term pseudolysogenic employed by these authors is defined differently from the same term as employed in Chapter 5 – Hariharan and Mitchell's meaning, essentially, is the equivalent of "unstable lysogen"). Consistent with dose dissemination, the nonlysogenized bacteria can be subsequently infected, thereby increasing the rate of botulinum toxin production, though in this case it may be lysogenic rather than lytic infections that are toxin producing.

By having VF genes encoded by prophages, VF release and dose dissemination can occur concomitant to phage-mediated killing of competitor bacterial strains, competitors which could potentially share in any benefits that may be derived from VF release. This phage-mediated killing of competitor bacteria is described by Stewart and Levin (1984) as "lysogen allelopathy" (mechanism 8 of Abedon and LeJeune, 2005). Remaining lysogenized cells are immune to this killing by virtue of a prophage-encoded immunity to infection (immunity can be considered a non-VF type of lysogenic conversion) (e.g., Allison *et al.*, 2003; Chapter 8). Lysogen allelopathy has been proposed to account for the rapid changes in culture composition observed for *Salmonella* (Bossi *et al.*, 2003), and will be discussed in Section 14.6.3.

14.3.3 Enhancement of phage fitness

"Direct enhancement of phage fitness" (mechanism 9 of Abedon and LeJeune, 2005) refers to a dual utility that can occur when a VF has a function that directly benefits the phage. The phage-benefiting function may be the same as the VF activity or completely different. One example is the hyaluronidase produced by some *Streptococcus* strains, possibly aiding in bacterial spread in tissue. This is a ϕVF that may be maintained as a prophage gene due to its supplying a direct enhancement of phage fitness, specifically degrading the bacterial hyaluronic acid capsule to gain access to the bacterial membrane (Hynes *et al.*, 1995). Supporting this hypothesis, Baker and colleagues (2002) have shown that the phage hyaluronidase can cleave bacterial hyaluronan. Other examples are *ace* and *zot* of phage CTXΦ, as discussed in Section 14.5.3.

"Indirect enhancement of phage fitness" (mechanism 10 of Abedon and LeJeune, 2005) describes a broad class of effects in which VFs enhance phage fitness by modifying the animal environment to promote phage growth and/or phage dissemination. VFs typically allow pathogenic bacteria to better infect, grow, or spread, often into environments where they could not otherwise grow. In this sense VFs can be considered what Woese (2004) described as "cosmopolitan genes," that is "special life style genes [that] allow adaptation to unusual environments." For lysogens, successful infection, growth, and spread within an animal is equivalent to achievement of this cosmopolitan nature by the resident prophages (Sections 14.2.2 and 14.2.3).

A prophage that has induced, however, may retain this adaptation to the animal environment only by finding new, replicating bacteria to infect, since phages typically display low or no productivity when infecting nondividing bacteria (Chapter 5). By generally enhancing bacterial growth within an environment, soluble VFs such as exotoxins could numerically or physiologically enhance the hosts available to phages released from VF-expression pathogens, thereby increasing likelihood of productive (or lysogenic) infection. Likewise, for bacteria whose ϕVFs promote dissemination of bacteria, e.g., via diarrhea, the same dissemination mechanisms could promote the spread of phage virions. Thus, certain ϕVFs may allow phage "adaptation" to animal environments via phage-mediated modification of the characteristics of those environments.

14.4 CONTEXT OF VF EXPRESSION AND UTILITY

From the above discussions we can conclude three key points. First, ϕVFs in principle can be of utility to bacteria, phages (directly or indirectly), or both bacteria and phages. Second, there can be considerable overlap in the interests of prophages and bacteria within uninduced lysogens, but this overlap in interests is almost completely absent during phage lytic infection. Finally, at least some of the conflict in interests between host and phage is alleviated given chronic rather than lytic phage release.

These statements are broad generalizations that serve as a means of organizing our thinking of how and why phages encode bacterial VFs. However, to understand the evolutionary ecology of specific VFs it is necessary to consider both the molecular details of their expression and the ecological context of that expression. So far (not surprisingly) clarifying molecular details has received more attention from researchers than the much more experimentally messy (and costly) analysis of *in situ* utility, especially of φVF utility to phage or bacterial populations rather than to individual bacteria (Abedon and LeJeune, 2005). In this section we provide an introduction to the "messy" latter, concentrating, as we will for much of the remainder of this chapter, on the evolutionary ecology of the phage encoding of virulence factors within the context of the animal colon.

14.4.1 VFs and the colon as a bacterial niche

It is common to think of the colon as a resource-rich environment. This may not be the case for all bacteria. Instead, competition between bacteria may limit access to nutrients, oxygen, and appropriate physical surfaces. The presence versus absence of these specific nutrients and other factors affecting bacterial growth affects what types of bacteria are capable of growing within a given intestinal environment (Ley *et al.*, 2006).

Takahashi and colleagues, for example, coinfected mice with *Clostridium butyricum* (which produces butyric acid that inhibits *E. coli* growth *in vitro*) and strains of *E. coli* O157:H7 and found that the mice suffered less Shiga-toxin-induced morbidity (Takahashi *et al.*, 2004). Likewise, a strain of *Bifidobacterium breve* that releases acetic acid (altering the pH of the intestine) inhibited *E. coli* O157:H7 growth in mice (Asahara *et al.*, 2004). *E. coli* O157:H7 EDL933, which were able to use gluconeogenic metabolites (mainly amino acids and fatty acids) as energy sources as well as glycolytic metabolites (sugars), were better able to compete with other bacteria at colonizing the mouse intestine than a 933 mutant strain that only used glycolytic metabolites (Miranda *et al.*, 2004). Intestinal mucosa and feces furthermore have different bacterial communities (Zoetendal *et al.*, 2002; Eckburg *et al.*, 2005) and similar bacteria can have different physiological states depending on their physical location, e.g., mucus lining the colon versus feces (Poulsen *et al.*, 1995).

From these examples we can envisage how VFs can aid bacteria in acquiring nutrients (e.g., iron), space (e.g., surfaces to colonize), or new environments (e.g., new animals). A subset of VFs, bacterial exotoxins, typically are soluble (that is, they are released from bacteria into the extracellular environment) and often damage eukaryote tissues, releasing nutrients in the process. Exotoxins also potentially aid bacteria in their acquisition of space by contributing to the invasion of previously uncolonized tissues (e.g., as would be found within bodies beyond physical barriers), by modifying bacteria-containing ecosystems such that existing bacteria are unable to continue colonizing, or by contributing to bacterial transmission to new locations (Brüssow *et al.*, 2004).

14.4.2 VFs and the colon as a phage niche

Phages are efficient vectors for transmission of genetic material between bacteria (Chapter 11). Over the years, however, we have encountered a number of individuals who have questioned the ability of phages to carry out their adsorption–infection–adsorption life cycle *in situ*, especially with regard to coliphages and colonic versus sewage growth. Phage *in situ* adsorption–infection, however, has been demonstrated for a variety of phage–bacterial systems (see Table 3 from Breitbart *et al.*, 2005 for additional references). Two studies have shown that, for bacteriophage T4, the colonic environment, but not sewage, contains the correct combination of salts and other co-factors needed for infection (Conley and Wood, 1975; Abedon, 1990; see also Kutter *et al.*, 1994). More recently, Muniesa and colleagues (2004) compared a variety of factors affecting the growth of undefined coliphages from sewage. They found that, for efficient propagation, coliphages required higher cell densities than would usually be expected to be found in aquatic environments including sewage and sewage-contaminated waters, reinforcing the idea that the intestinal tract is where most phage infection and lysogeny or lytic growth occurs.

As described previously (Section 14.4.1), different regions of the colon vary in terms of the ability of bacteria to grow (Poulsen *et al.*, 1994; Zoetendal *et al.*, 2002; Eckburg *et al.*, 2005; Ley *et al.*, 2006). For phages this means that presence of bacteria may not equate to the presence of infectable hosts. Chibani-Chennoufi and colleagues (2004b), for example, found that enteropathogenic *E. coli* treated with a lytic phage or mix of lytic phage strains were killed *in vitro* but not *in vivo* (using a mouse model). As they note, an intestinal lumen population of non-growing bacteria would be poor hosts for infecting virions although they would still presumably adsorb the phage (in essence inactivating the phage; at least temporarily, Chapter 5). Bacteria found in association with the intestinal mucosa, on the other hand, may have reduced exposure to phages due to limited diffusion through the surrounding mucus. These results contrast somewhat with earlier studies by Smith and Huggins (1983), who successfully treated enteropathogenic *E. coli* infections of calves, piglets, and lambs with a lytic phage, although they too noted that the effectiveness of the phage decreased as the *E. coli* colonized the intestine. Clearly, the colon represents a complex environment, and the growth of phages and bacteria is influenced by numerous factors (e.g., Smith *et al.*, 1987).

If the colon provides the correct environment for phage infection but contains a large fraction of non-growing bacteria that consequently make poor phage hosts (e.g., due to lack of a key nutrient such as iron), then nutrient enrichment is of obvious benefit to phages, as well as bacteria. Hemorrhagic lesions, for example, could increase the available nutrients within the intestine, and fecal thinning (i.e., diarrhea) could allow for better mixing of bacteria and nutrients. Increasing bacterial fitness, however, could tip the scales away from conditions in which a phage is better off displaying a bacterium-like state (i.e., as a prophage). For example, when bacteria are at lower densities, growing slowly, and/or have mucus-mediated partial protection from phage attack, then it is likely that phage lysogens are more fit than phage virions. However, high bacterial densities, faster bacterial growth, and a lack of physical protection for bacteria from phage attack should be conditions in which lytic infection is a more effective means of propagating phage population growth than a prophage's binary fission, especially since under these conditions a phage lysogen would be more susceptible to attack by heterologous lytic phages (Abedon, 2008). That is, an increase in the supply of readily infected bacteria, as a consequence of exotoxin expression, could directly benefit phage virions and thereby favor lysogen induction.

14.5 SURVEY OF EXOTOXIN-ENCODING PHAGES

Gaining an understanding of why phages might (and do) encode bacterial VFs, as well as the consequences of that encoding, requires extensive inference (i.e., hypothesis generation) from numerous molecular details. Organized by bacterial host, this is what we will attempt in this section. We do so, however, with two important caveats: (1) broad understanding requires consideration of large numbers of molecular details (particularly considering the number of unique VF–phage–bacterial–animal host permutations in existence), many of which are not yet known; (2) due to space limitations, there exist many more known molecular details than we will be presenting. For a fairly recent review of these many details, see Waldor *et al.* (2005).

We will focus much of our discussion on pathogens of the intestinal tract including *E. coli*, *Salmonella*, and *V. cholerae* systems, as the VF-encoding phages of these pathogens are particularly well studied. However, we will also review non-colonic organisms such as *Clostridium* and *Corynebacterium*.

In these organisms many ϕVFs play a similar role. They modify the bacterial environment so as to increase nutrient richness or to bias colonization within those environments toward VF-expressing bacteria at the expense of competing residents. For a summary of ϕVFs and their effects, see Table 14.1.

14.5.1 General considerations

VF expression can be considered at a number of levels. One of the better studied is their molecular genetics, i.e., regulation of transcription of mRNA and translation into proteins. Ecologically speaking, we may question whether this expression is regulated. If expression is regulated, then by what, how, and where within an ecosystem is such regulation imparted, and what impact does such regulation have on phage or lysogen Darwinian fitness. Next is the cell biology of protein transport, whether to cell envelopes or, in the case of exotoxins, secretion from the bacterium itself. To what degree does autonomy of VF transcription, translation, or transport from phage-progeny production imply phage versus bacterium ecological or evolutionary "ownership" of an exotoxin or its production?

Levels of production can also be considered in terms of dissemination of an effective toxin dose (Section 14.3.2). To achieve such dissemination, phages must be released, they must infect bacteria co-located within the same environment, and they must do so in a manner that results in additional toxin production. Relevant to such dissemination therefore are issues of phage lytic–lysogenic cycle decisions, phage potential to produce toxin with or without displaying lysogeny, and phage host range. Finally, there is effectiveness in terms of bacterial–animal interactions, including such things as bacterial attachment or exotoxin interference with host tissue function. Only given appropriate expression, transport, perhaps amplification, and action will a VF significantly impact animal physiology, and only by understanding the details of these processes can we ascertain the impact of ϕVFs on phage or lysogen ecology.

14.5.2 *Escherichia coli*

E. coli are a normal part of intestinal flora. Some strains are pathogenic, causing several different diseases depending on the site of colonization. These include systemic infections (sepsis or septicemia), meningitis, urinary tract and kidney infections, and a variety of diarrheic syndromes (Schaechter *et al.*, 1998). We will focus on strains causing diarrheic disease. Like most intestinal pathogens, the bacterium must be able to colonize the mucosal layer of the intestine (by invasion or on the surface) and then evade the immune system. These key abilities are mediated by a variety of VFs. For example, those VFs that remain as part of the bacterium, including adherence fimbriae, may be employed towards attaching to tissue surfaces, among other functions. Certain enterotoxins and myriad exotoxins are also responsible for diarrhea and additional pathogenic symptoms (see Nataro and Kaper, 1998, as a general reference for *E. coli* virulence factors). Among these we focus on the exotoxin, Shiga toxin (Stx).

14.5.2.1 STEC and Shiga toxin

E. coli strains that produce one or more types of Shiga toxin (Shiga-like toxins, also called verotoxins; Tyler *et al.*, 2005) are commonly grouped together as Shiga toxin-producing *E. coli* or STEC, and include numerous isolates of *E. coli* O157:H7. Unlike other strains of pathogenic *E. coli*, STEC do not adhere to or invade intestinal mucosal cells but instead are associated with disease that begins with diarrhea but progresses to more severe symptoms including hemorrhagic colitis (bloody diarrhea accompanied by intestinal cramping) and hemolytic uremia syndrome (a systemic disease that includes renal failure and intravascular coagulation) (O'Brien and Holmes, 1987). Central to these outcomes are one or both of the two major variants of Stx, Stx-1 and Stx-2. The genes encoding each variant are almost always located on separate prophages (Rietra *et al.*, 1989; Allison *et al.*, 2003). Stx-1 was later identified as having a nearly identical protein sequence to the toxin produced by *Shigella dysenteriae* 1.

Stx is a two-component toxin, the product of the *stxA* and *stxB* genes. In STEC these genes are encoded by highly mosaic lambdoid prophages (Neely and Friedman, 1998; Plunkett *et al.*, 1999; Unkmeir and Schmidt, 2000). The diversity of Stx-encoding phages is reflected in studies examining the flanking regions around Stx genes (Johansen *et al.*, 2001), in the differing levels of Stx expression seen when various Stx-encoding phages are placed in the same host bacterium (Wagner *et al.*, 1999), and in the differing distributions of STEC strains and Stx-encoding phages seen around the world (LeJeune *et al.*, 2006).

The two major types of Stx, Stx-1 and Stx-2, also differ in their means of regulation, especially in terms of transcription. Stx-2 genes are expressed with the late genes of their encoding prophages, and in particular with the expression of phage lysis genes (Muhldorfer *et al.*, 1996; Plunkett *et al.*, 1999; Wagner *et al.*, 2001 a). Consequently, Stx-2 production is intimately tied to and dependent on prophage induction. In general, treatments that are used to induce lambdoid prophages, such as mitomycin C and some classes of antibiotics, also induce Stx-2 production (O'Brien *et al.*, 1984; Muhldorfer *et al.*, 1996; Wagner *et al.*, 1999; Zhang *et al.*, 2000; Wagner and Waldor, 2002). Likewise, mutant prophages whose cI repressor is cleavage-resistant, and hence cannot induce, do not produce Stx-2 (Tyler *et al.*, 2004). The transcription of Stx-1 genes, by contrast, tends to be promoted by bacterial growth under low iron conditions. As a consequence, Stx-1 may be produced (though not released) independent of prophage induction. Furthermore, in most strains (but not all) levels of Stx-1 production are not affected by prophage induction (Ritchie *et al.*, 2003). This lack of association between phage induction and Stx-1 expression suggests that the function may be expendable, so far as phage fitness is concerned, when iron concentrations are high.

14.5.2.2 Stx and Stx-encoding phage induction and release

The mechanism of lambdoid phage induction is well understood and typically involves DNA damage triggering the bacterial SOS pathway (Little, 2005). Phages may enter the lytic phase in the absence of DNA-damaging treatment, a phenomenon referred to as spontaneous induction. Spontaneous induction likely occurs due to spontaneous DNA damage. For some phages, such as P22, this spontaneous induction occurs with the onset of host stationary phase (Ramirez *et al.*, 1999; Lunde *et al.*, 2003).

The amount of DNA damage needed for induction (the set point) varies between phage strains (Little, 2005). Set point can be modulated genetically by mutations in the phage repressor protein, repressor binding sites, or other regulatory proteins involved in the maintenance of the lysogenic state. Induction rates may also be modified by other global regulatory mechanisms. Sperandio and colleagues (2001) found that Stx expression, along with other SOS response genes and other non-phage-encoded VFs, are upregulated by quorum sensing.

The spontaneous induction rates of Stx-encoding phages have been shown to be much higher than that of phage λ. Livny and Friedman (2004) created a reporter system (SIVET) that could detect induction within STEC strains that had been modified so that induction did not lead to lethal production of phages. Using the SIVET system, they found that both H-19B (Stx-1 encoding) phage and 933W (Stx-2 encoding) phage were spontaneously induced at much higher rates than the lambdoid phages P21 and λimm 434, or λ itself. Livny and Friedman (2004) suggest that this increased rate of induction may have arisen due to selective pressure related to the requirement for phage induction to release Stx. As they also note, the rates of induction may be even higher during infection of the intestinal tract due to endogenous SOS-inducing agents. Hydrogen peroxide from phagocytic immune cells has been shown to induce Stx-2 production, for example (Wagner *et al.*, 2001b). Since the rate of Stx release from STEC is determined by the rate of phage induction, this increased rate of spontaneous induction (or a higher rate due to *in vivo* factors) means that STEC are constantly releasing Stx-encoding phages along with the exotoxin (Stx-2 for those strains encoding it, Stx-1 mainly under iron-deficient conditions). We can speculate from these observations that Stx-encoding phages are adapted particularly to intestinal propagation – in terms of both the colonic bioactivity of Stx and their potentially higher induction rate upon exposure to the colonic environment. A constant spontaneous production of Stx-encoding phages is supported by the results of Muniesa and colleagues, who found large numbers of Stx-2-encoding phages in samples of raw sewage obtained from urban wastewater treatment plants (Muniesa and Jofre, 1998) including many that propagated poorly on laboratory strains of *E. coli* (Muniesa *et al.*, 2004; Muniesa and Jofre, 2004).

The amount of extracellular Stx produced by an STEC appears to be correlated with the risk of severe disease (Ritchie *et al.*, 2003). Neither Stx-1 nor Stx-2 have secretory signals and they therefore accumulate within the cell until it is lysed (Tyler *et al.*, 2005). This is similar to the situation in *Shigella*, where Stx accumulates in the periplasmic space of the bacterium until the cell dies and the toxin is released (O'Brien and Holmes, 1987). In general, lysis in STEC is mediated by Stx-encoding bacteriophages as they complete their lytic cycle. For an Stx-2-encoding phage the timing of this lysis is tied directly to the expression of Stx-2, while for an Stx-1-encoding phage, lysis timing is controlled independently of control of toxin production (Section 14.5.2.1). In both cases it is the timing of phage-induced cell lysis that controls the timing of Stx release (Wagner *et al.*, 2002). The longer the time between the onset of Stx production (due to iron deficiency or phage induction) and subsequent cell lysis, the more Stx will accumulate for release. In addition, LeJeune and colleagues (2004) have found a correlation between the amounts of Stx-2 produced by different STEC strains and two different late gene control sequences.

14.5.3 *Vibrio cholerae*

Vibrio cholerae is the causative agent of cholera, an epidemic diarrheal disease. Over 150 serotypes of *V. cholerae* have been identified although only the O1 and O139 serotypes are associated with human disease (Kaper *et al.*, 1995). *V. cholerae* produces two major VFs: the toxin-coregulated pilus (TCP) and cholera toxin (CT). The TCP is the product of more than 15 genes clustered in the *Vibrio* pathogenicity island and is thought to be an essential colonization factor for vibrios to grow in the mammalian intestine (Boyd and Waldor, 2002). O'Shea and Boyd (2002) have found that the entire *Vibrio* pathogenicity island can move between *Vibrio* strains via generalized transduction by another vibriophage, CP-T1.

CT, which is encoded by a prophage, CTXΦ, affects a number of body systems but the overall effect is the production of a watery "rice-stool" form of diarrhea that seems ideally suited for dissemination of the bacterium, phage, or both (Kaper *et al.*, 1995). Unlike the VF-encoding phages of STEC (Section 14.5.2) or *Salmonella* (Section 14.5.4), which are related to phage λ, CTXΦ is a filamentous phage similar to the filamentous coliphages M13 and f1 (Russel and Model, 2006). CTXΦ is able to integrate into the *V. cholerae* genome and form a stable lysogen, as well as forming a plasmid-like replication form (Waldor and Mekalanos, 1996).

As with the lambdoid phages, treating *V. cholerae* with DNA-damaging agents, such as mitomycin C or UV radiation, leads to induction of the CTXΦ prophage. The integrated genome is not excised but is used as a template for new phage genome synthesis. Release of newly produced virions occurs without cell lysis, like other filamentous phages. Virions exit the cell through a pore created by the Eps secretory system (type II secretory system) of the bacteria. CT is also secreted from the cell via this same pore (Davis *et al.*, 2000; Davis and Waldor, 2005). Unlike Stx-2, CT expression is not coupled to phage induction. Quinones and colleagues have shown that for CTXΦ, prophage induction after UV irradiation depends on recA activity (Quinones *et al.*, 2005), while CT production does not (although SOS-inducing treatment does cause a small increase in CT production) (Quinones *et al.*, 2006). Instead, expression of *ctxA* and *ctxB*, the two genes encoding CT, is controlled as part of the bacterial ToxR regulon, a group of coregulated VFs. This regulon is controlled by three transcriptional activators, ToxR, ToxT, and TcpP, which respond to a variety of environmental factors allowing coordinated expression of the VFs (Reidl and Klose, 2002). The separation of CT expression and release provides a contrasting model to that of Stx-2, where the two processes are inherently linked.

CTXΦ has been found to encode two other exotoxins, Zot and Ace. Even in CT-deficient strains of *V. cholerae*, these are sufficient to trigger mild to moderate diarrhea (Fasano *et al.*, 1991). Both of these are examples of dual-utility VFs (Section 14.3.3). They act as exotoxins, increasing mammalian cell permeability (Fasano *et al.*, 1991; Trucksis *et al.*, 1993), but Ace is also a minor coat protein in the phage virion while Zot is part of the virion secretion protein complex (Waldor and Mekalanos, 1996; Davis and Waldor, 2005).

14.5.4 *Salmonella enterica*

The majority of the isolates of bacteria that are classified as *Salmonella* have been found to be a single species, *Salmonella enterica.* Within this species, however, are a vast number of antigenically distinct members or serovars. Over 2500 have been identified (Bossi and Figueroa-Bossi, 2005). These include the broad-host-range serovars Typhimurium and Enteritidis, which cause gastrointestinal disease, as well as the Typhi serovar, the causative agent of typhoid fever.

Over fifty years ago, Boyd (1950) recognized that lysogenic conversion was occurring in some strains of *Salmonella enterica* serovar Typhimurium (*S.* Typhimurium), which causes a watery diarrhea following tissue invasion (infectious bacteria may be shed for weeks). More recently, nearly a dozen different prophages with ϕVF genes have been identified in various *S.* Typhimurium serovars, although any particular serovar will have only four or five prophages integrated at different sites (Bossi and Figueroa-Bossi, 2005). In total, prophages account for about 5% of the DNA in the *Salmonella* genome. While prophages Gifsy-1 and Gifsy-2 are found in all enteropathogenic strains of *S. Typhimurium,* the remaining prophages vary. Like those of *V. cholerae,* the various *Salmonella* ϕVFs (along with bacterial chromosome- and plasmid-associated VFs) allow the bacteria to successfully colonize animal intestinal tracts. Also as in *V. cholerae,* all of the *Salmonella* ϕVFs are secreted, in this case by the bacterial type III secretion system (Wallis and Galyov, 2000).

The majority of VF-encoding prophages of *Salmonella* (both functional and defective) fall into two groups. These are lambdoid phages and a second group that includes phage P22 and P22-related phages (Bossi and Figueroa-Bossi, 2005). As with other temperate phages, non-defective *Salmonella*-associated prophages can be induced into lytic reproduction by a variety of DNA-damaging treatments. There is no good measure of the rate of spontaneous prophage induction in the wild, but in culture virions from lysogenized cells are detectable both with and without exposure to inducing agents (with hydrogen peroxide especially efficacious as an inducing agent; Figueroa-Bossi and Bossi, 1999; Frye *et al.*, 2005). Given the likelihood of bacterial engulfment by host phagocytes that use oxygen radicals to attack pathogens, prophage sensitivity to induction by hydrogen peroxide (like that of Stx-2-encoding phages; Section 14.5.2.2) is intriguing, particularly given that one of the VFs encoded by phage Gifsy-2 is a superoxide dismutase. Inducing this gene, at the same time the prophage is induced, may protect phage progeny from the same oxygen radicals potentially responsible for prophage induction.

14.5.5 *Corynebacterium diphtheriae*

C. diphtheriae, the causative agent of diphtheria, is a Gram-positive aerobe that is chiefly found in the upper respiratory tract, especially the throat. Encoded by the *tox* gene, the principal VF of *C. diphtheriae* is diphtheria toxin (DT) (Johnson, 2005). Analogous to the iron-dependent regulation of Stx-1 (Section 14.5.2.1), DT production is regulated by DtxR, a bacterial iron-dependent regulator of about 40 bacterial genes (Brüssow *et al.*, 2004). DT causes erythema (increased blood flow), induration (tissue stiffening due to inflammation), and necrosis (Holmes, 2000).

Conversion of nonpathogenic forms of *C. diphtheriae* to pathogenic forms following phage infection was first observed by Freeman (1951). The DT-encoding *tox* gene is now known to be located in the genome of one of several temperate bacteriophages, and represents the first VF mapped to a prophage genome. The encoding phages have been described as lambda-like, based on morphology and life cycle, and they can be induced by mitomycin C or UV. Compared to other phage groups, however, the corynebacteriophages have been relatively poorly studied (Johnson, 2005).

14.5.6 *Clostridium botulinum*

The clostridia – the causative agents of botulism, tetanus, and gas gangrene – are a group of spore-forming, obligately anaerobic species that include bacteria that live in soil, in the intestine, and which can invade tissue at wounds (Schaechter *et al.*, 1998). Perhaps best studied is *Clostridium botulinum*, which produces the neurotoxin botulinum (BoNT; Johnson, 2005). Seven antigenically distinct BoNTs have been identified (Schantz and Johnson, 1992). They are released from the bacteria as a complex that includes two phage-encoded exotoxins, a cytotoxin/hemagglutinin (C_2) and an ADP-ribosylating toxin (C_3), as well as several other non-toxin proteins (Johnson, 2005). Of the seven BoNTs, only the C_1 and D toxins have been demonstrated to be phage-encoded (Eklund and Poysky, 1974). Unlike the other VFs we have discussed, these neurotoxins do not appear to be involved in any modification of the immediate environment that could presumably benefit the lysogen or phage. However, the two other phage-encoded exotoxins in the toxin complex do have this role. Oguma and colleagues demonstrated that the C_2 hemagglutinin exotoxin could also be transferred from BoNT/C_1- and D-producing strains to nontoxigenic strains by phage conversion, although this linkage between the two toxins is not absolute as a few BoNT-plus/C_2-minus and BoNT-minus/C_2-plus strains were identified (Oguma *et al.*, 1976). Another bacteriophage gene, on the same phages that produce BoNT/C_1 or D, encodes the C_3 toxin (Popoff *et al.*, 1991). Both the C_2 and C_3 exotoxins have cytopathic effects like the VFs we have already discussed and, like cholera toxin, are part of the mechanism that allows the bacteria to move between extra-animal environments (in this case soil) and animals.

Little is known about the life cycle of the clostridia converting phages, although it appears that many toxin-encoding bacteriophages of clostridia are actually pseudolysogenic (Johnson, 2005; Chapter 5). Since the prophage is not a stably integrated prophage, pseudolysogenic strains can have a mix of prophage-containing and spontaneously cured cells. Strains with a higher fraction of prophage-containing cells would likely produce more toxin such that pseudolysogeny provides a different phage life-cycle-associated mechanism for controlling toxin production levels (Hariharan and Mitchell, 1976).

14.6 PHAGE PRODUCTION DURING PATHOGENESIS

To understand phage encoding of VFs it is helpful to explore how phage properties might influence VF utility or prevalence, especially during pathogenesis. In this section we focus on evidence for phage transmission within animals as well as the related potential for dissemination of an effective toxin dose (Section 14.3.2).

14.6.1 Phage transmission in animals

The ability of VF-encoding phages to infect and lysogenize bacteria within an animal host has been demonstrated for a number of species. Acheson and colleagues (1998) demonstrated *in vivo* transduction of both Stx-1- and Stx-2-encoding phages. For the Stx-1-encoding phage H-19B, they replaced portions of the *stxA* gene (Section 14.5.2.1) with a β-lactamase gene. *E. coli* MC4100 cells lysogenized with the altered phage were inoculated into mice along with ampicillin-sensitive *E. coli* AK16. Within 24 hours, ampicillin-resistant AK16 cells could be detected in stool. In a later study they created a similarly modified Stx-2-encoding 933W prophage (Zhang *et al.*, 2000). As in the previous study, mice were coinoculated with prophage-containing and target bacteria. Transductants were only seen in one of five control mice (no antibiotic induction) but if the mice were also given ciprofloxacin, transduced antibiotic-resistant target cells were readily detected in all mice. Gamage and colleagues (2003), using a differently modified 933W phage and the same mouse model, obtained similar results using coinoculation of purified phages instead of induced lysogenic bacteria, with C600 *E. coli* as the recipient cells.

Broudy and Fischetti (2003) have demonstrated *in vivo* lysogenic conversion of *Streptococcus pyogenes*. Using the lysogenic exotoxin-C-producing *S. pyogenes* strain CS112 (Tox-plus) and phage-naive strain CS24 (Tox-minus), they first demonstrated that the CS24 strain could undergo lysogenic conversion to Tox-plus when co-cultured with both the CS112 strain and cultured mammalian pharyngeal cells (although, interestingly, not in the absence of the pharyngeal cells). They next challenged mice oronasally with both strains. They detected lysogenic conversion of the CS24 strain in several but not all of the mice. Similar results were obtained when mice were challenged with CS24 (Tox-minus) bacteria and purified ϕCS112.

Kimsey and Waldor (1998) used a strain of *V. cholerae* containing a CTXΦ prophage that had been modified to carry a kanamycin-resistance gene to demonstrate *in vivo* transduction. When this strain (LAC-1) was coinoculated into suckling mice with either a CTXΦ lysogenized strain (E7946) or a CTXΦ-minus strain (BAH-2), kanamycin-resistant E7946 or BAH-2 cells could be recovered. The efficiency of transfer was about 10% for BAH-2 while it was about 1000-fold less for E7946 due to immunity to superinfection.

14.6.2 Host-range breadth of VF-encoding phages

In the absence of trade-offs (Breitbart *et al.*, 2005; but see also Chapter 2), a wide host range is of obvious utility to a phage since the more types of bacteria a phage can infect, the more bacteria available to it for infection. Nevertheless, while there are some phages with a broad host range (Jensen *et al.*, 1998; Chapter 11), most phages limit their host range to a single or limited number of species, although it has been suggested that this conclusion may be an artifact of phage isolation procedures (Jensen *et al.*, 1998; Weinbauer, 2004). In the case of VF-encoding phages, host range determines the limits of horizontal gene transfer, and several groups have characterized the host range of VF-encoding phages.

A number of studies have found, for example, that some of the Stx-encoding phages are able to lytically infect as well as lysogenize many strains of *E. coli* as well as some strains of *Shigella* (James *et al.*, 2001; Strauch *et al.*, 2001; Gamage *et al.*, 2004). Likewise CTXΦ, which encodes cholera toxin, has been shown by Faruque and colleagues to be able to lysogenize *V. mimicus* and convert this normally nonpathogenic strain to a CT-plus phenotype (Faruque *et al.*, 1999). This lysogenic conversion occurred in spite of the fact that the *V. mimicus* strains do not express TCP (the normal phage receptor), indicating that CTXΦ is able to infect via a second receptor, albeit at lower efficiency.

Clostridia phages are able to interconvert clostridia species associated with botulism and gangrene (*Clostridium novyi*) to produce the other strain's toxin (Eklund *et al.*, 1974). Given that the pseudolysogenic clostridia phages have been proposed to modulate toxin levels by altering the fraction of the bacterial population that contains a prophage (Section 14.5.6), an ability to infect other *Clostridium* species could expand the population of available bacterial hosts *in situ* for this purpose.

14.6.3 Contribution of phage transmission to pathogenesis

The association of Stx expression with virion production is suggestive of a potential for amplification of Stx production *in situ* via dissemination of an effective toxin dose (Section 14.3.2). Gamage and colleagues (2003) in fact showed that Stx toxin production can be amplified by commensal bacteria both *in vitro* and *in vivo*. Using the Stx-2-encoding phage 933W, they demonstrated that the production of Stx increased when a C600: :933W lysogen or the human isolate, *E. coli* O157:H7: :PT-32, were grown in the presence of phage-susceptible *E. coli* C600 cells but not with phage-resistant C600 (resistant due to the presence of a tox-minus 933W prophage). They obtained similar results using the C600 strains in a mouse intestinal model: greater Shiga toxin production in the presence of susceptible cells but not in the presence of phage resistant cells.

More recently, Gamage *et al.* (2006) examined the effect of an existing population of commensal bacteria on Shiga toxin production. A number of studies have shown that commensal bacteria interfere with colonization of the mouse intestine by *E. coli* O157 strains (Section 14.4.1). Gamage and colleagues found similar results using their phage-sensitive and phage-resistant strains of *E. coli* to measure bacterial colonization by a toxin-producing strain. They saw the same colonization interference with phage-sensitive and resistant commensals but a differential effect on Stx production. When they examined toxin production, they found Stx being produced in mice that had been coinoculated with phage-sensitive commensal bacteria but not in mice with phage-resistant commensal bacteria. Together these results suggest that STEC can use commensal bacteria as an amplification mechanism for toxin production.

Bossi and colleagues (2003) have suggested a similar role for toxin-encoding phages of *Salmonella* infecting commensal bacteria, based on co-culture experiments. In these studies they found that lysogenized strains of *Salmonella* quickly out-competed strains that did not contain prophages. This occurred through a combination of two effects. First, the released virions were able to infect and kill many of the sensitive, nonlysogenized cells (lysogen allelopathy; Section 14.3.2). Second, a few of the remaining sensitive cells were also infected but became lysogenized by the phages. In this way there was a rapid conversion of the mixed population of *Salmonella* to one in which almost all the bacteria were lysogenized. This observation is consistent with the sort of rapid horizontal gene transfer that has been seen among natural populations of pathogenic *Salmonella*.

14.7 CONCLUSIONS

In examining VF-encoding phages it is obvious that they are a highly variable group, ranging from λ-like to filamentous, and having in common only the ability to enter lysogeny. Rather than viewing this as a problem, however, diversity means that there are many different model systems available to the phage ecologist. For example, Stx-2 and cholera toxin both contribute to the development of diarrhea in humans but phage regulatory elements clearly control the expression of Stx-2 while bacterial regulatory elements are primarily responsible for regulating cholera toxin production (Waldor and Friedman, 2005).

Still, while this diversity of phages and expression mechanisms makes any all-encompassing conclusions difficult if not impossible, the roles of ϕVFs seem to fall into three broad categories. A few of the VFs are examples of dual-utility proteins (Section 14.3.3), having a role in the phage life cycle and a role in the bacterial host's life cycle. Others have no direct role in the phage life cycle but may instead be present as "morons" – more DNA that has been added to the phage genome by accident (Chapter 7). The majority of the VFs discussed here, however, seem to fall into a third group that can broadly be described as environmental modifiers (Table 14.1). Shiga toxin, cholera toxin, and diphtheria toxin, as well as some of the ϕVFs of *Salmonella* and the hemagglutinin toxin of certain clostridia, for example, cause or contribute to tissue necrosis and/or diarrhea in animal hosts of the lysogenized bacteria. In doing so they appear to be modifying the environment to supply more nutrients (regardless of location) or allow greater mobility (such as within the intestinal tract or even between intestinal tracts). This is in contrast to other VFs that are not usually found in phage genomes, such as attachment factors and host immune system evasion mechanisms. A nutrient-rich, high-mobility environment is of obvious benefit to the phage in promoting a healthier bacterial host, in supporting the growth of uninfected bacteria as hosts for induced virions, and in enhancing virion dissemination.

ACKNOWLEDGMENTS

We would like to thank Jeff LeJeune for bringing to our attention that the phage encoding of bacterial virulence factors could provide a deep well of interesting ecological questions.

REFERENCES

Abedon, S. T. 1990. The ecology of bacteriophage T4. Ph.D. thesis, University of Arizona.

Abedon, S. T. 2008. Ecology of viruses infecting bacteria. In B. Mahy and M. van Regenmortel (eds.), *Encyclopedia of Virology*. Oxford: Elsevier.

Abedon, S. T., and J. T. LeJeune. 2005. Why bacteriophage encode exotoxins and other virulence factors. *Evol. Bioinf. Online* **1**: 97–110.

Acheson, D. W. K., J. Reidl, X. Zhang, G. T. Keusch, J. J. Mekalanos, and M. K. Waldor. 1998. In vivo transduction with Shiga toxin 1-encoding phage. *Infect. Immun.* **66**: 4496–8.

Allison, H. E., M. J. Sergeant, C. E. James, *et al.* 2003. Immunity profiles of wild-type and recombinant Shiga-like toxin-encoding bacteriophages and characterization of novel double lysogens. *Infect. Immun.* **71**: 3409–18.

Asahara, T., K. Shimizu, K. Nomoto, T. Hamabata, A. Ozawa, and Y. Takeda. 2004. Probiotic bifidobacteria protect mice from lethal infection with Shiga toxin-producing *Escherichia coli* O157:H7. *Infect. Immun.* **72**: 2240–7.

Austin, B., A. C. Pride, and G. A. Rhodie. 2003. Association of a bacteriophage with virulence in *Vibrio harveyi*. *J. Fish Dis.* **26**: 55–8.

Baker, J. R., S. Dong, and D. G. Pritchard. 2002. The hyaluronan lyase of *Streptococcus pyogenes* bacteriophage H4489A. *Biochem. J.* **365**: 317–22.

Banks, D. J., S. B. Beres, and J. M. Musser. 2005. Contribution of phages to group A *Streptococcus* genetic diversity and pathogenesis. In M. K. Waldor, D. I. Friedman, and S. L. Adhya (eds.), *Phages: Their Role in Bacterial Pathogenesis and Biotechnology*. Washington DC: ASM Press, pp. 319–34.

Barksdale, L., and S. B. Arden. 1974. Persisting bacteriophage infections, lysogeny, and phage conversions. *Annu. Rev. Microbiol.* **28**: 265–99.

Barondess, J. J., and J. Beckwith. 1995. *bor* gene of phage λ, involved in serum resistance, encodes a widely conserved outer membrane lipoprotein. *J. Bacteriol.* **177**: 1247–53.

Bossi, L., J. A. Fuentes, G. Mora, and N. Figuero-Bossi. 2003. Prophage contribution to bacterial population dynamics. *J. Bacteriol.* **185**: 6467–71.

Bossi, L., and N. Figueroa-Bossi. 2005. Prophage arsenal of *Salmonella enterica* serovar Typhimurium. In M. K. Waldor, D. I. Friedman, and S. L. Adhya (eds.), *Phages: Their Role in Bacterial Pathogenesis and Biotechnology*. Washington DC: ASM Press, pp. 165–86.

Boyd, E. F. 2005. Bacteriophages and bacterial virulence. In E. Kutter and A. Sulakvelidze (eds.), *Bacteriophages: Biology and Applications*. Boca Raton: CRC Press, pp. 223–65.

Boyd, E. F., and M. K. Waldor. 2002. Evolutionary and functional analyses of variants of the toxin-coregulated pilus protein TcpA from toxigenic *Vibrio cholerae* non-O1/non-O139 serogroup isolates. *Microbiology (Reading)* **148**: 1655–66.

Boyd, E. F., B. M. Davis, and B. Hochhut. 2001. Bacteriophage–bacteriophage interactions in the evolution of pathogenic bacteria. *Trends Microbiol.* **9**: 137–44.

Boyd, J. S. K. 1950. The symbiotic bacteriophage of *Salmonella typhimurium*. *J. Pathol. Bacteriol.* **62**: 501–17.

Breitbart, M., F. Rohwer, and S. T. Abedon. 2005. Phage ecology and bacterial pathogenesis. In M. K. Waldor, D. I. Friedman, and S. L. Adhya (eds.), *Phages: Their Role in Bacterial Pathogenesis and Biotechnology*. Washington DC: ASM Press, pp. 66–91.

Broudy, T. B., and V. A. Fischetti. 2003. In vivo lysogenic conversion of Tox(−) *Streptococcus pyogenes* to Tox(+) with lysogenic streptococci or free phage. *Infect. Immun.* **71**: 3782–6.

Brüssow, H., C. Canchaya, and W. D. Hardt. 2004. Phages and the evolution of bacterial pathogens: from genomic rearrangements to lysogenic conversion. *Microbiol. Mol. Biol. Rev.* **68**: 560–602.

Chibani-Chennoufi, S., A. Bruttin, M.-L. Dillman, and H. Brüssow. 2004a. Phage–host interaction: an ecological perspective. *J. Bacteriol.* **186**: 3677–86.

Chibani-Chennoufi, S., J. Sidoti, A. Bruttin, E. Kutter, S. A. Sarker, and H. Brüssow. 2004b. In vitro and in vivo bacteriolytic activity of *Escherichia coli* phages: implication for phage therapy. *Antimicrob. Agents Chemother.* **48**: 2558–69.

Conley, M. P., and W. B. Wood. 1975. Bacteriophage T4 whiskers: a rudimentary environment-sensing device. *Proc. Natl. Acad. Sci. U.S.A.* **72**: 3701–5.

Davis, B. M., E. H. Lawson, M. Sandkvist, A. Ali, S. Sozhamannan, and M. K. Waldor. 2000. Convergence of the secretory pathways for cholera toxin and the filamentous phage, CTXΦ. *Science* **288**: 333–5.

Davis, B. M., and M. K. Waldor. 2005. Virulence-linked bacteriophages of pathogenic vibrios. In M. K. Waldor, D. I. Friedman, and S. L. Adhya (eds.), *Phages: Their Role in Bacterial Pathogenesis and Biotechnology*. Washington DC: ASM Press, pp. 187–205.

Eberhard, W. G. 1990. Evolution in bacterial plasmids and levels of selection. *Q. Rev. Biol.* **65**: 3–22.

Eckburg, P. B., E. M. Bik, C. N. Bernstein, *et al.* 2005. Diversity of the human intestinal microbial flora. *Science* **308**: 1635–8.

Eklund, M. W., and F. T. Poysky. 1974. Interconversion of type C and D strains of *Clostridium botulinum* by specific bacteriophages. *Appl. Environ. Microbiol.* **27**: 251–8.

Eklund, M. W., F. T. Poysky, J. A. Meyers, and G. A. Pelroy. 1974. Interspecies conversion of *Chlostridium botulinum* type C to *Chlostridium novyi* type A by bacteriophage. *Science* **186**: 456–8.

Faruque, S. M., M. M. Rahman, Asadulghani, K. M. N. Islam, and J. J. Mekalanos. 1999. Lysogenic conversion of environmental *Vibrio mimicus* strains by CTXΦ. *Infect. Immun.* **67**: 5723–9.

Fasano, A., B. Baudry, D. W. Pumplin, *et al.* 1991. *Vibrio cholerae* produces a second enterotoxin, which affects intestinal tight junctions. *Proc. Natl. Acad. Sci. U.S.A.* **88**: 5242–6.

Feil, E. J., and B. G. Spratt. 2001. Recombination and the population structures of bacterial pathogens. *Annu. Rev. Microbiol.* **55**: 561–90.

Figueroa-Bossi, N., and L. Bossi. 1999. Inducible prophages contribute to *Salmonella* virulence in mice. *Mol. Microbiol.* **33**: 167–76.

Freeman, V. J. 1951. Sudies on the virulence of bacteriophage-infected strains of *Corynebacterium diptheriae. J. Bacteriol.* **61**: 675–88.

Frye, J. G., S. Porwollik, F. Blackmer, P. Cheng, and M. McClelland. 2005. Host gene expression changes and DNA amplification during temperate phage induction. *J. Bacteriol.* **187**: 1485–92.

Gamage, S. D., J. E. Strasser, C. L. Chalk, and A. A. Weiss. 2003. Nonpathogenic *Escherichia coli* can contribute to the production of Shiga toxin. *Infect. Immun.* **71**: 3107–15.

Gamage, S. D., A. K. Patton, J. F. Hanson, and A. A. Weiss. 2004. Diversity and host range of Shiga toxin-encoding phage. *Infect. Immun.* **72**: 7131–9.

Gamage, S. D., A. K. Patton, J. E. Strasser, C. L. Chalk, and A. A. Weiss. 2006. Commensal bacteria influence *Escherichia coli* O157:H7 persistence and Shiga toxin production in the mouse intestine. *Infect. Immun.* **74**: 1977–83.

Hariharan, H., and W. R. Mitchell. 1976. Observations on bacteriophages of *Clostridium botulinum* type C isolates from different sources and the role of certain phages in toxigenicity. *Appl. Environ. Microbiol.* **32**: 145–58.

Holmes, R. K. 2000. Biology and molecular epidemiology of diphtheria toxin and the *tox* gene. *J. Infect. Dis.* **181** (Suppl. 1): S156–67.

Hynes, W. L., L. Hancock, and J. J. Ferretti. 1995. Analysis of a second bacteriophage hyaluronidase gene from *Streptococcus pyogenes*: evidence for a third hyaluronidase involved in extracellular enzymatic activity. *Infect. Immun.* **63**: 3015–20.

James, C. E., K. N. Stanley, H. E. Allison, *et al.* 2001. Lytic and lysogenic infection of diverse *Escherichia coli* and *Shigella* strains with a verocytotoxigenic bacteriophage. *Appl. Environ. Microbiol.* **67**: 4335–7.

Jensen, E. C., H. S. Schrader, B. Rieland, *et al.* 1998. Prevalence of broad-host-range lytic bacteriophages of *Sphaerotilus natans, Escherichia coli,* and *Pseudomonas aeruginosa. Appl. Environ. Microbiol.* **64**: 575–80.

Johansen, B. K., Y. Wasteson, P. E. Granum, and S. Brynestad. 2001. Mosaic structure of Shiga-toxin-2-encoding phages isolated from *Escherichia coli* O157:H7 indicates frequent gene exchange between lambdoid phage genomes. *Microbiology* **147**: 1929–36.

Johnson, E. A. 2005. Bacteriophages encoding botulinum and diphtheria toxins. In M. K. Waldor, D. I. Friedman, and S. L. Adhya (eds.), *Phages: Their Role in Bacterial Pathogenesis and Biotechnology*. Washington DC: ASM Press, pp. 280–96.

Kaper, J. B., J. G. Morris, Jr., and M. M. Levine. 1995. Cholera. *Clin. Microbiol. Rev.* **8**: 48–86.

Kimsey, H. H., and M. K. Waldor. 1998. CTXΦ immunity: Application in the development of cholera vaccines. *Proc. Natl. Acad. Sci. U.S.A.* **95**: 7035–9.

Krylov, V. 2003. The role of horizontal gene transfer by bacteriophages in the origin of pathogenic bacteria. *Rus. J. Genet.* **39**: 483–504.

Kutter, E., E. Kellenberger, K. Carlson, *et al.* 1994. Effects of bacterial growth conditions and physiology on T4 infection. In J. D. Karam (ed.), *The Molecular Biology of Bacteriophage T4.* Washington, DC: ASM Press, pp. 406–418.

Langley, R., D. T. Kenna, P. Vandamme, R. Ure, and J. R. W. Govan. 2003. Lysogeny and bacteriophage host range within the *Burkholderia cepacia* complex. *J. Med. Microbiol.* **52**: 483–90.

Lawrence, J. 1999. Selfish operons: the evolutionary impact of gene clustering in prokaryotes and eukaryotes. *Curr. Opin. Genet. Dev.* **9**: 642–8.

Lawrence, J. G. 1997. Selfish operons and speciation by gene transfer. *Trends Microbiol.* **5**: 355–9.

Lawrence, J. G. 2003. Gene organization: Selection, selfishness, and serendipity. *Annu. Rev. Microbiol.* **57**: 419–40.

LeJeune, J. T., S. T. Abedon, K. Takemua, N. P. Christie, and S. Sreevatsan. 2004. Human *Escherichia coli* O157:H7 genetic marker in isolates of bovine origin. *Emerg. Infect. Dis.* **10**: 1482–5.

LeJeune, J. T., D. Hancock, Y. Wasteson, E. Skjerve, and A. M. Urdahl. 2006. Comparison of *E. coli* O157 and Shiga toxin-encoding genes (*stx*) prevalence between Ohio, USA and Norwegian dairy cattle. *Int. J. Food Microbiol.* **109**: 19–24.

Levin, B. R., and C. T. Bergstrom. 2000. Bacteria are different: Observations, interpretations, speculations, and opinions about the mechanisms of adaptive evolution in prokaryotes. *Proc. Natl. Acad. Sci. U.S.A.* **97**: 6981–5.

Levin, B. R., and R. E. Lenski. 1985. Bacteria and phage: A model system for the study of the ecology and co-evolution of hosts and parasites. In D. Rollinson and R. M. Anderson (eds.), *Ecology and Genetics of Host–Parasite Interactions.* London: Academic Press, pp. 227–42.

Ley, R. E., D. A. Peterson, and J. I. Gordon. 2006. Ecological and evolutionary forces shaping microbial diversity in the human intestine. *Cell* **124**: 837–48.

Little, J. W. 2005. Lysogeny, prophage induction, and lysogenic conversion. In M. K. Waldor, D. I. Friedman, and S. L. Adhya (eds.), *Phages: Their Role in Bacterial Pathogenesis and Biotechnology.* Washington DC: ASM Press, pp. 37–54.

Livny, J., and D. I. Friedman. 2004. Characterizing spontaneous induction of Stx encoding phages using a selectable reporter system. *Mol. Microbiol.* **51**: 1691–704.

Lunde, M., J. M. Blatny, D. Lillehaug, A. H. Aastveit, and I. F. Nes. 2003. Use of real-time quantitative PCR for the analysis of ϕLC3 prophage stability in lactococci. *Appl. Environ. Microbiol.* **69**: 41–8.

Matthews, A. M., and R. P. Novick. 2005. *Staphylococcal* Phages. In M. K. Waldor, D. I. Friedman, and S. L. Adhya (eds.), *Phages: Their Role in Bacterial Pathogenesis and Biotechnology.* Washington DC: ASM Press, pp. 297– 318.

Miranda, R. L., T. Conway, M. P. Leatham, *et al.* 2004. Glycolytic and gluconeogenic growth of *Escherichia coli* O157:H7 (EDL933) and *E. coli* K-12 (MG1655) in the mouse intestine. *Infect. Immun.* **72**: 1666–76.

Muhldorfer, I., J. Hacker, G. T. Keusch, *et al.* 1996. Regulation of the Shiga-like toxin II operon in *Escherichia coli. Infect. Immun.* **64**: 495–502.

Muniesa, M., and J. Jofre. 1998. Abundance in sewage of bacteriophages that infect *Escherichia coli* 0157:H7 and that carry the Shiga toxin 2 gene. *Appl. Environ. Microbiol.* **64**: 2443–8.

Muniesa, M., and J. Jofre. 2004. Factors influencing the replication of somatic coliphages in the water environment. *Antonie van Leeuwenhoek* **86**: 65–76.

Muniesa, M., R. Serra-Moreno, and J. Jofre. 2004. Free Shiga toxin bacteriophages isolated from sewage showed diversity although the *stx* genes appeared conserved. *Environ. Microbiol.* **6**: 716–25.

Nataro, J. P., and J. B. Kaper. 1998. Diarrheagenic *Escherichia coli. Clin. Microbiol. Rev.* **11**: 142–201.

Neely, M. N., and D. I. Friedman. 1998. Functional and genetic analysis of regulatory regions of coliphage H-19B: location of Shiga-like toxin and lysis genes suggest a role for phage functions in toxin release. *Mol. Microbiol.* **28**: 1255–67.

Nesper, J., J. Blass, M. Fountoulakis, and J. Reidl. 1999. Characterization of the major control region of *Vibrio cholerae* bacteriophage k139: immunity, exclusion, and integration. *J. Bacteriol.* **181**: 2902–13.

O'Brien, A. D., and R. K. Holmes. 1987. Shiga and Shiga-like toxins. *Microbiol. Rev.* **51**: 206–20.

O'Brien, A. D., J. W. Newland, S. F. Miller, R. K. Holmes, H. W. Smith, and S. B. Formal. 1984. Shiga-like toxin-converting phages from *Escherichia coli* that cause hemorrhagic colitis or infantile diarrhea. *Science* **226**: 694–6.

O'Shea, Y. A., and E. F. Boyd. 2002. Mobilization of the *Vibrio* pathogenicity island between *Vibrio cholerae* isolates mediated by CP-T1 generalized transduction. *FEMS Microbiol. Lett.* **214**: 153–7.

Oguma, K., H. Iida, and M. Shiozaki. 1976. Phage conversion to hemagglutinin production in *Clostridium botulinum* types C and D. *Infect. Immun.* **14**: 597–602.

Paulsson, J. 2002. Multileveled selection on plasmid replication. *Genetics* **161**: 1373–84.

Plunkett, G., III, D. J. Rose, T. J. Durfee, and F. R. Blattner. 1999. Sequence of Shiga toxin 2 phage 933 W from *Escherichia coli* O157:H7: Shiga toxin as a phage late-gene product. *J. Bacteriol.* **181**: 1767–78.

Popoff, M. R., D. Hauser, P. Boquet, M. W. Eklund, and D. M. Gill. 1991. Characterization of the C3 gene of *Clostridium botulinum* types C and D and its expression in *Escherichia coli*. *Infect. Immun.* **59**: 3673–9.

Poulsen, L. K., F. Lan, C. S. Kristensen, P. Hobolth, S. Molin, and K. A. Krogfelt. 1994. Spatial distribution of *Escherichia coli* in the mouse large intestine inferred from rRNA in situ hybridization. *Infect. Immun.* **62**: 5191–4.

Poulsen, L. K., T. R. Licht, C. Rang, K. A. Krogfelt, and S. Molin. 1995. Physiological state of *Escherichia coli* BJ4 growing in the large intestines of streptomycin-treated mice. *J. Bacteriol.* **177**: 5840–5.

Pullinger, G. D., T. Bevir, and A. J. Lax. 2004. The *Pasteurella multocida* toxin is encoded within a lysogenic bacteriophage. *Mol. Microbiol.* **51**: 255–69.

Quinones, M., H. H. Kimsey, and M. K. Waldor. 2005. LexA cleavage is required for CTX prophage induction. *Mol. Cell* **17**: 291–300.

Quinones, M., B. M. Davis, and M. K. Waldor. 2006. Activation of the *Vibrio cholerae* SOS response is not required for intestinal cholera toxin production or colonization. *Infect. Immun.* **74**: 927–30.

Ramirez, E., M. Schmidt, U. Rinas, and A. Villaverde. 1999. RecA-dependent viral burst in bacterial colonies during the entry into stationary phase. *FEMS Microbiol. Lett.* **170**: 313–17.

Redfield, R. J. 2001. Do bacteria have sex? *Nat. Rev. Genet.* **2**: 634–9.

Reidl, J., and K. E. Klose. 2002. *Vibrio cholerae* and cholera: out of the water and into the host. *FEMS Microbiol. Rev.* **26**: 125–39.

Rietra, P. J., G. A. Willshaw, H. R. Smith, A. M. Field, S. M. Scotland, and B. Rowe. 1989. Comparison of Vero-cytotoxin-encoding phages from *Escherichia coli* of human and bovine origin. *J. Gen. Microbiol.* **135**: 2307–18.

Ritchie, J. M., P. L. Wagner, D. W. K. Acheson, and M. K. Waldor. 2003. Comparison of Shiga toxin production by hemolytic-uremic syndrome-associated and bovine-associated Shiga toxin-producing *Escherichia coli* isolates. *Appl. Environ. Microbiol.* **69**: 1059–66.

Russel, M., and P. Model. 2006. Filamentous phage. In R. L. Calendar and S. T. Abedon (eds.), *The Bacteriophages*, 2nd edn. Oxford: Oxford University Press, pp. 146–60.

Schaechter, M., N. C. Engleberg, B. I. Eisenstein, and G. Medoff. 1998. *Mechanisms of Microbial Disease*. Baltimore, MD: Williams and Wilkins.

Schantz, E. J., and E. A. Johnson. 1992. Properties and use of botulinum toxin and other microbial neurotoxins in medicine. *Microbiol. Rev.* **56**: 80–99.

Smith, H. W., and M. B. Huggins. 1983. Effectiveness of phages in treating experimental *Escherichia coli* diarrhoea in calves, piglets and lambs. *J. Gen. Microbiol.* **129**: 2659–75.

Smith, H. W., M. B. Huggins, and K. M. Shaw. 1987. Factors influencing the survival and multiplication of bacteriophages in calves and in their environment. *J. Gen. Microbiol.* **133**: 1127–35.

Sperandio, V., A. G. Torres, J. A. Giron, and J. B. Kaper. 2001. Quorum sensing is a global regulatory mechanism in enterohemorrhagic *Escherichia coli* O157:H7. *J. Bacteriol.* **183**: 5187–97.

Stahl, F. W., and N. E. Murray. 1966. The evolution of gene clusters and genetic circularity in microorganisms. *Genetics* **53**: 569–76.

Stewart, F. M., and B. R. Levin. 1984. The population biology of bacterial viruses: why be temperate? *Theor. Pop. Biol.* **26**: 93–117.

Strauch, E., R. Lurz, and L. Beutin. 2001. Characterization of a Shiga toxin-encoding temperate bacteriophage of *Shigella sonnei*. *Infect. Immun.* **69**: 7588–95.

Takahashi, M., H. Taguchi, H. Yamaguchi, T. Osaki, A. Komatsu, and S. Kamiya. 2004. The effect of probiotic treatment with *Clostridium butyricum* on enterohemorrhagic *Escherichia coli* O157:H7 infection in mice. *FEMS Immunol. Med. Microbiol.* **41**: 219–26.

Trucksis, M., J. E. Galen, J. Michalski, A. Fasano, and J. B. Kaper. 1993. Accessory cholera enterotoxin (Ace), the third toxin of a *Vibrio cholerae* virulence cassette. *Proc. Natl. Acad. Sci. U.S.A.* **90**: 5267–71.

Tyler, J. S., M. J. Mills, and D. I. Friedman. 2004. The operator and early promoter region of the Shiga toxin type 2-encoding bacteriophage 933W and control of toxin expression. *J. Bacteriol.* **186**: 7670–9.

Tyler, J. S., J. Livny, and D. I. Friedman. 2005. Lambdoid phages and shiga toxin. In M. K. Waldor, D. I. Friedman, and S. L. Adhya (eds.), *Phages: Their Role in Bacterial Pathogenesis and Biotechnology*. Washington DC: ASM Press, pp. 131–64.

Unkmeir, A., and H. Schmidt. 2000. Structural analysis of phage-borne *stx* genes and their flanking sequences in Shiga toxin-producing *Escherichia coli* and *Shigella dysenteriae* type 1 strains. *Infect. Immun.* **68**: 4856–64.

Vaca Pacheco S., Garcia Gonzalez O., and G. L. Paniagua Contreras. 1997. The *lom* gene of bacteriophage λ is involved in *Escherichia coli* K12 adhesion to human buccal epithelial cells. *FEMS Microbiol. Lett.* **156**: 129–32.

Wagner, P. L., D. W. Acheson, and M. K. Waldor. 1999. Isogenic lysogens of diverse Shiga toxin 2-encoding bacteriophages produce markedly different amounts of Shiga toxin. *Infect. Immun.* **67**: 6710–14.

Wagner, P. L., M. N. Neely, X. Zhang, D. W. K. Acheson, M. K. Waldor, and D. I. Friedman. 2001a. Role for a phage promoter in Shiga toxin 2 expression from a pathogenic *Escherichia coli* strain. *J. Bacteriol.* **183**: 2081–5.

Wagner, P. L., D. W. Acheson, and M. K. Waldor. 2001b. Human neutrophils and their products induce Shiga toxin production by enterohemorrhagic *Escherichia coli*. *Infect. Immun.* **69**: 1934–7.

Wagner, P. L., J. Livny, M. N. Neely, D. W. K. Acheson, D. I. Friedman, and M. K. Waldor. 2002. Bacteriophage control of Shiga toxin 1 production and release by *Escherichia coli*. *Mol. Microbiol.* **44**: 957–70.

Wagner, P. L., and M. K. Waldor. 2002. Bacteriophage control of bacterial virulence. *Infect. Immun.* **70**: 3985–93.

Waldor, M. K., and D. I. Friedman. 2005. Phage regulatory circuits and virulence gene expression. *Curr. Opin. Microbiol.* **8**: 459–65.

Waldor, M. K., D. I. Friedman, and S. L. Adhya. 2005. *Phages: Their Role in Bacteriophage Pathogenesis and Biotechnology*. Washington, DC: ASM Press.

Waldor, M. K., and J. J. Mekalanos. 1996. Lysogenic conversion by a filamentous phage encoding cholera toxin. *Science* **272**: 1910–14.

Wallis, T. S., and E. E. Galyov. 2000. Molecular basis of *Salmonella*-induced enteritis. *Mol. Microbiol.* **36**: 997–1005.

Weinbauer, M. G. 2004. Ecology of prokaryotic viruses. *FEMS Microbiol. Rev.* **28**: 127–81.

Woese, C. R. 2004. A new biology for a new century. *Microbiol. Mol. Biol. Rev.* **68**: 173–86.

Zhang, X., A. D. McDaniel, L. E. Wolf, G. T. Keusch, M. K. Waldor, and D. W. Acheson. 2000. Quinolone antibiotics induce Shiga toxin-encoding bacteriophages, toxin production, and death in mice. *J. Infect. Dis.* **181**: 664–70.

Zoetendal, E. G., A. von Wright, T. Vilpponen-Salmela, K. Ben-Amor, A. D. L. Akkermans, and W. M. de Vos. 2002. Mucosa-associated bacteria in the human gastrointestinal tract are uniformly distributed along the colon and differ from the community recovered from feces. *Appl. Environ. Microbiol.* **68**: 3401–7.

Part IV Modeling phage ecology

CHAPTER 15

Modeling bacteriophage population growth

David Stopar and Stephen T. Abedon*

15.1 INTRODUCTION

Different people have different views of what modeling biological phenomena is all about. For some, models are mathematical equations, also known as formal models (Haefner, 1996), that represent, for example, the growth of populations. For others, complex systems are studied in a simplified form in the laboratory, using experiments as physical models (Haefner, 1996). In other words, models, as broadly defined, are simplified representations of reality. This simplifying of the real world, though crucial to modeling success, nevertheless is a dangerous practice. Simplify too far, and a model's solutions, whether mathematical or empirical, will have nothing to do with reality. Simplify too little, and the model becomes too difficult to solve or too complex to derive useful conclusions from. A typical and ecologically very legitimate simplification, for example, is to ignore molecular or physiological details to concentrate instead on properties of whole organisms such as their fecundity or age of death. In this tradition, here we consider phage–bacterial, predator–prey interactions in fluid environments.

The environment of phages and bacteria may be modeled as unchanging over time, changing due to phage or bacterial actions (e.g., bacterial depletion due to phage growth or substrate consumption due to bacterial growth), or changing as a consequence of factors external to both phages and bacteria (e.g., due to chemostat outflow or protist grazing). Taking these various changes into account, in this chapter we consider the formal modeling of phage population growth as it occurs within well-mixed cultures. As we are

* Corresponding author

Bacteriophage Ecology: Population Growth, Evolution, and Impact of Bacterial Viruses, ed. Stephen T. Abedon. Published by Cambridge University Press.

aware that mathematics can be demanding, our goal will be to introduce phage ecological modeling in a way that is accessible to individuals with basic mathematical skills. We will mainly concentrate on discrete models, which are easier to intuitively grasp, but will also introduce continuous approaches to modeling phage growth in liquid cultures (i.e., as also employed in Chapters 2 and 17). Subsequent chapters will introduce complications such as spatial structure in the form of phage growth within plaques (Chapter 16) and phage growth and impact on bacteria within animals over the course of phage therapy (Chapter 17).

15.2 MODELING PHAGE GROWTH PARAMETERS

Most of the work on the modeling of phage population dynamics has considered growth by obligately lytic phages (often also described as virulent; Chapter 1). For models of the interaction between populations of virulent and temperate phages one is advised to consult Noack (1968), Stewart and Levin (1984), or Mittler (1996). In this section we consider basic mathematical algorithms employed to describe the various lytic phage growth parameters, such as rates of virion adsorption, infection latent period, and infection burst size. For convenience, in Table 15.1 we list symbols used as abbreviations for parameters or variables employed in modeling phage growth in this chapter. Techniques used for the experimental determination of phage growth parameters are reviewed in Adams (1959), Eisenstark (1967), Carlson (2005), and Hyman and Abedon (in press).

15.2.1 Overview of phage lytic growth

The phage replication cycle, for the purpose of modeling, may be split into several sequential steps (Abedon, 2006; see also Chapters 2, 3, and 4): (1) a diffusion-driven extracellular "search" for uninfected bacteria; (2) phage attachment to the host; (3) intracellular "growth" and phage-progeny maturation; and (4) release of phage progeny to the extracellular environment. The likelihood of adsorption is controlled by densities of phages (P) and bacteria (N) along with the phage adsorption constant (k). The duration of the resulting phage intracellular growth typically is described as the latent period (L). A lytic phage's latent period ends with the lysis of the cell, which releases B (for burst size) phage progeny into the extracellular environment.

15.2.2 Adsorption

Attachment is usually described as the first step of phage infection. The phage extracellular search is a time of free diffusion that delays the onset

Table 15.1 *Symbols used in models.*

Symbol	Description [units]: example parameter values
B	***Burst size*** [phage/bacterium]: (a) 98, (b) 100, (c) 80, (d) 12–60, (e) 71
F	***Flow of media*** in and out of the chemostat [fractional volume/time]: (a) 0.5–10.0, (b) 3.3, (c) 3.3, (d) 0.3–1.7, (e) 3.3 min^{-1}, all $\times\ 10^{-3}$
k	Phage ***adsorption constant*** [volume/time]: (a) 1.0, (b) 0.02, (c) 5.0, (d) 0.2, (e) 1.0 mL min^{-1}, all $\times\ 10^{-9}$
I_P, I_N, I_M	Phage and uninfected bacteria ***inactivation rates*** [inactivated/time]
K_s	***Substrate density for one-half μ_{max}*** [mass substrate/volume]: (a) n.e., (b) 4.0, (c) 0.073, (d) n.e., (e) 4 g mL^{-1}, all $\times\ 10^{-6}$
L	***Latent period*** [time]: (a) 30, (b) n.e., (c) 36, (d) 108–120, (e) n.e. min
M, M_t	***Infected-bacteria density*** [bacteria/volume]
N, N_t	***Uninfected-bacteria density*** [bacteria/volume]
P, P_t	***Free-phage density*** [phage/volume]
PFUs	***Plaque forming units*** [phage/volume]
S, S_t	***Limiting resource density*** (e.g., glucose) found within growth vessel of chemostat (usually initially set to S_r) [mass substrate/volume]
S_d	***Resource density that carries one bacterium through a single division cycle*** ("reciprocal of the yield of the bacteria"; Bohannan and Lenski, 1997) [mass substrate/bacterial division]: (a) 2.6, (b) 0.5, (c) 2, (d) n.e., (e) 2.6 g bacterium^{-1} division^{-1}, all $\times\ 10^{-12}$
S_r	***Resource density found in sterile media*** [mass substrate/volume]: (a) 0.1, (b) varied, (c) 0.1–0.5, (d) n.e., (e) 2.2 g mL^{-1}, all $\times\ 10^{-6}$
t, t_i	***Time*** and a single ***interval of incrementation***, respectively [time]
μ, μ_M	***Bacterial growth rate*** ($\mu = e^{\mu_M} - 1$, where μ_M is the Malthusian parameter) [time^{-1}]
μ_{max}	***Maximum bacterial growth rate*** [time^{-1}]: (a) 12.3, (b) 11.7, (c) 12.9, (d) 4.2, (e) 12.4 min^{-1}, all $\times\ 10^{-3}$

Parameter values are those presented in select studies by (a) Levin *et al.* (1977) for phage T2, (b) Mittler (1996) for a lytic/temperate phage, (c) Bohannan and Lenski (1997) for phage T4, (d) Middelboe (2000) for marine phage, and (e) Weitz *et al.* (2005). "n.e." means "not employed" or "no equivalent."

of virion attachment and, in so doing, serves to extend the phage generation time. Together this diffusion-mediated delay and subsequent attachment represent phage adsorption. Rates of phage adsorption to bacteria are known to be dependent on the properties and densities of phages and bacteria as well as environmental properties such as temperature, chemical make-up, and viscosity of the medium (Schlesinger, 1932; Stent, 1963; Hyman and Abedon, in press). See Chapter 17 (Table 17.1 and text) for consideration of how phage adsorption constants might vary between the laboratory and the real world. Though not considered here, a theory of phage particle adherence to inanimate surfaces or other planktonic organisms may also be described (e.g., Gerba, 1984).

15.2.2.1 Law of mass action

Interactions between bacteria and phage virions conform, approximately, to the law of mass action. In its original context this law states that the rate of molecular collisions of two chemical species in a dilute gas or solution is proportional to the product of the two concentrations. Mass action serves as a good approximation of phage behavior, especially in well-mixed habitats. For a stochastic treatment of phage attachment to bacteria one should consult Gani (1965) or Chang and Chang (1969). Non-diffusive movement such as bacterial motility is expected to increase phage–bacterial contact above that expected by diffusion alone (Stent and Wollman, 1952; Koch, 1960). We, like most phage modelers, will ignore these complications. All other factors affecting rates of phage adsorption – beside phage numbers, bacterial numbers, and bacterial movement – one traditionally combines into a single parameter called the phage adsorption constant, which, after Stent (1963), we call k.

The value of k is the likelihood that attachment will take place between a single phage and a single bacterium per some unit volume and unit time. Multiplying k by bacterial density (N) and by phage density (P), i.e., $k \cdot N \cdot P$, thus defines the rate of occurrence of phage–bacterial adsorption over a single unit time, t_i, (such as over 1 min) and within a single unit volume (such as within 1 mL). To predict the number of adsorptions over greater or lesser time lengths (t), multiply the above expression by t, i.e., $t \cdot k \cdot N \cdot P$. For the sake of simplification, note that we will consider in the main text of this chapter only the limiting case of unit time intervals and unit volume environments. That is, all specified time durations, for example, will be 1 min and all environmental volumes will be 1 mL. See Table 15.2 for how to introduce into equations incrementation intervals of other than unit time (shorter intervals can give better results though at the expense of computer processing time – as a rule of thumb, one should employ sufficiently short intervals that further reduction by one-half or more does not have a large impact on simulation results). Also, phages, bacteria, and other quantities, such as substrate, should be expressed as densities instead of as particle numbers. Absolute numbers within a given culture may be determined simply by multiplying density by culture volume. See Section 15.7.1, as found online (www.cambridge.org/9780521858458), for additional discussion of incrementation intervals.

15.2.2.2 Modeling phage adsorption assuming mass action

In modeling phage adsorption, the density of unadsorbed (i.e., free) phages remaining in a well-mixed environment, after a given time, may be expressed as a first-order difference equation. That is, phage density depends solely on the density of phages and bacteria in the immediately previous time step in the sequence:

$$P_{t+1} = P_t - k \cdot P_t \cdot N \tag{15.1}$$

where P_t is the density of free phages at time, t, and P_{t+1} is the free-phage density at the immediately subsequent time step, $t + 1$. The expression $k \cdot P_t \cdot N$ represents the fraction of adsorbed phage following a single time step while N is the density of bacteria that are available for phage attachment. In Equation 15.1, N is assumed to be constant in number. This constancy is indicated by a lack of subscript. Alternatively, the instantaneous change in phage density upon phage adsorption may be presented as $dP/dt = -k \cdot P \cdot N$. Since discrete models are intuitively easier to grasp than continuous models, we will discuss discrete models at length before providing a primer on non-discrete or "continuous" models (Section 15.5; see also Chapters 2 and 17).

15.2.3 Latent period and burst size

Once phage genetic material is taken up into the host cytoplasm, the phage infection begins. The duration of phage infection varies as a function of phage genes, general bacterial physiology, and even among genetically identical bacteria found within the same environment. The details of intracellular kinetics of growing viruses have been modeled for several phages (Srinivasan and Rangan, 1970; Gáspár *et al.*, 1979; Buchholtz and Schneider, 1987; Eigen *et al.*, 1991; McAdams and Shapiro, 1995; Endy *et al.*, 1997; Rabinovitch *et al.*, 1999, 2002). Nevertheless, in ecological models phage infection duration is usually modeled as a constant (e.g., Abedon *et al.*, 2001; also Chapters 2, 16, and 17). Here we describe this infection duration as the phage latent period, L. At the end of the lytic-phage latent period, phages are released into the extracellular environment with a characteristic burst size, B, which, like latent period, is typically described as an empirically derived constant.

15.3 SIMULATING PHAGE GROWTH IN BATCH CULTURE

Given a latent period of L during which phages infect bacteria but do not release progeny, a burst size of B phages that are released at the end of a phage latent period, and some algorithm describing phage adsorption such as that employed in Equation 15.1, we can simulate phage population growth in batch culture. Note, however, that in any circumstance simulations are only as good as the underlying mathematical model and should be interpreted with the same caution one brings to any experiment.

15.3.1 Algorithms for simulating phage population growth

In Table 15.2 we provide an overview of the mathematical algorithms employed in this chapter. Phage and bacterial growth-parameter values employed here as well as by various other authors are given in Table 15.1. In each step of the simulation, infected bacteria (M) that are one latent period old (M_{t-L}) instantaneously release a burst size (B) of free phage. The density of phages at the end of a given simulation time step (P_{t+1}) is therefore equal to the density of free phages initially present (P_t) minus the number lost to adsorption ($P_t \cdot k \cdot N$), plus the number released due to cell lysis ($B \cdot M_{t-L}$):

$$P_{t+1} = P_t - k \cdot P_t \cdot N + B \cdot M_{t-L} \tag{15.2}$$

Note that phage adsorption to infected bacteria is ignored but may be incorporated into Equation 15.2 simply by multiplying $k \cdot P_t$ by $(N + M_t)$ rather than solely by N. We will avoid consideration of phage adsorption to infected bacteria for now, however, but include this detail when considering chemostat modeling (Section 15.4). We also will ignore any possibilities for virus aggregation (Grant, 1994).

At the end of a given simulation time interval the number of new infected bacteria may be given by

$$M_{t+1} = k \cdot P_t \cdot N \tag{15.3}$$

as based on Equation 15.1. This algorithm, however, assumes that every phage adsorption to an uninfected bacterium gives rise to a new phage infection; in

Table 15.2 *Summary of algorithms employed in discrete modeling.*

Description	Algorithm	Comments
Phages lost to adsorption	$-P \cdot k \cdot (N + M) \cdot t_i$	t_i is an incrementation interval
Phages gained from burst	$B \cdot k \cdot N_{t-L} \cdot P_{t-L} \cdot t_i$ or $B \cdot e^{-L \cdot F} \cdot k \cdot N_{t-L} \cdot P_{t-L} \cdot t_i$ or $B \cdot e^{-L \cdot F} \cdot N_{t-L} \cdot (1 - e^{-P_{t-L} \cdot k \cdot t_i})$	N_{t-L} and P_{t-L} are the number of bacteria and phages present during an incrementation interval one latent period previous; the two latter expressions incorporate phage loss due to outflow
Infected bacteria gained from phage adsorption	$N \cdot P \cdot k \cdot t_i$ or $N \cdot (1 - e^{-P \cdot k \cdot t_i})$ or $N \cdot (1 - e^{-(P \cdot (1 - e^{-N \cdot k \cdot t_i})/N)})$	employ the first algorithm for $\sim P \cdot k \cdot t_i < 0.1$; the second for $\sim P \cdot k \cdot t_i < 0.1$; and the third when $\sim N \cdot k \cdot t_i > 0.1$
Uninfected bacteria gained from division	$N \cdot ((1 + \mu_{\max})^{t_i} - 1) . \frac{S}{K_s + S}$	this uses the Monod equation; note the placement of t_i
Loss to outflow (F) or other (I)	$-X \cdot F \cdot t_i - X \cdot I_x \cdot t_i$	where X is phages, bacteria, infected bacteria, or substrate density
Substrate gains and losses due to flow	$F \cdot (S_r - S) \cdot t_i$	S_r resource density in the reservoir, S resource density in the chemostat
Substrate consumed	$S_d \cdot (N + M) \cdot ((1 + \mu_{\max})^{t_i} - 1) \cdot \frac{S}{K_S + S}$	note that the metabolic impact of both infected and uninfected bacteria are shown and that $\mu_{\max}$ is *not* the Malthusian parameter

reality multiple phages often can adsorb individual bacteria. Thus, Equation 15.3 inherently overestimates the number of phage infections resulting from phage adsorption. One approach toward making sure that multiple adsorptions per time interval are only minimally ignored is to employ sufficiently short incrementation intervals such that approximately $t_i \cdot k \cdot P_t > 0.1$ is unlikely. Alternatively, especially if holding t_i constant at unit time, then one can employ an algorithm for infected-bacterium genesis that allows for a reduction in infection gains with higher phage multiplicities:

$$M_{t+1} = \left(1 - e^{-k \cdot P_t}\right) \cdot N \tag{15.4}$$

where $\left(1 - e^{-k \cdot P_t}\right) \cdot N \approx k \cdot N \cdot P_t$ for $k \cdot P_t < 0.1$.

Note that Equation 15.4 comes from assumptions of a Poisson distribution of phage adsorption to bacteria (e.g., Stent, 1963). The expression $t_i \cdot k \cdot P_t$ (above with t_i set equal to one) represents phage multiplicity of adsorption assuming phage adsorption with replacement (i.e., that phage density does not decline over the course of a simulation incrementation interval; Chapter 3). The quantity $e^{-t_i \cdot k \cdot P_t}$ is the fraction of bacteria that are adsorbed by zero phages. The fraction of bacteria adsorbed by one or more phages is therefore given by $1 - e^{-t_i \cdot k \cdot P_t}$. We can relax the assumption of phage adsorption with replacement by substituting $P_t \cdot (1 - e^{-t_i \cdot k \cdot N})/N$ for $t_i \cdot k \cdot P_t$ in $e^{-t_i \cdot k \cdot P_t}$ (Chapter 3). Here we mostly avoid making this substitution for the sake of avoiding excessive mathematical complexity and because doing so has a large impact only given high cell densities, lengthy incrementation intervals, or large adsorption coefficients. See, however, Fig. 15.1 for an exploration of the differences between these three approaches towards modeling phage-infection genesis. Note that Equation 15.4, which unless otherwise indicated we will employ instead of Equation 15.3, represents under many circumstances a not unreasonable compromise between simulation realism and simulation complexity. In particular, in Fig. 15.1 note the difference between approach one (based on Equation 15.3) and approaches two and three (the second approach is based on Equation 15.4 while the third approach employs a more complicated algorithm). See also the similarity, except at very high bacterial densities or long time intervals, between the second and third approaches. The algorithms employed in these three approaches are compared in the legend to Fig. 15.1.

15.3.2 Simulating phage population growth

In Fig. 15.2 we present the above-described simulation of phage population growth within batch culture assuming an absence of bacterial replication,

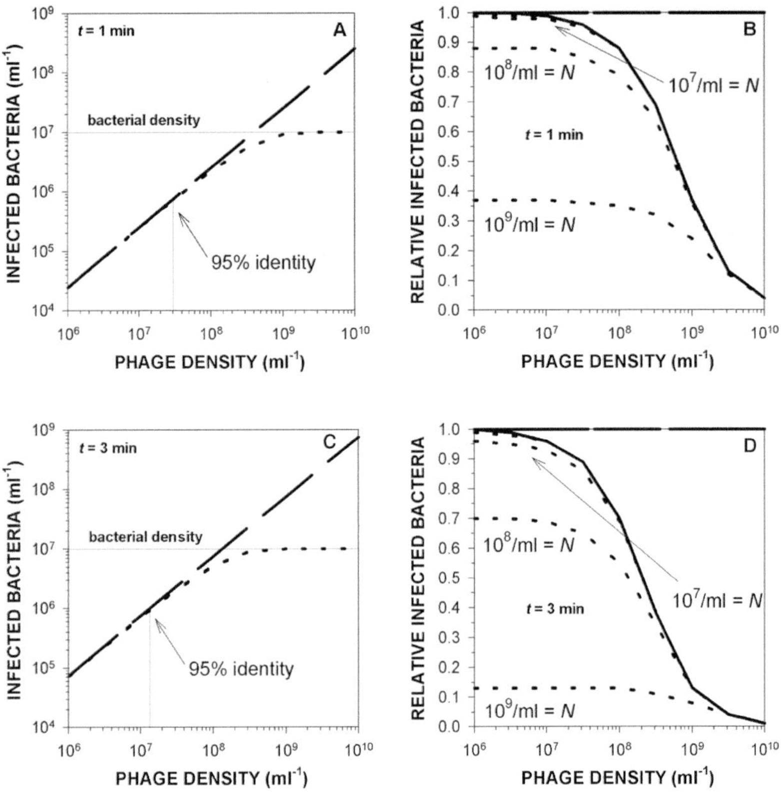

Figure 15.1 Three algorithms for infecting bacteria: (i) $M = t \cdot k \cdot P \cdot N$ (based on Equation 15.3), (ii) $M = (1 - e^{-t \cdot k \cdot P}) \cdot N$ (based on Equation 15.4), and (iii) $M = (1 - e^{-(P \cdot (1 - e^{-t \cdot k \cdot N})/N) \cdot}) \cdot N$ (which is an elaboration of Equation 15.4). Time (t), is indicated in the panels and $k = 2.5 \times 10^{-9}$ min mL^{-1}. The second approach ("ii") is used otherwise throughout this chapter. **Panels A and C**: The vertical lines indicate the phage threshold density at which the second approach (solid line) calculates ~95% as many infected bacteria as the first approach (dashed line). The third approach is graphed as a dotted line but is obscured in these panels by the second approach's solid line. Bacterial density, N, is 10^7 mL^{-1} in these panels. **Panels B and D**: The second approach (solid line) and third approach (dotted line) are presented relative to curves generated using the first approach (dashed line) at different bacterial densities, N, as indicated. Shown for the second and third approaches are $N = 10^5$, 10^6, 10^7, 10^8, and 10^9 bacteria per mL, though all but the third approach at $N = 10^7$, 10^8, and 10^9 bacteria per mL is obscured under the solid line.

keeping track of densities of free phages (P_t), of newly infected bacteria (M_{new}), and of plaque forming units (PFUs, i.e., what one would observe were one to directly plate phage-containing bacterial cultures; e.g., Hyman and Abedon, in press). The result is an exponential increase in free-phage density. Ignoring trade-offs, the rate of PFU increase will be greater the larger the phage burst size (Fig. 15.2A), the faster the phage adsorption rate (Fig. 15.2B), and the higher the bacterial density (Fig. 15.2C). This rate will also be greater given shorter phage latent periods (Fig. 15.2D).

15.3.3 Algorithms used to simulate phage–bacterial co-culture

Modeling assuming an absence of bacterial growth allows us to concentrate solely on the impact of phage characteristics on the dynamics of phage population growth (as in Fig. 15.2), but it has a major drawback in that it ignores phage–bacterial community dynamics. In this section we extend the above simulation to include bacterial growth as well as bacterial death due to phage infection. This approach represents an interim between simulation solely of batch-culture phage population growth and simulation of phage population growth within bacterial cultures actively growing in chemostats (Section 15.4).

The simplest approach toward simulating bacterial growth is to assume that bacteria replicate at rates that are influenced neither by bacterial nor by substrate densities. If we specify growth rate, μ, as the fractional increase in bacterial density per time interval, then the resulting first-order difference equations are

$$\begin{aligned} N_{t+1} &= N_t + \mu \cdot N_t = N_t \cdot (1+\mu)\,, \\ N_{t+2} &= N_{t+1} \cdot (1+\mu) = N_t \cdot (1+\mu) \cdot (1+\mu) = N_t \cdot (1+\mu)^2\,, \\ N_{t+x} &= N_t \cdot (1+\mu)^x = N_t + ((1+\mu)^x - 1) \cdot N_t \end{aligned} \tag{15.5}$$

where $\mu > 0$. We provide the additional equations as a primer on how one modifies μ to take into account non unit-time incrementation (Table 15.2). Note that μ in Equation 15.5 is *not* the Malthusian parameter (which as follows we will call μ_M, and which is otherwise known as the instantaneous growth-rate parameter or the specific growth rate). This is because we are not defining bacterial growth as a continuously compounding function but instead as a discrete, per-unit-time increase. We can easily convert from μ_M to μ, as employed here, with $\mu = e^{\mu_M} - 1$. For example, the maximum specific growth rate term of Bohannan and Lenski (1997) is $\mu_M = 0.7726$ per

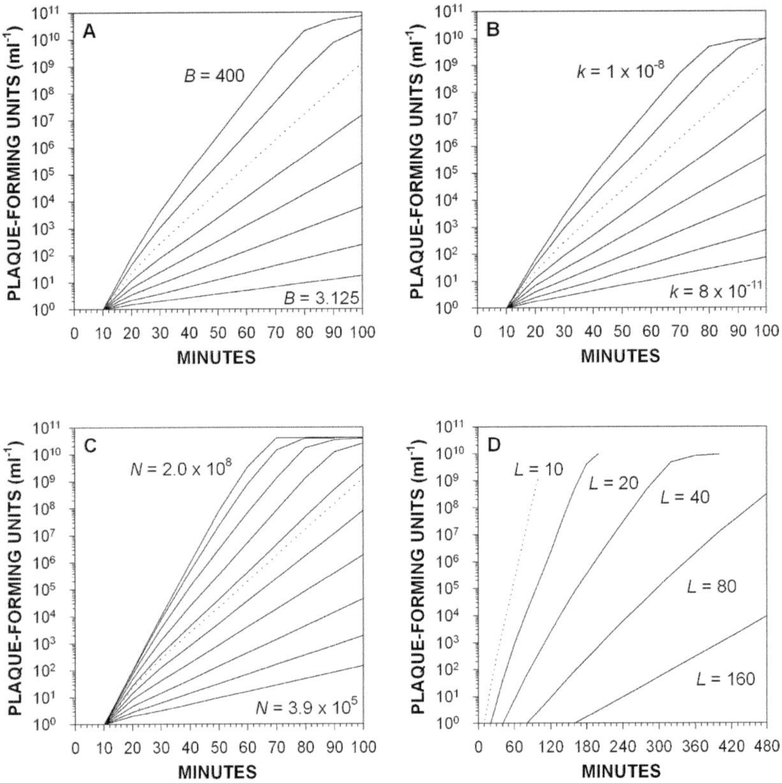

Figure 15.2 Recursive batch-growth, Excel-based simulations with constant bacterial density. Default parameters for all panels are $L = 10$ min, $B = 100$ phages/lysed bacterium, $k = 2.5 \times 10^{-9}$ mL min^{-1} (from Stent, 1963), $N = 1 \times 10^{7}$ bacteria mL^{-1}, and initial phage density is 1 per mL. Curves based on these default parameter values are shown as dotted lines. PFUs are indicated as observed at the beginning of simulation incrementation intervals. Rate of appearance of infected bacteria is modeled employing the algorithm presented in Equation 15.4. Adsorption to bacteria infected during previous rounds (secondary adsorption) is not considered. **Panel A**: Burst size (B) varies from top to bottom as 400, 200, 100, 50, 25, 12.5, 6.25, 3.125 phages/lysed bacterium. **Panel B**: Adsorption constant (k) varies from top to bottom as 10, 5, 2.5, 1.25, 0.625, 0.3125, 0.15625, and 0.078125 $\times 10^{-9}$ (i.e., 1.0×10^{-8} down to 7.8×10^{-11}) mL min^{-1}. **Panel C**: Constant bacterial density (N) varies for solid lines from top to bottom as 2.0, 1.0, 0.5, 0.25, 0.1, 0.125, 0.0625, 0.03125, 0.015625, 0.078125, 0.03906 $\times 10^{8}$ (bacteria mL^{-1}). **Panel D**: Latent period (L) varies (in min) as indicated.

hour, which corresponds to $\mu = 1.165$ per hour (which is a doubling time of ~54 min).

In addition to consideration of bacterial growth, we need to take bacterial death also into account, which in batch culture we will assume occurs only as a consequence of phage adsorption. The rate of phage adsorption to bacteria has already been considered for the formation of infected bacteria in Equation 15.4. Thus, we can account for both bacterial growth and phage-induced death by

$$
\begin{aligned}
N_{t+1} &= N_t + \mu \cdot N_t - M_{t+1}, \\
N_{t+1} &= N_t \cdot ((1+\mu) - (1 - e^{-k \cdot P_t})), \\
N_{t+1} &= N_t \cdot (\mu + e^{-k \cdot P_t})
\end{aligned}
\qquad (15.6)
$$

with the algebra resulting from substituting Equation 15.4 for M_{t+1}.

15.3.4 Simulating phage–bacterial batch co-culture

In Figs. 15.3A and 15.3B we present example simulations incorporating bacterial growth and death. Note in comparing these figures that final phage density is dependent on starting bacterial density and that, given sufficient phage densities, bacteria can be driven to extinction. In the following section we elaborate upon these simulations with bacterial growth and death in the guise of chemostat modeling. See Chapters 2 and 17 as well as Lenski (1988), Schrag and Mittler (1996), and Abedon (2006) for consideration of limitations on the phages' ability to drive bacteria to extinction.

15.4 CHEMOSTAT PHAGE–BACTERIAL CO-CULTURE

Well-mixed continuously flowing cultures known as chemostats (e.g., Smith and Waltman, 1995; Haefner, 1996; Chapter 9) provide the environmental basis for the majority of models of phage growth within communities of replicating bacteria. Chemostats differ from batch culture in that (1) nutrients continuously flow into chemostat growth chambers from a nutrient reservoir, thereby replenishing those lost in the course of bacterial growth, and (2) wastes, bacteria, and phages flow out. The latter has a number of consequences: wastes are limited in their ability to build up, bacteria never enter stationary phase, and phage densities readily decline. As phage densities decline due to a combination of reduced opportunity for bacterial infection and outflow, bacterial population densities can recover to levels that again are capable of supporting phage replication. Even ignoring the possibilities

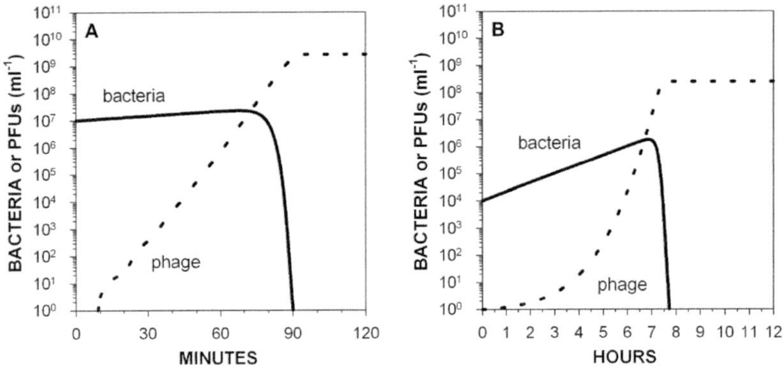

Figure 15.3 Recursive batch-growth, Excel-based simulations with incorporated bacterial growth. Default parameters are those used for Fig. 15.2 plus bacterial growth rate, $\mu = 9.6 \times 10^{-3}$ min^{-1}. Bacterial density is represented by solid lines and phage density is represented by dotted lines. **Panel A**: Initial bacterial density (N_o) is 1×10^7 bacteria per mL. **Panel B**: Same as panel A except that $N_o = 1.0 \times 10^4$ bacteria per mL. Note that the leveling-off of phage densities in the presented curves reflects limitations in bacterial availability. For similar effects of starting bacterial densities, see Fig. 17.1C, Chapter 17.

of inhomogeneities in chemostat cultures as spatial refuges for bacteria or the coevolution of phages and bacteria (Schrag and Mittler, 1996), bacteria and therefore phages can survive. See Chapter 2 for less phage-specific discussion of predator–prey oscillations and extinction.

15.4.1 Inflow and outflow in the chemostat

For a schematic representation of factors affecting phage, bacterial, and substrate densities within chemostats, see Fig. 15.4. Note that growth medium is added to the growth chamber at a constant flow rate, *F*. Stirring assures that incoming broth is rapidly dispersed and, furthermore, that the growth chamber contains minimal spatial heterogeneity (for a review of phage population growth in the absence of spatial heterogeneity, see Chapter 3; for discussion of phage growth with spatial heterogeneity, see Chapters 2, 4, and 16). Outflow of medium and organisms occurs at the same per-volume rate as inflow.

15.4.2.1 Modeling chemostat bacterium losses and gains

As a consequence of phage adsorption there is an associated loss of uninfected bacteria. Loss of uninfected bacteria additionally can occur via outflow from the chemostat as well as via other means of inactivation such as

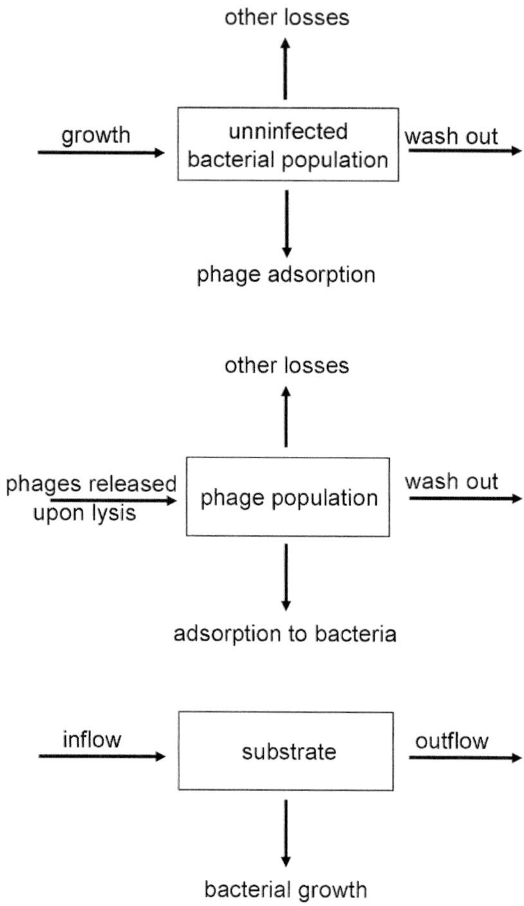

Figure 15.4 Schematic representation of factors affecting phage, bacterial, and substrate densities within chemostats. Not indicated are any "other" losses for substrate. The direction of the arrow indicates an increase (toward) or decrease (away) of uninfected bacterial population, phage population, or substrate concentration in the chemostat.

DNA damage or protist grazing (see Chapter 10, and Weinbauer, 2004). In addition, bacteria will increase in number as a consequence of cell division (Section 15.3.3). Taken together, the number of uninfected bacteria in the chemostat after one simulation incrementation may be given as:

$$N_{t+1} = N_t + \mu_{\max} \cdot \frac{S_t}{K_S + S_t} \cdot N_t - F \cdot N_t - k \cdot P_t \cdot N_t - I_N \cdot N_t \quad (15.7)$$

In words, this equation says that the density of uninfected bacteria at the end of an interval (N_{t+1}) is dependent on the density of the bacteria at the beginning of an interval (N_t). The bacterial density is increased by division in a bacterial density-dependent manner, which here we describe as $\mu_{\max} \cdot (S_t/(K_S + S_t)) \cdot N_t$. This expression is the Monod equation, where $\mu_{\max}$ is maximal bacterial growth rate, K_S is limiting substrate density at which bacteria grow at one-half maximum growth rate, and S_t is the limiting substrate density (Smith and Waltman, 1995; Haefner, 1996). In turn the bacterial density is decreased by outflow ($F \cdot N_t$), phage adsorption ($k \cdot P_t \cdot N_t$), and other mechanisms ($I_N \cdot N_t$). The latter we have arbitrarily indicated as occurring at a constant per-bacterium rate, I_N.

15.4.2.2 Modeling substrate losses and gains

In modeling substrate losses and gains, one generally considers only a single limiting resource, such as glucose, that controls bacterial growth rates. Substrate is gained via inflow from the chemostat reservoir and is lost due to media outflow and bacterial utilization. A simplifying assumption may be made that rates of substrate utilization by uninfected bacteria per bacterial division (S_d) is equivalent on a per-bacterium basis to that of phage-infected bacteria:

$$S_{t+1} = S_t + F \cdot (S_r - S_t) - \mu_{\max} \cdot \frac{S_t}{K_S + S_t} \cdot S_d \cdot (N_t + M_t) \qquad (15.8)$$

Here S_t is substrate concentration at the beginning of a time step while $S_r - S_t$ refers to the difference between incoming and outgoing substrate densities. The next term multiplies the total number of bacteria ($N_t + M_t$) by the "reciprocal of the yield of the bacteria" (Bohannan and Lenski, 1997), which is a measure of substrate utilization per bacterial division (S_d), along with an expression describing the number of new bacteria produced over the time interval per individual starting bacterium (which, as above, is the Monod equation). Note that an important modeling detail, not considered here, is substrate utilization by phage-resistant bacteria (Levin *et al.*, 1977; Bohannan and Lenski, 1997).

15.4.2.3 Modeling phage losses and gains

The modeling of free phages in chemostats, as well as the loss of infected bacteria, is complicated by delays (time lags) introduced by phage infection, i.e., the interval between phage adsorption and the phage-induced bacterial lysis that results in phage-progeny release (latent period, L). We therefore introduce phage gains and losses without consideration of this latent-period delay. We then improve modeling realism by introducing latent-period delays.

15.4.2.3.1 Phage losses and gains without latent-period delays

Ignoring for the moment phage multiplication, there are three routes by which phage concentration in chemostat growth chambers can change. The first is the loss of free phage due to adsorption to infected and uninfected bacteria (at rate, $k \cdot (N_t + M_t)$). The second is the loss of free phages due to chemostat outflow (at rate, $F \cdot P_t$). The third is by various additional means, such as adsorption to cell debris (Rabinovitch *et al.*, 2003), together at a per-free-phage rate of $I_p \cdot P_t$):

$$P_{t+1} = P_t - k \cdot (N_t + M_t) \cdot P_t - F \cdot P_t - I_p \cdot P_t \tag{15.9}$$

Note that one simplifying assumption is implicitly incorporated into the above equation, and that is that the same adsorption constant, k, holds for adsorption to both uninfected (N_t) and infected bacteria (M_t).

If phages are allowed to multiply, then gains in free phages occur as a consequence of phage-induced lysis of infected bacteria, which releases B viruses per infected cell. Total phage gains and losses thus are given by:

$$P_{t+1} = P_t + B \cdot k \cdot N_t \cdot P_t - k \cdot N_t \cdot P_t - F \cdot P_t - I_p \cdot P_t \tag{15.10}$$

Released free phages are then available during the next round for adsorption. We do not consider in Equation 15.10 any losses of free phages to infected-cell adsorption or infected-cell decay, because infected bacteria are at best transiently present when not allowing for any time delay between infection initiation and phage-mediated lysis. Simulations lacking time delays are presented in Figs. 15.5C and 15.5D. We will avoid discussion of these simulations, however, until we have introduced time delays, as employed in Figs. 15.5A and 15.5B.

15.4.2.3.2 Incorporating latent-period time delays

Incorporating time delays generally results in decreased system stability (Thingstad, 1974; Levin *et al.*, 1977; Chapter 2). For example, compare Figs. 15.5A and 15.5C. Consistently, greater likelihoods of bacterial extinction in chemostat models result when lysis does not occur immediately after phage adsorption. Despite this impact, a number of studies ignore time delays when formulating various chemostat models (e.g., Thingstad, 2000; Weitz *et al.*, 2005).

Two additions to Equation 15.10 are necessary to incorporate time delays. The first is to recognize, as seen in Equation 15.2, that the number of lysing bacteria during a given interval is a function of the number of bacteria that had been infected one latent period previously. The second is accounting for infected-bacterium losses to outflow over the course of a phage latent period. The latter is done via the term $e^{-L \cdot F}$, which is the fraction of a growth chamber's volume remaining following outflow (F) over the course of one latent period (L):

$$P_{t+1} = P_t + B \cdot k \cdot M_{t-L} \cdot e^{-L \cdot F} - k \cdot (N_t + M_t) \cdot P_t - F \cdot P_t - I_p \cdot P_t \tag{15.11}$$

Here M_{t-L} describes the density of bacteria that were infected one latent period (L) previously, that is, as by Equation 15.4, $M_{t-L} = N_{t-L} \cdot (1 - e^{-k \cdot P_{t-L}})$. Note that Equation 15.11 is otherwise equivalent to Equation 15.10 except that Equation 15.11 explicitly considers adsorption to (and inactivation by) already phage-infected bacteria (i.e., $k \cdot (N_t + M_t) \cdot P_t$ replaces $k \cdot N_t \cdot P_t$).

15.4.2.4 Infected-bacteria accounting with latent-period delays

Incorporation of phage latent periods complicates accounting for infected-bacteria losses due to phage-induced lysis, but significantly improves modeling realism. As above, we continue to model the increase in density of infected bacteria due to phage adsorption to uninfected bacteria ($k \cdot P_t \cdot N_t$) as well as losses to some combination of phage-induced lysis ($e^{-L \cdot F} \cdot M_{t-L}$), outflow ($F \cdot M_t$), and other means ($I_M \cdot M_t$). Thus,

$$M_{t+1} = M_t + k \cdot P_t \cdot N_t - e^{-L \cdot F} \cdot M_{t-L} - F \cdot M_t - I_M \cdot M_t \tag{15.12}$$

where $M_{t-L} = N_{t-L} \cdot \left(1 - e^{-k \cdot P_{t-L}}\right) \approx k \cdot N_{t-L} \cdot P_{t-L}$ when $k \cdot P_{t-L} \leq 0.1$ such that Equation 15.4 divided by Equation 15.3 is ≥ 0.95 (Fig. 15.1).

15.4.2.5 Summary of a chemostat model

We are now in a position to restate the four difference equations describing chemostat dynamics, ignoring for the sake of simplicity in presentation any phage or bacterial losses due to means other than chemostat outflow, phage adsorption, and phage-induced bacterial lysis:

$$S_{t+1} = S_t + F \cdot (S_r - S_t) - \mu_{\max} \cdot \frac{S_t}{K_S + S_t} \cdot S_d \cdot (N_t + M_t) \tag{15.13}$$

$$N_{t+1} = N_t + \mu_{\max} \cdot \frac{S_t}{K_S + S_t} \cdot N_t - F \cdot N_t - k \cdot P_t \cdot N_t \tag{15.14}$$

$$M_{t+1} = M_t + k \cdot P_t \cdot N_t - e^{-L \cdot F} \cdot M_{t-L} - F \cdot M_t \tag{15.15}$$

$$P_{t+1} = P_t + B \cdot M_{t-L} \cdot e^{-L \cdot F} - k \cdot (N_t + M_t) \cdot P_t - F \cdot P_t \tag{15.16}$$

The above algorithms have been generalized for non unit-time incrementation intervals in Table 15.2.

15.4.2.6 Stability analysis

An important issue in modeling is stability analysis of a multidimensional parameter space such as that defined by Equations 15.13–15.16. To perform a stability analysis of a system of difference equations, steady-state solutions (fixed points) are obtained. Next the system of equations is linearized in the vicinity of steady-state solutions and the eigenvalues of the Jacobian matrix are calculated. An interested reader should, for example, consult Rabinovitch *et al.* (2003) for stability analyses of a similar phage–bacterium system based on a set of difference equations. Although it is often very instructive to know the behavior of a system in the vicinity of the steady state, steady states are not a prerequisite for doing a simulation. Thus, stability analysis is a limited, though rather exact, means of analyzing chemostat models, while simulation inspection, as considered next, provides greater versatility.

15.4.2.7 Simulations of chemostat co-cultures

In Fig. 15.5 we present results of chemostat simulations using Equations 15.13–15.16 (all simulations shown were solved as spreadsheets using Microsoft Excel®; see Section 15.7.2, as found online (www.cambridge.org/9780521858458). Figure 15.5A shows the typical out-of-phase oscillations between phage and bacterial population densities where the presence of sufficient numbers of bacteria results in increases in phage densities, which results in decline in bacterial densities (due to increased predation; that is, "killing the winner," as reviewed in Chapter 10). After phage densities fall sufficiently, bacterial replication again occurs at rates in excess of those losses due to phage infection and outflow, resulting in recovery of the bacterial population and repeat of the process. Ultimately, however, the instability of the system is dictated by the densities achieved by the phage population along with the host's ability to evade phage attack (Abedon, 2006).

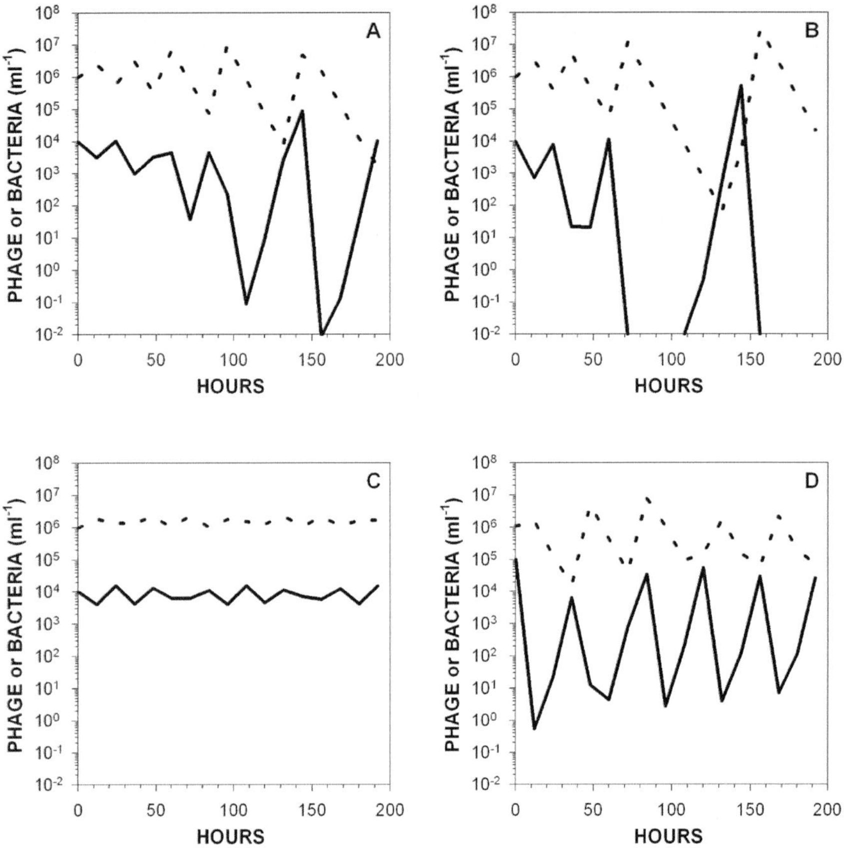

Figure 15.5 Recursive, Excel-based chemostat simulations using parameter values and inferred starting conditions of the 0.5 mg L^{-1} glucose chemostat simulation, 1×10^4 bacteria per mL, and 1×10^6 free phages per mL, approximately as employed by Bohannan and Lenski (1997). In all cases $t_i = 3$ min while dotted and solid lines are phage and bacterial densities, respectively. **Panel A** allows phage multiplicity to exceed 1.0 (i.e., Equation 15.4 is used). **Panel B** parameter values are the same as in panel A. However, instead of using the Monod equation we employ density-independent bacterial growth based on Equation 15.5 rather than Equation 15.7. **Panel C** shows the impact of ignoring time delays on chemostat stability and is equivalent to Panel A except that the latent-period time delay in the latter is ignored. **Panel D** indicates the importance of starting conditions and is equivalent to Panel C except that we use a starting bacterial density of 1×10^5 mL^{-1}.

In Fig. 15.5A one observes increasing instability as bacterial and therefore phage densities reach higher levels with each round. Depending on the volume of the environment, this increasing oscillation may eventually result in bacterial and then phage extinction.

Greater instability is introduced by removing density-dependent controls on bacterial growth (Fig. 15.5B). In Fig. 15.5C we present the effects of ignoring latent-period time delays, which, as noted, increases the stability of these systems. As shown in Fig. 15.5D, increasing initial bacterial density, while still ignoring latent-period time delays, increases the amplitude of the oscillations. However, a gradual damping of oscillations with increasing time of simulation is also seen. See Chapter 2 for more general considerations of the impact of initial phage and bacterial densities of oscillation amplitudes.

15.5 CONTINUOUS MODELS

With the discrete models described above most of the problems in phage growth modeling can be addressed. In general, however, for processes that change continuously over time, or which are discrete but with small enough time steps, continuous models using differential equations can provide better approximations of natural phenomena. Unfortunately, despite their naturalness, differential equations are frequently difficult to solve or, worse, have solutions that can only be numerically approximated (that is, solved using recursive modeling, as above). Mathematically continuous chemostat models usually consist of several coupled nonlinear differential equations describing the time evolution of different species such as virulent and temperate phages; infected, not-infected, sensitive, or resistant bacteria; and bacterial nutrients (Stewart and Levin, 1984; Lenski and Levin, 1985; Husimi, 1989; Schrag and Mittler, 1996; Middelboe, 2000; Carletti, 2002; Weitz *et al.*, 2005). Most of these models are based on the seminal work of Levin *et al.* (1977).

Here, before introducing phages into the system, we will first briefly review mathematical descriptions of bacterial and substrate steady-state densities. Note that our overall goal is to avoid ignoring continuous models in this chapter rather than to present a comprehensive review of their utility and mechanics. Chapters 2 and 10 also present continuous models of phage–bacterial co-culture.

15.5.1 Fixing steady-state densities without phages

Chemostat cultures allow one to experimentally obtain stable cellular (e.g., bacterial) growth rates that, at equilibrium, are usually lower than those seen during logarithmic growth in simple batch cultures. In theory it should be possible, by changing the chemostat flow rate, to fix the specific growth rate of a single type of bacterium in the chemostat (μ) at any value from zero to maximum (μ_{max}). Note that in continuous models both μ and μ_{max} refer to the Malthusian parameter (as discussed in Section 15.3.3).

15.5.1.1 Fixing the bacterial steady-state growth rate

In the absence of phages, the increase in bacterial density in the chemostat is given by a "biomass balance" of

$$\frac{dN}{dt} = (\mu - F) \cdot N \tag{15.17}$$

where μ is the Malthusian parameter (Section 15.3.3), F is the dilution rate of the chemostat, and N, cell density, is a measure of biomass. The equation states that the number of bacteria in the chemostat increases at a rate of $\mu \cdot N$ while at the same time it decreases at a rate of $N \cdot F$ due to washout. Note that in writing differential equations we indicate the instantaneous rate of change of bacterial density, so there is no need to indicate the bacterial density with a subscript of t. In the steady state when $dN/dt = 0$ (that is, there is no change in bacterial density over time) we have $0 = (\mu - F) \cdot N \rightarrow \mu = F$. That is, the per-capita rate of bacterial increase due to replication is equal to the per-capita rate of bacterial loss due to washout.

15.5.1.2 Fixing the substrate steady-state density

Gain in substrate occurs due to inflow from the chemostat reservoir, whereas loss occurs via both bacterial utilization and outflow. One can describe an instantaneous change of substrate density as

$$\frac{dS}{dt} = F \cdot (S_r - S) - S_d \cdot \mu \cdot N \tag{15.18}$$

where F is the chemostat dilution rate and S_d is the reciprocal of the growth yield, expressing the quantitative nutrient requirement of an organism. S_r is the concentration of the substrate in the nutrient reservoir, S is the concentration of the substrate in the chemostat growth chamber, and μ is the bacterial growth rate.

15.5.1.3 Simultaneous fixing of substrate and bacteria

One can solve Equations 15.17 and 15.18 for steady-state concentrations of substrate and biomass, where $dN/dt = dS/dt = 0$. By rearranging Equations

15.17 and 15.18, taking the Monod relation for specific growth rate (Section 15.4.2.1), and using $\mu = F$ we obtain non-trivial steady-state values for substrate concentration in the chemostat growth chamber ($\tilde{S}$) and biomass density ($\tilde{N}$)

$$\tilde{S} = \frac{K_S \cdot F}{\mu_{\max} - F} \tag{15.19}$$

$$\tilde{N} = \frac{1}{S_d}(S_r - \tilde{S}) = \frac{1}{S_d}\left(S_r - \frac{K_S \cdot F}{\mu_{\max} - F}\right) \tag{15.20}$$

Thus, in the absence of mechanisms of bacterial loss other than outflow and of substrate loss other than outflow and bacterial consumption, we have an expectation that both bacterial and substrate densities will reach steady-state equilibria (i.e., constant densities) as defined by these equations.

15.5.2 Continuous models with phage infection

The steady-state values for bacteria and substrate as given above are perturbed if phages are introduced into a chemostat. The basic theory of continuous models of phage growth within a chemostat was developed by Levin *et al.* (1977) and Lenski and Levin (1985). You will note that these models are essentially equivalent to those represented by Equations 15.13–15.16 for the limit where t_i goes to zero.

$$\frac{dS}{dt} = (S_r - S) \cdot F - S_d \cdot \mu \cdot N \tag{15.21}$$

$$\frac{dN}{dt} = \mu \cdot N - F \cdot N - k \cdot P \cdot N \tag{15.22}$$

$$\frac{dM}{dt} = k \cdot P \cdot N - F \cdot M - k \cdot \hat{N} \cdot \hat{P} \cdot e^{-L \cdot F} \tag{15.23}$$

$$\frac{dP}{dt} = B \cdot k \cdot \hat{N} \cdot \hat{P} \cdot e^{-L \cdot F} - P \cdot F - k \cdot P \cdot N \tag{15.24}$$

$\hat{N}$ and $\hat{P}$ refer to uninfected bacteria and free-phage densities, respectively, as observed one latent period, L, earlier (we employ notation that is different from that employed above – i.e., different from N_{t-L} and P_{t-L} – to avoid introducing t subscripts into continuous equations). In the simplifying case in the absence of consideration of the time delay associated with the phage latent period, without accounting for infected bacteria, and using the Monod type growth kinetics (Section 15.4.2.1) the corresponding densities can be written as (Weitz *et al.*, 2005)

$$\frac{dS}{dt} = (S_r - S) \cdot F - S_d \cdot N \cdot \mu_{\max} \frac{S}{K_S + S} \tag{15.25}$$

$$\frac{dN}{dt} = N \cdot \mu_{\max} \frac{S}{K_S + S} - F \cdot N - k \cdot N \cdot P \tag{15.26}$$

$$\frac{dP}{dt} = B \cdot k \cdot N \cdot P - F \cdot P \tag{15.27}$$

With the help of dimensional analysis, which reduces the number of original parameters in the equation, the system of Equations 15.25–15.27 may be further simplified. This is basically equivalent to saying that we have determined what are the minimal numbers of independent parameters in the equations. Bacterial and phage densities may therefore be given as

$$\frac{dN}{dt} = -N + \gamma \cdot \frac{(1 - N - P) \cdot N}{1 - N - P + \omega} - \kappa \cdot N \cdot P \tag{15.28}$$

$$\frac{dP}{dt} = -P + \kappa \cdot N \cdot P \tag{15.29}$$

using three dimensionless parameters: $\gamma = \mu_{\max}/F$, $\omega = K_S/S$, and $\kappa = S_d \cdot k \cdot B \cdot S/F$. For stability analysis of this simplified chemostat model, see Weitz *et al.* (2005). An alternative though more complex approach, performing the stability analyses with stochastic perturbation of some of the main parameters, has been proposed by Carletti (2002).

It is worth noting, in linking genetic change to community evolution, that the above basic model as presented in Equations 15.21–15.24 may be extended to include (1) incorporation of mutational events into the dynamics (Lenski and Levin, 1985), (2) the effect of resistant bacteria (Schrag and Mittler, 1996; Bohannan and Lenski, 1999, 2000a, 2000b), (3) community shift upon resource enrichment (Bohannan and Lenski, 1997), (4) the dynamics of lysogenic bacteria (Noack, 1968; Stewart and Levin, 1984), (5) coevolution branching leading to distinct quasipecies (Weitz *et al.*, 2005), and (6) phage selection upon periodic oscillations of the temperature in the chemostat (Aita and Husimi, 1994).

15.6 CONCLUDING REMARKS

The modeling assumptions one makes can determine the outcome of simulations. It therefore can be instructive to consider how modifying underlying assumptions can change results. As a consequence of oversimplification or other flaws in underlying assumptions, however, quantitative corroboration between experiment and theory is not necessarily easily obtained.

Nevertheless, it is only through an achievement of harmony between experiment and simulation that one can fully appreciate ecological processes.

15.7 APPENDIX

For the appendix to this chapter, see www.cambridge.org/9780521858458.

ACKNOWLEDGMENTS

We would like to thank Michael DuBow, whose 2000 Millennial Phage Meeting (in Montreal) marked the start of our collaboration.

REFERENCES

Abedon, S. T. 1989. Selection for bacteriophage latent period length by bacterial density: a theoretical examination. *Microb. Ecol.* **18**: 79–88.

Abedon, S. T. 2006. Phage ecology. In R.L. Calendar and S.T. Abedon (eds.), *The Bacteriophages*, 2nd edn. Oxford: Oxford University Press, pp. 37–46.

Abedon, S. T., T. D. Herschler, and D. Stopar. 2001. Bacteriophage latent-period evolution as a response to resource availability. *Appl. Environ. Microbiol.* **67**: 4233–41.

Adams, M. H. 1959. *Bacteriophages.* New York, NY: Interscience.

Aita, T., and Y. Husimi. 1994. Period dependent selection in continuous culture of viruses in a periodic environment. *J. Theor. Biol.* **168**: 281–9.

Bohannan, B. J. M., and R. E. Lenski. 1997. Effect of resource enrichment on a chemostat community of bacteria and bacteriophage. *Ecology* **78**: 2303–15.

Bohannan, B. J. M., and R. E. Lenski. 1999. Effect of prey heterogeneity on the response of a model food chain to resource enrichment. *Am. Nat.* **153**: 73–82.

Bohannan, B. J. M., and R. E. Lenski. 2000a. Linking genetic change to community evolution: insights from studies of bacteria and bacteriophage. *Ecol. Lett.* **3**: 362–77.

Bohannan, B. J. M., and R. E. Lenski. 2000b. The relative importance of competition and predation varies with productivity in a model community. *Am. Nat.* **156**: 329–40.

Buchholtz, F., and F.W. Schneider. 1987. Computer simulation of T3/T7 phage infection using lag times. *Biophys. Chem.* **26**: 171–9.

Carletti, M. 2002. On the stability properties of a stochastic model for phage-bacteria interaction in open marine environment. *Math. Biosci.* **175**: 117–31.

Carlson, K. 2005. Working with bacteriophages: common techniques and methodological approaches. In E. Kutter and A. Sulakvelidze (eds.), *Bacteriophages: Biology and Application*. Boca Raton, FL: CRC Press, p. 437–94.

Chang, M. L., and T. S. Chang. 1969. Direct solution of Markovian phage attachment to bacteria in suspension. *Math. Biosci.* **5**: 9–18.

Eigen, M., C. K. Bierbricher, M. Gebinoga, and W. C. Gardiner. 1991. The hypercycle: coupling of RNA and protein biosynthesis in the infection cycle of an RNA bacteriophage. *Biochemistry* **30**: 11005–18.

Eisenstark, A. 1967. Bacteriophage techniques. In K. Maramorosch and H. Koprowski (eds.), *Methods in Virology*, vol. 1. New York, NY: Academic Press, pp. 449–524.

Endy, D., D. Kong, and J. Yin. 1997. Intracellular kinetics of a growing virus: A genetically-structured simulation for bacteriophage T7. *Biotech. Bioeng.* **55**: 375–89.

Gani, J. 1965. Stochastic phage attachment to bacteria. *Biometrics* **21**: 134–9.

Gáspár, S., G. Rontó, and G. Müller. 1979. Determination of the biological parameters of bacterium-phage complexes. *Z. Allg. Mikrobiol.* **19**: 163–9.

Gerba, C. P. 1984. Applied and theoretical aspects of virus adsorption to surfaces. *Adv. Appl. Microbiol.* **30**: 133–68.

Grant, S. B. 1994. Virus coagulation in aqueous environments. *Environ. Sci. Technol.* **28**: 928–33.

Haefner, J. W. 1996. *Modeling Biological Systems*. New York, NY: Chapman & Hall.

Husimi, Y. 1989. Selection and evolution of bacteriophages in cellstat. *Adv. Biophys.* **25**: 1–43.

Hyman, P., and S. T. Abedon, in press. Practical methods for determining phage growth parameters. In M. Clokie and A. Kropinski (eds.), *Bacteriophages: Methods and Protocols*. Totowa, NJ: Humana Press.

Koch, A. L. 1960. Encounter efficiency of coliphage–bacterium interaction. *Biochim. Biophys. Acta* **39**: 311–18.

Lenski, R. E. 1988. Dynamics of interactions between bacteria and virulent bacteriophage. *Adv. Microbial. Ecol.* **10**: 1–44.

Lenski, R. E., and B. R. Levin. 1985. Constraints on the coevolution of bacteria and virulent phage: a model, some experiments, and predictions for natural communities. *Am. Nat.* **125**: 585–602.

Levin, B. R., F. M. Stewart, and L. Chao. 1977. Resource limited growth, competition, and predation: A model and experimental studies with bacteria and bacteriophage. *Am. Nat.* **111**: 3–24.

McAdams, H. H., and L. Shapiro. 1995. Circular simulation of genetic networks. *Science* **269**: 650–6.

Middelboe, M. 2000. Bacterial growth rate and marine virus–host dynamics. *Microb. Ecol.* **40**: 114–24.

Mittler, J. E. 1996. Evolution of the genetic switch in temperate bacteriophage. I. Basic theory. *J. Theor. Biol.* **179**: 161–72.

Noack, D. 1968. A regulatory model for steady-state conditions in populations of lysogenic bacteria. *J. Theor. Biol.* **18**: 1–18.

Rabinovitch, A., A. Zaritsky, I. Fishov, M. Einav, and H. Hadas. 1999. Bacterial lysis by phage: a theoretical model. *J. Theor. Biol.* **201**: 209–13.

Rabinovitch, A., I. Fishov, H. Hadas, M. Einav, and A. Zaritsky. 2002. Bacteriophage T4 development in *Escherichia coli* is growth rate-dependent. *J. Theor. Biol.* **216**: 1–4.

Rabinovitch, A., I. Aviram, and A. Zaritsky. 2003. Bacterial debris: an ecological mechanism for coexistence of bacteria and their viruses. *J. Theor. Biol.* **224**: 377–83.

Schlesinger, M. 1932. Adsorption of bacteriophages to homologous bacteria [translation]. In Anonymous, *Bacterial Viruses*. Boston, MA: Little, Brown and Co, pp. 26–36.

Schrag, S. J., and J. E. Mittler. 1996. Host–parasite coexistence: the role of spatial refuges in stabilizing bacteria–phage interactions. *Am. Nat.* **148**: 348–77.

Smith, H. L., and P. Waltman. 1995. *The Theory of the Chemostat: Dynamics of Microbial Competition*. Cambridge: Cambridge University Press.

Srinivasan, S. K., and A. Rangan. 1970. Age dependent stochastic models for phage reproduction. *J. Appl. Probab.* **7**: 251–61.

Stent, G. S. 1963. *Molecular Biology of Bacterial Viruses*. San Francisco, CA: W. H. Freeman.

Stent, G. S., and E. L. Wollman. 1952. On the two step nature of bacteriophage adsorption. *Biochim. Biophys. Acta* **8**: 260–9.

Stewart, F. M., and B. R. Levin. 1984. The population biology of bacterial viruses: why be temperate? *Theor. Pop. Biol.* **26**: 93–117.

Thingstad, T. F. 1974. Dynamics of chemostat culture, the effect of a delay in cell response. *J. Theor. Biol.* **48**: 149–59.

Thingstad, T. F. 2000. Elements of a theory for the mechanisms controlling abundance, diversity, and biogeochemical role of lytic bacterial viruses in aquatic systems. *Limnol. Oceanogr.* **45**: 1320–8.

Weinbauer, M. G. 2004. Ecology of prokaryotic viruses. *FEMS Microbiol. Rev.* **28**: 127–81.

Weitz, J. S., H. Hartman, and S. A. Levin. 2005. Coevolutionary arms races between bacteria and bacteriophage. *Proc. Natl. Acad. Sci. U.S.A.* **102**: 9535–40.

CHAPTER 16

Modeling phage plaque growth

Stephen M. Krone* and Stephen T. Abedon

16.1 INTRODUCTION

The modeling of bacteriophage growth is complicated by spatial structure. Spatial structure, for phages, consists of impediments to environmental mixing, impediments to phage diffusion, and impediments to bacterial motility. A spatially structured environment, however, is perhaps best described by what it is not: a well-mixed, fluid culture. Spatial structure as observed within the laboratory can consist simply of unstirred (or unshaken or not bubbled) broth cultures (Buckling and Rainey, 2002). More commonly, so far as phage growth is concerned, we find bacteriophages growing as plaques within bacterial lawns. Within agar, phages generally are free to diffuse, though not as readily as in broth cultures, while bacteria typically are somewhat immobilized. This spatial structure, in broad terms, replicates the spatial association seen within such naturally occurring microenvironments as biofilms, soil, and sediments as well as the various surfaces associated with plants, animals, and other multicellular organisms. Here we consider the mathematical modeling of phage growth within relatively homogeneous spatially structured environments as approximated by phage plaque formation within a soft agar overlay. We first consider a variety of models that are limited to describing the enlargement of plaques and then additional models that consider other aspects of phage plaque growth. Mathematically, the modeling results we discuss come from reaction–diffusion (RD) differential equations and stochastic cellular automata (CA).

* Corresponding author

Bacteriophage Ecology: Population Growth, Evolution, and Impact of Bacterial Viruses, ed. Stephen T. Abedon. Published by Cambridge University Press.

For consideration of factors affecting plaque growth in the laboratory, see Abedon and Yin (in press). Spatial structure in the guise of phage metapopulation ecology is introduced in Chapter 2 and the phage ecology of agar-based spatially structured environments is addressed in Chapter 4.

16.1.1 Plaque enlargement

Plaque enlargement occurs as a consequence of two distinct phenomena: a "diffusion" step that directly contributes to the growth of a plaque's diameter and an "infection" step that offsets diffusion-mediated declines in phage density. Virus infection is crucial additionally since it is the resulting phage-induced bacterial lysis that leads to the lawn clearing that defines plaque size and morphology. The majority of virus infection is thought to occur at the edge of plaque clearings, just outside of the region of visible clearing (Fig. 16.1).

16.1.2 Plaque morphology

Across its radius a plaque may be viewed as possessing at least four layers and sometimes a fifth (Fig. 16.1), which we describe starting outside and moving inward (and in reverse numerical order): (V) the bacterial lawn-proper, which contains neither virions nor phage-infected bacteria, (IV) a narrow "periphery" into which virions have diffused but within which phage infection has not yet occurred, (III) a "zone of infection" containing infected but predominantly not lysed bacteria, (II) a "zone of reduced turbidity," which gives the plaques of some phages a bull's-eye morphology but which otherwise is poorly understood (Chapter 4), and (I) a "zone of clearing," which is devoid of phage-infected bacteria. Thus, we envisage a wave of viral infection (Fig. 16.1, bottom) that leads with invasion by phages, via diffusion, into a plaque's periphery. In the wake of this invasion are increasing numbers of phage-infected bacteria going from the outside of the zone of infection inward. Numbers of infected bacteria thereupon reach a peak and then decline until one reaches the zone of clearing. It is this wave of viral infection that models of plaque enlargement seek to portray. See Chapter 4 for additional discussion of phage plaque morphology and formation.

16.2 MODELS OF PLAQUE-DIAMETER INCREASE

Models of plaque enlargement seek to predict rates of increase in the diameter of growing plaques (wavefront velocity; c). These models often have

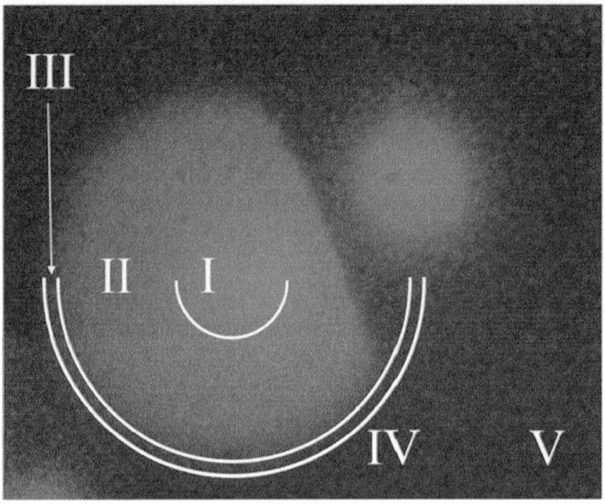

	I	II	III	IV	V
	Zone of Clearing	Zone of Reduced Turbidity	Zone of Infection	Periphery	Lawn
Virions	+	+	+ / –	+	–
Infections	–	?	+	–	–
Bacteria	–	?	+ / –	+	+

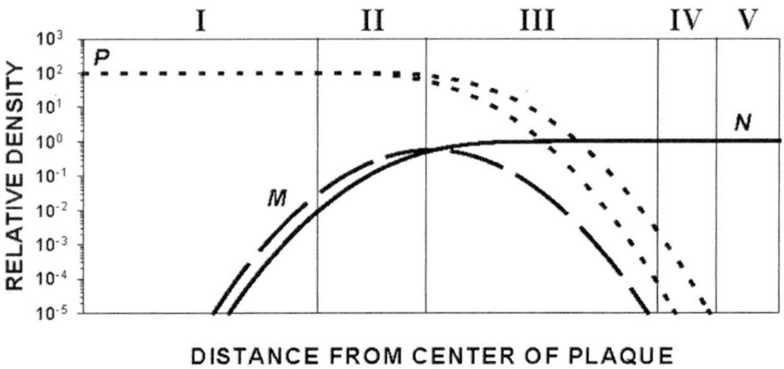

Figure 16.1 Three depictions of zones of a plaque. Top: a backlit digital photograph of two plaques grown on an *Escherichia coli* K12 lawn. On the right is the phage T4 mutant, *r48* (which does not display lysis inhibition; Chapter 3), while on the left is the phage T4 wild type. Only the plaque produced by the *r* mutant is discussed. The center or "bull's eye" (Chapter 4) is a region lacking in turbidity (I). This is surrounded by a relatively large region containing bacterial microcolonies where phage infection presumably is ongoing (II). The narrow band labeled III contains the portion of the bacterial lawn that is

more of a physical than biological flavor, describing large-scale patterns such as the outward, wavelike propagation of viral lysis initiated from a central focus.

16.2.1 Koch's model of plaque enlargement

Arthur Koch (1964) derived a heuristic model of what he describes as "the enlargement phase" of phage plaque development. We present Koch's model in its fully simplified form:

$$c = 10 \cdot \left(\frac{D}{L}\right)^{1/2} \tag{16.1}$$

where c is the constant "rate of increase of plaque diameter," D is a constant governing the rate of phage diffusion, and L is the phage latent period. Thus, in words, the diameter of a plaque is predicted to increase at a constant rate that is a function of the square root of a quantity consisting of the rate of phage diffusion divided by the length of the phage latent period, multiplied by a constant, which Koch approximates as equal to the number 10. The

Figure 16.1 (*continued*)
becoming infected. Note that this region and the next are inferred from the image rather than having been rigorously demonstrated. Just outside of this zone (IV) is a region that free phages have invaded but within which infection has not yet occurred. (V) indicates the phage-free bacterial lawn. Middle: a table describing the zones. "Bacteria" are phage-uninfected. Details are as described in the text. A "+" in the table indicates a density of greater than zero, a "–" indicates a density of zero, and a "+ / –" indicates densities of greater than or equal to zero. Presence of virions might vary across the width of the zone of infection, while uninfected bacteria should be essentially absent from the interior of the zone of reduced turbidity but perhaps abundant towards its exterior, as depicted in the lower figure. Bottom: an illustration of the traveling wave of bacterial infection and lysis similar to that presented by You and Yin (1999). P, M, and N are free phages (dotted lines), infected bacteria (dashed line), and uninfected bacteria (solid line), respectively (note that as depicted the y-axis does not necessarily cross the x-axis at zero). Two hypothetical P curves are shown, one indicating just phages generated by bacterial lysis (left) and the other, by shifting the curve slightly to the right, combines the idea of both lysis and net free-phage diffusion away from the center of a plaque. The M curve is a normal distribution (as rendered on a log scale). The N curve represents that fraction of bacteria remaining uninfected as the M wave travels left to right, and the zone marked "II" arbitrarily spans from $N = 0.01$ to $N = 0.5$. The leftward P curve represents the cumulative total of bacteria infected, going right to left, times a burst size of 100. Please keep in mind that this figure is presented to illustrate the concept of a traveling wave rather than representing a quantitatively accurate depiction of such a wave.

intuitive take-home message is that plaques will increase in diameter faster if phage diffusion occurs at a greater rate. Conversely, plaque diameter will increase more slowly if phage latent periods are longer.

Note that Equation 16.1 does not explicitly reference phage burst size, suggesting that the impact of phage burst size on increases in plaque diameter is not great. However, it is important to realize that this conclusion does not so much fall out of Equation 16.1, but rather was involved in its derivation, particularly of the constant, 10 (Koch, 1964). Note that bacterial density also is not explicitly found in Equation 16.1, but is described as not changing much over the course of lawn bacterial growth. Equation 16.1 thus represents an approximation of those constraints operating on plaque growth, but likely is less predictive than a more realistic (and thereby more complex) model could be. Subsequent to the derivation of Equation 16.1, Koch argues that the phage adsorption constant, k, also does not greatly impact the rate of plaque growth other than in terms of the embodiment of rates of virion diffusion that is part of that constant (see Chapter 15 for further discussion of the phage adsorption constant).

16.2.2 Yin and McCaskill's reaction–diffusion model

Yin and McCaskill (1992) provide a reaction–diffusion (RD) model of plaque enlargement that they derive from a model of RNA molecule propagation within a capillary tube. We provide in this section a detailed overview of the resulting model. Additional reaction–diffusion models of plaque formation are presented by Fort and Méndez (2002), Ortega-Cejas *et al.* (2004), and Fort *et al.* (2006), but these are not presented here due to space limitations. See Abedon and Culler (2007a) for further comparison expecially of the Ortega-Cejas *et al.* (2004) model with those of Koch (1964) and of Yin and McCaskill (1992).

16.2.2.1 Qualitative description of Yin and McCaskill's model

Yin and McCaskill's (1992) basic model is based upon various rate constants, which we introduce in a form that is based on that presented by Stent (1963):

$$\mathring{P} + \mathring{N} \xrightarrow{k_1} \mathring{P} : \mathring{N} \xrightarrow{k_{1+}} \mathring{M} \xrightarrow{k_2} \mathring{P} \cdot B \qquad (16.2)$$

where $\mathring{P}$ and $\mathring{N}$ are phage and uninfected bacteria, respectively, while $\mathring{P} : \mathring{N}$ represents a temporary complex between a phage virion and an uninfected bacterium, i.e., a post-collision, pre-irreversible adsorption state. A circle over a letter is to distinguish the state from the corresponding density using the same letter. The term B, by contrast, is the phage burst size, with $\overset{\circ}{P} \cdot B$ indicating that phage are produced in a ratio of B: 1 output (released) phage progeny to input (initially adsorbing) phage.

The state $\overset{\circ}{P} : \overset{\circ}{N}$ (phage–bacterium complex) is attained at a rate of k_1, which is the likelihood of phage reversible adsorption given an encounter with a bacterium. For a given complex formed upon this reversible adsorption, irreversible adsorption gives rise to a phage-infected bacterium, $\overset{\circ}{M}$, at a rate indicated by the constant k_{1+}. These infected bacteria are then converted into free phages at a rate of k_2 (in Equation 16.2), producing B phages each. Collision (as controlled by k_1) minus irreversible adsorption (as controlled by k_{1+}) thus indicates phage desorption, i.e., the decay of $\overset{\circ}{P} : \overset{\circ}{N}$ back to $\overset{\circ}{P} + \overset{\circ}{N}$ For the sake of consistency with the intent of the Yin and McCaskill (1992) model, we will define the associated desorption constant as k_{-1} and will treat $\overset{\circ}{P} : \overset{\circ}{N}$ and $\overset{\circ}{M}$ as identical states:

$$\left[\overset{\circ}{P} : \overset{\circ}{N} = \overset{\circ}{M}\right] \xrightarrow{k_{-1}} \overset{\circ}{P} + \overset{\circ}{N} \tag{16.3}$$

16.2.2.2 Quantitative description of Yin and McCaskill's model

Based on the summary presented as Equations 16.2 and 16.3, and assuming an absence of bacterial diffusion, Yin and McCaskill (1992) define a series of "reaction–diffusion equations" to define changes in P, N, and M (densities of phages, uninfected bacteria, and infected bacteria, respectively) as functions of time and spatial location along a fixed radial direction at the leading edge of the infection wavefront:

$$\frac{\partial N}{\partial t} = k_{-1} \cdot M - k_1 \cdot P \cdot N \tag{16.4}$$

$$\frac{\partial M}{\partial t} = k_1 \cdot P \cdot N - k_{-1} \cdot M - k_2 \cdot M \tag{16.5}$$

$$\frac{\partial P}{\partial t} = D \cdot \frac{\partial^2 P}{\partial r^2} + \frac{D}{r} \cdot \left(\frac{\partial P}{\partial r}\right) + k_{-1} \cdot M + B \cdot k_2 \cdot M - k_1 \cdot P \cdot N \tag{16.6}$$

where r indicates position along a vector starting at the focus of plaque initiation. Note that bacteria are assumed not to replicate. Instead, the bacterial lawn density, peripheral to the growing plaque, is set equal to a constant, N_o. This assumption of no bacterial growth simplifies modeling as well as model interpretation. It also is reasonable given a determination of plaque development as it occurs following maturation of the bacterial lawn to stationary phase, e.g., as is seen during much of the enlargement phase of phage T7 plaques (Yin, 1991). It also may be a reasonable basis for modeling phage growth as it may occur within spatially structured natural environments, though perhaps only to the extent that phage latent periods are very short relative to bacterial doubling times. We now explain in some detail what Equations 16.4–16.6 mean:

Equation 16.4 describes the instantaneous change in density of uninfected bacteria (N) as a function of time (t). The expression $k_{-1} \cdot M$ represents "births" of uninfected bacteria, which occur only via dissociation of temporary virion–bacterium complexes ($\mathring{P} : \mathring{N}$ in Equation 16.2, but here with density indicated as M). The reverse reaction is described by $k_1 \cdot P \cdot N$, the phage and bacterial association that results upon encounter between free phages (P) and uninfected bacteria (N). Thus, uninfected bacteria are lost only via phage adsorption and are gained only via phage desorption.

Equation 16.5 specifies the instantaneous rate of change in infected bacterial density (M) as a function of time. Note that loss of uninfected bacteria via phage adsorption in Equation 16.4 (as indicated by $-k_1 \cdot P \cdot N$) corresponds to gain of infected bacteria in Equation 16.5 by the same process (as indicated by $k_1 \cdot P \cdot N$). Similarly, whereas in Equation 16.4 uninfected bacteria are gained by dissociation of the phage–bacterial complex (as indicated by $k_{-1} \cdot M$), in Equation 16.5 they are lost via the same process (as indicated by $-k_{-1} \cdot M$). In Equation 16.5 infected bacteria are also lost as a consequence of phage-induced bacterial lysis (as indicated by $-k_2 \cdot M$). Infected bacteria as specified by Equation 16.5 therefore are transients, born and then lost as a consequence of phage–bacterial associations.

Equation 16.6 is most easily grasped by initially ignoring the first two terms. Thus, like uninfected bacteria, free phages are lost to bacterial adsorption and are liberated by bacteria–virion dissociation ($-k_1 \cdot P \cdot N$ and $k_{-1} \cdot M$, respectively). Unlike infected bacteria, however, phage-induced lysis results not in loss but in gain, with one burst size (B) of free virions added per infected bacterium lost ($B \cdot k_2 \cdot M$). These terms thus describe the interaction of phages in various ways with bacteria.

The first two terms of Equation 16.6, $D \cdot \frac{\partial^2 P}{\partial r^2}$ and $\frac{D}{r} \cdot \left(\frac{\partial P}{\partial r}\right)$, modify phage density as a function of both virion diffusion (with D the diffusion constant) and radial position within the plaque (as described by r). In understanding Equation 16.6 it is important to realize that what is ultimately being described is the one-dimensional rate of increase in plaque size, and that this is done by modeling conditions at the front of this increase (Fig. 16.1), i.e., at the periphery and zone of infection of a plaque. Phage density thus declines as a consequence of phage adsorption to bacteria and then increases as a function particularly of the release of phage progeny from infected bacteria.

However, phage density also can increase as a function of phage diffusion (at rate, D) from the interior of the plaque. The first term, $D \cdot \frac{\partial^2 P}{\partial r^2}$, thus indicates that so long as phage densities are moving from the interior to the exterior of a plaque, then phage densities will increase as a function of rates of phage diffusion from the interior. The second term, $\frac{D}{r} \cdot \left(\frac{\partial P}{\partial r}\right)$, on the other hand, declines as the plaque enlarges, and so is ignored by Yin and McCaskill except as it applies to very small plaques. Note that these two terms, when added together, simply amount to the polar coordinate representation of the usual diffusion operator (given by D times what in mathematics and physics is known as the Laplacian) that describes radial diffusion in two dimensions. The fact that the second term becomes small as the radius, r, increases means that, as the plaque gets larger, its boundary becomes more and more like a straight line and the wavefront locally looks like a constant-speed "planar wave."

16.2.2.3 Calculating wavefront velocity as plaques spread

Yin and McCaskill consider traveling-wave solutions to Equations 16.4–16.6 and, assuming that viral adsorption and desorption processes are fast relative to the death rate of infected hosts, arrive at an "equilibrated adsorption" approximation for the expected constant wavefront velocity (c) observed as plaques spread:

$$c = 2 \cdot \left[\frac{D \cdot k_2 \cdot (B-1) \cdot f \cdot K_{\mathrm{max}}}{(1 + f \cdot K_{\mathrm{max}})^2}\right]^{1/2}$$

$$= 2 \cdot \left[\frac{D \cdot k_2 \cdot (B-1) \cdot N_o \cdot k_1/k_{-1}}{(1 + N_o \cdot k_1/k_{-1})^2}\right]^{1/2} \tag{16.7}$$

where f is the ratio of bacterial density (N_o) to a theoretical maximum bacterial density (N_{max}; that is, $f = N_o/N_{\mathrm{max}}$) and $K_{\mathrm{max}} = k_1 N_{\mathrm{max}}/k_{-1}$. The first expression is converted to the second expression by making these substitutions for f and K_{max}. Breaking up the second expression into parts: $k_2 \cdot (B-1)$ represents an increase in P, i.e., phages added to the population as a consequence of infected-bacterium lysis, with the –1 indicating that one free phage was lost to initiate this infection. These phages diffuse at a rate that is indicated by D. The rate of phage acquisition of bacteria is a function of phage prevalence ($B-1$), bacterial density (N; which at a plaque's periphery is equal to N_o), and the various constants controlling phage adsorption and desorption: k_1/k_{-1}.

In the denominator, the term $(1 + N_o \cdot k_1/k_{-1})$ indicates that rapid irreversible adsorption to bacteria can also slow down plaque growth, since this value increases as a square function. The expression, however, approaches

1.0 for $N_o \ll k_{-1}/k_1$. Thus, some bacterial adsorption is important for the occurrence of plaque spread (since otherwise the bacterial lysis that defines a plaque will not occur), but too much adsorption, particularly at higher bacterial densities, ultimately could hinder that plaque spread. You and Yin (1999), in a numerical exploration of the Yin and McCaskill (1992) reaction–diffusion model, suggest however that in fact no declines in c will occur as a function of increasing phage adsorption.

16.3 MODELS OF PLAQUE-FECUNDITY INCREASE

An alternative approach to modeling plaque formation is to define c, the rate of radial plaque-size increase, to be a constant, and to then concentrate on aspects of plaque formation other than wavefront velocity. Kaplan *et al.* (1981b) present such a model, where their goal was to determine the number of bacteria that become infected over the course of plaque formation, assuming a replicating bacterial lawn. If we multiply the number of bacteria infected by B (phage burst size per infection) then we can, in principle, predict the overall fecundity (total number of phages produced) per plaque, or at least what the maximum fecundity might be (Abedon and Culler, 2007b). In this section we discuss the Kaplan *et al.* model and in the subsequent section, 16.4, we combine their model with the c-estimating models presented above (in Section 16.2).

16.3.1 Modeling a three-dimensional plaque

Kaplan *et al.* (1981a) screened plaques for the presence of a specific phage-encoded antigen and concurrently developed a model for plaque formation which they presented as a companion paper, though designated as an "appendix" (Kaplan *et al.*, 1981b). Kaplan *et al.* were interested primarily in the potential for plaques to produce phage-encoded antigen, which to a first approximation should vary as a function of number of bacteria that have been lysed over the course of plaque formation. Accordingly, they developed their model as a means of predicting average plaque sizes and number of bacteria lysed, both as functions of initial indicator-bacteria density. Plaque size is relevant to total number of bacteria lysed since, all held constant, a larger plaque should result in greater bacterial involvement in phage population growth. Furthermore, Kaplan *et al.* were able to calculate the number of bacteria lysed both as an endpoint of plaque development and as a function of time prior to that end point (i.e., prior to the bacterial lawn reaching a productive-infection-refractory physiological state such as stationary phase).

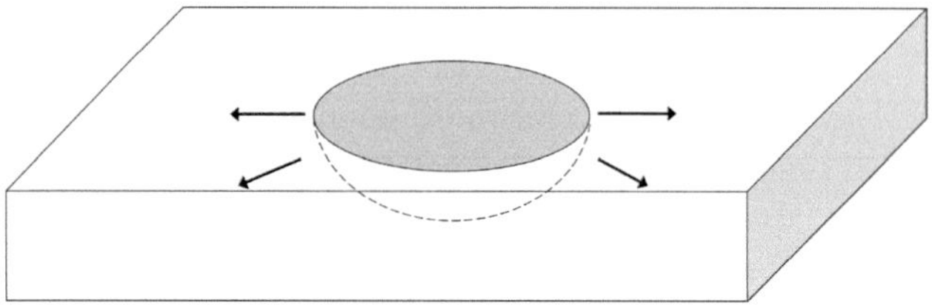

Figure 16.2 Depiction of a three-dimensional plaque as a truncated sphere. Shown is a rectangular section of top agar (soft agar) containing a bacterial lawn. This plaque was initiated approximately at the air–agar interface (up) so is truncated at that interface but is otherwise spherical. The arrows represent a sampling of the outward directions of plaque enlargement in three dimensions. Were this plaque larger, or had it been initiated closer to the top agar–bottom agar interface (down), then it may have become truncated at both interfaces at this point in its development, or alternatively just at the lower interface had it been initiated even closer to that plane. Smaller plaques initiated sufficiently distant from both interfaces could be spherical without truncation.

Kaplan *et al.* (1981b) begin the motivation of their model with a discussion of plaque shape. Unlike those models discussed in Section 16.2, the Kaplan *et al.* model emphasizes plaque three-dimensionality rather than a single vector of plaque development (by necessity, given their justification for developing their model, but also reasonably realistic given the relative smallness of their plaques). As such, their initial assumption is that plaques represent truncated spheres (Fig. 16.2). The truncation occurs at the agar–air interface (a second truncation could be assumed to occur at the soft agar–hard agar interface, but this is explicitly not considered by Kaplan *et al.*). While the perceived width of plaques is the most common measure of plaque size, for Kaplan *et al.* (1981b) it is plaque surface area, especially that bordered by soft agar and therefore by bacteria, that is the more relevant measure since it is at that surface that the wave of phage infection occurs.

Because otherwise spherical plaques may be truncated (Fig. 16.2), the surface area (i.e., the periphery; Fig. 16.1) of a given plaque is a function of the overall thickness of the soft-agar layer, the depth within the soft agar at which the plaque-initiating phage infection occurred, and the plaque's subsequent radius. That is, greater distances from air (or hard agar), such that truncation is minimal, equates with greater ratio of plaque surface area in contact with the bacterial lawn to plaque radius. With greater plaque size, on the other hand, there is a greater likelihood of truncation, resulting in the average plaque surface area in contact with the bacterial lawn not increasing as quickly as a function of plaque radius. Greater soft-agar thickness, by increasing the separation of hard agar and air, decreases the impact of these hard-agar and air interfaces on plaque surface area. Practically speaking, however, greater soft-agar thickness comes at a cost in plaque visibility as well as bacterial access to nutrient sources, oxygen supply, and waste sinks.

The surface area of a spherical (not truncated) plaque is $4 \cdot \pi \cdot (\text{radius})^2$. The area of a maximally truncated plaque, that is, one with focus at the soft-agar–air interface, is $2 \cdot \pi \cdot (\text{radius})^2$, i.e., the surface area of the bottom half of a putative spherical plaque. Aiming at something intermediate between these two extremes, Kaplan *et al.* (1981b) assume an average plaque surface area of $3 \cdot \pi \cdot (\text{radius})^2$, though in fact this was accomplished via a surface area integration rather than via simple division (B. Rothschild, personal communication). Note, as described above, that this approximation clearly would vary both as a function of soft-agar thickness and with plaque size, with thicker agar and smaller plaques resulting in an approximation that is closer to $4 \cdot \pi \cdot (\text{radius})^2$ while thinner agar and larger plaques should be better approximated by $2 \cdot \pi \cdot (\text{radius})^2$, or even less if truncation occurs at both top and bottom. It is an open question the degree to which spherical progression of phage population growth occurs within spatially structured natural environments without truncation, though presumably truncation should vary at least as a function of environment thickness.

16.3.2 Estimating final plaque radius

Like Koch (1964) and Yin and McCaskill (1992), Kaplan *et al.* (1981b) assume that plaque formation occurs with radius increasing at a constant rate (see their Figure 1 for supporting evidence as determined via time-lapse photography). Unlike these earlier models, however, the Kaplan *et al.* model explicitly accounts for ongoing division of uninfected bacteria throughout plaque development. Accounting for host division is important (1) for keeping track of total numbers of bacteria lysed, (2) in determining plaque virion productivity, (3) toward addressing potential for bacteria physiological change, and (4) in defining the end point of plaque enlargement. Host division, however, does not appear to impact rates of plaque enlargement during much of plaque development (Mayr-Harting, 1958; Koch, 1964; Kaplan *et al.*, 1981b), at least prior to maturation of the bacterial lawn.

The equation in Kaplan *et al.* for final plaque radius, R, is

$$R = (c/\mu) \cdot \ln(N_f/N_o) \qquad (16.8)$$

where all terms are either empirically determined parameters (c, μ, and N_f) or initial conditions (N_o). The constant, c, is the rate of increase of the plaque radius as a function of time, i.e., as employed in Equations 16.1 and 16.7;

μ is the growth-rate constant of those bacteria making up the lawn; N_f is the maximum (or final) density of bacteria in the lawn (i.e., as at the point of lawn maturation); and N_o is the initial density of bacteria making up the bacterial lawn. Note that an explicit assumption of Equation 16.8 is that phage replication ceases once the lawn reaches a density of N_f bacteria per mL, thus explaining the absence of time, t, in the equation. This model, therefore, is not applicable to phages such as T7 which can continue to replicate following lawn maturation (Yin, 1991; Robb and Hill, 2000).

16.3.3 Estimating plaque productivity

The total number of virions produced per individual plaque is a function of the total number of bacteria infected times the average phage burst size (per bacterium infected) and minus phage losses. Losses can be due to multiplicities of infection that are greater than one (Chapter 3, 5, and 8), or, alternatively, phage infections that do not lead to bacterial lysis (e.g., of stationary-phase cells or, for temperate phages, infections that result in lysogeny; Chapter 5). To a first approximation, the total number of virions produced per unit time should initially increase as both plaque surface area and lawn bacterial density increase, since both factors result in a presence of greater bacterial numbers at the periphery of plaques. On the other hand, physiological changes in bacteria might result in reduced infection burst sizes as bacterial densities increase, thereby to some degree offsetting the positive effects of increased bacterial numbers on per-plaque rates of virion production. A resulting model of plaque productivity thus would be the sum of the instantaneous products of phage burst size, plaque soft-agar surface area, and infected bacterial density within that "surface." The latter two terms increase with time, while the phage burst size at least potentially decreases as a function of time.

By concentrating on number of bacteria lysed rather than on number of virions liberated, Kaplan *et al.* (1981b) are able to disregard these complications on per-bacterium productivity (i.e., burst size). They proposed the differential equation

$$dM/dt = N_o \cdot e^{\mu \cdot t} \cdot 3 \cdot \pi \cdot (c \cdot t)^2 \qquad (16.9)$$

for the density, M, of newly infected bacteria (here considered to be equivalent to newly lysed bacteria). The term $N_o \cdot e^{\mu \cdot t}$ indicates bacterial density within the uninfected region of the bacterial lawn and $3 \cdot \pi \cdot (c \cdot t)^2$ is a measure of the surface area of a truncated spherical plaque (i.e., with $c \cdot t$ equal to the *radius* of the semi-spherical plaque – at least if we assume that the radius of the initial phage inoculation is negligible). Note that an implicit assumption made by Kaplan *et al.* (1981b) in deriving this equation is an occurrence of instantaneous infection and then lysis of all bacteria found at "the thin shell at the surface of the growing plaque." See Chapters 2 and 15 for discussion of problems with assumptions of instantaneous infection and then lysis in modeling phage growth in broth culture.

Estimation of cells lysed from the point of plaque initiation until a given time, T, is simply a matter of integrating Equation 16.9 from $t = 0$ to $t = T$. Doing so, however, requires empirical determination of both μ and c. We avoid explicitly presenting the resulting equation, as published by Kaplan *et al.* (1981b), due to the relative lack of facile biological inference that may be gained from close analysis.

Subsequent estimation of plaque productivity requires an empirical determination of phage burst size, and could also require some estimation of how phage burst size can change throughout plaque development, though an obvious simplification is to treat burst size as a constant (Abedon and Culler, 2007b). With the latter simplification, a determination of overall plaque productivity involves simply solving an integrated Equation 16.9 from $t = 0$ to t_f, where t_f is the total duration of plaque growth, and then multiplying the resulting estimation of cumulative numbers of phage-infected bacteria by the phage burst size. Doing so, however, requires an empirical determination of this total duration of plaque development. Determination of this parameter may be achieved in conjunction with the also-necessary determination of c.

Of interest, Kaplan *et al.* (1981b) estimate that in their system approximately 80% of phage-induced cell lysis during plaque development occurs during the final 20% of time required for complete plaque development. This makes intuitive sense since the area of affected lawn should be greatest near maximum plaque diameter, as should also the bacterial lawn density. It means, however, that most phage liberation during plaque development likely also occurs at this time, though with the caveat that increasing lawn maturation may impact phage burst size. It remains to be determined therefore whether or not our extension of Kaplan *et al.*'s work to model plaque productivity can retain adequate predictive power, given potential deviations from assumptions of burst size constancy.

16.4 CELLULAR AUTOMATA MODELS

In this section we describe a spatially explicit mathematical description of phage/bacteria dynamics using a class of individual-based lattice models known as interacting particle systems or stochastic *cellular automata* (CA). These models incorporate stochasticity and spatial structure, in the form of a two-dimensional lattice that can display resolution down to the level of individual cells and phages, and can be thought of as providing a microscopic picture of a microbial community. CA models have been applied widely to various biological systems (Durrett and Levin, 1994a, 1994b, 1998; Tilman and Kareiva, 1997), including some that involve microbial communities (Durrett and Levin, 1997; Kerr *et al.*, 2002; Kreft and Bonhoeffer, 2005; Wei and Krone, 2005). The reader can consult these references for more details on such models, and for further applications.

16.4.1 Simple CA model for plaque growth

Perhaps the simplest CA model that can be applied to phage plaque modeling is a version of the usual SIR (susceptible – infective – removed) models from epidemiology. Here, within the above spatial framework, one considers the possible states of lattice sites to be susceptible (S), infective (I), and removed (R) individuals; in the phage setting, these correspond to individual uninfected bacteria (susceptibles), infected bacteria (infectives), and dead bacteria (removed). The latter are represented as vacant sites. There is no consideration of free phage particles or subtleties of the phage life cycle. Instead, uninfected bacteria become infected by infected bacteria at a rate that is proportional to the number of neighboring individuals infected by that pathogen (which on a typical square lattice ranges from zero to eight infected neighbors, corresponding to eight directions – S, N, E, W, SE, SW, NE, and NW – as indicated in Fig. 16.3). Infected individuals lyse at a constant rate. Despite the simplicity of such a model, it captures many of the essential features of plaque formation, including the roughly circular shape and constant rate of plaque radial spread (see the first frame of Fig. 16.4). More complicated CA models that include details of phage bursts, etc. (Section 16.4.2) yield very similar overall behavior.

To get an idea of how a CA model can provide insight into the dynamics of plaque formation (and other things!), let us consider the above simple model and see how it leads to a microscopic explanation for why circular plaques are so natural even when plaque formation is modeled as a stochastic process. Imagine a situation in which we place a small number of infected bacteria in the middle of a homogeneous environment composed of uninfected bacteria, thus initiating the formation of a "plaque" of removed sites (dead bacteria). Since the infection spreads by direct contact between susceptibles and infectives, the rate at which an uninfected bacterium found on the boundary of the wave (wavefront) of viral infection becomes infected is proportional to the number of neighboring lattice sites that contain an infective. Alternatively,

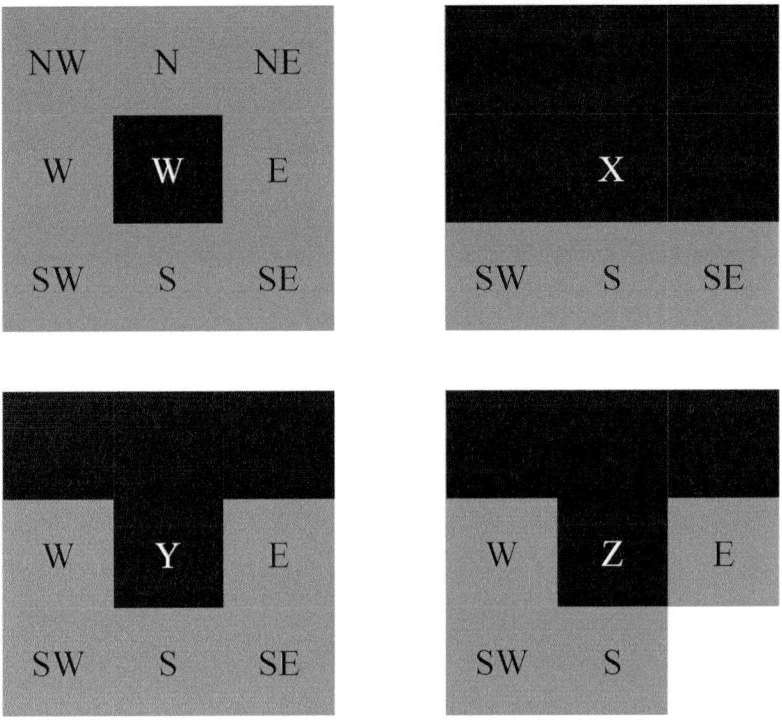

Figure 16.3 Close up of SIR-type cellular automata lattice showing susceptibles (uninfected bacteria, black shading) and infectives (infected bacteria, gray shading). Neighboring sites are named in this figure according to the directions of the compass, with north being up. The susceptible indicated by "W" is surrounded by the maximum number of possible infectives. The susceptible indicated by "X" is found adjacent to the periphery of a smooth wavefront. The susceptible indicated by "Y" is found within a dent in the wavefront. "Z" is equivalent to "Y" except that one of the infectives surrounding it has been converted to a dead bacterium (removed, no shading). During a small time increment of the model, note that the likelihoods of susceptible conversion to infective may be expressed as $W \propto 8$, $X \propto 3$, $Y \propto 5$, and $Z \propto 4$. This arrangement indicates one limitation of strict SIR modeling vis-à-vis plaque formation, where dead bacteria are described as no longer infective, whereas in plaques the dead bacteria behind wavefronts would be expected to contain free phages and therefore remain infective (though not necessarily exposed to uninfected bacteria).

one can think of the infectives on the wavefront as racing to infect susceptible sites. Uninfected bacteria that lag behind the wave (corresponding to dents in the wavefront) will typically be in contact with more infected bacteria than a typical susceptible at a smooth part of the wave, and hence will have a higher infection rate (Fig. 16.3). This acts to fill in the dents along the wavefront.

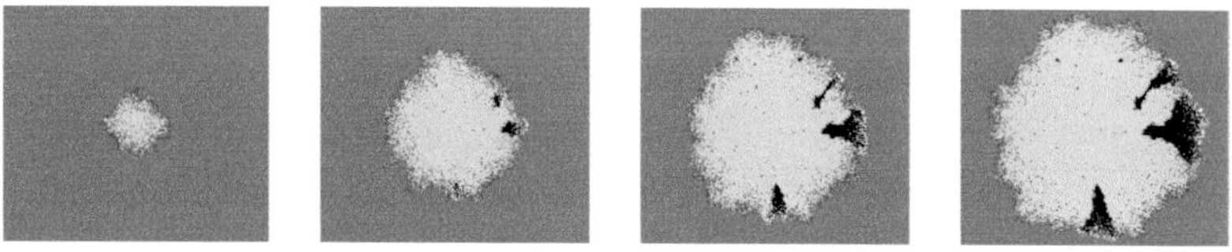

Figure 16.4 Four successive observations of an SIR-type CA simulation showing the development of a bacteriophage plaque. In the first frame, a small plaque begins to form. Shown are bacteria infected by wild-type phages (darkest gray), which are scattered about much of the plaque peripheries (except in front of the black wedges), uninfected bacteria (gray background), and lysed bacteria formerly infected by wild-type phages (light gray), which make up the bulk of the plaque interiors. Beginning in the second frame, mutant phages appear within the zone of reduced turbidity of the plaque (see Fig. 16.1). Bacteria infected by these mutant phages are light gray and are found in front of wedges of lysed bacterial cells formerly infected by mutant phages (black). These bacteria formerly infected by mutant phages are leaving tell-tale sectors within the overall plaque (the wedges). In this simulation, mutant phages have a slight fitness advantage over wild-type phages.

Similarly, protrusions in the wave lead to susceptibles that are in contact with fewer neighboring infectives. Thus the average speed of such a part of the wavefront will be smaller than the smooth part of the wave. In other words, although stochasticity results in a wavefront that is irregularly shaped, the afore-mentioned differences in local transmission rates result in the wavefront being self-correcting, and thus maintaining the global circular shape of the wave.

16.4.2 CA model with mutation

Wei and Krone (2005) used a slightly more complicated SIR-type CA model to examine the fate of mutant phages in a radially expanding plaque (Fig. 16.4). This model includes a mutant strain of the phage that differs from the wild type in its infection rate and/or the death rate of infected bacteria. They found that spatial structure affects the likelihood that certain mutant phages can invade, with successful mutants forming clonal sectors (seen as black wedges in Fig. 16.4) originating at the edge of the growing plaque. Early experimental work indicating the presence of these mutant sectors appeared in Yin (1993; see also Chapter 4). These CA model results differ significantly from phage–bacteria dynamics in chemostats (Chapters 2 and 15), for example, where the basic reproductive ratio determines whether a mutant pathogen can out-compete and eventually replace the wild-type pathogen (May and Anderson, 1983; Bremermann and Thieme, 1989; Murray, 1989).

Simulations in Wei and Krone (2005) showed that conditions leading to the successful invasion of a mutant pathogen in the CA model are quite different from those in an ordinary differential equation model (which describe phage growth within well-mixed environments) and rather close to those in a reaction–diffusion model (Sections 16.2 and 16.3). Other quantities, such as mean invasion time (i.e., the time it takes for a mutant phage to appear and successfully propagate in the expanding plaque; Fig. 16.4), are strongly influenced by the nature of the randomness acting on small scales and can only be studied with stochastic models. These results demonstrate that spatial structure and stochasticity can have a pronounced effect in shaping the evolution of pathogens in addition to affecting plaque morphology.

16.4.3 A more elaborate CA model

Coberly *et al.* (unpublished: Spatial structure and host evolution facilitate coexistence of competing bacteriophages) developed an experimental model system and a corresponding CA model to investigate the role played by spatial structure in determining the outcome of competition between two phages (wild-type isolates of DNA bacteriophages ϕX174 and α3) on a single bacterial host (*E. coli* C).

16.4.3.1 Between-plaque competition experiments

The experiments were carried out on agar plates for a total of 15 incubation periods, with parallel sets of experiments using two different incubation times (short = 5 h and long = 18 h). After each incubation period, a reduced sample of phages was transferred to a fresh lawn of bacteria using a 384-prong replicate picker (essentially, a rectangular "bed of nails") that preserved the spatial locations of the sampled phages. The beginning of each such transfer initiates the formation of multiple small plaques with foci separated by several millimeters. Within the incubation period, these plaques collide, resulting sometimes in head-to-head competition between two phage types. To determine the effects of spatial structure, each treatment was also carried out with phages mixed prior to transfer with the picker. In this case, each transfer initiates the formation of plaques that typically contain both phage types – at least while the densities of both phage types are sufficiently large to be represented in the picker sample.

In the case of mixed transfer, short-incubation experiments led to phage ϕX174 competitively excluding phage α3, and in the long-incubation experiments ϕX174 drove α3 to low densities and sometimes to extinction. When spatial structure was preserved at each transfer, however, ϕX174 and α3 were able to coexist in the long-incubation experiments, while short-incubation experiments still led to ϕX174 competitively excluding α3. The persistence of the weaker competitor in the spatially structured experiments when incubation times are long seems to be accounted for by the evolution of hosts resistant to the more competitive virus but susceptible to the weaker competitor (Coberly *et al.*, unpublished).

16.4.3.2 CA competition modeling

Simulations of the CA model were consistent with these experiments, including the ability of the weaker competitor, α3, to coexist with the dominant competitor, ϕX174, if incubation times were sufficiently long (Fig. 16.5). Mutant bacteria that were resistant to ϕX174 but susceptible to α3 were able to arise and spread in ϕX174-dominated regions, but only if the resistant bacteria were given sufficient time to become established; these "percolating" clusters of mutant cells provided a spatial refuge for the weaker competitor, leading to coexistence. In the absence of any within-host competition that might occur if we allowed for superinfection or coinfection (Chapter 8), the simulations also suggested that shorter latent periods were much more beneficial than larger burst size in spatial environments. This is consistent with a similar finding by You and Yin (1999) that was based on a reaction–diffusion equation model.

The CA model was even able to account for curious oscillations in the densities of the two phages; these were shown to be due to a "quantization" effect that was a function of the sampling area of a picker prong and the spacing between prongs. Other oscillations, much larger and temporary, were apparently caused by mutations of the phages; this feature could not be captured in simulations since the only mutations modeled were those of the host bacteria.

16.4.3.3 The competition CA model

We now describe the basic ingredients of the CA model presented by Coberly *et al.* The simulations take place on a 600×600 square lattice of sites, where each site can be vacant, occupied by one of two uninfected bacterial strains – the wild-type strain (susceptible to both ϕX174 and α3) or the resistant mutant strain (resistant to the dominant phage ϕX174 but susceptible to α3). A site – whether containing no cell, an uninfected cell, or an infected cell – can also contain free phages. To keep things relatively simple, the model does not include coinfection or superinfection (Chapter 8). A susceptible cell at a site containing a free phage particle will become phage-adsorbed at a rate that depends on the phage type. Similarly, an infected cell displays a

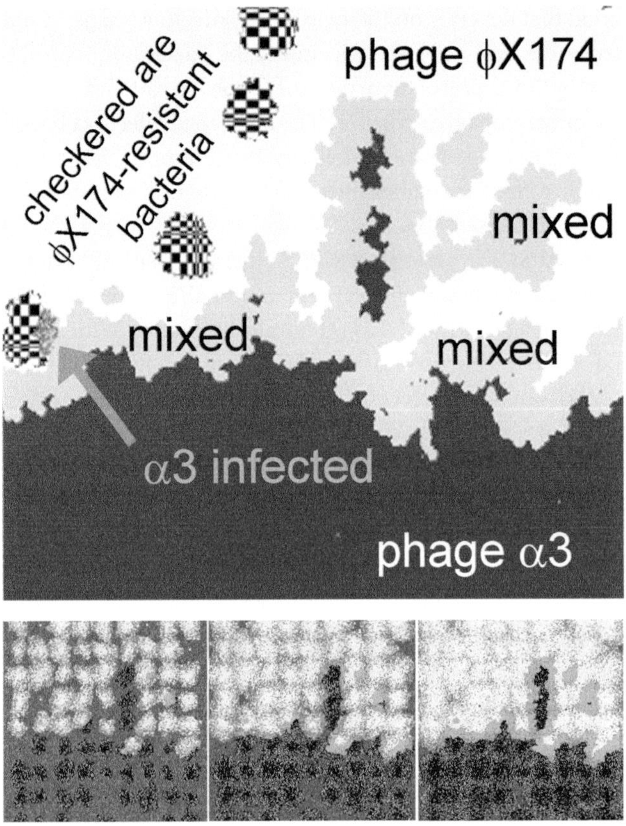

Figure 16.5 Snapshot from partial view of a CA simulation showing early stages of bacteriophage competition on an agar plate. Top: final time point during a long-incubation period. Light and dark regions indicate presence of phages ϕX174 and $\alpha 3$, respectively, while the regions of intermediate darkness are hybrid zones containing both phage types. The checkered regions are colonies (four of them) of ϕX174-resistant bacteria, with a region of the lower-left colony showing infection by phage $\alpha 3$. Bottom: three prior time points during the same incubation period, with early, middle, and later progressing from left to right. Dark indicates phage-$\alpha 3$-containing regions while light indicates regions containing phage ϕX174. Gray surrounding the dark or light regions are infected bacteria. The darker gray outside of the regions containing infected bacteria – that is, the background – represents the uninfected bacterial lawn. The gray surrounding black, noticeable especially in the third time point, are regions containing both phage types.

latent period that depends on the type of the infecting phage. If both phage types occupy the site, only one will infect the bacterium, with the winner determined by the timing of the arrivals of the phage particles and the random delay until one or the other adsorbs. The combined effects of burst size and diffusion of phage progeny are modeled by distributing phage progeny among all sites within a circular neighborhood of the lysed cell, with some prescribed burst "reach" or "range."

The simulations are initiated with an initial density of bacteria that are randomly distributed (0.05 of sites) and an initial density of 0.0005 (also of total sites) for phages, with φX174 randomly distributed in the top half of the lattice and α3 on the bottom half. Death of bacteria in the model is assumed to result only from phage infection. Bacteria reproduce by depositing a single daughter cell at a randomly chosen neighboring site that does not already contain a bacterial cell; if there are no neighboring vacant sites, the reproduction event is suppressed. Thus, the model does not allow for the "piling up" of bacteria on our 2-dimensional lattice. Nutrients are not explicitly modeled and bacteria retain their original susceptibility to phage infection, even when they can no longer reproduce due to crowding.

At the end of each incubation period, a "sample" from the spatial configuration of phage is transferred to a "fresh lawn" of bacteria, again beginning with a fraction, equal to 0.05, of the lattice sites occupied by bacteria. For spatial transfers (i.e., those preserving the spatial locations of the phage), the sample "prongs" in the simulation form a 9×9 grid with each prong consisting of a 5×5 square of neighboring sites. Phages are picked up by a total of 25 sites per prong ($25 = 5 \times 5$) with each able to carry α3 phage, φX174 phage, or both or neither, depending on phage presence at the sites on the parent plate. See Coberly *et al.* (unpublished) for further details.

Although the CA model makes a number of simplifying assumptions, including the fact that the length scales for the lattice and sampling prongs are smaller than those for the experiments (e.g., think of a simulation as corresponding to about a 1 mm square in the experiment, and the experimental plate being about 100 mm on a side), it does account for key ingredients of spatial structure at the level of individuals, including the stochastic nature of the dynamics. This led to the simulations capturing and helping to explain many aspects of the behavior of the experimental system, including some that were quite subtle, as described above. Most importantly, the CA model was able to demonstrate how a combination of spatial structure and sufficiently long incubation periods could allow clusters of resistant bacteria (resistant to the dominant phage competitor) to spread enough to enable the weaker competitor to persist. Since bacteria produce phage-resistant mutant offspring with high probability in large populations, these dynamics captured by the CA model are likely to be quite prevalent in real populations.

16.5 CONCLUSION

Reaction–diffusion equations of plaque growth (Sections 16.2 and 16.3) are lacking in at least two ways. First, they do not explicitly consider individual phage–bacterial interactions. Second, they do not easily incorporate inhomogeneities within plaques, as surely must occur during phage growth in nature (Chapter 4), including, in all likelihood, over the course of phage-mediated treatment of bacterial disease (i.e., phage therapy; Chapter 17). Still, these models are predictive of bulk properties of phage-plaque development within soft agar overlays. As such, they lay the foundation for a more sophisticated modeling of phage growth – and phage impact on bacteria – within spatially structured environments, such as those achieved using cellular automata (CA) models. Since CA models are capable of incorporating stochasticity and other features of bacterial and phage dynamics down to the level of individual bacterial cells, they can generate additional patterns that appear to play important roles in plaque formation and other spatial aspects of phage populations. These include the formation of microcolonies in the bacterial host population and inhomogeneities within plaques.

There are connections between reaction–diffusion (RD) equations and CA models. Such connections between these different types of spatial models can be obtained via "hydrodynamic" or "fast-stirring" limits of properly scaled CA models (Durrett and Neuhauser, 1994; Krone, 2004). Roughly speaking, this means that certain large-scale behaviors of a given CA model can be approximated by an appropriate RD equation. The RD equation in fact can be thought of as being somewhere between a CA model and a mass-action (ordinary) differential equation (e.g., Chapters 2, 15, and 17). At one extreme, the CA model is the more detailed and can account for the random, spatially explicit behavior of individual bacteria. At the other extreme, the mass-action model only considers globally averaged, non-spatial densities. The RD equation is still based on averages, but these average densities are computed locally, thus preserving some large-scale spatial structure. As an example of the connection between RD and CA models, the "shape theorem" of interacting particle systems (Durrett, 1988) describes the same large-scale constant-speed traveling wave behavior for the spatial SIR model that is known to occur for the corresponding RD equations.

One can learn a great deal about phage plaque formation and other aspects of microbial population biology through mathematical models. No model can completely capture all the details of the dynamics; in fact, models that attempt to come too close to reality are often of little use. Insight comes from finding the proper blend between reality and abstraction, and in comparing the predictions from different models (and with empirical data).

ACKNOWLEDGMENTS

We thank Wei Wei for help with the CA simulations and figures. Thank you to John Yin, who read and commented on an early draft of the manuscript (covering Sections 16.1–16.3). S.M.K. was supported in part by National Institutes of Health grant P20 RR 16448 from the COBRE program of the National Center for Research Resources.

REFERENCES

Abedon, S. T., and R. R. Culler. 2007a. Bacteriophage evolution given spatial constraint. *J. Theor. Biol.* **248**: 111–19.

Abedon, S. T., and R. R. Culler, 2007b. Optimizing bacteriophage plaque fecundity. *J. Theor. Biol.* **249**: 582–92.

Abedon, S. T., and J. Yin, in press. Bacteriophage plaques: theory and analysis. In M. Clokie and A. Kropinski (eds.), *Bacteriophages: Methods and Protocols.* Totowa, NJ: Humana Press.

Bremermann, H. J., and H. R. Thieme. 1989. A competitive-exclusion principle for pathogen virulence. *J. Math. Biol.* **27**: 179–90.

Buckling, A., and P. B. Rainey. 2002. Antagonistic coevolution between a bacterium and a bacteriophage. *Proc. Biol. Sci.* **269**: 931–6.

Dieckmann, U., R. Law, and J. A. J. Metz. 2000. *The Geometry of Ecological Interactions: Simplifying Spatial Complexity.* Cambridge: Cambridge University Press.

Durrett, R. 1988. *Lecture Notes on Particle Systems and Percolation.* Belmont, CA: Wadsworth.

Durrett, R., and S. A. Levin. 1994a. Stochastic spatial models: a user's guide to ecological applications. *Philos. Trans. R. Soc. Lond. B Biol. Sci.* **343**: 329–50.

Durrett, R., and S. A. Levin. 1994b. The importance of being discrete (and spatial). *Theor. Pop. Biol.* **46**: 363–94.

Durrett, R., and S. A. Levin. 1997. Allelopathy in spatially distributed populations. *J. Theor. Biol.* **185**: 165–72.

Durrett, R., and S. A. Levin. 1998. Spatial aspects of interspecific competition. *Theor. Pop. Biol.* **53**: 30–43.

Durrett, R., and C. Neuhauser. 1994. Particle systems and reaction diffusion equations. *Ann. Probab.* **22**: 289–333.

Fort, J., and V. Méndez. 2002. Time-delayed spread of viruses in growing plaques. *Phys. Rev. Lett.* **89**: 178101

Fort, J., J. Pérez, E. Ubeda, and F. J. García. 2006. Fronts with continuous waiting-time distributions: Theory and application to virus infections. *Phys. Rev.* E **73**: 021907.

Kaplan, D. A., L. Naumovski, B. Rothschild, and R. J. Collier. 1981a. Chromogenic detection of antigen in bacteriophage plaques: a microplaque method applicable to large-scale screening. *Gene* **13**: 211–20.

Kaplan, D. A., L. Naumovski, B. Rothschild, and R. J. Collier. 1981b. Appendix: a model of plaque formation. *Gene* **13**: 221–5.

Kerr, B., M. A. Riley, M. W. Feldman, and B. J. M. Bohannan. 2002. Local dispersal promotes biodiversity in a real-life game of rock–paper–scissors. *Nature* **418**: 171–4.

Koch, A. L. 1964. The growth of viral plaques during the enlargement phase. *J. Theor. Biol.* **6**: 413–31.

Kreft, J. U., and S. Bonhoeffer. 2005. The evolution of groups of cooperating bacteria and the growth rate versus yield trade-off. *Comment Microbiol.* **151**: 637–41.

Krone, S. M. 2004. Spatial models: stochastic and deterministic. *Math. Comp. Mod.* **40**: 393–409.

May, R. M., and R. M. Anderson. 1983. Epidemiology and genetics in the coevolution of parasites and hosts. *Proc. R. Soc. Lond. B Biol. Sci.* **219**: 281–313.

Mayr-Harting, A. 1958. Die Entwicklung von Phagenloechern und der Mechanismus der Phagenwirkung in festen Naehrboeden. *Zbl. f. Bakt. Paras. Infek. u. Hyg.* **171**: 380–92.

Murray, J. D. 1989. *Mathematical Biology*. New York, NY: Springer-Verlag,

Ortega-Cejas, V., J. Fort, V. Méndez, and D. Campos. 2004. Approximate solution to the speed of spreading viruses. *Phys. Rev.* E **69**: 031909.

Robb, F. T., and R. T. Hill. 2000. Bacterial viruses and hosts: Influence of culturable state. In R. R. Colwell and D. J. Grimes (eds.), *Nonculturable Microorganisms in the Environment*. ASM Press, Washington, DC: ASM Press, pp. 199–208.

Stent, G. S. 1963. *Molecular Biology of Bacterial Viruses*. San Francisco, CA: W. H. Freeman.

Tilman, D., and P. M. Kareiva. 1997. *Spatial Ecology: the Role of Space in Population Dynamics and Interspecific Interactions*. Princeton, NJ: Princeton University Press.

Wei, W., and S. M. Krone. 2005. Spatial invasion by a mutant pathogen. *J. Theor. Biol.* **236**: 335–48.

Yin, J. 1991. A quantifiable phenotype of viral propagation. *Biochem. Biophys. Res. Com.* **174**: 1009–14.

Yin, J. 1993. Evolution of bacteriophage T7 in a growing plaque. *J. Bacteriol.* **175**: 1272–7.

Yin, J., and J. S. McCaskill. 1992. Replication of viruses in a growing plaque: a reaction–diffusion model. *Biophys. J.* **61**: 1540–9.

You, L., and J. Yin. 1999. Amplification and spread of viruses in a growing plaque. *J. Theor. Biol.* **200**: 365–373.

CHAPTER 17

Modeling of bacteriophage therapy

Jason J. Gill

17.1 INTRODUCTION

Bacteriophage therapy, the treatment of bacterial infections with bacteriophages, is a topic that has received increasing attention in recent years. While the primary practice of phage therapy has been conducted with an eye towards the treatment of bacterial infections in humans, the concept of eliminating undesirable bacterial populations using phages can be extended to agriculturally important animals, plants, and even finished foodstuffs. While the potential benefits of phage therapy have been well documented, the mechanisms of phage therapy are less well understood, except in general terms. There is little in the way of standard criteria for selecting dose size, timing, or frequency when treating bacterial infections with phages. One approach toward addressing such concerns attempts to gain a better understanding of the *in vivo* reality of phage therapy through the development of theoretical mathematical models. Most models have comprised a series of differential equations, with theoretical parameters selected to examine the impacts of various phage therapy constraints. In this chapter I give a basic overview of the mathematical modeling approaches which have been applied to simulate phage therapy regimens. Simulations will be provided, where appropriate, to illustrate the behavior of these models. Finally, I briefly explore the current limitations of modeling given our understanding of how phages and bacteria interact *in vivo*.

For more on modeling phage–bacterial interactions, see Chapters 2, 10, 15, and 16; Chapter 13 provides a primer on phage interactions with animals,

Bacteriophage Ecology: Population Growth, Evolution, and Impact of Bacterial Viruses, ed. Stephen T. Abedon. Published by Cambridge University Press.

including in terms of phage therapy, while Chapter 12 discusses the use of phages as bacterial biocontrol agents in food.

17.1.1 The growing antibiotic crisis

In the years before the development of antibiotics, bacterial infections of all kinds were a common cause of morbidity and mortality in humans. More recently, the major killers of the past, such as *Mycobacterium tuberculosis*, *Staphylococcus aureus*, and *Vibrio cholerae*, once controllable via the employment of chemical antibiotic agents, have re-emerged in forms that are resistant to multiple antibiotics. Compounding this problem is the apparent dearth of novel antibiotics under development (Clarke, 2003), and an aging population cohort in the West which is generally more susceptible to many kinds of bacterial infections (Yoshikawa, 1997). In response to these trends, concerns have been voiced that we are moving towards a "post-antibiotic era" in which human populations will be vulnerable to morbidity and mortality caused by bacterial pathogens in a manner which has not been seen since the 1930s (Alanis, 2005).

17.1.2 Phage therapy

Bacteriophage therapy, or the use of bacteriophages to treat bacterial infections, has been studied in one form or another for almost as long as phages have been known to exist (Summers, 2001). Phage therapy was studied intensively around the world through the 1930s and 1940s as a means of combating bacterial infections in humans. This research was halted in the West, however, with the discovery of broad-spectrum chemical antibiotics that were effective and reliable, in contrast to early phage therapy research, which produced inconsistent results (Summers, 2001). While effectively abandoned in the West, development of phage therapy continued in other parts of the world, notably Eastern Europe and the former Soviet Union, and remains a part of the standard treatment repertoire in these regions in the modern era (Slopek *et al.*, 1987). Recent Western interest in phage therapy was sparked by the work of Smith and Huggins (1983), and phage therapy has proven successful against a number of pathogenic bacteria in limited trials (Smith and Huggins, 1983, Biswas *et al.*, 2002, Cerveny *et al.*, 2002, Matsuzaki *et al.*, 2003). See Chapters 12 and 13 for additional references and considerations of phage therapy.

A fundamental understanding of the dynamics of phage therapy is required in order to better predict treatment outcomes. As an ecological concept, phage therapy presents an extremely complicated system. The target pathogenic bacterium must be driven to extinction by the lytic action of one or more phages; this interaction proceeds in the context of a living host, which adds spatial structure (e.g., Chapters 2, 4, and 16), a variety of nutritionally diverse sub-environments, and an immune response to the system. Empirical work to gain understanding of the phage–bacterium–host interaction has been conducted in several recent studies, either in the form of *in vitro* model systems (O'Flaherty *et al.*, 2005, Gill *et al.*, 2006a) or in mouse models (Bull *et al.*, 2002). Another method for exploring the *in vivo* phage–bacterial interaction is the use of mathematical models, which simulate some aspects of the phage therapy dynamic.

Throughout this chapter we will be examining the interactions of bacteriophages with their host bacteria. In the context of phage therapy against pathogenic bacteria, the bacteria are host to the phages and yet are themselves hosted by the animal they are infecting. In order to avoid confusion, I will refer to the bacteria, where appropriate, as the phage's host or host cells; the animal or human which is colonized with a pathogenic bacterium shall be referred to as the subject.

17.1.3 Why do we model?

Mathematical models allow us to define and explore the dynamics of complex systems. As discussed elsewhere in this volume, the accurate predictive modeling of even the most controlled of experimental systems of chemostats or batch cultures is not a trivial matter, although approximations of reality are possible (Chapters 2, 15, and 16). Unlike these highly controlled *in vitro* systems, phage therapy in an *in vivo* system is fraught with complexities which are only beginning to be understood. Capturing every nuance of the *in vivo* phage therapy reality is beyond the scope of even the most complex models, and likely will be so for some time to come. It is a valid question, therefore, to ask why one should even attempt to construct such models when so much of the *in vivo* reality of phage therapy remains inscrutable. One could counter that mathematical modeling represents a reasonably tractable means of exploring the behavior of a complex system in order to generate predictions based on the interactions of known, predicted, or even strictly hypothetical variables in a complex system. While reliable *a priori* predictions of general phage therapy practice are not possible at this time, models can provide a rational basis for the design of phage therapy experiments and a context for the meaningful analysis of experimental outcomes.

There are a number of models examining idealized phage–bacterial interactions that can be applied to our understanding of phage-therapy dynamics, and many of these models have been discussed elsewhere in this volume (Chapters 2, 15, and 16; see also Chapter 12). Models have also been used to explicitly examine several different aspects of potential *in vivo* phage–bacterial interactions, such as the evolution of bacterial resistance to the phage (Chapter 2), within-subject competition between pathogen strains, the effects of spatial refuges (Chapters 2 and 16), and the effects of clearance of the phage by the subject's immune system (Chapter 13).

17.2 MODELING PHAGE THERAPY

A number of mathematical models have been proposed that examine various aspects of the phage–bacterial interaction. The standard approach to phage modeling in general has been to develop deterministic models (as opposed to stochastic ones; e.g., Chapter 16). These models consist of a set of several differential equations, each of which tracks a component population within the model. The interactions of the model components are governed by a series of (often hypothetical) rate constants and are solved simultaneously with respect to time (for a primer on continuous modeling of phage–bacterial interactions, see Chapter 15 as well as Chapter 2). In the presentation of differential equations in this chapter, all rate parameters govern an exponential function such that, for example, a bacterial growth rate, μ_N, can be expressed as $N_0 e^{\mu_N t} = N_t$, where N_0 is the density of cells at time 0, t is the number of units of elapsed time, and N_t is the density of cells extant after the time t has elapsed. The doubling time may be calculated as $\ln(2)/\mu_N = T_d$, where T_d is the doubling time. In cases where rates are negative (e.g., rates of free-phage decay), this same relationship can be used to calculate half-life.

The models all assume phage–bacterial interactions that occur on a basis of mass action (Chapter 15), which appears to offer adequate simulation as long as the output of model components (e.g., the calculated density of cells present at a given time, t) do not fall below a level at which fractional results may lead to the generation of artifacts. Individual-based modeling approaches, which track the interactions of each cell as an individual entity, have been employed to a limited extent (Chapter 16) but may come into greater prominence in the future as computational resources become more readily available. Some of the models developed cover phage–bacterium coevolution or coexistence (see Lenski and Levin, 1985; Crowley *et al.*, 1980; Schrag and Mittler, 1996; Chapter 2). Other models have been developed with an eye specifically towards phage therapy (Levin and Bull, 1996; Payne *et al.*, 2000;

Weld *et al.*, 2004). It is these latter models which are of primary interest in this chapter.

17.2.1 Model overview

The examination of phage therapy by modeling is a quantitative approach for gaining what is essentially a qualitative answer: the survival of the subject after phage treatment. While each of these models has its own unique characteristics and perspectives, they all use the same basic framework for examining phage–bacterial interactions. The phages and their bacterial hosts are assumed to exist in a dimensionless, thoroughly mixed habitat. Bacteria begin each simulation run at a set inoculum level and divide at a given rate. The phages similarly begin each simulation at a set number and then successfully adsorb to and infect their host cells at a given rate. Each infection results in the subsequent release of a number of progeny phages and the removal of the host cell from the bacterial population, with progeny phages free to adsorb to other susceptible host cells. The phages are presumed to be completely lytic (there is no lysogen formation; Chapters 1 and 5), and effects such as lysis inhibition are not considered (Chapter 3).

Throughout the simulation, both phages and bacteria are removed from the system by the subject's own response against these foreign bodies. This removal may take the form of a specific immune response mediated by antibodies, innate immunity such as macrophage grazing or complement (Chapter 13), or removal from the site of infection by some other mechanism, such as dilution. It is worth noting here that, in an actual *in vivo* system, the exact form of response is likely to differ in both type and magnitude between the phages and the bacteria.

This simulation continues for a given amount of time and generally concludes with one of two outcomes: (1) the phage eliminates the bacterial population, which is considered a treatment success; or (2) the bacterial population increases until it reaches some predetermined level, or persists in the subject for longer than some predetermined time limit. In this case, the subject is presumed to be unable to recover, which is considered a treatment failure. Levin and Bull (1996) used a terminal threshold of 1×10^{10} CFU (colony-forming units) per mL and ran simulations to 10 hours; if the bacterial population exceeded this threshold before this time, then the treatment was considered a failure. Work by Payne and Jansen (2001, 2003) presented simulations run for 20 or 25 hours. Excessively long simulation runs tend to exhibit cyclical behavior, which here will be considered to have little illustrative power in terms of the general model dynamics that are covered. In this chapter, simulations of 18 hours will be presented in all cases, which is long enough to observe the general behavior of the model in question. Furthermore, model parameters have been selected on the basis of their ability to illustrate some of the interesting features of the model dynamics, rather than as perfectly realistic estimates of these parameters in actual *in vivo* treatment. Indeed, such parameter estimates would, in any case, be expected to vary widely between pathogens, phages, subjects, and the various possible sites of infection. The importance of parameter selection is discussed in Section 17.3.3.

17.2.2 Basic model

Let us first consider a basic form of the model, with a single population of phage-susceptible bacteria, *N*, treated by a single phage type, *P*. Despite its apparent simplicity, this basic model illustrates many of the interesting dynamics of the theoretical phage-therapy system. The relationship can be described in Model I, below, which is a simple two-compartment model:

$$dN/dt = (\mu_N \cdot N) - (I_N \cdot N) - (k \cdot P \cdot N) \tag{17.1}$$

$$dP/dt = (k \cdot P \cdot N \cdot B) - (I_P \cdot P) \tag{17.2}$$

where the bacteria multiply according to a specific growth rate, μ_N, and are removed from the system by actions of the attacking phages, at the rate governed by $k \cdot P \cdot N$, where k is the phage adsorption constant, and also by the subject's immune response, at rate I_N. In cases where I_N is constant with respect to time, this term is somewhat redundant as it simply reduces the net bacterial growth rate, such that if $\mu_N = 2$ per hour and $I_N = 0.5$ per hour, the effect is the same as setting $\mu_N = 1.5$ per hour and $I_N = 0$ per hour. In general, if $I_N > \mu_N$ (rate of non-phage bacterial removal is greater than the rate of increase in bacterial number due to multiplication), then the bacteria will be cleared by the subject's immune response without any intervention by the phage. If the impact of the combination of phage-mediated lysis and immune response on bacterial density is less than the bacterial division rate, then a net increase in the bacterial population will be observed.

Here it is assumed that the subject's immune response is constant over time, irrespective of the size of the bacterial population. Other modeling approaches, such as presented in Levin and Bull (1996), implement a variable immune response, which increases in magnitude proportionally to the size of the bacterial load. It is also assumed that the bacteria will not reach a stationary phase of growth in this system. In the case of the phage-therapy models presented here, this issue is essentially abrogated by examining relatively low bacterial densities over short periods of time, situations in which a bacterial stationary phase is unlikely to come into play. The topic of limits on bacterial growth is also treated in the discussion of the Monod equation, Chapter 15.

Phages adsorb to bacteria in Equations 17.1 and 17.2 according to the term $k \cdot P \cdot N$, which is the derivative of the second-order rate equation. This model represents a simplified version of the phage life cycle (Fig. 2.1, Chapter 2), in that there is no accounting for the latent period; that is, infected bacteria are assumed to burst instantaneously and release B number of phages (Chapter 15). Phages are removed from the system by the subject's immune response (Chapter 13), I_P, which, like I_N, here is assumed to be constant over both time and space. In this respect, this model is similar in principle to the chemostat model presented in Chapter 15, in that there is a constant phage loss, though loss rates of phages and bacteria are governed by two different parameters, I_P and I_N, which are not necessarily equal. If the rate of phage replication falls below the rate of phage removal by the immune response, then a net decline in free phages will be observed, potentially resulting in phage extinction.

As shown in Fig. 17.1, simulations conducted using the basic Model I (Equations 17.1 and 17.2) can be used to illustrate the effects of altering key parameters on the behavior of the system. The rate of phage removal from the system, represented by I_P, for example, affects the treatment outcome (Fig. 17.1 A). As one might expect, a phage removal rate of 0 is preferable in terms of increasing the success of treatment. This observation would imply that an important factor in obtaining a cure by phage therapy may lie in the ability to keep viable phages at the site of the infection for as long as possible (Chapters 12 and 13). In other ways, the system behaves in a manner which is counterintuitive based on what is observed when treating bacterial infections with antibiotics. For example, alterations in the phage adsorption constant, k, also affect treatment success, as shown in Fig. 17.1B. However, contrary to what one may expect, the simulation suggests that a phage with a lower adsorption constant (in this case, 1×10^{-10} mL per hour) would be more successful in curing the subject than would a more rapidly adsorbing phage. This is a consequence of slower phage adsorption allowing the accumulation of higher bacterial densities, which in turn results in higher phage densities and thereby greater eventual depression in bacterial densities. Similarly, the simulation shown in Fig. 17.1 C suggests that delaying the administration of the phage treatment actually improves treatment success, which is contrary to what would be expected from most clinical experience. This observation is a result of the increased peak phage density as a consequence of the

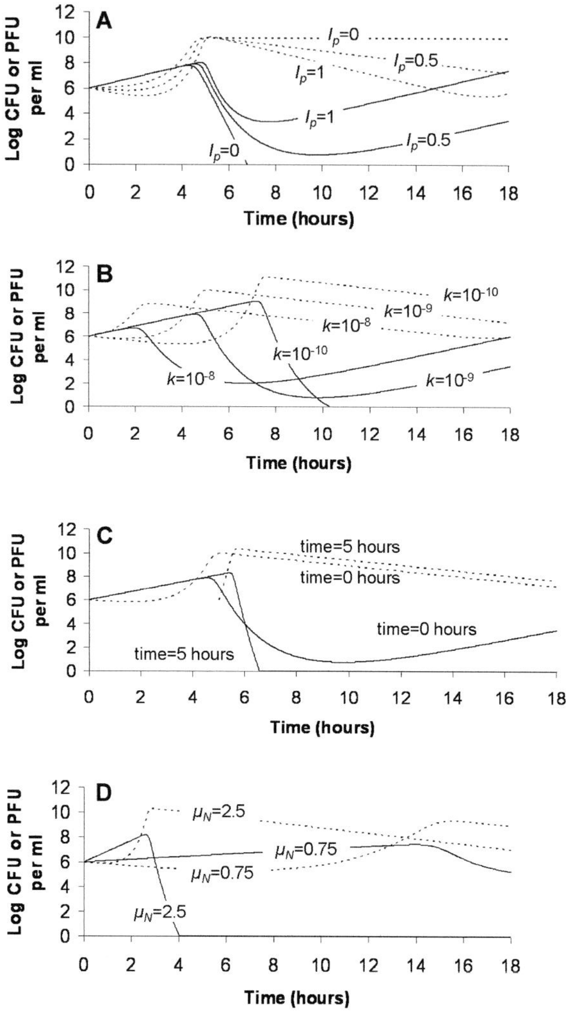

Figure 17.1 Simulation of phage therapy using Model I, showing the effects of altering certain parameters. Unless otherwise noted, $\mu_N = 1.5$ per hour, $I_N = 0.5$ per hour, $I_P = 0.5$ per hour, $k = 10^{-9}$ mL per hour, $B = 100$ new virions per lysed cell; the starting values of P and N are 1×10^6 PFU per mL and CFU per mL, respectively. (A) The effects of altering the phage removal rate, I_P; (B) the effects of altering the phage adsorption constant, k; (C) the effects of delaying phage treatment by 5 hours; (D) the effects of altering the growth rate of the bacterial population, μ_N. Dotted lines represent simulated phage densities, P, and solid lines represent bacterial densities, N.

bacterial density being allowed to grow to a higher level. The effect of bacterial population size is also illustrated in Fig. 17.1D, which shows the effects of altering the bacterial growth rate parameter μ_N. In this case, slowly growing bacteria ($\mu_N = 0.75$ per hour) are cleared poorly in comparison to a rapidly growing population ($\mu_N = 2.5$ per hour) (see also Fig. 15.5, Chapter 15).

These somewhat paradoxical observations are peculiar to an antimicrobial therapy that relies on a density-dependent process affected by the concentration of not only the antimicrobial (i.e., phage) but also by the density of the pathogen, as governed by the term $k \cdot P \cdot N$. Unless extremely high initial phage doses are applied (e.g., see Chapter 12, appendix), replication is required for the phage to reach sufficient levels to eliminate the pathogen population; phage growth is supported by higher bacterial densities, where the rate $k \cdot P \cdot N \cdot B > I_P$. In this case, bacterial growth rate would also be expected to influence therapeutic outcome, as shown in Fig. 17.1D. A variety of factors which may reduce net bacterial growth, such as an increased immune response or the addition of a chemical antibiotic to the system, may be expected to reduce the efficacy of the phage therapy regimen by reducing the size of the phage's potential host pool. A model suggesting the possible antagonism of antimicrobials to the success of phage therapy has been proposed (Payne and Jansen, 2003).

It should also be noted that such acutely host-dependent behavior is only observed in simulations where phage decay occurs at a rate that is greater than zero, i.e., $I_P > 0$. Alternatively, in a system where phages are never removed, they are free to accumulate in the medium until they reach such a concentration that, given enough time, their hosts are eliminated, regardless of phage or system properties. This observation was predicted by Chao and Levin (1977), and also presumes the lack of a bacterial stationary phase, which otherwise confers reduced susceptibility of the bacterium to the phage. As shown in Fig. 17.1 A, the phage replicates extremely slowly when first introduced into the system, as the densities of phages and bacteria are too low to allow adsorption at an appreciable rate given the adsorption constant used in this simulation (see also Fig. 15.3, Chapter 15). As the bacteria multiply, phage replication increases due to the increased density of available hosts for phage attachment; increased phage density contributes further to the rate of productive phage infection (see also Fig. 15.2, Chapter 15). In the absence of phage removal (i.e., where $I_P = 0$), this situation leads to the extinction of the bacterial population. If the phage is also subject to removal, however (e.g., where $I_P = 1$ per hour, Fig. 17.1 A; Chapter 13), then the bacteria may persist at low levels and eventually recover as more phages are removed from the system. In model simulations, this leads to a cyclical rise and fall of the bacterial and phage populations (see, for example, Fig. 2.4, Chapter 2, and Fig. 15.5, Chapter 15). The qualitative interpretation of this outcome is as a treatment failure.

In the case of Model I, the phage replication rate is determined by the adsorption constant, k, the densities of free phages and bacteria (P and N, respectively), and the burst size, B. The concentration of bacterial hosts in a given system is suggested to be an important determinant in the success of a phage-therapy treatment, in cases where phage replication is required to generate or maintain adequate phage titers at the site of infection. As noted by Kasman *et al.* (2002), phage adsorption and replication will still occur even at extremely low bacterial densities. However, this replication rate is far outstripped by the rate of phage decay, I_P, resulting in a net loss of free phages (see also Chapter 3). At the same time, the bacterial replication rate, μ_N, which is independent of bacterial density and greater than the rate of decay, I_N, allows the bacterial population to recover, unless the phage density is sufficiently high (e.g., Chapter 12) such that the phage adsorption rate exceeds the net rate of bacterial replication. Bacterial concentrations below this threshold will not be bound by free phages rapidly enough to support phage growth that is faster than phage decay.

17.2.3 The Payne and Jansen model

The basic Model I shown above (Equations 17.1 and 17.2) makes several simplifications to the replication of phages on their bacterial hosts. By removing the assumption of a latent period, for example, an important component of the phage life cycle is ignored (Chapters 2 and 15). Model I may be elaborated upon by adding a term which allows for a simulation of the phage latent period, as shown in Model II, adapted from a model presented by Payne and Jansen (2001):

$$dN/dt = (\mu_N \cdot N) - (I_N \cdot N) - (k \cdot P \cdot N) \tag{17.3}$$

$$dM/dt = (\mu_M \cdot M) - (I_M \cdot M) - (\lambda \cdot M)(k \cdot P \cdot N) \tag{17.4}$$

$$dP/dt = (B \cdot \lambda \cdot M) - (k \cdot P \cdot N) - (I_P \cdot P) \tag{17.5}$$

Bacteria are partitioned into two groups: infected cells, denoted M, and uninfected cells, denoted N. As free phages (P) adsorb to uninfected host cells (as governed by the rate term $k \cdot P \cdot N$), they are transferred to the second compartment in the model, M. It is assumed here that phages only adsorb to uninfected hosts (that is, infected cells, M, are not available for phage adsorption in Equation 17.5), though in reality a single host may be adsorbed

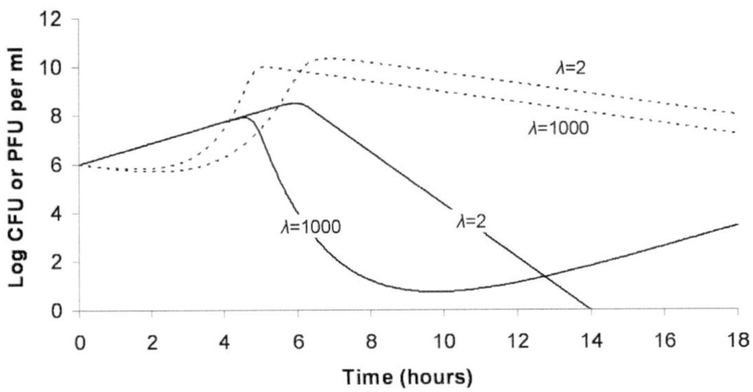

Figure 17.2 Simulation of phage therapy using Model II, showing the effects of the lysis rate parameter λ, measured as the number of latent periods per hour. Results are the product of two separate simulation runs. In both cases, $\mu_N = 1.5$ per hour, $I_N = 0.5$ per hour, $I_P = 0.5$ per hour, $k = 10^{-9}$ mL per hour, $B = 100$ new virions per lysed cell; the starting values of the phage and bacterial densities, P and N, are 1×10^6 PFU per mL and CFU per mL, respectively. Dotted lines represent simulated free-phage densities, P, and solid lines represent densities of uninfected bacteria, N. A lysis rate (λ) of 1000 per hour represents essentially instantaneous lysis.

by multiple phages (Chapter 8). It is presumed that the subject's immune response does not distinguish between infected and uninfected bacteria, and so both are removed at the same constant rate, I_N. The division rate of infected cells is governed by its own parameter, μ_M, which is distinct from the division rate of uninfected cells; this rate may be set to zero if it is presumed that phage-infected cells cease division upon infection.

The latent period in this model is simulated by the lysis rate term, λ, which is equivalent to the reciprocal of the phage latent period, which here is measured in hours. In this case, the lysis of infected bacteria is simulated as a continuous process, rather than as a discrete burst. Modeling the latent period in this manner has the effect of increasing phage population growth rates because some fraction of infecting bacteria lyse very early, while still releasing a full burst size of phage virions (see, for example, Weld *et al.*, 2004). See Chapter 16 for phage plaque models employing a similar treatment of phage latent period.

The addition of a latent period in the form of the lysis rate in Model II predicts a markedly different form of bacterial lysis than shown in Model I, even when all other parameters are kept equal. In this case, the rate, μ_M, has been set to zero, presuming that infected hosts, M, do not divide. As shown in Fig. 17.2, adjusting the lysis rate to 1000 (latent period $= 1/\lambda = 1/1000$

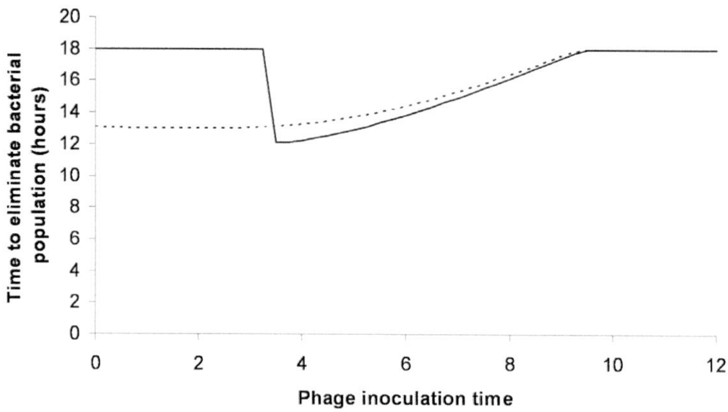

Figure 17.3 The effects of treatment time and phage dose size on phage-therapy treatment efficacy in simulations of Model II. In both cases, $\mu_N = 1.5$ per hour, $I_N = 0.5$ per hour, $I_P = 0.5$ per hour, $k = 10^{-9}$ per hour, $B = 100$ new virions per lysed cell, $\lambda = 2$ per hour; the starting value of N is 1×10^6 CFU per mL. The dotted line represents a phage dose of 1×10^6 PFU per mL, and the solid line represents a dose of 1×10^8 PFU per mL. Simulations were run for 18 hours; a line at 18 hours indicates that bacteria were not eliminated during the course of the simulation run. Note that the lower phage dose is generally more effective, and that, in the case of the higher dose, delaying treatment results in greater efficacy.

= 0.001, effectively yielding instantaneous lysis) generates results similar to those shown in Model I (Fig. 17.1). However, the simulation including a lysis rate of 2 per hour leads to slower phage growth and also a higher peak phage density, with subsequent elimination of the pathogen population (Fig. 17.2).

Payne and Jansen (2001) have noted, in the context of this model, that the size of the phage dose and the timing of administration have a major effect on treatment success. This effect is seen in the simulation of Model I (Fig. 17.1 C), where a delay in treatment results in clearance of the pathogen population by phages due to the greater accumulation of potential phage hosts. The overall effects of phage dose size (density) and treatment delay on treatment efficacy are illustrated for Model II in Fig. 17.3, where the time required for clearance of the total bacterial population ($N + M$) is plotted against the delay in phage application. Multiple simulations were run with delay (changes in dose time) from 0 to 12 hours at intervals of 0.25 hours. In the case of a higher phage dose (1×10^8 PFU per mL; PFU are plaque-forming units), early administration appears to be contraindicated by this model, as no clearance is observed before the simulation endpoint of 18 hours, for reasons discussed in the following paragraph. Clearance is first observed at 12.1 hours when phage administration is delayed by 3.5 hours and efficacy gradually wanes up to 9.5 hours, after which phages are unable to clear the pathogen population before the 18-hour endpoint. When a lower phage dose (1×10^6 PFU per mL) is used, pathogen clearance is observed at 13.1 hours when phages are administered at times ranging from immediate (at time 0) up to 9.25 hours. It is important to note here that these results are dependent upon requiring that phage replication within the subject generates sufficient phage numbers to eliminate the pathogen population. Had phages been added at high enough densities at the beginning of the simulation, then dramatic bacterial killing would be observed independent of the timing of phage addition (Chapter 12, appendix).

Long treatment delays may also result in eventual clearance of the pathogen. In simulations, however, this occurs after the set threshold of 18 hours, and also requires the accumulation of improbably high pathogen densities due to the lack of a maximal growth function in this model (Chapter 15). In simulations where an early phage administration results in efficacy loss (i.e., Fig. 17.3, 1×10^8 PFU per mL phage dose), the initial killing of the pathogen halts when its population is reduced to a density below which new hosts no longer are adsorbed, lysed, and release progeny more rapidly than phages are removed; by the time the bacterial population recovers, the phage population decay (controlled by I_P) has occurred to such an extent that there are not enough free phages remaining in the system to effect cure within 18 hours.

It should be noted here that this interesting phenomenon is not commonly observed in empirical studies of phage therapy. Work involving *S. aureus* in a mouse peritonitis model showed that early (time 0) phage doses at a variety of concentrations yielded high cure rates (Matsuzaki *et al.*, 2003). Similarly, phage therapy against *Vibrio vulnificus* in iron-dextran-treated mice found decreased efficacy when phages were administered more than 3 hours after bacterial challenge (Cerveny *et al.*, 2002). Bull *et al.* (2002) also noted a decrease in efficacy with delayed treatment by phages.

17.2.4 Modeling resistance of bacteria to phage

Another common factor that is often of interest in studies of phage therapy is the acquisition of resistance by bacteria against phage attack (Chapters 2 and 10). In their examination of work conducted by Smith and Huggins (1982), Levin and Bull (1996) modeled the effects of mutational resistance of the bacteria against phage attack. In this case, the pathogenic *Escherichia coli*

K-1 bacterium is able to acquire resistance to the phage by losing the ability to produce capsule, which is the receptor used by the K-1 capsule-specific phage used in these studies. However, the K-1 capsule is also a major pathogenicity determinant for the *E. coli* cells, making phage-resistant bacterial mutants considerably less pathogenic than the wild type (Gross *et al.*, 1977), and presumably more susceptible to clearance by the subject. This factor may be conceptually modeled by adding a fourth term to Model II, which tracks a subpopulation of phage-resistant bacteria within the subject, denoted here as *R*. A model depicting these factors, conceptually based on models presented in Payne and Jansen (2001) and Levin and Bull (1996) is shown below as Model III:

$$dN/dt = (\mu_N \cdot N) - (I_N \cdot N) - (k \cdot P \cdot N) \tag{17.6}$$

$$dR/dt = (\mu_R \cdot R) - (I_R \cdot R) \tag{17.7}$$

$$dM/dt = (\mu_M \cdot M) - (I_M \cdot M) - (\lambda \cdot M)(k \cdot P \cdot N) \tag{17.8}$$

$$dP/dt = (B \cdot \lambda \cdot M) - (k \cdot P \cdot N) - (I_P \cdot P) \tag{17.9}$$

In this model, bacteria are again transferred from the uninfected (*N*) to phage-infected (*M*) populations by phage adsorption. Because bacterial mutation to phage resistance is a stochastic process and in reality does not occur reliably, a set population of phage-resistant bacteria is presumed to exist at the start of the simulation. The growth rate of this resistant population, *R*, is governed by its own parameter, μ_R. Because this population is considered to be completely invulnerable to phage attack, it is removed only by the subject's immune response, I_R. However, for capsule-deficient bacteria, removal by the subjects' immune response is presumed to occur more easily, so the parameter I_R may be set considerably higher than its counterpart affecting the phage-sensitive (e.g., encapsulated) population, I_N.

This model predicts that, in addition to the factors discussed in Sections 17.2.2 and 17.2.3, the development of bacterial resistance to phage attack results in treatment failure only if the mechanism of resistance confers a net selective advantage to the bacterium. In this case, phage resistance is only advantageous to the pathogen if cells displaying the resistant phenotype are not excessively susceptible to removal by the subject's immune response, described by I_R; if $I_R > \mu_R$, then this population will decline within the subject. This is also a satisfying explanation of the results of Smith and Huggins (1982), as noted by Levin and Bull (1996). The selection pressures against the loss of capsule present in the *in vivo* setting make phage-resistant mutants considerably less viable *in vivo* and treatment by phages more successful. This attenuation of bacterial virulence associated with phage resistance has been noted in some plant pathogenic bacteria as well (Hendrick and Sequeira, 1984; Schoonejans *et al.*, 1987), suggesting it may be an attractive target in implementing phage therapy in other situations. For more on bacterial mutation to phage resistance and the associated fitness costs, see Chapter 2.

17.2.5 Non-mutational resistance

As noted above, the development of mutational resistance by bacteria to phages is not an insurmountable obstacle in the application of phage therapy, particularly if such resistant mutants also display reduced virulence. However, aside from mutational resistance, the pathogen population may resist phages by other means. It has been noted in some animal studies of phage therapy that delays in treatment greatly reduce treatment success (Cerveny *et al.*, 2002; Bull *et al.*, 2002); in work presented by Bull *et al.* (2002), there is strong empirical evidence that mutational resistance was not a likely cause of treatment failure. It is possible that these delays represent the time required for the pathogen to adapt to the *in vivo* environment and reach a state which is genetically sensitive to the phage but phenotypically resistant, either through physiological adaptations or through migration to protected sites within the subject.

In Model III, bacterial resistance is presented as an absolute barrier to phage-mediated killing. Model IV, described below, attempts to examine the effect of hidden or phenotypically resistant bacteria in the course of phage therapy, by allowing a limited amount of phage adsorption to the resistant bacterial population, and, importantly, by allowing bacterial cells to move from the resistant to the sensitive population. This effect is similar in concept to the population-stabilizing properties of spatial refuges as described in Schrag and Mittler (1996), and the modeling of "persister" cells by Roberts and Stewart (2005). The model is modified from Model III with the assumption that the bacteria are able to move from the resistant bacterial population, *R*, into the normal population, *N*, at a rate that is governed by the parameter, *T*; this would represent the shedding of bacteria protected in a biofilm or intracellular compartment into the "sensitive" or planktonic population.

$$dN/dt = (\mu_N \cdot N) - (I_N \cdot N) - (k \cdot P \cdot N)(T \cdot R) \quad (17.10)$$

$$dR/dt = (\mu_R \cdot R) - (I_R \cdot R) - (k_R \cdot P \cdot R) - (T \cdot R) \quad (17.11)$$

$$dM/dt = (\mu_M \cdot M) - (I_M \cdot M) - (\lambda \cdot M)(k \cdot P \cdot N) \quad (17.12)$$

$$dM_R/dt = (\mu_{MR} \cdot M) - (I_R \cdot R) - (\lambda \cdot M_R)(k_R \cdot P \cdot R) \quad (17.13)$$

$$dP/dt = (B \cdot \lambda \cdot M)(B \cdot \lambda \cdot M_R) - (k \cdot P \cdot N) - (k_R \cdot P \cdot R) - (I_P \cdot P) \quad (17.14)$$

The subpopulation of bacteria, R, may be less susceptible to phage attack due to incorporation into a biofilm, phenotypic changes, intracellular survival, or any combination of these or other factors. Bacteria in this protected state are presumed to multiply, adsorb phage, and be removed by the subject's immune response at rates which are lower that those used for the general bacterial population.

Bacteria enter the sensitive bacterial population, N, from the protected bacterial population, R, at a rate governed by the term, T. In this model, bacteria do not transfer from the N population into the R population; while this transfer does likely occur in reality, here it is assumed that the contribution of this movement to the growth of the population R is insignificant when compared to the division of cells already in this compartment. While in the protected state, bacteria divide at the rate, μ_R, and are removed by the subject's immune response at the rate, I_R. Phage are capable of adsorbing to the protected bacteria at a rate governed by k_R, which is presumed to be lower than k. There are therefore two populations of phage-infected bacteria, M and M_R; the only difference between the two is their rate of removal, governed by I_N and I_R, respectively. For the sake of simplicity, it is assumed here that the phage latent period and burst size are the same for hosts in both the N and R populations. Phage-infected bacteria are again assumed to stop multiplying upon phage infection, thus the term, μ_M, and its corresponding term governing the growth rate of phage-infected resistant cells, μ_{MR}, have both been set to zero.

In this model, μ_R is presumed to be lower than μ_N and has been set at 0.5 per hour, as opposed to 1.5 per hour for μ_N. Due to their protected status, the rate of removal of the R population, I_R, is also presumed to be low, and has been set to 0.1; the R population is therefore maintained as a slow-growing population of protected bacteria. The transfer-rate parameter, T, which governs the rate at which cells move from the R to the N population, is 0.01 per hour. The simulation is established with an initial sensitive pathogen load, N, of 1×10^6 CFU per mL and a population of protected cells, R, of 1×10^4 CFU per mL, 1% of the sensitive population. As with the other models presented here, the phage burst size, B, is 100, λ is 2 per hour, and the adsorption constant, k, is 10^{-9} mL per hour. The adsorption rate against cells in the R population, k_R, is presumed to be lower, at 10^{-10} mL per hour. The initial phage dose, administered at 5 hours, is 1×10^8 PFU per mL. In this model, the protected pathogen population, R, survives phage administration due to its low adsorption constant, k_R (Fig. 17.4 A), and replenishes N over time via the function $T \cdot R$, allowing the overall infection to persist. If the protected state of the R population is removed by adjusting k_R to 10^{-9} mL per hour (the same value as k), then the phages are able to

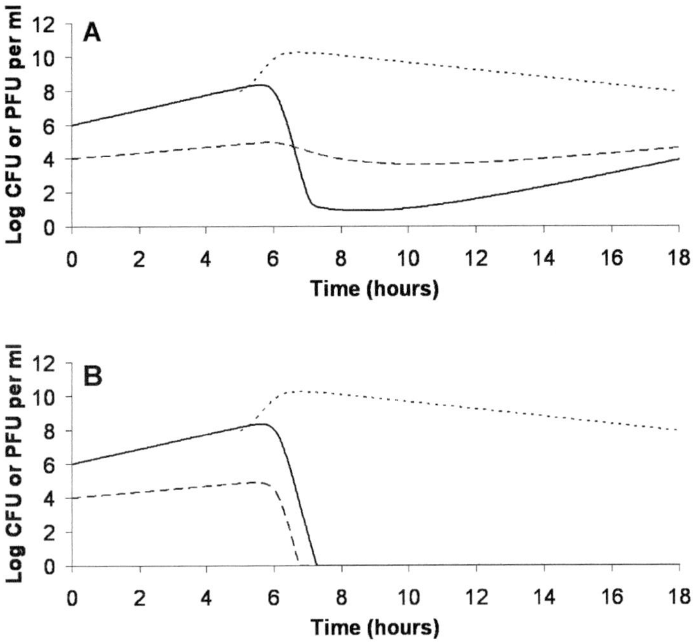

Figure 17.4 Simulation of phage therapy using Model IV, showing the effects of altering the rate, k_R, of phage adsorption to hosts in the resistant, R, population. In all cases, $\mu_N =$ 1.5 per hour, $\mu_R = 0.5$ per hour, $I_N = 0.5$ per hour, $I_P = 0.5$ per hour, $I_R = 0.1$ per hour, $k = 10^{-9}$ mL per hour, $\lambda = 2$ per hour, $B = 100$ new virions per lysed cell, $T = 0.01$ per hour; the starting values of N, sensitive bacteria, and R, resistant bacteria, are 1×10^6 and 1×10^4 CFU per mL, respectively. A phage dose of 1×10^8 PFU per mL delayed to 5 hours is used in both cases. (A) $k_R = 10^{-10}$ mL per hour; (B) $k_R = 10^{-9}$ mL per hour. Dotted lines represent simulated phage densities, P, solid lines represent bacterial densities, N, and dashed lines represent densities of resistant bacteria, R. Note the survival of bacteria, in their resistant form (R, dashed lines) when R forms are partially protected (panel A) versus unprotected (panel B), and the associated recovery of sensitive bacteria in panel A (solid line).

eliminate the pathogen population shortly after the inoculation of the phages (Fig. 17.4B).

17.3 MODELS IN THE REAL WORLD

As mentioned at the beginning of this chapter, mathematical models are a theoretical tool for understanding the interaction of phages and bacteria using a set of predetermined assumptions. When using these models to describe phage therapy, however, there is little in the way of current knowledge as to how well the models can predict treatment outcomes in an *in vivo* setting. Perhaps the greatest value of phage therapy modeling in its current state, therefore, is to underline how little is truly understood about the interaction of phages and their hosts in a system as complex as a living subject. Even though phage adsorption constants, latent periods, and burst sizes may be precisely determined *in vitro* for use in mathematical simulations of phage therapy, the reality is that the phages or their bacterial hosts are unlikely to behave the same *in vivo* as they do *in vitro*. There are numerous additional factors which could potentially be addressed when constructing models of phage therapy, a few of which will be considered in this section.

17.3.1 Predictive value of current models

Work by Weld *et al.* (2004) has shown that models of phage therapy based on parameters obtained from *in vitro* experiments were unable to predict the behavior of phage and pathogen in a mouse model. The model employed in this study was a three-compartment system similar to Model II, with the exception that the phage latent period was simulated by a more complicated time-delay system which is effectively equivalent to that described in terms of differential equations in the chemostat model presented in Chapter 2 (Equation 2.8) and the discrete time-delay chemostat model employed in Chapter 15 (Equation 15.11). While the modeling approach of Weld and colleagues was shown to accurately predict phage growth *in vitro*, the introduction of phages and their hosts into an *in vivo* rat model produced results which were not in agreement with the model predictions. These results are echoed in those found by Chibani-Chennoufi *et al.* (2004). *E. coli* introduced into the gastrointestinal tracts of axenic mice was found to be lysed but not eliminated by an inoculated phage; in standard mice with an extant gut flora, this phage was found to have little effect on the *E. coli* population. In all cases, the bacteria were found to be sensitive to the phage *in vitro*, and the occupation of protected sites within the gut by the *E. coli* cells was suggested as a possible cause of these observations (Chibani-Chennoufi *et al.*, 2004).

In contrast, the models employed in Levin and Bull (1996) do possess some explanatory power for the results observed by Smith and Huggins (1982), wherein the resistance of the pathogen to its phage was coupled to a major virulence factor, the *E. coli* K-1 capsule. This observation again highlights the potential utility of phage therapy modeling in its current form: as a qualitative tool for unraveling the potential interactions between bacteria and phages in what is still largely the "black box" of the *in vivo* system. The extensive work presented by Weld *et al.* (2004) confirms that, at this stage, it is premature to expect quantitative predictions from mathematical modeling of phage therapy.

17.3.2 Spatial structure and heterogeneous populations

All phage therapy models explored to date make the explicit assumption that the phages and bacteria exist in a spatially unstructured, homogeneous, and well-mixed environment. This implies that each individual phage particle, upon its release from a lysed host cell, has an equal opportunity to adsorb to any new host contained in the model population. While this assumption may hold at least partially true in the case of certain clinical syndromes, in many cases the pathogen population (or a subpopulation thereof) is localized to a surface-attached or otherwise immobilized state within the subject. Funk *et al.* (2005) have proposed a model of viral growth in a spatially structured environment. In this model, the environment is divided into a two-dimensional array of sites, and free virions are able to diffuse only into sites which are adjacent to their site of release. The introduction of spatial coupling into the model was found to suppress the appearance of cyclical growth and decay as found in models with assumptions of homogeneity. Phage propagation within a biofilm-type environment may well resemble phage propagation through an agar medium as found in plaque growth (Chapter 4), in which case methods covered in Chapter 16 may need to be incorporated into models of phage therapy (see also descriptions of phage metapopulation ecology, Chapter 2).

The particular microenvironment colonized by a pathogen is unique to each pathogen and infection site. *S. aureus*, which can cause a diverse array of clinical syndromes including toxic shock syndrome, bacteremia, mastitis, and osteomyelitis, may exist in a surface-attached state and also intracellularly within macrophages, neutrophils, or epithelial cells (Lowy, 1998). *Pseudomonas aeruginosa* is believed to exist primarily as a biofilm when colonizing wounds and the lungs of cystic fibrosis patients (Costerton *et al.*, 1999). Other pathogens, such as *Listeria monocytogenes*, are able to migrate to and travel within the intracellular environment (Hamon *et al.*, 2006). All of these modes of bacterial pathogenesis, and a plethora more not mentioned here, would be expected to severely limit or completely eliminate the ability of a free bacteriophage to bind to the surface of its host until the host leaves such a protected state. As shown in simulations of Model IV, which models a heterogeneous but dimensionless host population, protected bacterial subpopulations are able to persist in the subject relatively unaffected by free phages. This suggests that greater knowledge of the mode of bacterial pathogenesis and its effects on phage-mediated lysis are crucial for a systematic application of phage therapy *in vivo*. Furthermore, modeling approaches will need to be modified to take into account the pathogenesis and disease progression of different pathogen–subject systems.

17.3.3 Parameter estimation

Throughout this chapter, the parameters used for model simulations were selected based on their ability to illustrate various interesting features of the models. These parameters satisfy a basic requirement of biological plausibility, and are similar in many respects to the parameters selected for use by Levin and Bull (1996) in their simulations. However, "biologically plausible" could define a relatively broad range of possible parameter estimates, and would certainly be highly dependent upon the particular bacterial species and phage strain under study. Typical parameter estimates used in this chapter and other studies of phage-therapy modeling are shown in Table 17.1.

Growth-rate parameters of 1.5 per hour (equivalent to a doubling time of 0.46 hours) were used throughout most of this chapter and also in Levin and Bull (1996). Payne and Jansen (2001) selected a growth rate of 0.5 per hour in their study (doubling time of 1.39 hours). Weld *et al.* (2004) measured an *in vivo* net bacterial growth rate of 0.36 per hour in rats. However, the actual growth rate of the bacterial population in this case cannot be precisely determined as this measurement also includes any losses of bacteria within the system. Bacterial growth rate has a strong effect on phage therapy treatment outcomes, as shown in Fig. 17.1D.

Phage growth parameters have also been shown to be closely tied to the host cell division rate, which is an indicator of cell physiology. In the case of *E. coli*, for example, low bacterial growth rates have been associated with lower phage adsorption rates, longer latent periods, and smaller burst sizes (Hadas *et al.*, 1997). In *P. aeruginosa*, phage infection under starvation conditions may lead to the maintenance of a pseudolysogenic state (Ripp and Miller, 1998; Chapter 5) which in terms of model parameter estimation would translate into a dramatically extended latent-period parameter. These examples illustrate the importance of understanding pathogen physiological status under *in vivo* conditions when simulating phage therapy in a mathematical model.

The decay rate of free phages in the system, represented in this chapter by the parameter I_P, also plays an important role in simulation outcomes, as shown in Fig. 17.1 A. As shown in Table 17.1, a range of parameters have been used in phage therapy simulations. Free-phage decay rates of approximately

Table 17.1 *Selected parameter estimates reported in previous studies of phage therapy modeling, and in this chapter.*

Parameter description	Symbol (as used here)	Units (as used here)	Levin & Bull, 1996	Payne & Jansen, 2001	Weld *et al.*, 2004	This work
Bacterial division rate	μ_N	per hour	1.5	0.5	0.09 to 0.36	1.5
Phage-infected bacteria division rate	μ_M	per hour	NA	0.5	0	0
Phage burst size	B	per bacterium	100	100	150	100
Phage adsorption constant	k	mL/hour	1×10^{-9} to 2×10^{-11}	1×10^{-7}	1.2×10^{-6}	1×10^{-9} to 1×10^{-10}
Phage latent period	λ	per hour	0	5	2.3	2
Free-phage decay rate	I_P	per hour	0	5	0.36 to 1.2	0.5

Note the range of parameter estimates used, and also the variation in choice of measurement units (minutes vs. hours). "NA" denotes the absence of a term for this parameter in the model.

0.36 per hour and 1.2 per hour were calculated for *E. coli* phages K1—5 and T4, respectively, following their introduction into the rat gastrointestinal tract (Weld *et al.*, 2004). A free-phage decay rate of approximately 0.4 per hour was observed following the introduction of *S. aureus* phage K into the mammary gland of lactating cattle (Gill *et al.*, 2006b). Merril *et al.* (1996) measured a circulating free-phage decay rate of approximately 0.37 per hour for wild-type λ*vir* phage following intraperitoneal administration in mice, and decay rates of approximately 0.05 per hour for their long-circulating mutant derivatives (see also Chapter 13). As shown in Fig. 17.1 A, low phage decay rates are expected to increase phage-therapy efficacy, and this prediction was shown empirically by Merril *et al.* (1996), where the long-circulating phage exhibited greater efficacy than its wild-type parent phage. This empirical support for the model prediction illustrates a role for such models in the development of novel hypotheses surrounding phage therapy.

As shown in the previous mathematical simulations of phage therapy (Fig 17.1B), the phage adsorption constant, *k*, is of major importance in these models. This effect is also illustrated in Fig. 3 of Levin and Bull (1996). The adsorption constant governs the rate at which free phages adsorb to their hosts at any given concentration of phages and bacteria. This constant represents the aggregate effects of the phage diffusion rate (which is, in turn, governed by such factors as virion size and shape, and viscosity of the medium); the size, diffusion rate, and motility of the host cells; the type and amount of phage receptor expressed on the host cells; and the affinity of the phage's attachment proteins to the bacterial cell surface. As shown in Table 17.1, values of *k* used by various workers range over several orders of magnitude. All of these values are biologically plausible: Schrag and Mittler (1996) reported adsorption constants ranging from 2×10^{-8} to 4×10^{-10} mL per hour for phage T1X under different growth conditions, while Weld *et al.* reported average rates of 3×10^{-9} mL per min (equivalent to 1.8×10^{-7} mL per hour) for phage T4 under optimal conditions. There is little guarantee, however, that such constants measured *in vitro* will accurately reflect the phage–bacterial interaction *in vivo*.

Environments which contain turbulent flow would be expected to increase adsorption rate and thereby increase the effective value of *k* in these environments (Koch, 1960). Similarly, particles with a larger effective radius, either through extension of the phage receptor from the cell surface (as may occur when capsule or flagella are a primary phage receptor) or through cellular aggregation, are expected to increase the encounter efficiency of a phage and its host and hence the apparent value of *k* (Koch, 1960). Factors which inhibit phage adsorption to its host cell have also been observed in some systems. Production of a bacterial capsule which is capable of inhibiting phage adsorption has been documented in *E. coli* (Scholl *et al.*, 2005) and *S. aureus* (Wilkinson and Holmes, 1979), and capsule expression has been found to be highly strain-dependent in *S. aureus* (Herbert *et al.*, 2001). Exposure of *S. aureus* to milk or serum proteins also strongly promotes the aggregation of the cells and induces a significant reduction in the phage adsorption rate (O'Flaherty *et al.*, 2005, Gill *et al.*, 2006a). In addition, this rate reduction was found to be highly dependent on the subject, ranging from complete inhibition of phage adsorption to adsorption rates comparable to optimal conditions (Gill *et al.*, 2006a).

Throughout this chapter, all rate parameters are given in units of hours. The issue of measurement units may lead to confusion when attempting to track a large number of these parameters, particularly when such parameters are reported in different unit scales. Bacterial growth rates, for instance, are typically measured and expressed in rates per hour. A specific bacterial growth rate of 1.5 per hour translates into a doubling time of 27.6 minutes, a biologically plausible rate of growth. Phage adsorption constants, however, may be expressed as units per minute (e.g., Weld *et al.*, 2004; Chapter 15) or as units per hour (e.g., Schrag and Mittler, 1996). In some important contributions to the phage-therapy modeling literature, such as Levin and Bull (1996), parameters are given as units per hour but are not clearly delineated, which could potentially lead to confusion when attempting to repeat these simulations. It is therefore recommended and encouraged that measurement units be explicitly stated in all descriptions of mathematical models.

17.4 CONCLUSIONS

There is a current and growing need for the development of novel antimicrobial treatments, as pathogenic bacteria become resistant to conventional antibiotics. One such treatment is the application of phage therapy, which holds significant promise in the treatment of a diverse array of bacterial infections. Being by its nature a complicated predator–prey interaction (see Chapter 2), phage therapy appears to be more difficult to implement *in vivo* than are conventional antibiotics. Mathematical models of phage therapy are one method of achieving a greater understanding of the interactions involved, which in turn could lead to more successful phage therapy implementation in a clinical setting. The models proposed to date apparently have limited quantitative predictive ability, but are useful for the generation of novel hypotheses and, possibly, for qualitative predictions of treatment outcomes. However, the utility of these models would be extended immensely by a more thorough understanding of the *in vivo* subject–pathogen–phage interaction.

ACKNOWLEDGMENTS

I would like to thank Robin McKellar and Parviz Sabour of Agriculture and Agri-Food Canada for their inspiration, ideas, and support in the writing of this chapter. This work was supported by Agriculture and Agri-Food Canada, the Dairy Farmers of Ontario and US Public Health Service grant AI064512.

REFERENCES

Alanis, A. J. 2005. Resistance to antibiotics: are we in the post-antibiotic era? *Arch. Med. Res.* **36**: 697–705.

Biswas, B., S. Adhya, P. Washart, *et al.* 2002. Bacteriophage therapy rescues mice bacteremic from a clinical isolate of vancomycin-resistant *Enterococcus faecium*. *Infect. Immun.* **70**: 204–10.

Bull, J. J., B. R. Levin, T. DeRouin, N. Walker, and C. A. Bloch. 2002. Dynamics of success and failure in phage and antibiotic therapy in experimental infections. *BMC Microbiol.* **2**: 35.

Cerveny, K. E., A. DePaola, D.H. Duckworth, and P.A. Gulig. 2002. Phage therapy of local and systemic disease caused by *Vibrio vulnificus* in iron-dextran-treated mice. *Infect. Immun.* **70**: 6251–62.

Chao, L. and B. R. Levin. 1977. A complex community in a simple habitat: an experimental study with bacteria and phage. *Ecology* **58**: 369–78.

Chibani-Chennoufi, S., J. Sidoti, A. Bruttin, E. Kutter, S. A. Sarker, and H. Brüssow. 2004. In vitro and in vivo bacteriolytic activity of *Escherichia coli* phages: implication for phage therapy. *Antimicrob. Agents Chemother.* **48**: 2558–69.

Clarke, T. 2003. Drug companies snub antibiotics as pipeline threatens to run dry. *Nature* **425**: 225.

Costerton, J. W., P. S. Stewart, and E. P. Greenberg. 1999. Bacterial biofilms: a common cause of persistent infections. *Science* **284**: 1318–22.

Crowley, P. H., S. C. Straley, R. J. Craig, *et al.* 1980. A model of prey bacteria, predator bacteria, and bacteriophage in continuous culture. *J. Theor. Biol.* **86**: 377–400.

Funk, G. A., V. A. Jansen, S. Bonhoeffer, and T. Killingback. 2005. Spatial models of virus-immune dynamics. *J. Theor. Biol.* **233**: 221–36.

Gill, J. J., P. M. Sabour, K. E. Leslie, and M. W. Griffiths. 2006a. Bovine whey proteins inhibit the interaction of *Staphylococcus aureus* and bacteriophage K. *J. Appl. Microbiol.* **101**: 377–86.

Gill, J. J., J. C. Pacan, M. E. Carson, K. E. Leslie, M. W. Griffiths, and P. M. Sabour. 2006b. Efficacy and pharmacokinetics of bacteriophage therapy in the treatment of subclinical *Staphylococcus aureus* mastitis in lactating dairy cattle. *Antimicrob. Agents. Chemother.* **50**: 2912–18.

Gross, R. J., T. Cheasty, and B. Rowe. 1977. Isolation of bacteriophages specific for the K1 polysaccharide antigen of *Escherichia coli.J. Clin. Microbiol.* **6**: 548–50.

Hadas, H., M. Einav, I. Fishov, and A. Zaritsky. 1997. Bacteriophage T4development depends on the physiology of its host *Escherichia coli. Microbiology* **143**: 179–85.

Hamon, M., H. Bierne, and P. Cossart. 2006. *Listeria monocytogenes*: a multifaceted model. *Nat. Rev. Microbiol.* **4**: 423–34.

Hendrick, C. A. and L. Sequeira. 1984. Lipopolysaccharide-defective mutants of the wilt pathogen *Pseudomonas solanacearum. Appl. Environ. Microbiol.* **48**: 94–101.

Herbert, S., S. W. Newell, C. Lee, *et al.* 2001. Regulation of *Staphylococcus aureus* type 5 and type 8 capsular polysaccharides by CO_2. *J. Bacteriol.* **183**: 4609–13.

Kasman, L. M., A. Kasman, C. Westwater, J. Dolan, M. G. Schmidt, and J. S. Norris. 2002. Overcoming the phage replication threshold: a mathematical model with implications for phage therapy. *J. Virol.* **76**: 5557–64.

Koch, A. L. 1960. Encounter efficiency of coliphage–bacterial interaction. *Biochim. Biophys. Acta* **39**: 311–18.

Lenski, R. E. and B. R. Levin. 1985. Constraints on the coevolution of bacteria and virulent phage: A model, some experiments, and predictions for natural communities. *Am. Nat.* **125**: 585–602.

Levin, B. R. and J. J. Bull. 1996. Phage therapy revisited: the population biology of a bacterial infection and its treatment with bacteriophage and antibiotics. *Am. Nat.* **147**: 881–98.

Lowy, F. D. 1998. *Staphylococcus aureus* infections. *N. Engl. J. Med.* **339**: 520–32.

Matsuzaki, S., M. Yasuda, H. Nishikawa, *et al.* 2003. Experimental protection of mice against lethal *Staphylococcus aureus* infection by novel bacteriophage phi MR11. *J. Infect. Dis.* 187: 613–624.

Merril, C. R., B. Biswas, R. Carlton, *et al.* 1996. Long-circulating bacteriophage as antibacterial agents. *Proc. Natl. Acad. Sci. U.S.A.* **93**: 3188–92.

O'Flaherty, S., A. Coffey, W. J. Meaney, G. F. Fitzgerald, and R. P. Ross. 2005. Inhibition of bacteriophage K proliferation on *Staphylococcus aureus* in raw bovine milk. *Lett. Appl. Microbiol.* **41**: 274–9.

Payne, R. J. and V. A. Jansen. 2001. Understanding bacteriophage therapy as a density-dependent kinetic process. *J. Theor. Biol.* **208**: 37–48.

Payne, R. J. and V. A. Jansen. 2003. Pharmacokinetic principles of bacteriophage therapy. *Clin. Pharmacokinet.* **42**: 315–25.

Payne, R. J., D. Phil, and V. A. Jansen. 2000. Phage therapy: the peculiar kinetics of self-replicating pharmaceuticals. *Clin. Pharmacol. Ther.* **68**: 225–30.

Ripp, S., and R. V. Miller. 1998. Dynamics of pseudolysogenic response in slowly growing cells of *Pseudomonas aeruginosa*. *Microbiology* **144**: 2225–32.

Roberts, M. E. and P. S. Stewart. 2005. Modelling protection from antimicrobial agents in biofilms through the formation of persister cells. *Microbiology* **151**: 75–80.

Scholl, D., S. Adhya, and C. Merril. 2005. *Escherichia coli* K1's capsule is a barrier to bacteriophage T7. *Appl. Environ. Microbiol.* **71**: 4872–4.

Schoonejans, E., D. Expert, and A. Toussaint. 1987. Characterization and virulence properties of *Erwinia chrysanthemi* lipopolysaccharide-defective, phi EC2-resistant mutants.*J. Bacteriol.* **169**: 4011–17.

Schrag, S. J., and J. E. Mittler. 1996. Host–parasite coexistence: the role of spatial refuges in stabilizing bacteria–phage interactions. *Am. Nat.* **148**: 348–77.

Slopek, S., B. Weber-Dabrowska, M. Dabrowski, and A. Kucharewica-Krukowska. 1987. Results of bacteriophage treatment of suppurative bacterial infections in the years 1981–1986. *Arch. Immunol. Ther. Exp. (Warsz)* **35**: 569–83.

Smith, H. W. and M. B. Huggins. 1982. Successful treatment of experimental *Escherichia coli* infections in mice using phage: its general superiority over antibiotics. *J. Gen. Microbiol.* **128**: 307–18.

Smith, H. W. and M. B. Huggins. 1983. Effectiveness of phages in treating experimental *Escherichia coli* diarrhoea in calves, piglets and lambs. *J. Gen. Microbiol.* **129**: 2659–75.

Summers, W. C. 2001. Bacteriophage therapy. *Annu. Rev. Microbiol.* **55**: 437–51.

Weld, R. J., C. Butts, and J. A. Heinemann. 2004. Models of phage growth and their applicability to phage therapy.*J. Theor. Biol.* **227**: 1–11.

Wilkinson, B. J. and K. M. Holmes. 1979. *Staphylococcus aureus* cell surface: capsule as a barrier to bacteriophage adsorption. *Infect. Immun.* **23**: 549–52.

Yoshikawa, T. T. 1997. Perspective: aging and infectious diseases: past, present, and future. *J. Infect. Dis.* **176**: 1053–7.

Index

Abiotic 15, 18, 35, 40, 221
Abundance of bacteria
Grazer impact (see Grazers)
Normal flora (see Flora)
Viral impact 260
Worldwide 1
Abundance of phages 14, 119, 260–1
Aquatic 14, 252, 261
dominance of viruses by phages 14, 251
historical 12, 119–20, 252, 261
infections per second 178
relative abundances (time, space) 252
turnover time 178
Blue whale unit 14
Declines with storage 252
Earth 14, 147–8, 178, 333
Phage-to-bacterium ratios
observed 268
rule of thumb 1
Sediments 14, 289–90
Soils 14, 289
bulk soil vs. rhizosphere 289
seasonal variation 123, 290–1
Tailed phages 6, 178–9
Theory 268
Adaptation 225
Experimental 219, 224–5
(see also Experimental evolution)
Facilitation of 226, 227
mutation rate, high 226
population size, large 226, 227
progressive environmental change 221
speed limits 223, 224
Limits (see Adaptive walks)
Adaptive landscape 157
Drift, importance of 158
Fitness landscape (synonym) 224
Fitness (peaks) 157–8, 225
environmental change, impact 158
landscape constantness 221
local (closest) peaks 158
local-peak trapping (see Epistasis – Fitness)
phage-affecting parameters 225
Genotype lattice rigidity 224–5
Mapping phage landscapes 169
Shifting balance theory (Wright's) 158
Topology/shape/ruggedness 225–6
Adaptive walks 224–5, 227
Alternative pathways 230–1
parallel evolution (see Evolution – Parallel)
Fitness
compensatory mutations (see Compensatory mutations)
peaks, ascent of 225, 227, 230, 231–4
recovery 153, 228, 236
Limits to adaptation 227–9, 236–7
clonal interference 223, 228
Problems with two-stage chemostats 239
Quantification 227
fitness 231

Adaptive walks (*cont.*)
number of substitutions 231
polymorphism 231
Shape 231, 233–4
earlier increases greatest 231, 233
fitness plateauing 231
identifying limits 229
large steps 234–5
theory 231–4
Adsorption 32, 196, 390–2
Bacterial resistance (see Resistance, partial)
"Capture" equivalent 35–6
Constant/coefficient/rate 87, 392, 445, 448, 460–1
estimating virus productivity 254
Constants, various phages 391, 459
approximations 399, 446, 449, 450, 454, 455
phage T1X 460
phage T4 460
phage not indicated 57
Stent's (1963) value 88, 323, 397, 399
Dispersal time (see Extracellular search)
Free phage (see Free phage – Adsorption)
Initial in environment 82–3
plaques 96
Mean time 321–4
Modeling 390, 392–3, 395
bacteria adsorbed 324, 394–7, 400, 404
plaques (see Plaques, modeling – Adsorption)
secondary adsorption 394
Multiplicity of (see Multiplicity – Adsorption)
Preadsorption 96–7
Primary phage/adsorption 196
Rates, influences on 48, 83, 392, 456
antagonistic pleiotropy 162
bacterial capsule (see Capsules, bacterial)
cofactor requirements (phage T4) 364
dynamic variation in (assertion) 272
less than ideal 8, 271
motility, bacteria 392
target radius 460
turbulent flow 460
Secondary 70, 196
burst, reduced contribution 71
ignoring 394, 399, 432, 448
likelihood 86
lysis-inhibition (see Lysis inhibition)
modeling 394
multiple adsorption (see Multiplicity)
plaque, phage losses 426
superinfection exclusion (see Superinfection – Exclusion)
superinfection not synonym 197
Steps of
(1) reversible 420
(2) irreversible 8
Adult period (phage maturation) 45, 64, 390
Maturation rate 45
bacteria quality impact on 48
enhancement of (unsure how) 84
values of 58
Observation *in situ* 255
Productive latent period (syn) 71
Release of phage 65, 115
Alleles/mutations
Beneficial (few) 155–6, 221
fitness and 222–3, 230, 235
Deleterious (most) 155, 209
fitness and 159
Neutral (many) 155, 158, 221
fitness and 152, 155–6
see also Genes
Animal–phage interactions
Antiquity 333, 346
DNA uptake 344, 346
Gene therapy 339, 343–5
Immune system, interaction with (see Immune system)
Interference, viral infections and tumors 345
Phage administration, nasopharynx
amyloid plaque monitoring 338
brain, transport to 338–9
vaccination 339, 342
Phage administration, oral 337–8
circulatory system, appearance in 338
mammalian cells, appearance in 346
spleen and liver, appearance in 346
Safety of 336, 341
Vaccination 341–2
gene delivery vectors 342–3

Vasculature mapping and targeting 343–4
Virulence factors (see Virulence factors)
Antagonistic pleiotropy 160
Adsorption rates, phage 162
Host range 161–2
trade-offs 160–2
Antibiotics
Impact on phage therapy (historical) 440
Post-antibiotic era 440
Resistance to 440, 461
Aquatic environments, phage ecology of 12, 251, 258
Abundance of phages (see Abundance of phages – Aquatic)
Antibacterial activity by phages
per day kill rate 255
partitioning between grazers and viruses 255, 258, 261, 265, 267–8
susceptibility to multiple phages 255
variation in as function of time 255
Diversity of phages 257–8
Lysogeny in 121–2
frequency 122, 125, 286
induction 122–3, 255
seasonal variation 122
Nutrient cycling 12, 253, 269
Pelagic
abundance, phage 14, 261
bacterial diversity 266
food web microbial ecology 272
prokaryote-to-eukaryote ratios 251
trade-offs in shaping food web 262
viral activity 253, 258
viral diversity 257
Phage production/activity (see Production of phages)
Sediments (see Sediments)
Temperate phages (see also Lyosgeny, temperate) 121
Transduction (see Transduction – Aquatic)
Autotrophy 257
Photosynthesis (see Carbon, organic – Photosynthetically)

Bacteria
Abundance (see Abundance of bacteria)
Biocontrol (see Biocontrol of bacteria)
Coliforms 306
Competition specialist (see Specialization, bacterium – Competition)
Culturability 133
Cyanobacteria (see Cyanobacteria)
Detection (see Bacterial detection)
Diversity (see Diversity of bacteria)
Feast–famine existence 114, 116, 121
Flora (see Flora)
Growth rates (see Bacterial growth)
Heterotrophic (see Heterotrophic)
Indicator, for plaquing (see Indicator, plaquing)
Lactic acid bacteria (see Lactic acid bacteria)
Outcrossing (see Horizontal gene transfer – Bacterial)
Phage typing of 288, 308–9
Predation by phages (see Predation)
Species (see Bacterial species)
Whole cultures 2–3, 79
Bacterial detection using phages 302, 308
Foodborne bacteria 308–9
Escherichia coli 306, 309, 311–12
Escherichia coli O157:H7 309, 311–13
Listeria monocytogenes 310–11
Mycobacterium bovis 310
Mycobacterium tuberculosis 310
Salmonella spp. 309–10
Salmonella Enteritidis 312–13
Salmonella Typhimurium 310
Staphylococcus aureus 310
Food matrix problems 302, 309
Phage ecology of 308
Technologies, general
fluorescently labeled virions (FBA) 308, 312
phage amplification 308–9, 312–13, 322
reporter-labeled phage 311
reporter phage 308–12
Technologies, specific
cocktails, phage 309
enrichment of bacteria 308–10, 312–13
flow cytometry 312
immunomagnetic separation (IMS) 312
most probable number method (MPN) 310–11

Bacterial detection, using phages (*cont.*)
phage therapy, use in 340
qualitative detection 309–10
quantitative detection 310–11
reporter genes (see Reporter genes and proteins)
virion inactivation post adsorption 309
Bacterial growth
Losses, non-phage related 389, 401, 443
grazing (see Grazers, eukaryotic)
modeling 395
Modeling 391, 395, 398, 444
density-independent 398
Malthusian parameter 391, 395, 398–400, 409
maximum 57, 391, 398, 403
Monod equation 395, 403, 407, 410–11
phage infected 459
reciprocal yield 57, 403, 409
substrate, half-maximum 57, 391, 403
Rate 41, 49, 267–8, 271–2, 398, 401, 403, 409–10, 426, 442, 444, 446–7, 452, 454, 458, 461
increase, factors causing 283, 290
reduction, factors causing 95, 444, 447
Rate, values of 458, 459, 461
maximum 57, 391
problems estimating *in vivo* 458
reciprocal yield: 57, 391
substrate, half-maximum 57, 391
Wall population 36
Bacterial species
Acholeplasma laidlawii 130
Agrobacterium tumerfaciens 334, 344
Azotobacter vinlandi 129
Bacillus spp.
phage SPO1 (see Bacteriophage isolates)
Bacillus megaterium
phage G 179
Bacillus pumilus 130
Bacillus subtilis
phage ϕ29 180
pseudolysogeny 125–6, 131
Bacteroides fragilis
phages as fecal indicator 305
pseudolysogeny 130
Bradyrhizobium japonicum 295
Brochothrix thermospacta
biocontrol 319
Campylobacter jejuni 316
biocontrol (see Biocontrol of bacteria)
foodborne pathogen 316
Clostridium spp. 365, 372, 375
Clostridium botulinum (see *Clostridium botulinum*)
Clostridium butyricum
E. coli O157:H7 inhibition 363
Clostridium novyi
exotoxin introconversion 375
Corynebacterium spp. 365
Corynebacterium diphtheriae 353, 372
diphtheria toxin (see Diphtheria toxin)
phage β (see Bacteriophage isolates)
Escherichia coli (see *Escherichia coli*)
Halobacterium spp.
phages Hh1 and Hh3 (see Bacteriophage isolates)
Halobacterium salinarium
phage Hs1 (see Bacteriophage isolates)
pseudolysogeny 129
Lactobacillus plantarum
phages ϕJL-1 and LP65 (see Bacteriophage isolates)
Lactococcus lactis (see Dairy phages)
Listeria monocytogenes (see *Listeria monocytogenes*)
Listonella pelagia 130
Mycobacterium bovis 310
Mycobacterium tuberculosis 310
detection
major killer, past 440
Myxococcus fulvus 130
Myxococcus virescens 130
Pasteurella multocida
virulence factors, phage-encoded 345, 358
Pseudomonas spp.
biocontrol 320
lysogeny 124, 286
plant pathogenic 201
pseudolysogeny 127–30
Pseudomonas aeruginosa (see *Pseudomonas aeruginosa*)

Pseudomonas fluorescens (see *Pseudomonas fluorescens*)
Pseudomonas syringae
biocontrol 320
pathovar *phaseolicola* 202, 204
phage φ6 (see Bacteriophage isolates)
Pythium ultimum 292
Rhizobium leguminosarum
biovar *viceae* transduction 296
Rhizobium meliloti 295–6
Rhodopseudomonas capsulata 131
Salmonella spp. (see *Salmonella* spp.)
Serratia sp. CP6 296
Serratia liquefaciens 123, 289–90
Shigella spp. 369
Shigella dysenteriae
Shiga toxin 367
Sphaerotilus natans 296
phages SN-1 and SN-T (see Bacteriophage isolates)
Staphylococcus spp.
phages, antibodies against 336
phages, clearing of from blood 335
Staphylococcus aureus (see *Staphylococcus aureus*)
Streptococcus spp.
phages as fecal indicator (see Fecal indicators)
virulence factors, phage-encoded 353
Streptococcus pyogenes 361
lysogenic conversion 374
pseudolysogeny 130
strains CS24 and CS112 374
virulence factors, phage-encoded 353, 357, 358
virulence factors, phage regulation 360
Streptococcus thermophilus (see Dairy phages)
Streptomyces spp.
pseudolysogeny 130
soil ecology 282
soil microcosm experiments 124
Synechococcus spp. (see Cyanobacteria)
Vibrio cholerae (see *Vibrio cholerae*)
Vibrio fischeri 309
Vibrio harveyi 358
Vibrio mimicus (see Bacteriophage CTXΦ)
Vibrio vulnificus 451
Xanthomonas campestris 320
Bacteriophage 2, 333
Abundance (see Abundance of phages)
Advantages as model organisms 55, 147, 148, 163, 166–7, 169, 217, 219, 235
limitations 217
microbes in general 224
whole-genome sequencing 218
Animal interactions (see Animal–phage interactions)
Antiquity 185, 333
Categories
coliphages (see Coliphages) (see Fecal indicators)
corynebacteriophage 132
cyanophages (see Cyanophages)
dairy phages (see Dairy phages)
filamentous (see Bacteriophages, filamentous)
F^+ (see Fecal indicators)
isolates (see Bacteriophage isolates)
lambdoid (see Bacteriophages, lambdoid)
large, exceptionally so 179
listeriophages (see *Listeria*)
long-circulating (see Bacteriophages, long-circulating)
mycobacteriophages 188
obligately lytic (see Obligately lytic)
rhizobiophages 291, 295
roseophages 119
somatic (see Fecal indicators – Coliphages)
T-even (and T4-like; see Bacteriophage T2/T4) 69, 184
tailed (see Bacteriophages, tailed)
temperate (see Lysogeny, temperate phages)
virulent (see Bacteriophage virulence)
Cocktails (see Cocktails, phages)
Contamination, dairy 302–3, 320
Conversion (see Conversion, phage)
Display 341–3, 345
Diversity (see Diversity of phages)
Ecology (overview) 15

Bacteriophage (*cont.*)
Evolution (see Evolution)
Free phages (see Free phages)
Genes (see Genes, phage)
Genomes (see Genomes, phage)
Growth parameters (see Growth parameters, phage)
Host range (see Host range)
Immunity against (see Immune system)
Infection types (see Bacteriophage infection types)
Lysogen/lysogeny (see Lysogen)
Multiplicity (see Multiplicity)
Population growth (see Growth, population, phages)
Production (see Production of phages)
Release (see Release)
Specialization (see Specialization, phage)
Superinfection (see Superinfection)
Therapy (see Phage therapy)
Trade-offs (see Trade-offs, phage)
Transduction (see Transduction)
Typing 288–9
Viral nature of (historical) 2
Virion characteristics (see Bacteriophage and other virus families) 5
Virulence (see Bacteriophage virulence)
Virulence factors, phage-encoded (see Virulence factors, phage . . .)
Bacteriophage and other virus families 6, 178–9
Algal viruses (not phages)
Phycodnaviridae 190
Animal and plant 190
Adenoviridae 190
Reoviridae (reovirus) 190
Archaeal viruses (not phages)
Fuselloviridae 179
Guttaviridae 179
Lipothrixviridae 179
Rudiviridae 179
Salterprovirus 179
Cubic (icosahedral) phages 6
Corticoviridae 179
Leviviridae 7, 178
Microviridae 6, 179, 189–90
Tectiviridae 179, 190
Enveloped phages
Cystoviridae 6, 7, 178, 190, 199, 208
Filamentous and helical phages 6, 56
Inoviridae 6, 7, 179
Pleiomorphic phages
Plasmaviridae 179
Tailed (Caudovirales) phages 6, 178
Myoviridae 6, 305
Podoviridae (tailed) 6, 290
Siphoviridae (tailed) 6, 290, 304–5
Bacteriophage infection types 5–10, 64, 115
Non-productive 9, 64, 116
abortive infection 8, 9, 83
reductive (see Reductive, below)
restricted infection (see Restriction)
Plasticity 131
Productive 9, 64, 115, 121, 126, 129, 288, 390
chronic infection (see Release – Chronic)
ecological aspects 65–6, 116–17, 119, 120, 123, 128–9, 362, 447
lytic cycle (see Lytic cycle)
obligately lytic (see Obligately lytic)
Reductive 114, 120–1, 123
lysogenic cycle (see Lysogenic cycle)
pseudolysogenic state (see Pseudolysogeny)
(see also Physiology, bacterial)
Bacteriophage isolates (and host) 6–7
Phage 12 (*S.* Enteritidis) 315
Phage 933W (*E. coli*)
infection *in situ* 373, 375
spontaneous induction 368
Phage 936 (type) (*L. lactis*) 304
Phage α3 (*E. coli*) 431–4
Phage A21 (*A. vinlandii*) 129
Phage A511 (*L. monocytogenes*) 305
mosaicism 305
wide host range 310
Phage AS-1 (*Synechococcus* spp.) 105
Phage β (*C. diphtheriae*) 132, 334
Phage c2 (type) (*L. lactis*) 6, 304
Phage CP-T1 (*V. cholerae*) 370
Phage CTXΦ (*V. cholerae*) (see Bacteriophage CTXΦ)
Phage DT1 (*S. thermophilus*) 304

Phage ES18 (*Salmonella* spp.) 182
Phage ϕ6 (*P. syringae*) (see Bacteriophage ϕ6)
Phage ϕ29 (*B. subtilis*) 6
structural relatedness 190
Phage ϕCS112 (*S. pyogenes*) 374
Phage ΦJL-1 (*L. plantarum*) 304
Phage ϕR2f (*P. fluorescens*) 119
Phage ϕR73 (*E. coli*) 157
Phage ϕX174 (*E. coli*) (see Bacteriophage ϕX174)
Phage f1 (*E. coli*) 7, 370
virulence evolution 54
Phage F2 (*E. coli*) 7
Phage F116 (*P. aeruginosa*)
host physiology impact 118, 124, 128
lysogen induction 128
pseudolysogeny 127–8
Phage fd (*E. coli*) 7
vaccine display 342
Phage Ff (*E. coli*) 253
Phage G4-like (*E. coli*) 230
Phage G (*B. megaterium*)
largest phage genome 179–80
Phage Gifsy-1 (*S. typhimurium*) 371
Phage Gifsy-2 (*S. typhimurium*) 371
Phage GP100 (*P. fluorescens*) 291
Phage H19-B (*E. coli*) 368
Phages Hh-1 and Hh-3 (*Halobacterium* spp.) 305
Phage HK97 (*E. coli*)
mosaicism 182
relatedness to other phages 182, 187, 190
Phage Hs1 (*H. salinarium*) 129
Phage K (*S. aureus*) 305, 460
Phage K-1 (*E. coli*) 451–2
bacterial resistance 456
Phage K1-5 (*E. coli*) 460
Phage K139 (*V. cholerae*) 357
Phage KVP40 (*Vibrio* spp.) 119
Phage λ (*E. coli*) (see Bacteriophage λ)
Phage L5 (*Mycobacterium* spp.) 6
Phage LG1 (*E. coli* and other enterics) 312
Phage LH7 (*L. monocytogenes*) 317
Phage LP65 (*L. plantarum*) 305
Phage M13 (*E. coli*) 7, 370
oral feeding/mammal uptake of DNA 346
mutation rate 151
Phage MS2 (*E. coli*) 7
adaptive landscapes 169
clonal interference 155
compensatory mutations 164
first genome sequenced 169
salinity impact 285
Phage Mu (*E. coli*) 6
increased lysogen fitness 119
mosaicism 187
Phage P-SSM4 (*Prochlorococcus* spp.) 180
Phage P1 (*E. coli* and other enterics) 6
host-range breadth 288
increased lysogen fitness 119
Phage P2 (*E. coli*) (see Bacteriophage P2)
Phage P21 (*E. coli*) 368
Phage P22 (*Salmonella* spp.) (see Bacteriophage P22)
Phage P4 (*E. coli*) 207–8
hyperparasitism of phage P2 207–8
defective but not DI particle 207–8
induces phage P2 208
related phages 157, 208
Phage P100 (*L. monocytogenes*) 318
Phage P335 (type) (*L. lactis*) 304
Phage ϕ29 (*B. subtilis*) 180
Phage PP01 (*E. coli*) 311
Phage PRD1 (*E. coli*)
attachment protein diversity 168
environmental transport 285
highly stable in environment 285
virion internal membrane 285
Phage Q1 (*S. thermophilus*) 304
Phage Qβ (*E. coli*) 7
compensatory mutations 164
mutation rate 150–1
synergistic epistasis 164
Phage S-PM2 (*Synechococcus* spp.)
phosphate depletion 118
photosynthesis genes 132
Phage SfV (*S. flexneri*) 186–7
Phage SIO1 (*Roseobacter* spp.) 119
Phage SJ2 (*S. enteritidis*) 312
Phages SN-1 and SN-T (*S. natans*) 296
Phage SP6 (*Salmonella* spp.) 294

Bacteriophage isolates (and host) (*cont.*)
Phage SPO1 (*B. subtilis*) 6
mosaicism 305
Phage T1 (*E. coli*) 6
bacterial resistance 150
Phage T1X (*E. coli*)
adsorption rate 460
chemostat experiments 39
Phage T2 (*E. coli*) (see Bacteriophage T2)
Phage T3 (*E. coli*) 6
chemostat experiments 39
mitochondrial RNA polymerase 334
Phage T4 (*E. coli*) (see Bacteriophage T4)
Phage T5 (*E. coli*) 6
chemostat experiments 39, 41
"no" cost to resistance 41
secondary adsorption block 71
two bacterial-species community 293
Phage T7 (*E. coli*) (see Bacteriophage T7)
Phage Tuc2009 (*L. lactis*) 197
Phage UT1 (*P. aeruginosa*)
host physiology impact 118
pseudolysogeny 127–8
Bacteriophage CTXΦ
Characteristics 370
ssDNA genome 115
temperate 115, 130, 297, 370
Evolutionary ecology ("interests") 356
Exotoxins encoded
Ace 358, 370
Cholera toxin (see Cholera toxin)
Zot 358, 370
Hosts
Vibrio cholerae (see *Vibrio cholerae*)
Vibrio mimicus 374
Infection *in situ* (suckling mice) 374
Lysogeny 370
lysogenic conversion 297, 374
Receptors
primary, TCP 269–70, 379
secondary 374
Superinfection immunity 374
Bacteriophage Ecology Group (BEG) 3
Bacteriophage ϕ6 7, 201
Adaptation
Adaptive landscapes 169
Adaptive robustness 210
Population size importance 226
Biogeography 208
Cheating (see Cheating)
Coinfection (see Coinfection – Phage ϕ6)
Epistasis 163–4, 166, 206–7
Host shifting 208
Muller's ratchet (see Muller's ratchet)
Mutation
ease of acquisition 166
rapid evolution 213
rate 151
Recombination, genetic
molecular, limited 203
reassortment 199, 212
template switching 199
Structural homology among dissimilar viruses 148, 190
Bacteriophage ϕX174 6
Adsorption proteins, faster evolution in 168
DI particles (see DI particles)
Epistasis 163
Experimental phylogeny 189
ϕX174-like
experimental adaptation 235
extreme value theory 232
Fitness
changes (see Fitness – Phage)
trade-offs from host switching (see Trade-offs, phage – Host range)
Immune response to 336
Mutation
adaptive landscape mapping 169
compensatory 164
rate 151
single-step adaptation 230
single mutation effects 162
temporal pattern of substitutions 236
Plaques, serial passage
bacterial resistance (see Resistance, bacteria)
cellular automata simulation 432–4
competition between 431–2
Second genome sequenced 169
Selective sweeps 154, 155

Structural homology among dissimilar viruses 190
Two bacterial species community 161
Bacteriophage λ 6
Chemostat experiments 39
Decay free phages 460
Genome 178, 187
λ-like phages 188, 305
Lysogeny
highly understood 115, 116
increased lysogen fitness 119
induction rate 368
lytic–lysogenic decision 117
Mutation
lysis timing 47
mutational robustness 165–6
rate 151
vir 39, 460
Mouse circulation
clearing without adaptive immunity 337
decay rate 460
prolonged survival 337
Oral feeding/mammal uptake of DNA 346
Order Caudovirales 178
Receptor 117
Recombination (historical) 151
Reporter-gene engineered 309
Vaccination vector 342
Bacteriophage P2 6, 207
Induction, not by SOS 207
Increased lysogen fitness 119
Phage P2-like prophages 207
Phage P4 helper phage 157, 207, 208
Phage T4, interference with replication 197
Superinfection immunity 197
Bacteriophage P22 6
Bacterial detection 309
Induction by stationary phase 368
Relatedness to other phages 190
analogous gene exchange 186
nonhomolgous recombination 182
Superinfection immunity 197
Virulence-factor encoding 371
Bacteriophage T2
Adsorption, two different receptors 41
Bacterial resistance (see also Resistance, partial) 39, 264
Chemostat experiments 39
Host-range mutants 42
Mutation rate 151
Bacteriophage T2 and T4 co-experiments
Genetic divergence, barriers to hybridization 167
Host-range mixing following coinfection (see Host range – Mixing)
Recombination rate, homology dependence 168
Bacteriophage T4 6
Adsorption (see Adsorption)
Bacterial resistance 40, 264
Chemostat experiments 39–41
Decay free phages 460
Gene clustering, tail fibers 355
Mouse circulation removal 335
Mutation rate 151
Oral administration, humans 338
Order Caudovirales 178
Phenotypic plasticity 49
Plaque visualization 417
Structural homology 190
T4-like phages 167–8
Bacteriophage T7 6
Bacterial resistance 40, 42
Epistasis 163, 168
Evolution, strain (historical) 166
Experiments
Arms race 43, 169
Serial passage for high growth rate 229
Two-bacterial species community 293
Growth mutational robustness 166
Host-range mutants 42–3
Mitochondrial RNA polymerase 334
Origin (historical) 166
Plaque formation 99, 240
stationary-phase lawn 98–9, 109, 420, 426
Rodent circulation survival 337–8
complement-mediated inactivation 338
T7-like phages 188
Bacteriophage virulence 11, 66, 79
Different perspectives 78
Evolution of
horizontal transmission 54–5

Bacteriophage virulence (*cont.*)
short-sightedness 79
vertical transmission 54–5
well-mixed environments 80
Temperate phage virulence 11
Virulence factors, phage-encoded (see Virulence factors)
Virulent phages (see Bacteriophages, virulent)
Whole cultures, against 78–9
multicellularity and 78–80
Bacteriophages, filamentous 6, 7, 370
Chronic release 56
Examples: Phages CTXΦ, f1, fd, M13, etc. (see Bacteriophage isolates)
Miniphage (see DI particles)
Nasal administration 338–9
Parasites (ecological role) 56
Bacteriophages, lambdoid
Induction 367–8, 370
Mosaicism 182
Super group 305
STEC
diversity in 369
Virulence factors encoding 357–8, 371
Bacteriophages, long-circulating
Phage λ 337, 460
Phage T7 338
Bacteriophages, tailed
Abundance 6, 178–9
Antiquity 191
Binary (synonym) 6
Evolution 179
capsids 186
deep phylogenetic connections 190–1
flamboyance 179, 189
non-homologous recombination 191
Genetic
divergence 185, 190
diversity 188
Genomes 179–81, 191
Lifestyles 120
Order Caudovirales 6, 178, 290, 304–5
Plaques more dramatic 179
Structural homology 190
Bacteriophages, virulent 11
Lysogeny loss 12, 304
potential for mutation back to temperate 12
Problems with usage 11
obligately lytic as synonym 390
Temperate, not 11, 78, 126, 318, 356, 390, 408
Barriers to gene flow between phages 208
Genetic divergence 166–8
Post-zygotic barriers (see Barriers, post-zygotic)
Pre-zygotic barriers
geographic barriers 208
host-range barriers 43, 167, 208–9
Surmounting barriers 167
Barriers, post-zygotic 169
Hybrid incompatibility 168–9
selection against recombinants 178, 183–4, 191
Recombination incompatibilities 167, 168–9
Sequence homology, rate control of 168
Biases
From pre-enrichment 5
In broth-culture growth rates (virulence) 5, 288
In completed genome sequences 187
In electron microscopy 179, 252–3
In enumeration 5, 252–3
In host type 5, 179
In host-range breadth 5, 288
In phage diversity understanding 179
In plaquing 5, 179
Toward measure of obligately lytic growth 256
Virulence factor-encoding by temperate phages 360
Biocontrol of bacteria in food 302–4, 313–14, 320
Active vs. passive 315–16
Control of human pathogens
Campylobacter spp. 314–17
Escherichia coli O157:H7 314–15
Listeria monocytogenes 302, 317–18, 320
Salmonella spp. 310, 314–16
S. Enteritidis 315–16
Control of spoilage of food 318, 319
Brochothrix thermospacta 319

Pseudomonas spp. 319
Pseudomonas tolaasii 318
Impact of
food matrix 315
phage density/adsorption 321–4
virion stability 316, 319–20
Regulatory issues 302, 314, 320, 341
Biocontrol of plant pathogens 318, 320
Pseudomonas syringae 320
Xanthomonas campestris 320
Biofilm
Bacteria
heterogeneity 95, 288
Pseudomonas aeruginosa 457
source 453
Control
biocontrol 318, 323
phage therapy 454
Lysogeny, importance to 121
Phage propagation within 457
Plaques as model for 94, 101, 109, 415
Sealed-off microenvironments (speculation) 77
Spatial structure (see Spatial structure – Environments)
Ubiquity 121, 281, 287, 297
disease 109, 457
food 309, 318
Virion stabilization or trapping 66, 290
Biogeochemistry
Movement of energy and materials 265, 268, 272
parasitism: larger to smaller forms 258
predation: smaller to larger forms 258
Sinking of material 269–70
dissolving of material (virues) 258
Virus impact on 269
Biogeography 208
Biotic 15, 18, 55, 287
Blooms, microbial 122, 256, 286, 293
Burst size 45, 390
Effective burst size 67, 83
insufficient 70
Factors affecting 48, 83
adsorption timing 71
bacterial physiology (see Physiology, bacterial)
coinfection impact (see Coinfection – Consequence – burst size)
lysis inhibition 69, 105
lysogeny 70, 120
multiplicity dependence 69
Modeling 57, 390–1, 394–5, 445
cellular automata 432, 434
latent period, dependence on 45, 46, 57, 58
plaques 419–21, 427
Measures of 118
difficulty *in situ* 456
importance of measuring 7
uncertain value within natural communities 253
values for 459
Plaques
minute plaques 105
plaque fecundity/productivity 423, 426–7
reduced during formation 100, 426–7
wavefront velocity 103, 432
Population
fecundity, impact on 69, 75
growth rate, impact on 67, 153, 398–9
trade-offs with latent period (see Trade-offs, phage – Burst)
Related concepts
cooperation and cheating 72
economy equivalent 74, 76–7, 85
energetic gain equivalent 47
estimating virus productivity 254
fecundity equivalent 67, 69
fitness surrogate 77, 153–4
marginal value theorem 46–7
stability 58
Selection for larger 69, 109, 153, 160

Capsules, bacterial
Costly to produce 265
Phage defense mechanism 265, 460, 461
Phage receptor 265
phage K-1 451–2
Specific bacteria
Escherichia coli 461
Staphylococcus aureus 461

Carbon, organic
Dissolved (see Dissolved organic material)
Global sink 12
Movement to deeper waters 269–70
Nutrient/energy 117, 121–2, 124, 128–9, 256–7, 259, 265
Photosynthetically fixed 12, 257
Scarcity 114
Trophic movement 12, 259
Cellular automata (see Plaques, modeling, cellular automata)
Cheating/cheaters
Coinfection (phage φ6) 198, 201, 204
costs of 205
intracellular cheating 201, 204
reducing population fitness 205
segmentation and 211–12
selfish genes/segments 209–10, 212
"tolerated" in populations 205
wild-type-infected specialists 204
Cooperation 72
Defection/defector (synonym) 73, 81
Exclusion of 77–8
co-replication 354–5
mutations, limiting 78, 84
Expedient growth (see Economy)
cheating in 73
large volumes and 78
latent period 72
Genes, cheater (see Genes, phage – Cheater)
Lysis inhibition (see Prisoner's dilemma – Superinfection exclusion)
Multicellularity, avoidance in 76, 84
(see also Cooperation–defection)
(see also D1 (defective interfering) particles
(see also Spatial structure)
Chemostat 35, 400
Batch culture, differences from 400
One-stage (see Chemostat, one-stage)
Two-stage 236, 239
adaptive walks, problems with 239
coinfection (see Competition – Intracellular . . .)
Chemostat, one-stage 240, 401
Experimental
bacterial refuges 36, 401
bacterial resistance 40, 42
coexistence 36, 39, 41, 442
community dynamics 40
few phage-beneficial mutations 240
lysogens and pseudolysogens 119, 128–9
phage-bacterial 36, 39, 161
Modeling 57–8, 402, 405–7, 410–11
bacteria, infected 405, 410
bacteria, uninfected 57, 401–3, 409–11
burst size (see Burst size – Modeling)
dimensional analysis 411
extensions of models 411
fixing densities 408–10
flow, media 57, 391, 401
latent period (see Latent period – Modeling)
limit cycles 58
Lotka–Volterra 33–5, 262, 265
lysogeny 124–5, 411
phage therapy (see Phage therapy, modeling)
phages, free 57, 404–5, 410–11
substrate 57, 395, 403, 409–11
Stability/instability
burst size 58
density control of bacterial growth 407–8
enrichment 39, 41, 57
host susceptibility 406
negative density dependence 34–5, 57–8
phage density 406–7
stability analysis 406, 411
time lag (see Time lags)
Cholera 369
Phage therapy of 340
toxin (see Cholera toxin)
Vibrio cholerae (see *Vibrio cholerae*)
Cholera toxin (CT) 369, 370
Diarrhea 370
Dissemination, extra-organismal 373
Environmental modification 376
Expression
bacteria regulated 360, 376
CTXΦ, not coupled to induction 370
environmental factors 370
genes *ctxA* and *ctxB* 370
secretion 370

SOS, small increase in 370
ToxR regulon 370
Phage CTXΦ 130, 297, 358, 370, 374
Clonal interference 154–5, 223
Complicates experiment interpretation 230
Confused with epistasis 224, 237
Evolution
parallel evolution 231
step-wise evolution 227, 233
Example: overcoming host resistance 155
Fitness and 154–5
Genome sequence analysis 238
Impediment to allele fixation 223
adaptation biases 230, 234
close together mutations 238
multiple beneficial alleles 230
mutation disappearance 231
Rate of adaptation
population size 227, 238
speed limits 223, 228
Recombination interferes with 223, 235, 237–8
Clostridium botulinum 372
Phage conversion 372
Pseudolysogeny 361, 373, 375
Virulence factors, phage-encoded 353, 372
botulinum toxin (BoNT, C_1 and D) 358, 361, 372
C_2 hemagglutinin exotoxin 358, 372–3
C_3 ADP-ribosylating exotoxins 358, 372–3
dissemination roles 373
dose dissemination 36
environmental modification 372
linkage between 372
Co-replication 354–9
Cooperation and 354
Prophage-bacterial 115, 120, 355–6
Virulence factor gene-phage 356
(see also Linkage (genetic))
Cocktails, phage
Detection, bacterial 309
Phage therapy 314–15, 317, 320
Coexistence
Bacteria, aquatic 270
Hutchinson's paradox 261–3
phage impact 268
Phage, plaques 431–2
Phage and bacteria
chemostat 36, 39, 41, 442
Coinfection 70, 151, 195–6
Avoidance 71
Competitive exclusion (see Competition – Exclusion)
Consequences 195, 197–8, 201
arms race 236
cheating (see Cheating – Coinfection)
competition within hosts (see Competition – Intracellular)
complementation (see Complementation)
gene exchange (see Recombination – Coinfection)
phenotypic mixing (see also Host range – Mixing) 198–9, 207
polyploidy 199–200
Fitness impact 196–7
complementation (see Complementation – Fitness)
intracellular competition (see also Competition – Intracellular) 198, 203–5
more rapid selection 206
Muller's ratchet reversal 203
mutation accumulation 211
selective environment modification 221
trade-offs 204
Historical 196
Limiting experimentally 202
Phage ϕ6 195, 201, 211
complementation 201, 206, 210–11
intracellular competition 201, 204, 237
limits to coinfection 201–2
phenotypic mixing 201
Scenarios for (between)
among resident prophages 151
between "heterologous" phages 195, 207
between "homologous" phages 207
newly infecting phages 151, 196
newly infecting phages and prophages 151, 196–7
superinfection 197

Coliphages 4, 195
- Ecological concepts 55
- Examples: phages 933W, α3, f1,F2, fd, Ff, αR73, αX174, G4, H19-B,HK97, K-1, K1-5, LG1, M13, MS2, Mu, P1, P2, P21, P4, PP01, PRD1, Qβ, T1, T1X, T2, T3, T4, T5, T7, etc. (see Bacteriophage isolates)
  - term usage 155, 197, 207, 312, 370
- Fecal indicators (see Fecal indicators)
- Filamentous 370
- Infection, colonic 364
- Mycobacteriophage comparison 188
- Propagated, where? 364

Community structure 31, 37–8, 55, 294

Compensatory mutations 164
- Epistasis, example of 164–5
- Fitness
  - impact 164–5, 228
  - lower if not compensating 164–5
  - recovery 228
- Genetic technique 242
- Genome stabilization 164
- Muller's ratchet impediment 160
- Pseudoreversion (synonym) 164
- Reversion, more likely than? 164
- Second-site suppressors 225

Competition
- Apparent 41–2
- Competition specialist, bacterial (see Specialization, bacterial – Competition)
- Economy vs. expediency 74–5
- Exclusion, competitive
  - chemostats with partial resistance 42
  - coinfection, during 203
  - Hutchinson's paradox 261
  - plaques, within 431–2
- Exploitative 41, 56, 66–8, 70
  - multiplicity component 68
  - "strictly clonal" 201
  - termination of 69
  - time component 68
- Final yield 75
- Interspecific 73
- Intracellular 200–1, 203, 241
  - arms race 236
  - between heterologous phages 207
  - cheating (see Cheating – Coinfection)
  - exploiting coinfecting phages 203–4
  - phage φ6 201, 204, 237
  - two-stage chemostats 239
- Plaques
  - avoidance of competition 237
  - bacterial resistance to phages 432
  - between plaques 431–2
  - modeling 432–5
- Spatial structure 431
- Temperate vs. obligately lytic 123, 290, 390

Complementation 198, 200, 211
- DI particles 200
- Fitness 200, 206, 210
- Genetic dominance 199–200
- Negative consequence, evolutionary 200
- Phage φ6 201, 206, 210–11
- Weakened robustness selection 210–11

Conversion, phages 132
- Lysogenic (see Lysogenic conversion)
- Lytic 132

Cooperation–defection 72
- Cooperation facilitation 76
  - cheater exclusion (see Cheating – Exclusion of)
  - co-replication (see Co-replication)
  - defector avoidance 77, 84
  - environmental maturation 77–8
  - founder effect 76
  - replicative fidelity 78, 84
  - temporary sequestration (see Spatial structure – Cooperation)
- Cheater (defector synonym) 73
- Coinfection and cooperation 204
- Defection facilitation 80
- Latent-period evolution (see Trade-offs, phage – Burst size)
- Multicellularity and cooperation 76, 84
- Phage RB69 81
- Related concepts 72
  - cheating (see Cheating)
  - economy vs. expedience (see Economy . . .)
  - husbanding 73
  - prisoner's dilemma (see Prisoner's . . .)

tragedy of the commons (see Tragedy . . .)
virulence (see Bacteriophage virulence)
Core genes 184
Critical mass metaphor 83
Culturing, batch 394, 396–8, 400–1
Bacteria, infected 394–6
Bacteria, uninfected 400
Phages, free 394, 398
Culturing continuously (see Chemostat)
Cyanobacteria
Aquatic
large fraction photosynthesizers 12
minimal lysogeny 122
Phages
lysis susceptible 12
multiple types 255
Prochlorococcus spp.
phage S-PM2 133
Synechococcus spp.
lysogeny 122
phage S-PM2 132–3
phages of 133
Transduction 133
Cyanophages 4
Abundances 122
Examples: phages AS-1, P-SSM4, S-PM2, etc. (see Bacteriophage isolates)
Genes (see Genes, phage – Cyanophage)
Phosphate limitation 118, 257
Temperance 121

Dairy phages 303
Difficult to inactivate 303
Fermentation failure 302–3, 320
Lactococcus lactis phages 304
examples: phages 936, c2, P335, Tuc2009 (see Bacteriophage isolates)
Resistant starter cultures 303
Streptococcus thermophilus phages 304
examples: phages DT1 and Q1 (see Bacteriophage isolates)
Decay, virion 83, 117
Acidity 285–6, 316
Balance against replication 83, 253, 445, 448
Effective burst size 83
Food, stability in 316, 319–20
Mechanisms of 285–6
Modeling 391, 404
phage therapy 442, 445, 447–8, 451, 458
Plaques 109
Protection from 116
Rates of 114, 117, 253
aquatic 117, 260–1
phage K1-5 460
phage λ*vir* 460
phage T4 460
values of 117, 458–60
Resistance to 67
Soils 285–6
Survival 66, 71, 242
Defense specialist, bacterium (see Specialization, bacterium – Defense)
Definitions
Bacteria
conjugation 294
horizontal (lateral) gene transfer 294
lawns 107
lysogen 120
lysogenic 11
lysogenic conversion 13
partial phage resistance 8
phage therapy of 439
restriction by 9
Bacteriophages
unicellular-organism parasites 3
viruses of bacteria 2, 333
Bacteriophages, infection types
abortive 8
chronic 9
coinfection 70, 196
lysogeny 9, 115
lytic 9, 64
productive 115
pseudolysogenic 8, 9, 115–16, 125–6
restrictive 9
superinfection 70
Bacteriophages, life cycle aspects
adsorption 196, 390–2
adsorption, primary 196
adsorption, secondary 70, 196

Definitions (*cont.*)
adsorption constant 392, 460
adult (maturation) period 45
burst size 45, 390
burst size, effective 67, 83
eclipse 44–5
exclusion, superinfection 197
extracellular search (dispersal time) 390–2
F^+ 306
growth parameters 7
holin 49
immunity 197
infection initiation 196
latent period 390, 403
lysis inhibition 105
lysis rate 449
somatic 306
valence 5
Bacteriophages, population growth
carrier state 8
plaque 95
plaque fecundity/productivity 423
plaque-forming unit (PFU) 97
plaque zones 416
plaques, mean invasion time 431
virulence 11
Bacteriophages, types
binary 6
cubic 6
filamentous 56
helical 6
host-range mutants 42
lytic 11
obligately lytic 11, 56
temperate 11, 56
virulent 11
Ecology 1, 15, 31
community 37
evolutionary 15, 18, 52, 195, 360
landscape 16
metapopulation 49
organismal 15
population 16
Ecology, bacteriophage 15
behavioral 15
community 16
ecosystem 16
exploitative competition 66
killing the winner 13
landscape 16
physiological 15
population 16
subject 441
Ecology, ecosystem
abiotic component of 15
biotic component of 15
phyllosphere 282
rhizoplane 282
rhizosphere 282–3
spatial structure 94, 415
spatially unstructured 415, 457
Ecology, microbial
osmotrophs 258
viral loop 258–9
Genetics and evolution
adaptive landscape 157
adaptive walk 227
antagonistic pleiotropy 160
bottlenecks 158
clonal interference 154–5, 223
coevolved genes 354
compensatory mutations 164
complementation 198
dominance 199–200
epistasis 162
epistasis, sign 224
epistasis, synergistic 163, 206
evolution 149
evolvability 165
experimental evolution 17
fitness 153, 200
fixed allele 152
linkage disequilibrium 354
linkage equilibrium 354
moron 185–6
mosaicism 180–1
Muller's ratchet 159
mutation 149–50
natural selection 152
neutral alleles 152
population genetics 149
selective sweep 154

transduction 13, 294
transformation 294
Models 389
cellular automata, stochastic 427
chemostat 35, 400
formal 389
mass action 33, 392
physical 389
Other
epifluorescence microscopy 252
exotoxins 363–4
neoantigens
tragedy of the commons 53
Recombination 151
heterologous 152
homologous 151, 181
illegitimate 295
non-homologous 181
Detection of bacteria, using phages (see Bacterial detection)
DI (defective interfering) particles
Cheating and 204–5, 211
Complementation 200
Examples (phage)
phage φX174 200
phage, filamentous (miniphage) 200
Incompletely defective viruses (see Incompletely defective viruses)
Mechanism of interference 211
Replication advantage 201
Diphtheria toxin (DT)
Encoding
first VF mapped to prophage (historical) 372
phage β encoded 357, 372
phage-encoded 132, 334
tox gene encoded 372
Principal VF of *C. diptheriae* 372
Regulation of expression 372
Toxicity 347, 372, 376
Dispersal period (see Extracellular search)
Dissemination (see Extracellular search)
Dissolved organic material/matter (DOM)
Bacterial assimilation 12
Diversion of carbon and energy 259
Viral-lysis created 12, 258, 266
Diversity of bacteria 12, 13, 253
Killing the winner (see also Killing the winner)
hypotheses 267
Lysogeny impact
induction of prophage 260
lysogenic conversion (see Lysogenic conversion)
Niches 266
Transduction (see also Transduction) 13
Diversity of phages (and viruses)
Aquatic 257–8
Environmental 4
expectations (theory) 267–8
without culturing 5
Genome types 6
Infection types (see Bacteriophage infection types)
Isolates 188
Metagenomics (see Metagenomics)
Spatial diversity 188–9
Types of phages 6
bacteriophage families (see Bacteriophage and other virus families)
bacteriophage isolates (see Bacteriophage isolates)
Virion morphologies 6
Dormancy 114
Lysogeny (see Lysogeny)
Pseudolysogeny (see Pseudolysogeny)
Reservoirs of phages 120
Drift, genetic 17, 157
Adaptive peaks
epistasis 162
history of drift 157–8
meandering down 158
Allele fixation (stochasticity) 158, 165
time needed 158, 222
Bottlenecks 158, 202
effective population size 158–9
fitness reducing 152–3, 15–9, 202, 226
Muller's ratchet (see Muller's ratchet)
mutation accumulation 159, 206, 210, 226, 236
mutation fixation (see Substitution)
mutation loss 239

Drift, genetic (*cont.*)
plaque to plaque 159, 202
selection overwhelming 158
Change in allele frequency 17, 149
non-beneficial accumulation 160
phenotype independence 152
population size dependence 156, 158–9, 202, 231
randomness 202
wild-type loss 159
Founder effect 77, 83–4
Inbreeding depression 158
Model for studying 166, 219
(see also Fitness, Darwinian)

Ecology 1, 15, 31
Behavioral 15, 43
optimal foraging (see Optimization – Foraging)
Community (see Ecology, community)
Ecosystem 16
Evolutionary ecology 15, 17–18, 52, 64, 195, 360
coinfection 195, 198, 213
hyperparasitism 208
latent-period evolution 55
multiple adsorption 71, 196
phage–phage interactions 195
phenotype robustness 209
source and sinks 149
trade-offs (see Trade-offs) 54
virulence factors, phage-encoded 353, 360, 362–3
Landscape 16
Mathematical 32
Metapopulation 49–50
community modules 56
phage–bacterial metacommunity 50–2, 54–5
Microbes as model organisms 31–2
Organismal 15
Physiological 15
Population 16
Ecology, community 16, 37
Community modules 37
Phage–bacterial 35, 38
Predators 37, 56
Resource enrichment 42
Stability
neutrally stable oscillations 33
stable fixed point 36
stable limit cycle 36
Economy vs. expediency 73
Burst size 74, 76–7, 85
Competition 74–5
migration restriction 53–4
productivity vs. competitive ability 52–3
prudent vs. rapacious 53
Growth 72–3, 84
economy 72–6, 81
expediency 72–6, 81
Prisoner's dilemma (see Prisoner's dilemma – Exponential)
Trade-offs 73–4
productivity vs. competitive ability 53, 54
yield vs. growth rate 76
Effective population size 158–9
Electron microscopy
d'Hérelle and 2
Environmental
aquatic 12, 252, 289
diversity of phages 5
filamentous phages, excluding 253
phages, distinctive shapes 179, 252
soil (see Abundance of phages – Soils)
visibly infected cells 254
Virion characterization 6
Enrichment, resource
Chemostat 38, 42
densities of organisms, impact on 39–41
destabilizing 39, 41, 57
phage T4 39–41
predation importance 42
stabilizing 41
Paradox of 39, 41
Environmental modification by phages 12, 14
Bacterial diversity (see Diversity of bacteria)
Nutrient cycling (see Nutrient cycling – Viral lysis)
Virulence factors 362, 376
bacterium, ongoing presence not required 358
diarrhea 376
exotoxins 13, 363, 372

fitness and (see Fitness, Darwinian – Virulence factors)
immediate environment not modified 372, 376
nutrient release 363, 366
tissue damage 363, 366, 376
Environments
Animals (see Animal–phage interactions)
Aquatic (see Aquatic environments)
Food
dairy (see Dairy phages)
non-dairy (see Fermented foods – Non-dairy phages)
Host organism 31
Modification by phages (see Environmental modification)
Oligotrophic 125, 128, 252, 262
higher lysogen frequencies 122
Terrestrial 12, 281–2
plants (see Plants)
soil (see Soils)
EPA (Environmental Protection Agency) 320
Epistasis 162
Adaptation, impediment to 224
Adaptive peaks
genetic drift 162
local-peak trapping 162
Antagonistic epistasis 162–3, 224
Compensatory mutations (see Compensatory mutations)
Examples
phage φ6 163–4, 166, 206–7
phage φX174 163
phage Qβ 164
phage T7 163, 168
P. fluorescens and phage 164
Examples, counter
Phage MS2 164
Fitness 162–3, 168, 206
antagonistic/diminishing returns 224, 235
Linkage
prone to epistasis 237–8
virulence factors, phage-encoded 360
Recombination and
clonal interference vs. epistasis 224, 237
hybrid incompatibility 168
Second-site suppressors 225
Sign epistasis 224
Synergistic epistasis 163, 206
advantage of sex 206
positive epistasis 163
small effect vs. complementation 206–7
Escherichia coli
Biocontrol (see Biocontrol of bacteria)
Butyric acid inhibition 363
Capsule (see Capsules, bacterial)
Detection (see Bacterial detection)
Examples of phages infecting: 933W, α3, f1, F2, Ff, φR73, φX174, HK97, K1-5, φX174, LH7, λ, M13, MS2, Mu, P1, P2, P4, P21, PP01, PRD1, Qβ, T1, T1X, T2, T3, T4, T5, T6, T7, etc. (see Bacteriophage isolates)
Fecal indication 306
Genetic model 196
"is true for the elephant" 148
Killing the winner (see Killing the winner) 264
Lysogeny 119
Pathogeneic 366–7
enteropathogenic 353, 360, 364–5
Shiga-toxigenic (see Shiga-toxigenic *E. coli*)
Phage-encoded
phage ΦX174, fitness on 161
proteins 357
Shiga toxin (see Shiga toxin)
virulence factors 353, 357–8, 367
Phage host 8, 55, 99, 118, 150, 168, 178, 188–9, 271, 417, 460
chemostat 32, 39–40, 287
two bacterial species community 161–2, 293–4
Phage resistance 39–40
Phage therapy (see Phage therapy)
Phages as fecal indicator (see Fecal indicators)
Physiology impact on phages (see Physiology)
Pseudolysogeny (see Pseudolysogeny)
Resistance to phages (see Resistance, bacteria partial)

Escherichia coli strains
15597 306
AK16 373
B 40, 41, 42, 50, 169
C 307, 431
C600 373, 375
C-3000 307
CR63 75, 79
K-1 451, 456
capsule as virulence factor 452
K12 54, 311, 417
MC4100 373
O157:H7 363
biocontrol (see Biocontrol of bacteria)
detection (see Bacterial detection)
foodborne pathogen 314
intestinal colonization 375
medical implications 314
O157:H7PT-32 375
O157:H7 EDL933 363
O157:H7 Sakai 116
S/6/5 67
Evolution/evolutionary biology 17, 149
Common descent (homology) 148
Convergent evolution 148
Experimental (see Experimental evolution)
Fitness, Darwinian (see Fitness, Darwinian)
Fixation (see Substitution)
Genetic variation 149, 152
mutation (see Mutation)
recombination (see Recombination)
selection (see also Selection, natural) 209
Genomics 6, 17, 147, 177–8, 180, 182, 189, 229
Horizontal gene transfer (see Horizontal gene transfer)
Macroevolution 167
Microevolution 167
Parallel 230–3
alternative beneficial mutations 230, 233
Phage 17, 177
evolutionary ecology (see Ecology – Evolutionary)
genomics (see Genomics)
phage relatedness (see Phylogeny, phage) 17
Microbial, history of 147–8
Natural selection (see Selection, natural)
Population genetics 149
Evolvability 165–6
Exaptation 156
Exclusion 8
(see also Superinfection – Exclusion)
Exotoxins 363
(see also Virulence factors, phage-encoded)
Expediency (see Economy vs. expediency)
Experimental evolution 17, 18, 219–20, 238
Adaptation (see Adaptation – Experimental)
Constant conditions, problems maintaining 221, 225
Extremes, problem of 220
Hitchhiking (see Hitchhiking)
Large-effect mutation biases 221–2
Multiplicity, problem of 221
Mutational equilibrium, lack of 221–2
Propagation methods
chemostats (see Chemostat)
phage T7 plaques 240
serial passage (see Serial passage)
Recombination difficult to eliminate 223
Experimental microbial communities 31
Broth
batch culture 394, 396–8, 400–1
continuous culture (see Chemostat)
three-species interactions 293–4
More complex model communities/ecosystems
mesocosm 118, 264, 267, 271
metapopulation (see Ecology – Metapopulation)
microcosm (see Microcosms)
Solid and semi-solid substrate
agar-based (see Plaques . . .)
biofilm (see Biofilm)
Extracellular search (dispersal period) 390
Bacterial density impact 68
constant dispersal time 48
exponentially distributed 48
Dispersal period/time 390–2
adsorption rate inverse 44, 47
humming birds 44–5
optimal "phoraging" 47
Dissemination 71–2, 80

Marginal value theorem 47
Plaques
stage 1 (see also Plaques, stages) 96

FDA (Food and Drug Administration) 302, 314, 320, 341
Fecal indicators (mostly food)
Animal-virus surrogates 307–8
Bacteria 305–6
Phages 305–7
Coliphages 303, 305–7
Water 305–7
Fermented foods, phages of
Dairy phages (see Dairy phages)
Lactic acid bacteria 304
Non-dairy phages 304–5
bacterial ecological succession 304
examples: phages ΦJL-1, Hh-1, Hh-3, LP65, etc. (see Bacteriophage isolates)
Filamentous (see Bacteriophages, filamentous)
Fitness, Darwinian 152–3
Alleles (see Alleles/mutations)
Bacterium
horizontal gene transfer 133, 297
phage virulence impact 78
DI particles 200
Evolutionary ecology 360
Landscape, fitness (see Adaptive landscape)
Limits (see Adaptive walks – Limits)
Lysogeny
fitness goals, divergent 157
lysogenic conversion (see Lysogenic conversion – Fitness)
superinfection immunity (see Superinfection – Immunity)
Measuring fitness 153–4, 210, 238–9
difficulties 239, 241
growth rates 46, 228–9, 241
individual substitutions 242
plaques 240–1
Mutations
large effects 222, 235
site-directed mutagenesis 226
small effects 223
Natural selection (see Selection, natural – Fitness and)
Payoff matrix (Prisoner's dilemma) 81–2
Phage 225
adsorption rate 162
burst size 77, 153–4
coinfection/multiple adsorption (see Coinfection – Fitness)
competitive ability 53, 203
growth capacity 195
harnessing cellular machinery 203
historical 218
latent period, impact on 47, 69, 77
population growth rate 72
Rates of fitness increase 235
population size impact 227, 238
Recovery 153, 228, 236
Relative fitness 200–1
Resource acquisition 43
energetic gain 44
Statistical fitness 217–19
Trade-offs (see Trade-offs – Fitness)
Transposons 157
Virulence factors 366
bacterium fitness 297, 362, 365, 452–3
phage fitness 356, 361–2, 365–6
Flora
Food 303, 312, 320
competing (for detection) 311
normal 313
phage modulation of 304
spoilage 313, 319–20
Human, normal 332–3
disruption 13
gastrointestinal 366, 456
Flow 95
Fast 95
Turbulent 460
Food webs, microbial (aquatic, pelagic, planktonic) 263
Biogeochemical aspect 265
Ecological theory of 260
Lytic viruses and 258–9, 261, 265, 272
Modeling (see Modeling – Food webs)
Organic material
release and reincorporation 259
Role of predatory processes 259

Food webs, microbial (*cont.*)
grazers (see Grazers – Consumers)
Trade-offs 262
(see also Trade-offs, bacterial – Competitive)

Free phages or viruses
Abundances 291
Adsorption 89
bacterial adsorption rates 323–4
bacterial density issues 321, 323
mean free time 321, 323
multiplicity issues 88
Avoidance of coinfection 71
Environmental
occupancy 287
survival issues 283–4, 286
trapping 290
virus community dominance 251
Existence conditions 288
Inertness, behavioral 290
Less complex than lysogens 359
Loss to superinfection 85
longer latent periods 85
Modeling (broth culture)
adsorption 393, 404
bacterial adsorption 404
batch growth 398
chemostats and time delays 403
density 391, 410
gain via lysis 394, 404
inactivation 404
initial density 407
outflow 404
Phage therapy (modeling)
adsorption 448, 460
bacteria, protection from 457
decay 442, 445, 448, 451, 458–60
densities 449
diffusion, two-dimensional 457
gain 443
Plaque formation (including modeling)
adsorption 432
density 418
gain via lysis 420–1
ignoring free phage 428
loss to infection 422
periphery 418
phage–bacterial association 421
pre-plaque 96
site occupancy 432
virion movement 97, 418
wavefronts 429
Storage as 219
Virulence factors
isolation biases 360
less beneficial to 356

Genes, bacteria
Cosmopolitan genes 362
Pathogenicity islands 355, 369–70
Receptor, phage 117
Reporter genes (see Reporter genes and proteins)

Genes, phage
Cheater genes 355
coinfection 209–10, 212
prophage 355
selfish genes 156, 210
Common attributes and types
average size 180
coevolved genes 184, 354–5
core genes 184
co-transcription 180
direction of transcription 180
duplication 152, 180, 186, 212, 221
head genes/proteins 184, 186–7, 207–8, 337, 342
intragenic mosaics (see Joints)
lysis genes/proteins 207
lysogeny 187, 197
lytic infection 187
novel associations 191
pathogenicity gene (see also Virulence factors) 297
RNA (e.g., tRNA) genes 180
tail genes/proteins 180, 182–4, 199, 207, 354
tightly packed into genomes 180, 187
Core genes 184
non-core group genes 184
Cyanophage genes 119, 132–3
hli/hliP (HLIP) 119, 133
petE (plastocyanin) 133

petF (ferredoxin) 133
psbA (D1) 119, 132
psbD (D2) 132
Functionality of genes 184, 187, 200
Morons (see Morons)
gene *15* 182
Non-coding genome regions 180
Phosphate metabolism genes 119
Pseudolysogeny
rI 129–30
Selection for useless genes 187
Technology
gene therapy 339, 343–5
reporter genes (see Reporter genes and proteins)
Virulence factor genes
ctxA, *ctxB* 370
tox 372
stxA, *stxB* 373
Genetic elements
Inteins 185
Introns 185
Mobile 185, 355
Morons (see Morons)
Niceness of 355
Plasmids 342, 355
conjugation 202, 294
experimental adaptation of 227
phage P4 208
phage CTXΦ plasmid-like replication 370
Ti plasmid 334, 344
vaccination with 342
virulence-factor associated 371
Transposons 157, 185–6, 355
Genomes, phage 6
DNA
double-stranded 6, 18, 151, 167, 178, 179, 180
single-stranded 6, 7, 151, 179, 189, 228, 232
RNA 199, 228
double-stranded 6, 7, 148, 151, 178, 189, 201
single-stranded 7, 151, 178
Topology
circular 179, 189
linear 178, 179
segmented 6, 7, 178, 199, 201, 212
Genomics 180
Metagenomics (see Metagenomics)
Phage evolution 6, 17, 147, 177–8, 180, 182, 189, 229
Grazers, eukaryotic
Bacterial community
affects size 260, 263
bottom-up control 266
dominant predator 260
top-down control 266
Bacterial prey
competition specialist susceptibility 263
existence conditions 266, 267–8
grazing resistance 263
instantaneous removal 254
non-selective (relatively) among 260, 265–6, 268
partitioning between grazers and viruses 255, 258, 261, 265, 267–8
Consumers, grazers as (trophic level) 260
connection to higher trophic levels 12, 266, 268
of DOM assimilators 12
Density of grazers 260
Grazing quantification 254
Modeling (see Modeling – Food webs)
Growth
Bacterial (see Bacterial growth)
Logistic
carrying capacity 34, 72, 77
Parasite
intracellular 4
within-host 4
Phage (see Growth, population, phages)
Predator 33
Prey 33
Growth parameters, phage 7, 15
Adsorption constant (see Adsorption – Constant)
Burst size (see Burst size)
Decay, infected bacteria 404
Decay, virion (see Decay, virion)
Desorption 420

Growth parameters, phage (*cont.*)
Diffusivity 460
impediments to 415
Eclipse period 44–5, 48, 160
measures of 58
Generation time 46, 68, 392
Host physiology dependence (see Physiology, bacterial)
Latent period (see Latent period)
Modeling (see Modeling – Algorithms)
Phenotypic plasticity 49
Physiology, impact on (see Physiology)
Rise 393
variance in waiting times 48
Growth, population, phages 16
Behavior and game theory
economy vs. expediency (see Economy – Growth)
later-offspring discounting 80
lysis inhibition (see Lysis inhibition)
prisoner's dilemma (see Prisoner's dilemma)
virulence (see Bacteriophage virulence)
Broth
batch culture 73, 394
chemostat modeling 400, 410
comparing plaque growth 95–6, 100
fitness measure 154, 239
isolation biases 179
modeling 46, 390, 396–8
selection on 228–9, 239, 241
Coinfection, leading to 202, 207
Initiation of growth
broth 82–3
plaques (see Plaques, stages)
Limits to growth
bacterial density 83–4
carrying capacity 77
distance between environments 80
plaques (see Plaques, stages)
virion loss 83
volume 76–7
(see also Ecology – Metapopulation)
Lytic growth 64–5, 67
exponential and faster 67–9, 84
multiplicities low 65–7, 221
multiplicities near one 65–7, 69–70
multiplicities over one 65–7, 221
sustaining growth 83–4
T-even phages (model for) 82
Phage attributes affecting 46, 228–9, 399
adsorption rate: 84, 398–9
bacterial density 398–9, 447
burst size 67, 153, 398–9
eclipse period 84
genome size 229
historical factors 229
latent period, dependence on 46–8, 84
maturation rate phage 84
Spatial structure
diffusion-only limit 95
plaques, descriptive 96, 97, 98, 100
plaques, modeling 100, 415, 423, 425
wild 94, 101, 105, 106, 109, 435
Temperate phage growth 11

Heterotrophic
Carbon and heterotrophic bacteria
availability variable over time 257
dissolved assimilation 12, 259
increases mediate collapse 256
recalcitrant materials less available 12
reduced trophic movement 12
Phage
lysis susceptibility 12
lysogeny prevalence 122
phosphate limitation impact 257
Predator–prey relationships 264
History of phages 3, 177
d'Hérelle, F. 2, 3
discovery 2
Environmental abundance 12, 119–20, 252, 261
Molecular genetics 148–9
Twort, F. W. 2
Hitchhiking (natural selection)
Genetic 160, 161, 222
reduced emphasis 222
(see also Lysogeny, temperate – Virulence factors)
Horizontal gene transfer (HGT) 17
Bacterial 13, 148, 294, 355
conjugation 202, 294
fitness 133, 297

transduction (see Transduction)
transformation 294
Phage
fossils of HGT 182
morons (see Morons)
mosaicism (see Mosaicism)
phages to eukaryotes 334
Photosynthesis genes 133
Host range
Adaptive peak example 158
Breadth 288
advantages 167, 288
broad host range, phages with 119, 288, 296
disadvantages 43, 168
eu- and prokaryote spanning (speculation) 333–4
narrow expected 260, 284, 288
presumptive treatment problems (phage therapy) 340
underestimation, reasons for 288
valence 5
Cocktails to broaden ranges (see Cocktails, phage)
Coevolution, phage–host 36, 42, 169
arms race 43, 168–9
lopsided 43
Red Queen 168
Complete receptor replacement 43
Determinable only experimentally 5, 304, 333
Ecology of 167
mutational "entrance" into environment 82
niche expansion 167
reproductive isolation 167
specialization 167–8
trade-offs (see Trade-offs, phage – Host range)
Host shifting 43, 167
Mixing, phenotypic 199
Mutants, phage 38, 42–3, 55, 218, 230
antagonistic pleiotropy (see Antagonistic pleiotropy)
attachment protein variability 167–8
large mutational effects 234
plaque visualization 104–5
Viruses
broad, some 333
defining characteristic 5, 17
narrow, mostly 333, 344
phages (in general) 2
Human–phage interactions (see Animal–phage interactions)
Hutchinson's paradox 261–2
Reversed 270–1
Viral solution to 262
killing the winner (see Killing the winner)
Hyperparasitism 207–8

Immune system, anti-phage 332, 334, 442–3
Adaptive phage interactions 336–7, 341
Innate phage interactions 334–8
Immunity (see Superinfection – Immunity)
Incompletely defective viruses
Cheating and 200, 211–12
(see also Cheating – Coinfection)
Phage φ6 200, 204–5, 211
mechanism, undetermined 211–12
reassortment impact 212
tolerating genotypes 205
Prisoner's dilemma (see Prisoner's dilemma – Coinfection)
Indicator, plaquing
Mixed indicator 104
evolution during 105
resistant indicator 104
Phage isolation 5
Induction, prophage (see Lyosgeny, induction)
Interacting particle systems 427
Cellular automata (see Plaques, modeling, cellular automata)
Interspecific interactions 16
Chemostat, phage–bacterial (see Chemostat)
Control
biocontrol (see Biocontrol of bacteria)
bottom-up 37, 41–2, 266
top-down 37, 41–2, 266
Eukaryote-involving trophic relationships 264

Interspecific interactions (*cont.*)
Horizontal gene transfer (see Horizontal gene transfer)
Host range, phage (see Host range)
Immunity, anti-phage (see Immune system)
Killing the winner (see Killing the winner)
Mutualistic 16
Phage therapy (see Phage therapy)
Physiological bacterial impact on phage (see Physiology)
Red Queen hypothesis 168
Restriction (see Restriction)
Selection, levels of (see Selection, natural – Levels of)
Trophic interactions (see Trophic)
Victim–exploiter 16, 32–3
grazing (see Grazers)
infection (see Bacteriophage infection types)
inherently antagonistic 52
parasite 16
parasites of unicellular organisms 3–4
parasitoid 16
predator 16, 33, 36–9
prey 33, 36–7, 39
Virulence factors (see Virulence factors)

Joints, novel 181
Creation of 181, 183
Gene boundaries 181–2
Marker reassorting 183
Restricted locations, argument against 183
Within genes 181–3

Killing the winner
Bacterial diversity enhancement 13, 261–2
phage-mediated bacterial decline 406
Competition vs. defense (see Specialization, bacterial) 263
Host specificity of viruses 260
Hypotheses from 267
(H1) few dominant host organsms 265, 267, 270–1
(H2) comparable virus-host richness 267–8, 271
(H3) host changes result in virus changes 267–8, 271
(H4) high host growth differentials support high virus:host ratios 267–8, 271–2
Substantiation 264, 270
complications toward 270
Viruses, no killing the winner for 268

Lactic acid bacteria (LAB) 304
(see also Dairy phages)
(see also Fermented foods)
Lambdoid phages (see Bacteriophages, lambdoid)
Latency 14, 115
Latent period (see Latent period)
Lysogeny (see Lysogeny)
Latent period 32, 390, 403
Burst size 45–6, 57–8
Evolution 242
longer 69
minimization of 160
optimization 46–8, 84
shorter 68–9, 73, 432
Factors impacting on
host 48, 393
environment 48–9, 114, 118, 393, 456, 458
Generation time component 68–9
Impact on
fitness, phage (see Fitness, Darwinian – Phage)
population growth, phages 398
stability (see Time lags)
Lysis rate surrogate (k_2)
plaquing 420–1, 430
phage therapy 449
Lysogeny as extended 70, 117, 286
Measures of 459
Modeling 57, 391, 393–4, 445, 448
chemostats 58, 404–5
Optimal 46–8, 84
Plaque modeling 418–19
cellular automata 432
incorporating time delays 99–100
Pseudolysogeny as extended 458

"Residence time" equivalent 47
patch residence time 44–5
optimal residence time 44
Time lags (see Time lags)
Trade-offs (see Trade-offs)
Lateral gene transfer (LGT) (see Horizontal gene transfer)
Lawn, bacterial 107
Bacteria immobilization 415
Clearing of 416
confluent lysis of 101
holes in lawn 98
"lysins" 102, 106
pre-clearing 98
Density of
increasing 97–8, 107, 425–7
modeling as constant 420
modeling as replicating 423, 426
phage-resistant bacteria 107
Microcolony 107–8
Physiology of
burst reduction 426
lysis delay 108, 426
maturation 98
microcolony impact (speculation) 108
standard bacterial growth curve 98
Plaque formation 99–100, 415, 417
phage-free (see also Plaques, zones) 433
race with lawn maturation 103
replica plating (plaque transfer) 431, 434
Standard bacterial growth curve 98
Turbidity 3, 99
Life-history strategies 56, 65
Coexistence 56, 123–4
Optimal life-history theory 49, 55
Selective modification of 242
Trade-offs (see also Trade-offs, phage – Burst size) 160
Transition difficulties 49, 56
Linkage, genetic 354, 355
Epistasis (see Epistasis – Linkage)
Genes 354–5
Linkage disequilibrium 208, 354–5
Linkage equilibrium 354
Virulence factors 344–5
Listeria monocytogenes
Biocontrol 302, 317–18, 320
Detection 310–11
Listeriophages 310–11, 317–18
examples: phages A511, LH7, P100, etc. (see Bacteriophage isolates)
Properties
fetal morbidity and mortality 317
foodborne pathogen, common 314, 317
intracellular pathogen 457
Long-circulating bacteriophages (see Bacteriophages, long-circulating)
Lotka–Volterra 33–5, 262, 265
Lysis
Bacterial loss/death 12, 359, 405–6, 421
Community-wide 65–7, 86
lysis from without 86–7, 324
lysis-inhibition collapse 67, 86–7
optimizing timing 71
Holin proteins 49
Infection termination 9, 32, 65, 390
Instantaneous 404, 427
Lytic cycle (see Lytic cycle)
Lytic phages (see Lytic phages)
Plaque formation
bacteria lysed, number 423, 425–7
clearing (see also Lawn, bacterial 416, 423
traveling wave of lysis 418
virion release 418, 421, 422
Viral loop and lysis (see Viral loop)
Virion release 64, 362, 390, 394, 403–4, 418, 421
Virulence-factor release 357, 359–60
Shiga toxin (see Shiga toxin – Expression)
Lysis inhibition 85, 105
Collapse 67, 86–7
Defective (*r*) mutants 105, 130, 417
model simplification 443
Dilemmas (see Prisoner's dilemma)
Growth parameter impact
burst size 69, 105
latent period 67, 105
Induction 85
multiple/secondary adsorption 85–6, 105, 108, 196
Phage population growth aspect 82

Lysis inhibition (*cont.*)
Plaques
bull's eye 108
microcolony retention 108, 417
mottled plaques 106
smaller plaques 105
Why not in all phages? 70
Lysogen 120
Community composition, influences 116
Experimental dynamics 411
Fitness considerations (see Lysogenic conversion)
Frequency/prevalence/occurrence 288, 291
aquatic environments (see Aquatic environments)
oligotrophic environments (see Environments – Oligotrophic)
soils 286
Identification/recognition
DNA probing 122
immunity to superinfection (see also Superinfection – Immunity) 121
induction 122
sequencing bacterial genomes 115–16
Lysogenic, property of bacterium 11
Multiple genetic entities 354
bacterial exploitation, long-term 70
interests conflicting 121, 157, 286, 354, 356, 362
interests overlapping 156, 286, 356, 362
related lysogens 359
symbiotic relationship 116
Plaques, within 104
Success, evolutionary 116
Superinfection (see also Superinfection)
coinfection of 207
immunity (see Superinfection – Immunity)
lytic infection 255
phage P4 of P2 lysogens 207–8
Lysogen allelopathy 361
Virulence factors and 359, 361
Salmonella spp. 361, 375–6
Lysogenic conversion 13–14, 131–2
Bacterial diversity increasing 13, 121, 132–3
Examples
Phage CTXΦ 297, 374
Streptococcus pyogenes 374
Fitness impacts 116, 131, 157
division rate advantages 118–19
niche expansion/opening 116, 131
virulence factor expression 356, 359, 362, 366, 371–2
why necessary? 355–6
Phenotypes acquired 132, 356, 361
(see also Superinfection – Immunity)
(see also Virulence factors, phage-encoded)
Lysogenic cycle (see also Lysogeny) 8, 9
DNA phages only 115
Immunity (see Superinfection – Immunity)
Induction (see Lysogeny, induction)
Lysogen (see Lysogen)
Lysogeny (see Lyosgeny)
Lyosgenic conversion (see Lysogenic conversion)
Prophages (see Lysogeny, prophages)
Selection, for and against 365
Lysogeny 9, 114–15
Advantages
reproduction enhancement 69–70
survival enhancement 66, 116, 120, 123, 284, 286
Alternative descriptors
dormant state 114
non-productive infection 64
reductive infection 120
Establishment
animals, within 364, 373
integration 9
requires energy 125
Historical 2, 11
Induction (see Lysogeny, induction)
Latent-period extension 70, 117, 286
burst enhancement 70, 120
generation time, impact on 70, 85
Lysogen allelopathy (see Lysogen allelopathy)
Lysogenic conversion (see Lysogenic conversion)
Lytic–lysogenic decision 117, 120, 124, 129, 366

bet hedging 125
host physiology impact 120, 123, 124, 129
microenvironment impact 124
modeling (see Modeling – Temperate phages)
Plaques, impact on 104, 426
Polylysogeny 115–16, 151, 353
Escherichia coli O157:H7 116
Salmonella Typhimurium 371
Temperate phages (see Lysogeny, temperate phages)
Well defined in phage λ (see Bacteriophage λ – Lysogeny) 115
Lysogeny, induction
Controlling factors 120, 260, 368
DNA damage 120–1, 368, 371
hydrogen peroxide 371
ignorance, natural conditions 260
mitomycin C 122–3, 367, 370, 372
modeling (see Modeling – Temperate phages)
nutrients 124, 129
spontaneous 121, 368, 371
ultraviolet (UV) light 122–3, 128, 370, 372
uninducibility 122
Examples
phage CTXΦ 370
phage λ, rate of 368
Salmonella prophages 371
Phage production 121, 255
lysogeny adaptation loss 362
virulence factors and 356
Repression of induction 121
bacteria survive 8
dormant lysogens 123
immunity (see Superinfeciton – Immunity)
Termination of lysogeny 120–1, 129
Timing optimization 286
Upon induction
selective killing of bacteria 260
Stx-1 production 368
Stx-2 production (see also Shiga toxin) 368
unlinking phage and bacterial genes 354
virulence factor release 359
Lysogeny, prophages 115
Bacteria-like state, phage in 365
Co-replication with bacteria (see Co-replication – Prophage)
Cosmopolitan genes 70, 117, 286
Costs of being a prophage 365
Curing 8–9
Defective/vestigial 115, 353, 360
Duplication, selective constraints on 355
Exotoxin encoding 353
Induction (see Lysogeny, induction)
Lysogen allelopathy (see Lysogen allelopathy)
Mosaicism in (see Bacteriophages, tailed – Genomes – mosaicism)
Protection from decay (see Lysogeny – Advantages)
Superinfection of (see Superinfection – Immunity)
Transposons in 157
Virulence factors (see Virulence factors – Phage/prophage encoding)
Lysogeny, temperate phages 11, 56, 78
Ecology
advantages of temperance 286, 355–6
bacterial density depression 286, 292
bacterial properties impact on 123
competition with lytic phages 123, 290, 390
durability of virions 256
host nutritional state 117–18
Examples: phages CTXΦ, λ, P2, P4, P22, Tuc2009, etc. (see Bacteriophage isolates)
additional examples 304, 309
Stx-phages (see Stx-encoding phages)
Genomes
DNA phages 115
Immunity (see Superinfection – Immunity)
Infection (see also Lysogenic cycle) 120
Isolation as lysogens 360
Lysogeny module defects (see Bacteriophages, virulent)
Lytic cycles and 11, 126
obligately lytic, not 11
virulence and 11

Lysogeny, temperate phages (*cont.*)
Modeling (see Modeling – Temperate phages)
Nomenclature
"lysogenic phage", no such thing 11
virulent phages (see Bacteriophages, virulent)
Parasitic 56
Plaques 104
Pseudolysogeny 129–30
Virulence factors 359–60
Lytic cycle 8–9, 11, 32, 64, 390, 394
Environmental importance 254–5, 261
Gene expression
lytic conversion 132
morons 186
photosystem proteins 133
Lysogens
lytic following induction 121, 255
lytic infection of 255, 365
Nutrient limitation
carbon/energy 118
inorganic factors 118–19
Productive infection 115, 119
Relation to reductive infections 126, 365
Synchronization of lytic cycles 255
Virulence factors benefiting 356, 359–60
Lytic phages
Environmental
carriage 267
factors controlling 260, 272
historical 261
impact (see also Killing the winner) 262, 268
protozoan grazing 265–6
viruses, lytic 258
Evolutionary goal 195–6
Obligately lytic (see Obligately lytic)
Shunting material down to dissolved state 266, 270
see also Dissolved organic material
Temperate phages, includes most 11, 117, 121
Virulent (not fully synonymous) 11
(see also Bacteriophages, virulent)
Malthusian parameter (see Bacterial growth – Modeling)
Marginal value theorem 44, 47–9, 80
Marker genes and proteins
β lactamase 373
Kanamycin resistance 374
Mass action 33, 49, 239, 392–3, 435, 442
Further considerations 49, 392, 435
Mesocosms 118, 264, 267, 271
Metagenomics 133–4, 187, 189
Human microbiome 332
Limitations
gene association identification 189
phenotype determination 5
Strengths
culturing not required 5, 188
diversity indication 188
reduced sampling discrimination 188
Microbial loop 12, 259
Microcolonies (during plaque formation)
Bull's eye formation 107–8
Cellular automata 435
Delays in elimination 107
Lawns 107–8
Lysis inhibition (see Lysis inhibition – Plaques)
Lysogens 104
Time dependence impact 110
Persistence within plaques 107–8
Microcosms
Aquatic
freshwater and pseudolysogeny 128–9
marine and phosphate 256
marine and UV exposure 123
Broth
chemostat 39
feast–famine 124
resistant bacteria 39
victim–exploiter interactions 31
without stirring 293
Microbial module 37
Soil 291, 293
Bacillus subtilis 286–7, 292
streptomycetes 124
Microscopy
Electron (see Electron microscopy)
Epifluorescence 252

bacterial counts (DEFT) 312
viral counts 252
Modeling 389
Algorithms employed 390
adsorption (see Adsorption – Modeling)
bacteria (see Bacterial growth)
burst, phage gained due to (see Burst size – Modeling)
latent period (see Latent period – Modeling)
multiplicity (see Multiplicity – Adsorption, of)
Batch culture (see Culturing in batch)
Chemostat (see Chemostat, one-stage – Modeling)
Continuous time 408
chemostat model 57–8, 408–11
continuously compounding 398
instantaneous rate of change 409
Lotka–Volterra (see also Lotka–Volterra) 33
Differential equations 408–9
bacteria infected per plaque 426
chemostat model 408
mass action (ordinary) 431, 435
phage therapy 439, 442
reaction diffusion (plaques) 415
time delay (see also Time lags) 36, 456
Discrete time 390, 393, 398
algorithm summary 395
difference equations 393, 398, 405–6
incrementation intervals 392–3
parasites and hosts 32
recursive modeling 399, 401, 407–8
utility 408
Food webs 263
grazing 266
idealized flow scheme 259
Lotka–Volterra 262, 265
Phage therapy (see Phage therapy, modeling)
Plaques (see Plaques, modeling)
Predator–prey 33–5
Substrate 57, 395, 403, 409–11
Temperate phages 390
induction 125
lytic–lysogenic decision 124–5
Why model? 441
Modules
Chemostats (see Chemostat)
Community 37, 56
phage–bacterial 37–9, 41–2, 56–7
spatially continuous surfaces (see also Plaques) 56
Metapopulations (see Ecology – Metapopulation) 56
Microbial 37, 56
Microcosms 37
Molecular techniques (see Techniques, molecular)
Morons 13, 185–6, 295, 376
Acquisition of 186, 295
Benefits
gain of new genes 186, 221
invention of novel properties 221
to prophage 186
toward lytic growth 186
Example 182
Mosaicicsm 180–1
Example: phages A511, HK97, K, LP65, Mu, SfV, and SPO1 (see Bacteriophage isolates)
Tailed phages (see Bacteriophages, tailed – Genomes)
Muller's ratchet 159, 238
Bottlenecks 238
Demonstration of 238
phage ϕ6 159, 202, 226
phage MS2 159–60
Impediments of Muller's ratchet
compensatory mutations 160
sex 159, 203
Multiplicity 88
Adsorption, of (MOA) 65–6, 88–9
algorithms, defining 87–9, 396
infected bacteria creation 396–7
Adsorption-rate multiplicity dependence 321
of individual bacteria 321, 324
of individual phages 321, 323
Infection, of (MOI) 88
algorithms, defining 87–8, 323

Multiplicity (*cont.*)
discrepancies 315
experiments 205, 311, 314–16, 407
Multiple adsorption consequences 195–7
coinfection (see Coinfection)
historical 196
lysis inhibition (see Lysis inhibition)
multiplicity reactivation 196
plaques 426
secondary adsorption (see Adsorption, Secondary)
superinfection 70, 195, 197
Phage population ecology, impact on 65–7
low multiplicity 66–7
multiplicities approaching one 69–70
multiplicities exceeding one 77
Mutation 149–51, 221
Categories
deletion 152–3, 164, 166, 180, 191, 198, 221, 222, 304
duplication 152, 180, 186, 212, 221
insertion 152, 157, 164, 180, 186, 189, 191, 304
missense 236
non-synonymous 185
point 166, 180, 185, 189, 191, 222, 225, 304
rearrangement 180, 221–2
silent 236
synonymous 155
Compensatory (see Compensatory mutations)
Environment interactions 150–1, 162
Fitness (see Alleles)
Jackpot distribution 150
Luria and Delbrück (1943) experiment 150
Mean invasion time (plaques) 431
Mutation rates 150–1
different apparent 181
high in RNA phages 18, 166, 228
increasing using mutagens 166

Natural selection (see Selection, natural)
Negative density dependence/regulation 34–5, 57–8
Normal flora (see Flora)
Nutrient cycling 12, 259
Environments
aquatic 12, 253, 269
soils 282
Predation effect on 258–9
microbial loop 12, 295
Viral lysis effect on 258–9
viral loop (see Viral loop)
Nutrients (typically limiting)
Carbon (see Carbon, organic)
Glucose
chemostats carbon source 40, 57–8, 264, 391, 407
laboratory community 264
lysogen growth advantage 119
phage T5 resistance 41
Nitrogen
bloom in microcosm 256
burst size impact 118–19
mixed algal community 264
photic zone balance 269
Phosphorous
adsorption rate 118
burst size 118, 257
DNA production 119, 253
food web sharing 263
laboratory communities 256–7, 264
latent period 257
phosphate-limitation gene 119
photic zone balance 269

Obligately lytic 32, 56
Employment of
ecological analysis 56, 390
fitness (example) 153
isolate descriptor 290–1, 295, 305
propagation and fitness 238
Host impact
soils 286–7, 292
static microcosm 293
virulence against 78
Parasitoid-like 56, 64
Population growth 66
Survival
pseudolysogeny 127–9
soil 286
Virulence factor not encoding 359–60
Usage of term 11

Oligotrophic (see Environments – Oligotrophic)
One-step growth 100
Optimization 55
Foraging 43–5
patch residence time 44–5
"phoraging" 44
Lysis timing 46–8, 84
exponential adsorption impact 48, 69
host density 47–8, 67
host quality 47–8
optimization model system 49, 56
phage-multiplicity dependence 65–6
plasticity and 49
Lysogen-induction timing 286
Marginal value theorem 44, 47–9, 80
Oscillations 33
Damped 34, 408
Increasing 407
Neutrally stable 33
Out of phase 406
Osmotrophs 258–9

Packaging
Genomes 32, 187, 203, 208, 212
Phenotypic mixing 199
Transduction 294–5
Paradox of enrichment 39, 41
Phage display 341–3, 345
Phage therapy 340–1, 439–41, 461
Active vs. passive 315–16
Antibiotic alternative 340, 461
Biocontrol, phages-mediated (see Biocontrol of bacteria)
Cocktails of phages 314–15, 317, 320
Detection (see Bacterial detection)
Escherichia coli 456
enteropathogenic 364
History 339–40, 440
host range narrowness 340
modern use 440
phage fate in animals 335
Host vs. subject (definitions) 441
Impacts of
bacterial growth rates 458
bacterial pathogenesis, modes of 458
decay rates, phage 460
delayed treatment 451, 453
host range narrowness 340
immune system 336
pathogen physiology (see Physiology, bacterial)
phage-protected sites/conditions 453, 456–7
resistance of bacteria to phage attack 451
Regulation of 341
Phage therapy, modeling 439, 460
Bacteria
bacterial resistance to phage 451, 453
density 442, 447–8
immobilized/surface attached 457
stationary phase 444
Modeling behavior or results
cycling behavior 443
modeling limitations 441
outcomes, possible 443
overview of approaches 442–4
poor predictor 455–7
treatment delay 445, 450–1
utility of 441, 456
Phage
adsorption rate, impact of 445, 460
decay of phage 445, 448
dose size 450
immune response 443, 445
latent period, ignoring 445, 448
lysis rate incorporation 448–9
parameter determination 456, 458–61
replication within subject 448, 451
Unstructured environment assumptions 457
Phage typing 288, 308–9
Phenotypic plasticity 49, 131
Phylogeny, phage
Deep 189–91
tailed phages 190–1
Experimental 189
Tailless phages 189
Physiology, bacterial, impact on phages 117–18
Bacterial hosts in experiments
Escherichia coli 458
Pseudomonas aeruginosa 118
Pseudomonas fluorescens 119

Physiology, bacterial, impact on phages (*cont.*)
Environment impact 48–9, 118, 256–7, 456
aquatic environments 256–7
Limiting factors
amino acids 256–7
carbon and energy 256–7
phosphate 256–7
Parameters affected 117–19, 458
burst size, impact on 49, 114, 117–18, 124, 256
eclipse 49
latent period 49, 117, 118
lysogen replication 117, 119
lytic–lysogenic decisions 117
phage receptor (adsorption) 117
Phage therapy, importance of 458
Plaques (see Plaques – Bacterial)
Phytoplankton 256, 259, 261, 263, 264, 266, 269–70
Planktonic 119, 259, 392, 453
Plants
Biocontrol of plant pathogens 318, 320
Growth, phage impact on 291
Phyllosplere 282
Phizoplane 282
Rhizosphere 282–3, 289, 291
Seasonal phage variation 292–3
Spatial structure 415
Soils, dead plant material impact 283
Plaques 94–110, 415–36
Bacterial physiology 100
growth parameters impact 100, 426
lawn (see Lawn, bacterial)
vs. natural assemblies of bacteria 101
Bottlenecking 159, 202, 226
Broth, differences from 95, 100–1
Competition in (see Competition – Plaques)
Evolution within 240
Existence conditions 5
Historical 3
Inhomogeneities
lysogen microcolonies 104
microcolonies (see Microcolonies)
natural selection 109
phage density 108–9
turbidity (see Plaques, morphology – Bull's eye)
Lawn (see Lawn, bacterial)
Productivity/fecundity 154, 240, 423, 426
Pure-culture technique 77
Serial transfer 159, 240–1
Spatial structure 415
Stages of formation (see Plaques, stages)
Virion loss
adsorption (see also Plaques, modeling – Adsorption) 95–6
secondary 426
unproductive infection 426
Zones of (see Plaques, zones)
Plaques, modeling 415
Adsorption 419, 421–3
desorption 420–2
diffusion independent 420
equilibrated 422
Bacteria
infected/lysed (see Lysis – Plaque formation)
density 419–22
division, assumption of none 420–1
Burst size (see Burst size – Plaques)
Cellular automata (see Plaques, modeling, cellular automata)
Diffusion 418, 419, 421–2
adsorption constant component 419
Enlargement 415, 416, 421, 422
wavefront velocity 99, 418, 422
wavefront velocity, constancy in 423, 425
Homogeneity requirement 435
Productivity/fecundity (see Plaques, modeling, productivity)
Reaction–diffusion 415, 419–20, 423, 431–2
cellular automata similarities 431, 435
predictive of bulk properties 435
shortcomings 435
Release of phage progeny 421
Sensitivity analysis 100
Time dependence 109–10
Traveling wave 416, 418, 422, 424
Plaques, modeling, cellular automata 415, 427–8
Bacteria

microcolonies 435
resistant to phages 432, 434
Modeling details, results, and extensions
experimental testing 431–2, 434
inhomogeneities 435
initiation 434
length scales 434
plaque morphology 428, 431
stochasticity 427–8, 431, 435
Phage
burst size 432, 434
coexistence 431–2
competition 432–5
diffusion 434
latent period (see Latent period – Plaque)
mean invasion time 431
mutation 430, 432
oscillations in densities 432
Phage refuge, spatial 432
Reaction–diffusion connections 435
SIR (Susceptible–Infective–Removed) 428–30
Spatial structure 427, 434
Traveling wave 435
stochasticity in 430
wavefront 429, 430
Plaques, modeling, productivity/fecundity 423
Bacterial density 423, 425–6
Productivity 426–7
Wavefront velocity, constant 425
Plaques, morphology 102
Bull's eye 106–7, 416, 417
lysis inhibition 108
microcolony resistance to lysis 107–8
mutation to phage resistance 107
Lysis inhibition (see Lysis inhibition)
Mutants 103–4
clear 104
halo size (lysin production) 106
host range (mixed indicator) 104
minute (small) 105
mottled (mosaics) 106
rapid lysis (large) 105
sectored/star (overgrowth) 106
turbidity increases (adsorption deficient?) 105–6
Pseudolysogeny (see Pseudolysogeny – Plaque impact)
Shape 424–6
Size 103, 154
large 240
mutation affecting 79
small 105, 179
vs. per-plaque productivity 103
Temperate phages 104
Turbidity 106, 417
Zones of (see Plaques, zones)
Plaques, plaque assay 5, 240
Bacterial detection use 308, 312
Biases 5, 179
Low agar concentrations 179
Plaque-forming unit (PFU) 97, 240, 398
Plaques, stages of formation 96
(1) Extracellular search 96
(2) First adsorption/infection 96–7
(3) Phage population growth 97–9
constancy of 98–9, 425
diffusion of phage 95, 415–16
enlargement phase 418
infection–diffusion 97, 416
rates of enlargement 99
selection for rapidity 109
spreading stage 97
(3a) Pre-visibility growth 95, 98
(3b) Visible plaque growth 99
(3c) Post-lawn-maturation growth 99, 426
stationary-phase lawn 98–100, 109, 420, 426
(4) Termination of phage population growth 100–1
stationary-phase bacteria 100–1, 423
(5) Post-termination modification 101–2
adsorption to stationary-phase bacteria 101–2, 109
continued phage spread 101–2
enzymatic lawn modification 102
lysins (see Lawn, bacterial)
Plaques, zones 416–18
(1) Clearing 416
(2) Reduced turbidity 416, 418, 430
(3) Infection 416, 418, 421
(4) Periphery 102, 416, 418, 421–2
plaque's edge 102, 106–9, 241, 429–30

Plaques, zones (*cont.*)
surface area 424, 426
(5) Phage-free bacterial lawn (see also Lawn, bacterial) 418
Poisson distribution 88, 396
Polymorphism 231
Population size, effective 158–9
Predation
Biogeochemistry (see Biogeochemistry – Movement)
Chemostat enrichment importance (see Enrichment – Chemostat)
Diversity of bacteria impact (see Killing the winner)
Nutrient cycling (see Nutrient cycling – Predation)
Sinking of material (see Biogeochemistry – Sinking)
Prevalence
Bacteria (see Abundance of bacteria)
Phage (see Abundance of phages)
Prisoner's dilemma (and PD-like dilemmas) 81
Coinfection dilemma 205, 220
Selfish segment interpretation 212
Cooperation–defection 81–2
Exponential growth dilemma 81
cooperation–defection 81, 84, 87
deviation from PD 82, 84
resolution 84–6
(see also Economy vs. expediency)
(see also Trade-offs – Growth rate)
(see also Tragedy of the commons)
Illustration of 81, 82
Superinfection exclusion dilemma 85–6
cooperation–defection 85–7
deviation from PD 85–6
resolution 87
Productivity
Photosynthesis 12, 269–70
Primary 12
Production of phages/viruses
Aquatic
obligately lytic infection 255, 261
role of induction (see Lysogeny, induction – Phage production)
synchronization 256–7
Measures of 253, 256
(1) measures of change in viral abundance 253, 256
(2) measures of virus decay 253
(3) rate of incorporation of radiotracer 253–4, 256
(4) viral tracer change in abundance 254, 256
(5) estimation of host–virus contact rates 254, 256
(6) frequency of visibly infected cells 254
(7) contact-rate reduction by dilution 254, 256
(8) virus-production measures 254–6
Prophages (see Lysogeny, prophages)
Protozoa/Protists (see Grazers, eukaryotic)
Pseudolysogeny 9, 114–16, 125–31
Carrier state and 2, 8
Ecology of
alternative infection strategies 115, 127, 131, 134, 255
chemostat experiments 128
dormant state 114
environmental relevance 125–6, 131
host growth inhibition and 127, 129
host inviability and 129
nutrients reverse 116, 126, 128–9, 131
phenotypic plasticity 131
starvation response 116, 126, 128–9, 131, 458
survival strategy 117, 120, 125, 128–9, 131, 133
Epidemiology and 130
Examples, environments
feces 130
fresh water 125, 128–30
salt water 129
soil 126, 130–1
sewage 130
Examples, hosts 127, 129–30
Acholeplasma laidlawii 130
Archaebacteria 131
Azotobacter vinlandii 129
Bacillus pumilus 130
Bacillus subtilis 125–6, 131
Bacteroides fragilis 130
Clostridium botulinum 361, 373, 375

Escherichia coli 129
Halobacterium salinarium 129
Listonella pelagia 130
Mycobacterium spp. 130
Myxococcus fulvus and *virescens* 130
Pseudomonas spp. 127–30
Pseudomonas aeruginosa 125, 128–9, 458
Rhodopseudomonas capsulate 131
Streptococcus pyogenes 130
Streptomyces spp. 130
Vibrio cholerae 130
Examples, hosts 127, 129
Examples, phages
phage A21 129
phage CTXΦ 130
phage F116 127–8
phage Hs1 129
phage T4 129–30
phage UTI 127–8
Genes (*rI*) 129-30
Infection aspects 8–10
host survival 127
immunity, lack of 127
latent period extension 458
long-term survival of pseudolysogen 127
lysogeny, reduction to 116, 129
no preprophage replication/unilateral inheritance 126–8
poorly understood 116, 127
reductive infection 120
subcultured, can't be 129
unproductive 115, 126–8
Persistent infection 8
Phages
absence from RNA phages 115
obligately lytic 127–9
temperate 129–30
Plaque impact 108, 127
turbid 127–7
Preprophage 116, 126–8, 129, 131
Virulence factors (see Virulence factors, phage-encoded – Evolutionary ecology)
Pseudomonas aeruginosa
Biofilm, existence as 457
Lysogeny
experimental 124
freshwater isolates 122, 125, 286
Phages F116, UT1, etc. (see Bacteriophage isolates)
Physiology impact on phages (see Physiology)
Pseudolysogeny of 125, 128–9, 458
Virulence factors, phage-encoded 353
Pseudomonas fluorescens
Biocontrol agent 291–2
Epistasis 164
Phages ΦR2f, GP100, etc. (see Bacteriophage isolates)
Phenotypic diversification 293
Physiology impact on phages (see Physiology)
Pure-culture technique 77, 166

Reaction diffusion (RD) (see Plaques, modeling – Reaction–diffusion)
Recombination 151–2
Bacterial genes into phages 295
Coevolved genes
core genes 184
no functional interchangability 184
not split apart by 181, 184, 355
Coinfection and 151, 195–6, 198, 202–3, 223, 237
clonality and 202, 220
Cooperation and 354–6
Difficulties, experimental
high levels, introducing 237
preventing altogether 243
varying levels 237
Evolution and genetics
adaptation and 220, 226–7, 235, 242–3
advantages of, attempts to evaluate 220, 237
barriers to gene flow (see Barriers, post-zygotic)
clonal interference in absence of (see Clonal interference)
epistasis and (see Epistasis – Recombination)
evolutionary change 149
fine structure of the gene/fine-scale mapping 103, 230
fixation beneficial alleles 222–3

Recombination (*cont.*)
Muller's ratchet (see Muller's ratchet)
phenotypic mixing, distinguishing from 199
selection against, lack of 187
selective sweep 154
sex (see Sex)
significance of 151, 237
variation generation by 149, 152, 184, 196, 198–9, 209
Heterologous recombination 152
gene duplication 152
template switching 199
Historical 198–9
Homologous recombination 151, 181, 183, 199
continuum of homology 181
higher frequency occurrence 183
intermediate rates of 223
micro homologous recombination 183
near homologous recombination 181
no signature of 183
transduction and 295
(see also Bacteriophage T2 and T4)
Illegitimate recombination 152, 295
Intragenic 237–8
Joints, novel (see Joints, novel)
Known for many phages 223
Metagenomics ("*in silico*" recombination) 189
Molecular 151
limited 203, 212, 354
Non-homologous recombination 181
creation of novel joints (see Joints, novel – Creation of)
evidence of 184
frequency 178, 183
importance 191
leaves signature 183
moron insertion 186
not positional 183
variation, novel/pervasive 183, 184, 187
Phage density and 237
Reassortment 199
Red Queen hypothesis 168
Refuge (for bacteria from phages)
Non-mutational 453–5
spatial 401, 432, 442
wall populations 36
Release 390
Chronic 7–9, 16, 115, 356, 362
budding 8
continuous 8, 49
extrusion 8, 370
selective interests 359
Lysis (see Lysis)
Reporter genes and proteins
β galactocidase (*LacZ* gene) 338
Bioluminescence (*lux* genes) 309–11
Green fluorescent protein (GFP) 311, 342–3
antibodies against 343
Ice nucleation (*ina* gene) 310
Reporter phages 208–12
Reproductive isolation (see Barriers, post-zygotic)
Resistance, bacteria to phages 168
Adsorption blocked 8–9
capsule loss 452
form of exclusion 8
Arms race (see Host range – Coevolution)
Colonies, phage resistant 150
Cost to
experimental 39–40, 43
theory (see Trade-offs, bacterial)
Ecology
chemostats 36, 39–40, 271
community module 38
bottom-up control 41
Phage therapy 451, 453
Plaques 432
Resistance to phages K-1, T1, T2, T4, T5, T7 (see Bacteriophage isolates)
Resistance, partial, bacteria to phage 8, 38
Chemostat experiments 40, 42
Cost to 42, 43
Evolution in the wild (speculation) 271
Phage T2 40–2, 48, 264
Sensitive strain impact 41–2
Restriction 8
Anti-phage bacterial immunity 9
Effective burst size and 83
Retron Ec73 157
Robustness, mutational/genetic 165, 209

Brittleness (antonym) 209
Complementation 210
reduced selection for robustness 210
Empirical support
Phage ϕ6 210
Phage λ 165–6
vs. Evolvability 165–6
Fitness/natural selection 165, 209
Growth mutational robustness 166
Ruminants, inefficient food use by 13

***Salmonella* spp.** 371
Biocontrol 310, 314–16
Detection (see Bacterial detection)
Food, contaminated with 306, 309, 316
Horizontal gene transfer; rapid 376
Lysogen allelopathy 361, 375–6
Lysogen induction 371
Phages as fecal indicator (see Fecal indicators)
Phages 12, P22, SJ2, SP6 (see Bacteriophage isolates)
Prophage 353, 371
Salmonella enterica 371
phage ϕX174, fitness on 161
three-species interactions 161, 293, 294
Salmonella Enteritidis
biocontrol 315–16
detection 312–13
Salmonella typhi 371
Salmonella Typhimurium
detection 310
lysogenic conversion 371
Virulence factors, phage-encoded 353, 357, 365, 371
disease 376
lambdoid phage-encoded 370–1
phage P22-like encoded 371
secreted, all are 371
Sediments
Abundance of phages 14, 289–90
Generalized transducing phages 296
Phage PRD1 dispersion 285
Spatial structure 415
Selection, natural 152–4
Adaptation, key to understanding 152
beneficial alleles 196
(see also Ecology – Evolutionary)
Clonal interference (see Clonal interference)
Deleterious alleles 156
Evolvability (see Evolvability)
Fitness and 152, 154, 209
complications 242
Hitchhiking (see Hitchhiking)
Large population sizes 156
Levels of
prophage vs. bacteria 156–7
prophage vs. prophage genes 157
prophage vs. transposon 157
selfish/cheater genes 156–7, 355
Propagation rate increases 241
Purifying selection
recombinant elimination 183
sex (see Sex – Purifying selection)
synergistic epistasis 206
synonymous mutations 185
Robustness (see Robustness)
Selection coefficient 156, 163
Selective sweep
hitchhiking and 154–5, 160
phage-resistant bacteria 293
recombination 223
vs. single-step adaptation 227
Serial passage/propagation/transfer 158, 221
Broth 76, 228, 239, 241
Metacommunity 50–1
Mice 337
Plaques 159, 240–1
Sex 202
Adaptation 205–6, 223
Bacterial (see Horizontal gene transfer – Bacterial)
Counters Muller's ratchet (see Muller's ratchet)
Evolution of sex 159, 206–7
F (sex) pili 7
Facultative sexuality 213
Horizontal gene transfer (see Horizontal gene transfer)
Mixis 199, 205
Phage ϕ6 201, 206

Sex (*cont.*)
Recombination (see Recombination)
Selfish gene evolution 210–12
Purifying selection 160, 206–7
Shifting balance theory (see Adaptive landscape)
Shiga-toxigenic *E. coli* (STEC) 367
Adherence, lack of 367
Escherichia coli O157:H7 (see also *Escherichia coli* strains) 367
Induction, prophage 368–9
Lambdoid phage 369
Phage-encoded toxin (see also Shiga toxin) 370
Phage-mediated lysis 369
Syndromes, associated 367
Shiga toxin, importance to 369
Toxin-dose dissemination (see Toxin-dose dissemination)
Shiga toxin (Stx)
Escherichia coli associated 353, 358, 360
(see also Shiga-toxigenic *E. coli*)
Expression
butyric acid, blocking by 363
colonic 369, 375
disease 369, 376
following prophage induction (Stx-2) 368
low iron conditions (Stx-1) 368–9
lysis gene association 355
multisubunit proteins 355
quorum sensing 368
release, lysis dependent 360, 368–9
stxA and *stxB* genes 373
toxin-dose dissemination (see Toxin-dose dissemination)
Phages (see Stx-encoding phages)
Shigella 369
periplasmic accumulation 369
Variants of 358
Stx-1 368, 369, 372, 373
Stx-2 368, 369, 370, 371, 373, 375
Verotoxin 367
Soft agar overlay 98, 415, 435
Spatial structure model system 109
Soils
Abundance of phages (see Abundance of phages)
Bacteria
high densities 289
nutrient density impact 287
phage impact 293–4
Bacterial species
Bacillus subtilis 287
Bradyrhizobium japonicum 295
Clostridium botulinum 372–3
Pseudomonas fluorescens 291–3
Pythium ultimum 292
Rhizobium leguminosarum 296
Rhizobium melioti 295–6
Serratia sp. CP6 296
Serratia liquefaciens 123, 289, 290
Streptomyces spp. 282
Decay, virions, soils 285–6
clay interaction 290
Experiments
in situ 293
microcosm (see Microcosm – Soil)
Lysogeny 286
Obligately lytic phages 287
interfering with biocontrol 291
Plant impact on
decay products 282–3
rhizosphere (see Plants – Rhizosphere)
roots 282
Pseudolysogeny 126, 130–1
Similarities to other environments 281
Spatially structured environment 94, 282
plaques as model for 415
sealed-off microenvironments 77
Temperate phages 123–4
Transduction 295–6
Transport through 285
Variation in 282–3
animals, action of 282
microorganisms, action of 283
nutrient cycling (see Nutrient cycling – Soils)
Spatial structure 56, 94, 415
Bacteria and 96
Refuge for bacteria from phages (see Refuge)
Cooperation, evolution of 53, 76–8, 84
absence of 74

limiting volumes 77–8
temporary sequestration 76–8
Environments 95
biofilm 94
bodies 441, 457
naturally occurring 94, 415, 421, 457
General properties 95
fast movement 95
heterogeneity 56
minimizing 431
slow movement 95
Models of
cellular automata 427, 434
laboratory 415
plaques 94, 109
metapopulations (see Ecology – Metapopulation)
microcosms (see Microcosms – Soil)
Phages and 56, 94
bacterial impact 94, 101, 293
coexistence (phage–phage) 431–2
competition 431
competitive exclusion (see Competition – Exclusion)
mutant invasion 430
stochasticity and 431
Sources and sinks 101
Spatially unstructured 95, 415, 457
Specialization, bacterium
Competition (specialist/type) 263
conditions favoring, oligotrophic 262
nutrient affinity, higher 266
nutrient uptake specialist (synonym) 264
phage susceptible 263
predicted abundance lower with phage 263
virus–host ratios, predicted high 267–8
Defense (specialist/type) 263–4
conditions favoring, eutrophic 262
nutrient affinity, lower 266
phage resistant 263
predicted abundance higher with phages 263
virus–host ratios 267–8
Spatial niche 293
Trade-offs (see Trade-offs, bacterial – competitive)
Specialization, phage
Coinfection/cheating 201, 204
Extreme environments 257
Host range 167–8
coevolution 168
see also Transduction – Specialized
Speciation (see Barriers, post-zygotic)
Staphylococcus aureus 457
Aggregation in milk and serum 461
Capsule (see Capsules, bacterial)
Detection (see Bacterial detection)
Foodborne pathogen 310
Major killer, past 440
Mouse peritonitis model 451
Phage K (see Bacteriophage isolates)
Phage therapy 451
Reduction in phage binding 461
Virulence factors, phage-encoded 353
toxins, phage-encoded 357
Structural complexity 38
Stx-encoding phages
Colonic adaptation 369
Host-range breadth 369, 374
Induction and release 368–9
Lambdoid 358
Substitution (fixation) of alleles 152
Drift impact (see Drift, genetic – Allele fixation)
deleterious alleles 226
mutant allele accumulation 236
neutral alleles 152, 154–5, 158, 221–2
Interference with 155, 223
clonal interference (see Clonal interference – Impediment)
failure to fix 155
Rates of 235–7
Selection impact
beneficial alleles 221–3, 226, 235–6
large benefit for initial fixed 155
recombination impact 222–3
Substrate (resource), limiting
Differential growth impact 41, 262–3
Dynamic variable 35
Modeling 57, 395, 403, 409–11
Multiple 36
Values of 57

Superinfection 70, 197
Coinfection and 197
Exclusion 70, 197
altering receptors 197
blocking DNA injection 197
free-phage antagonist 71, 85–7
heterologous 197, 207
lacking in pseudolysogens 127
lysis dilemma 85
Immunity 8, 70, 121, 157, 197
examples: phages CTXΦ, Hh-1, Hh-3, P2, P22, etc. (see Bacteriophage isolates)
free-phage antagonists 85
heterologous 197, 207
homologous 121, 197
lacking in pseudolysogens 127
lysogenic conversion 132, 361
lysogeny associated 72, 197, 288, 361
Multiple adsorption consequence 70, 195, 197
(see also Adsorption – Secondary)

Techniques, molecular
Denaturant gradient gel electrophoresis (DGGE)
dominant hosts number 267, 270, 271
not measure of activity 265
PCR-step biases 271
sources of error 271
Polymerase chain reaction (PCR)
16S rRNA gene amplification 296
phage presence determination 346
quantitative PCR (qPCR) 154
Pulsed field gel electrophoresis (PFGE)
dominant phage numbers 267, 271
size genome phage differentiation 257–8
Temperate phages (see Lysogeny, temperate phages)
Time lags (delays)
Destabilizing 35–6, 58, 404, 408
Ignoring 403–4, 408, 410, 448
Latent period 36, 57, 403, 410, 456
accounting for infected bacteria 404–5
in viral production estimation 254
increasing destabilizes 57–8, 404
linking to burst size greatly destabilizes 58
Plaques 99, 100
ignoring 427
Prey to predator-progeny lag 35–6
Transiently infected bacteria 404
Toxin-dose dissemination/amplification 361
Clostridium botulinum 361
Lysogen allelopathy association 361
Shiga toxin 361, 375
Toxins (see Virulence factors)
Trade-offs 54, 169
Diversity maintenance, general theme 43
Fitness 160–2, 204
Genetic mechanisms
antagonistic pleiotropy (see Antagonistic pleiotropy)
neutral mutation accumulation 160–1
Life-history traits 68, 160
Pathogen virulence evolution 54
spatial structure impact 54
Trade-offs, bacterial
Competitive ability vs. phage resistance 39–41, 262, 263, 272
bacterial capsules (see Capsules, bacterial)
porins 262–3
shaping pelagic food webs 262
Winnie-the-Pooh strategists 262
Phage protection vs. phage invitation
bacterial capsules 265
Virulence vs. phage resistance 452–3
Phenotypic 38
Trade-offs, phage
Burst size vs. latent period/generation time 68–9, 73–5, 160
cooperation–defection 72, 85
higher bacterial density costs 72, 74
life-history trade-off 160
lower bacterial density costs 68, 72, 74
lysogeny and 70
modeling 46
plaques 109
Growth rate vs. growth yield 74
finite volumes 74
reducing trade-off 85
(see also Economy vs. expediency)

(see also Prisoner's dilemma – Exponential growth dilemma)
(see also Tragedy of the commons)
Host range 43
antagonistic pleiotropy 160-1
coevolution, limts to 168
phage ϕX174 161
Neutral mutation accumulation
phage models for testing 162
Reproductive output vs. virulence 54
horizontal vs. vertical transmission 54–5
Within vs. between-cell competition 204
Tragedy of the commons 53, 55, 73
Evolution of restraint 54
Growth rate vs. growth yield 73
Husbanding commons 73
(see also Economy vs. expediency)
(see also Prisoner's dilemma – Exponential growth dilemma)
(see also Trade-offs – Growth rate)
Transduction 13, 294, 295
Abortive 128
Aquatic 253, 296
(see also Cyanobacteria – Transduction)
Beneficiaries 295
Generalized 13, 132, 294–5
16S RNA gene, phage encapsulated 296–7
examples (and null results) 295–6, 370
frequency 296, 297
transfer of functional units 355
Mediated *in situ* 374
Morons (see Morons)
Packaging 294–5
Sex, bacterial 14
Specialized 294–5
biases in genes transferred 295
definition, broad 131
definition, narrow 13, 295
examples 296
Transposon 157
Increase in copy number 157
Parasitism of phages 157
Trophic interactions
Movement up trophic levels 12, 259–60
Viruses as "process" 260
(see also Food webs)
(see also Grazers – Consumers)

Unicellular-organism parasites (UOPs) 3–4

Vaginosis, bacterial 13
Vibrio cholerae 365, 369
Intestinal colonization 371
Latent infection
frequency in 130
lysogenic (phage) conversion 297
pseudolyosogeny, frequency in 130
Major killer, past 440
Phage CTXΦ (see Bacteriophage CTXΦ)
Serotypes 369
Transduction 370, 374
Vibrio pathogenicity island 369, 370
Virulence factors, phage-encoded 353
Ace 358, 370
cholera toxin (see Cholera toxin)
Glo 357
secretion of 371
toxin-coregulated pilus (TCP) 369–70, 374
Zot 358, 370
Viral loop 12, 258–9
Respiratory losses 259
(see also Nutrient cycling – Viral lysis)
Virulence
Phage virulence (see Bacteriophage virulence)
Trade-off model 80
Virulence factors (VFs)
Bacterial growth enhancement 363–4
Colonization enhancement 371
Cosmopolitan genes 362
Phage receptor as 456
Virulence factors, phage-encoded (ϕVFs) 148, 354
Bacteria associated with: *Clostridium botulinum*, *Corynebacterium diphtheriae*, *Escherichia coli*, *Pasteurella multocida*, *Salmonella* spp., *Salmonella enterica*, *Salmonella* Typhimurium, *Staphylococcus aureus*, *Streptococcus pyogenes*, *Vibrio cholerae*, *Vibrio harveyi*
Ecology 363, 366, 376

Virulence factors, phage-encoded (*cont.*)
bacterial growth enhancement 362
cosmopolitan genes 362
ecological niche opening 116
environmental dissemination 356, 362
environmental modification (see Environmental modification – Virulence factors)
hitchhiking to appropriate environments 359
host range limitations, phage 374
Evolutionary ecology 356, 360, 362, 365–6
bacterial availability to phage 362, 365
colon, expression in 363
dual utility 361–2, 370, 376
enhancement of phage fitness (see Fitness, Darwinian – Virulence factors)
enhancement of productive infection 356, 359
expressed only during lytic cycle 360
growth enhancement, lysogen 356, 359
interests, phage lytic cycle 359
lysogen allelopathy (see Lysogen allelopathy)
phage properties, impact of 373
selection on/benefiting entity 157, 353–4, 356, 360–2, 366
toxin-dose dissemination (see Toxin-dose dissemination)
Examples of phage-encoded virulence factors
botulinum toxin (BoNT) 358, 361, 372
C_2 hemagglutinin toxin 358, 372–3
C_3 ADP-ribosylating toxins 358, 372–3
cholera toxin (see Cholera toxin)
diphtheria toxin (see Diphtheria toxin)
cell growth and carcinogenesis 345
hyaluronidase 358, 360–2
neurotoxins (*Clostridium botulinum*) 372
Shiga toxin (see Shiga toxin)
superoxide dismutase 371
Genetics
epistasis 360
expression regulation by bacteria 360
expression regulation by phages 360
gene-phage linkage arguments 354, 355
quorum sensing and expression 368
VF function-improving phage genes 360
Phage/prophage encoding 353, 357, 373, 376
bacterial infection within animals 373–4
defective prophage 353
gene-phage co-replication 356
helper-phage requirement 360
multiple prophage (polylysogeny) 353
phage therapy and 341
temperate phage 359–60
Mechanisms toward retention
direct enhancement of phage fitness 361
dissemination of an effective toxin dose 361
epistasis linking of VF and phage genes 360
indirect enhancement of phage fitness 362
lysogen allelopathy (see Lysogen allelopathy)
Obligately lytic phage, not encoding 359–60
Release 358
component of intact lysogen 356, 367
following prophage induction 353
secreted with phage virions 356–9
secreted without phage virions 356, 366
upon phage-induced lysis 353, 359–60

Whole cultures 2–3, 79

Printed in the United Kingdom by TJ Clays Ltd.